SURVIVING MOLD

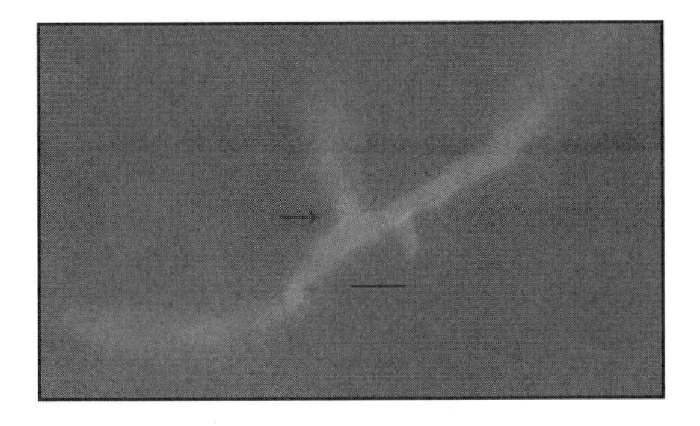

Life in the Era of Dangerous Buildings

RITCHIE C. SHOEMAKER, MD

Otter Bay Books
BALTIMORE, MD 2010

Please direct all correspondence and book orders to:
Ritchie C. Shoemaker, MD
500 Market Street
Pocomoke, City, MD 21851

Library of Congress Control Number 2010940299
ISBN 978-0-9665535-5-0

Published for the author by
Otter Bay Books, LLC
3507 Newland Road
Baltimore, MD 21218-2513

Printed in the United States of America

Dedication

Surviving Mold is dedicated to all Mold Warriors who have pulled together to help others in the ongoing search for a cure. The advances in clinical medicine and research you will read about here are a result of a group effort that brought support from those who have needed support themselves, participation from those who were too ill to join in at one time and hard work from those who were disabled before. To those who joined in the effort, I salute your courage and tenacity.

To those who want to be a part of this exciting field of medicine, politics, public opinion and legal battles, the road for you has been blazed. Now that so many scientific mysteries about the devastating illness caused by exposure to Dangerous Buildings are unveiled no one needs to suffer without hope and no one needs to stay ill with no one to turn to. Since 2005 when *Mold Warriors* was published, there is now fulfillment of old dreams and attainment of new goals. The role for those affected by Dangerous Buildings can now turn to opening the legal injustices to the sunshine of truth and revealing to all that we will not let a readily recognizable and readily treatable condition be either ignored or untreated again.

This book is dedicated to the life and work of Frank Fuzzell, my friend, and Wells Shoemaker, my father, men I will not forget.

Acknowledgments

Any book this long has a long list of those due thanks. From Editors Patti Schmidt and Matthew Hudson to guest authors Laura Mark MD, Pierre and Susan Belperron, Jerry and Loretta Koppel, Erin Culpepper, Tom Jones, Erik Johnson, Greg Weatherman, Dr. Tom Harblin, Lee Thomassen and David Lasater, *Surviving Mold* became a group effort. Comments from Margaret Maizel, Jonathan Wright, Cheryl Wisecup, Bruce Nichols, Don Malloy and JoAnn Shoemaker have strengthened this book immensely. Ellen Shoemaker provided a poignant sidebar for "Blood in the Sink." Dennis House was always willing to lend a statistical ear to knotty problems.

Debbie Waidner, Debbie Bell and Barbara Gray continue to set the professional standard for what patient care in a mold illness practice requires. While mold patients can be demanding and difficult, the emotional rewards of seeing them return to a normal life creates a satisfaction different from the joys of primary care medicine.

A special thanks to Matthew Hudson, the Professor, a man of extraordinary depth of understanding and competence.

My greatest thanks are for my bride, JoAnn, who has supported this effort for two years.

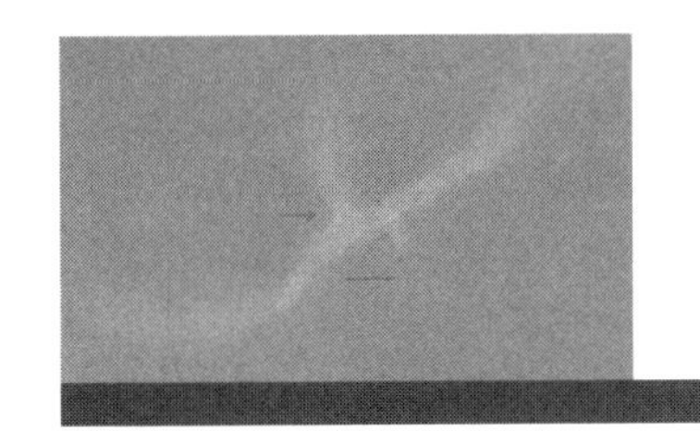

Table of Contents

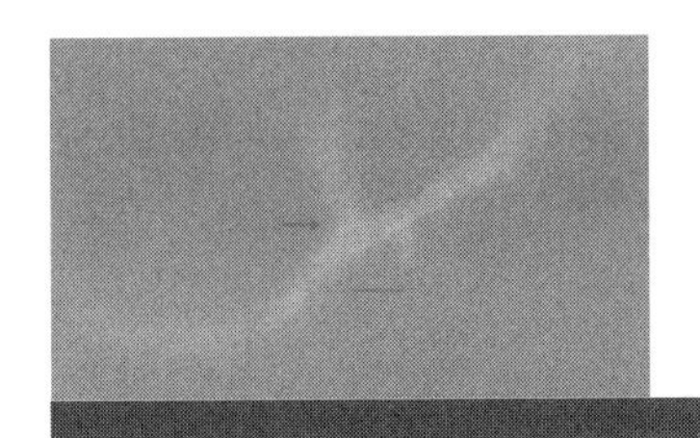

Foreword

We are not born defenseless. We arrive into the outer world already well equipped with the most sophisticated bio-hardware known to man. Indeed the most advanced military programs and our best scientists and billions of dollars cannot buy anything approaching this technology. The design of our inflammation-producing immune response system has survived intact for billions of years; it permits us to survive during those first hours in a hostile world - and for decades thereafter. Our personal nano-army is not passive. Once activated by news of intruders it is proactive and so aggressive that if its targeting software has a design flaw or has been programmed incorrectly then it doesn't stand down if it cannot find the enemy. Instead it begins targeting friendlies. The "friendlies" thus targeted are then attacked by the immune system's warriors, the monocytes, macrophages and T-killer cells. But the host is the only friendly that this aggressive system can harm – and so it does. That is us. As the attack continues, vasoactive intestinal polypeptide (VIP) and alpha melanocyte stimulating hormone (MSH) our self-defenses to this aggression are overwhelmed and their production is destroyed. By then our health is likewise in ruins. As Dr. Shoemaker observes in a later passage - like Pogo - "we have seen the enemy and he are us."

This is our innate immune system. It is very sophisticated, very complex and neither sleeps nor relaxes for as long as we live. It is a self-starter and is fully automatic. We cannot coach or control it. We can help it teach because it enables the acquired immune system (think vaccination). The recognition and targeting equipment (immune response genes) used by this guardian is so important that we have 2 sets of it, one from each parent. There are not so many of these software packages available to humans. Only 54 main versions (with some sub-groups) have been discovered so far. They have model numbers such as 4-3-53 and 11-3-52B and a few have only 2 components such as 1-5. Sometimes we get two identical models. For instance, I am 4-3-53 and 17-2-52A, while my son is 17-2-52A and 17-2-52A.

When it comes to dealing with the effects of inhaling the toxic stew that usually prevails in water damaged buildings (WDB), some targeting mechanisms are far more effective than others at calling in the Delta Force of T-killer cells and other specialized warriors. The targeting equipment is called HLA DR and model numbers 15-6-51 and 10-5 and 8-3 are but three of the list of 48 main groupings you really want to have. Sadly about 25% of the human population appears to have the targeting machinery that malfunctions when faced with WDB exposure. There are six main groups of these. The worst models of all are 4-3-53, 11-3-52B and 14-5-52B. 17-2-52A, 7-2-53 and 13-6-52C are not good either. Sadly for the unlucky, HLA DR models are original equipment and no trade-ins are possible.

This book is not about a theory. It is about biological facts that have been established by tens of thousands of lab tests on humans, not mice. This book lets you hear from the trailblazing scientist who has mapped this hitherto unknown territory and who as a treating physician has helped thousands of patients. All this because, starting from a moment of serendipity (think Fleming and penicillin), he painstakingly thought and reasoned and experimented his way from the darkness of ignorance into the light of knowledge. This book also lets you hear from some of those patients about their own journeys from dark despair, surrounded by the ignorance of the medical profession, to a place where they have regained their ability to enjoy life. Ritchie Shoemaker is that scientist physician. His patients will introduce themselves as they appear. Some of their personal stories are full of detail that will interest some, but not all readers, so feel free to dig deeply into some chapters and skim others.

So now, as Sherlock might say, "The game's afoot."

Matthew C. Hudson

September, 2010

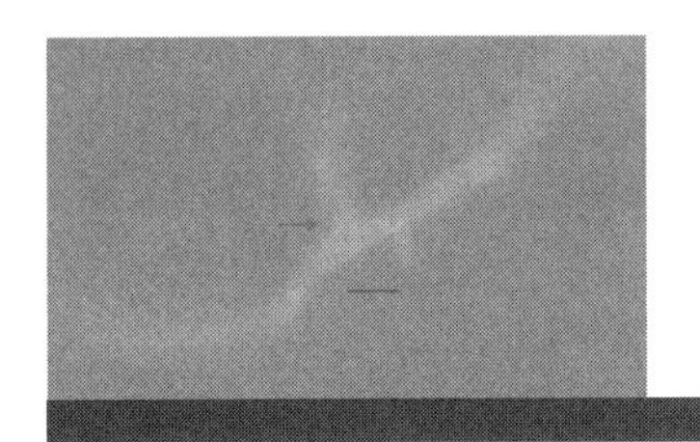

Preface

Surviving Mold tells the stories of people who've been made ill after they were exposed to WDB (water-damaged buildings). These illnesses reflect a growing societal problem: dangerous buildings.

There are many ways buildings become home to a toxic mix of microbes, fragments of microbes and harmful chemicals. Buildings can host fungi, bacteria, mycobacteria and actinomycetes as a result of construction defects like inappropriate ventilation and HVAC systems; faulty construction of crawl spaces or inadequate building design; using failed technologies like flat roofs or fake stucco cladding without adequate caulking; incomplete basements exposed to saturated ground water conditions; or not correcting water intrusion; or remediation that doesn't *clean* as its final requirement. It's no wonder that the Occupational Safety and Health Administration (OSHA) tells us that more than a quarter of U.S. buildings are water-damaged.

And if you've got water-damaged buildings, you've got patients with mold illness. After the illness is identified and corrected, and sadly, that doesn't happen often in the United States, the *goal for patients is to stay well.* And that doesn't happen often in the U.S. either. If you have an illness acquired following exposure to the interior environment of a WDB, an illness I will call "mold illness," then *Surviving Mold* mandates a change in your life style since the mold illness has changed your immune responses and your life.

For a minute, imagine you're Lindsay Mather. Ever since you moved into the cute cottage at the bottom of the hill, you're ill all the time. The basement leaks after it rains, it smells musty and you just found out that the flexible fiberglass ducts - the source and delivery method of heated and cooled air to every room in the house - are filled with mold. Apparently, someone didn't seal the ductwork very well, as if any taping job could ever be airtight forever. Air from the basement leaks into the ductwork each time the HVAC system blows air past the failed tape jobs, so now the microbes that *were* growing *only* in the basement are found in *every room of your house*. Your homeowner's policy most likely excludes correcting mold problems. The mold company you hired to look at the biological stew that inhabits your new home charges you more than $50,000 to fix the problem, but will not guarantee your safety when they're finished.

You can't work as a computer company supervisor any more, because you can't stay focused. Your last performance review was so terrible, you agreed to take an unpaid leave of absence until you felt better. The disability policy the company provided will carry you through until you return to work.

Sounds bad, right? But you're lucky compared to many others. Your physician knows more than his peers about mold, human health and exposure to WDBs. He told you to move out right away, before it was too late. But where can you go? How do you know the next place won't be moldy, too? You looked, but just walking into some of those apartments made you immediately more ill. His medications aren't much help, although the sleeping pills and breathing meds get you through the week.

Once again, you read over the registered letter that arrived today from the disability insurance company. More bad news: your short-term disability ran out last month and your application for long-term disability was denied after the doctor who "reviewed" your chart for the insurance company said you didn't have any significant lung impairment, that your cognitive impairments were stress-related, and the aches and pains are just fibromyalgia that is pre-existing because your medical record plainly shows an automobile accident twelve years ago. He never even examined you, yet he has the power to give or withhold the measly amount of disability insurance you paid for, month after month, thinking it would guarantee you help if you ever needed it.

Oh, his opinion is called "independent." What is independent about some physician with *never-disclosed* credentials siding with an insurance company so he will get a check from his employer, the middle man company the insurance company paid? No opinion, no check. And since the insurance company itself didn't pay the doctor for the opinion, somehow the words become unbiased? Yes, such appearance of independence is a sham, with deniability all around. Try applying for long term disability if you think the contracted independent opinion won't work against you.

It is a simple prescription for outrage against truth. And it gets worse. When the "independent" doc is an Occupational Medicine (OccMed) practitioner, I can essentially guarantee you he won't know anything about mold illness and will cite idiotic opinions from the ancient and fallacious ACOEM report (contrived and published in 2002) to back up his malarkey. More on the politics of deception for hire from the ACOEM statement about human illness will follow in these pages.

Whatever made you think a system that stays in business by more often *denying* disability would actually agree you were disabled? Saying that you needed psychiatric care, and the independent doc used that masterful ploy as they all usually try to, was the cruelest stab of all. Imagine an adminstrative judge reading the opinion that you are a wacko; will his opinion of you as a person be enhanced?

You want to confront this "independent" person, accuse him of - What? Negligence? Ignorance? - yet he is surrounded by the reputation of his esteemed institution and all that entails. You learn that you could hire a lawyer to fight for you, but you'd have to pay roughly $15,000 to get on his case list, $500 per hour for his time and then pay 25% of your first five years of disability income to him for his work. Even then, you have no idea if he knows anything about mold illness. Without a lawyer, though, you don't have much chance of getting disability. But *with* a lawyer, the disability company bleeds you dry financially with legal fees, making one motion after another and postponing again and again. Justice postponed is justice that's more expensive to achieve, if ever, and their lawyers know that people filing disability cases have little money to spare after months or years of illness and financial difficulties.

OK, now imagine you're a physician. If Lindsay called you for help, how would you answer her questions? Where can she go that might be safer? Of all the things she owns, what can she take with her, or is it all contaminated? How can she know if her new residence will make her ill? Where will she find the money to start over again? All she has left to her name is the damaged home and its mortgage. She's too tired to even look up this information for herself.

Now imagine the same case, but our example patient has two children who aren't doing well in school. This family rents instead of owns their home and the husband is tired of his wife being sick. She just isn't the same person he married ten years ago, he complains to you. *His* doctor said mold doesn't hurt anyone (his opinion has nothing to back it up, but the physician's opinion isn't questioned). His wife tells you that her husband said he was working late last night at the office, but when she called there, he didn't answer the phone. What can she do, she asks crying? Where can she go?

Now our example patient is a famous TV personality whose melodious voice has sounded like a chorus frog with laryngitis ever since she moved into the fancy condo down the street from the White House. If she can't be the voice that America trusts, she's out but her big mortgage won't stop. She's convinced that she's ill from the mold that's growing all over the condo, and so are scores of her fellow condo-owners. They've demanded the company that built the condos remediate the mold problem, and the condo's attorneys are fighting them tooth and nail. Their strategy includes hiring an experienced professional witness (another "independent," we might note, who, as expected, hasn't even seen a patient in the practice of medicine for more than twenty five years) to say *under oath* that mold couldn't hurt a flea. You heard the insurance company has already paid this witness more than $500,000 and you haven't even been to court yet. If you sue and lose, and the case would be heard in the most conservative court in the United States, you might be forced to pay the defense consultant's fees, even though he never saw or examined you.

Imagine going bankrupt because a witness didn't tell the truth under oath. Does it happen in the U.S.? Of course it does and not just in mold cases. Makes you wonder about the legal system, doesn't it? All it would take would be for one judge to stand up and tell a professional witness that he is a liar, as they usually are and cite him for perjury. Doesn't happen. And how about the ridiculous items the defense bar makes up in motions of one kind or another: usually just outright lies. Isn't it time to get out the "sanction sticker gun," labeling their brief cases with official "liar" stickers all the way off the bar? That behavior certainly appears to be accepted as standard and expected behavior. Just tell the truth, for once, will you?

Now imagine you're a wealthy Florida entrepreneur who owns homes in many locations. Your high-rise view of the Fort Lauderdale beach has been ruined by the wrecking crew taking down the other wing of your building because it's full of mold. You have to vacate, but the condo association is balking at your demands to give you the replacement value of your Oriental rugs and ancient museum-quality Egyptian textiles. Do you just throw them away? Since they're full of mold, too, and if you take them with you, you'll contaminate your new environment.

Or imagine you're one of a group of Air Traffic Controllers who work at a major U.S. airport. As determined by the U.S. Department of Labor in a landmark decision that's setting precedents all over the nation (Haefner v FAA, 3/09, Appendix 6), you're now 100% disabled from mold exposure. What do you have to do to return to work at another airport tower two years from now, so you can support your three young children?

Or let another example patient come to my Pocomoke, Maryland office for treatment of her mold illness. She finds out her genetic make-up is the kind that ensures she'll become ill with even minor exposure to a water-damaged building. Where can she go? Where is the most likely place *not* to run into a WDB? Is there such a place? Although she's asked her landlord repeatedly to correct the water intrusion, visible mold and musty smells, he isn't doing it. When the landlord ignores her repeated requests, is that a form of personal injury caused by the landlord's negligence? Can she sue him because his behavior is hurting her?

Now suppose in our hypothetical case that our patient's exposure was at her workplace, where the water comes in from roof leaks and broken pipes in the ladies rest room. There are two others ill. Their request to have medical bills reimbursed, sick time covered and vacation time reinstated was denied by the employer. The group of three hired an attorney to help them with their claim for compensation, but the business owner counter-attacks by firing all three of them. The three ladies banded together to sue the employer for wrongful firing and now personal injury from the exposure to a WDB, but the employer cleverly brings in NIOSH inspectors to evaluate the building just after the massive clean-up of the building is completed. What can NIOSH say? In their inspections, they didn't find anything scary and as far as the massive mold lake under the building, well, they didn't look there. The employer then trumpets the clean bill of health from NIOSH as evidence that the three ladies are just wackos or lazy good-for-nothings looking for a free ride.

The ladies argue that NIOSH didn't look where they should have; didn't sample like they should have; didn't interview the entire work force like they should have; didn't ask the right questions of those they interviewed like they should have; and didn't have any basis in actual medical practice to say what they did (see Chapter 19, New Orleans, for a review of the similar approach that NIOSH used in evaluation of sick people in St. Bernard Parish). Make this scenario more interesting, though, as each of the three ladies performed a classic re-exposure trial that proved the workplace made them ill. Add in the EPA scientist who walked around with NIOSH, quietly sampling areas of the building that NIOSH didn't see, like the boiler room, finding *Stachybotrys* growing like Spanish moss on old oaks in Mobile, Alabama. Then really make the employer look bad by having a judge agree to let the ladies' experts examine the building, finding lakes of standing water under the building and mold growing everywhere. Does one believe (1) the ladies; (2) the actual medical data that proves causation of illness; (3) the EPA; or (4) the newer inspections done when the building wasn't white-washed? Or will the employer hire bull-dog attorneys to make us think he acted in good faith, all the while hiding behind the flawed work of NIOSH? And here the case really gets interesting as one must wonder if NIOSH would be willing to stake its name and reputation on shoddy medical work and superficial reporting of environmental conditions.

I'd be remiss to leave out the flat-roofed buildings, especially those so often seen in public schools. If Lindsay is a teacher, what recourse does she have if her school made her sick? She proved that beyond any doubt, but since she can't sue the school district, who will cover her losses? She lost hundreds of hours of sick time, not to mention her chance for promotion, because of her illness.

And what about the children? What do we do with kids in public schools when the teachers are sick, the maintenance staff says mold is an ongoing problem but the principal and the superintendent say mold never hurt anyone? Predictably, the administration trundles out an indoor hygienist who took trivial five-minute air samples that showed nothing wrong. He says the school is so safe he would want his children to attend, but unfortunately won't be able to do so. Based on multiple studies, we know that on average, 24 percent of all kids are at risk for illness if they're exposed to a WDB. And based on average, negative air samples aren't reliable to assess the absence of the potential for illness.

Finally, let's say our example patient works in the main office building of the State of New York at 10 State Street in Albany. Let's say this patient is proven to be ill (proof comes from the repetitive exposure protocols, like the ladies above did, that I'll talk about often in the chapters that follow), and the building managers quietly make individual accommodations for her. But what's done for the other 1500 workers in the building? Is there a screening process that the building or the employees should go through, to see if there's a reasonable concern for illness?

If you were this patient's out-of-state physician, surely you'd refer the case to a New York-licensed physician for follow-up. And when nothing happened, would you contact the Union for the State Workers? Sure, and then when still nothing happened, whose problem is it that people are unknowingly being exposed to hazardous working conditions? If you called NIOSH, would your duty have been met? Do you also have a duty to the developing fetus of the administrator on the second floor? What about to the asthmatic on the eighth floor whose illness flared after he took it upon himself to clean the soot and black stuff out of the duct work near his work station?

And when the asthmatic ended up on a respirator for five days, would the original patient who quietly was retired, the in-state physician, the building managers or the Union (or all!) be financially responsible for their negligence of "failure to inform" that allowed the asthmatic patient to face death? Was there not a duty on all parties involved to inform the at-risk people working in the building that it was dangerous to some?

Finally, let's try to imagine what circumstances could possibly have arisen to create such a Gordian Knot of conflicting issues.

Envision yourself hypothetically as the president of one of the largest trade organizations in the country, an organization that prioritizes lobbying efforts in our nation's capital above the rest of its broad activities. This organization has no qualms about spending millions of dollars to influence the outcome of political campaigns across the country. Now imagine that you have been presented with evidence that a new class of public health threat is directly tied to

the core of your membership's primary vehicle for the creation of wealth -- that of the ownership of Property. This new and deadly threat can be traced to the very business model that has supported the unprecedented expansion of the housing market in this country over the last 30 years, where bigger, faster and cheaper means of building have equated to bigger and faster profits for everyone concerned, from builders and bankers to real estate and insurance agents.

Your role at the helm of this trade organization requires a response from you, both public and private. In the face of potential losses for your membership on the order of hundreds of billions of dollars, and your multinational corporate constituents screaming for a solution, what do you do? Act in the best interest of the people (who are ultimately your customers), and take a huge long term loss that might threaten entire industries to save suffering and lives? Or with almost unlimited power and influence at your disposal, do you solely embrace the bottom line? Once the decision is made to trade lives for dollars, what are the limits as to what is fair in the game of business, national politics, and public health policy?

There are so many variations on the theme of human suffering initiated by exposure to water-damaged buildings and each story I've heard is a little bit different. As I write this in summer of 2010, I've seen and treated well over 6000 patients since 1998 with an illness caused by exposure to the interior environment of water-damaged buildings. Each patient has a syndrome that's readily identified by blood tests performed in standard medical labs and paid for by insurance companies. Each has a multi-system, multi-symptom illness and each can be helped with the protocols our group has published in refereed scientific journals, books and magazines.

Now that we know so much more about "mold illness" than when my first book on this topic (*Mold Warriors)* was published in 2005, the goal of this book is not just to lead the way to truth. Now the goal is to *heal* these patients' of their incredible inflammatory injuries and to *prevent* subsequent relapse. *Surviving Mold* is the beginning of what we'd all like to see: an illness defeated and lives restored, all over the world.

We're entering a new era of acceptance of mold illness as an inflammatory disease. The self-serving arguments against mold illness we heard in the early 2000's have been exposed as nonsense by U.S. government agencies and the World Health Organization. Maybe those who deny the existence of mold illness are acting in good faith. Maybe the moon is truly made of green cheese with an outer core of moon rocks.

I have no doubt that the matter of the integrity, honesty and thoroughness of witnesses for the insurance industry will be decided by the courts and in the court of public opinion as were the same issues decided for Big Tobacco (Chapter 13). When that time will arrive? Perhaps tomorrow.

In the meantime, it's time to tell these patients' stories of how they survived.

CHAPTER 1 Survival: There Is So Much to Learn

Surviving Mold isn't just a fanciful idea. For someone injured by exposure to the interior environment of a water-damaged building (WDB), survival demands a new lifestyle, a necessary adaptation to a life forever diminished. Not everyone needs to learn about what water-damaged buildings do to the 25% of exposed people who become ill, but certainly those people do. And so do their families, their employers, co-workers and classmates.

This illness has strong genetic roots and in the end the injuries these patients suffer won't go away without intervention unless their very genetic make-up is altered. And of course, that doesn't happen.

After "mold illness" begins—usually insidiously—it quietly and silently takes away energy, cognition and easy breathing; it leaves behind pain, fatigue and often weight gain. *Surviving Mold* begins by teaching its victims what's wrong with them and continues by explaining how to get appropriate treatment and prevent re-exposure.

And when there *is* an exposure, *Surviving Mold* means rapidly accessing definitive help.

As you'll see, *Surviving Mold* means getting past uninformed physicians, unreliable, even devious insurance companies and (if involved in litigation) defense attorneys who distort truth as easily as they breathe. *Surviving Mold* can mean leaving the job you need or the home you love. *Surviving Mold* divides families, breaks apart relationships and erodes its victims' self-esteem.

Surviving Mold begins with knowing what's wrong and *developing a plan* to defeat the abnormal physiology of innate immunity that would otherwise inexorably progress to take the life out of living. Having the right plan means that mold illness patients need to both listen and learn, two functions they often no longer perform well. Having the right plan means being patient, because a step-by-step treatment process doesn't allow anyone to skip steps or be creative and do things differently. It means ignoring what you heard from a brother-in-law that he heard in a talk on the radio; or read in a blog written by some self-anointed expert from deserted sands; or read on a web page written by a scheming shyster trolling for your money. All those pitfalls are out there.

Living this plan demands attention to detail, too, but in the end, all this listening, learning and doing, slowly brings back a life filled with meaning and fullness, vitality and insight, as the new therapies (all tested in IRB-approved clinical trials) correct the immunologic disease often called mold illness.

The greatest sense of loss in *Surviving Mold* comes to people who've recovered a vital life, only to lose it suddenly when they're re-exposed to a water-damaged building ("WDB"). The feeling approaches what one might experience when leaving behind an Asian tiger cage prison after five years surviving insane isolation and the torture of deprivation, now to taste freedom, fresh fruit and a cold glass of homemade lemonade, only to be cruelly thrown back into the tiger cage. Can you imagine how devastating that is? That feeling steals hope, breeds depression and erodes the desire to live. Defeating that desolation is part of *Surviving Mold,* too.

Let me be clear. When I say 'mold illness,' that is a jargon term for a chronic inflammatory response syndrome caused by exposure to all kinds of bad things - toxins, chemicals, enzymes, fragments of cell walls and more - all routinely found inside WDB. Those of us who treat mold illness have learned a lot about the disease over the years. I began the life as a mold illness-treating physician in 1998. The science in this field has come a long way since my first mold patient was confirmed to have a biotoxin illness. 6,000 mold patients later (and counting) among a total of 8,000 biotoxin patients, along with other treating physicians I feel that we're entering our first "Golden Age" of diagnosis and therapy.

Along the way, applying new-found knowledge of the physiology of mold illness provided unexpected benefits. Delaney Liskey of Virginia Beach, Virginia, doesn't have demyelinating lesions in her spinal cord and brain any longer; Ben Allen of New Orleans has tossed aside his beta-seron and all of his inhalers. We can reasonably ask, what happened to their multiple sclerosis when they were treated for a mold illness? Read on to learn more about their true stories.

Surviving Mold will share some science and some stories of real people. I'd like *Surviving Mold* to tell you about some of the friends on our side who are striving to make dangerous buildings safe for all. I'll also mention some of the foes. I'd like *Surviving Mold* to bring hope if you're ill and understanding of those who *are* ill, if you're not.

You'll meet Erin Culpepper and Laura Mark, two people who mold couldn't keep down even after it nearly destroyed their lives. Erin can now have a family because her treatment didn't come too late. Laura can be a grandmother since her daughter Shannon now has her life back. You'll meet Jerry Koppel, a man with an illness that therapy couldn't fix, yet he still sacrificed everything to make sure his son's genius was protected and allowed to shine.

One of the most incredible chapters in *Surviving Mold* is the saga of the Belperrons (Chapter 8). Susan and Pierre Belperron literally fought the medical establishment in Boston and the Massachusetts bureaucracy to find a way for their three brilliant sons to obtain the education they needed. Educations that they did finally obtain despite moldy public schools and universities as well as the moldering mentalities of academics, academies and statehouses alike.

Liana Jones has her life back after her odyssey through inflammation and a sinus infection that was mistakenly misdiagnosed as a brain cancer. The other day, she sang her version of "I Feel Good," as if she were James Brown re-incarnated. I joke with her that if James Brown had graduated from Harvard, the song would have been titled, "I Feel Well."

Mystery Diagnosis was intrigued by her story; they will start filming this fall.

Not all days are as lighthearted as Liana's last visit with me. And there are struggles ahead, as you'll read in the chapters on government agencies, lawyers and Dr. Tom Harblin's chapter on policy.

Erik Johnson has been an observer of mold illness since he was sickened in Truckee High School, the scene of the beginnings of the outbreak of 'Chronic Fatigue Syndrome' (CFS) in Incline Village, near Lake Tahoe in Nevada. He's learned that avoiding exposure to water damaged indoor environments is crucial to his survival. Thus, situational awareness is a sixth sense he can't live without. He argues that mold exposure as part of day-to day life is the missing factor in what is called CFS. Only a few listen despite the obvious mountains of data that confirm Erik's observations.

You can listen to Greg Weatherman tell us what all people exposed to moldy conditions need to know about cleanup and remediation. Susan and David Lasater will take Greg's comments one step further as they tell you what they did to Survive Mold.

You'll meet two men who've helped me more than any others. Frank Fuzzell isn't with us any longer, but his teachings about chemical alteration of habitat, deception and duplicity of agencies, and the power of large corporations lives on. Professor Matthew Hudson has the last word in *Surviving Mold*. The Professor is a true leader, inspiring confidence and loyalty in those who know him. I won't be offended if you read his words first.

Yet for all our advances in identifying and developing treatments for the innate immune responses gone haywire (the source of the many symptoms of the illness), there are more advances on the horizon. Soon we'll be unraveling the fundamental basis for all these inflammatory immune responses by showing the differential gene activation - and suppression - that'll show us the genomic basis for immune responses gone wild. Soon we'll be able to provide targeted gene therapy for those who have become ill and possibly, targeted gene therapy for those at risk. Surely, the funding for this research will come.

It may seem like a simple idea, but once we know which genes are bad actors, then we should target our therapies to "shutting off" the noise made by these genes. But until that day comes, *Surviving Mold* means learning what's in this book. The problems faced by mold survivors stem from a dense injury to quality of life. We'll meet a clutch of people who've succeeded in surviving mold; we need to learn from what they've done.

You'll read compelling stories of real people; you'll read some fairly dense science jargon that might make your eyes glaze over at times. But give it a try. When the physiology I share becomes your illness, I feel you're empowered by knowing what's wrong. You might feel that reading the science twice is a good idea, but once the ideas sink in, you'll know more than your physician. And I think that's a good thing, because it's important to understand your illness.

If you are a teacher, you need to learn about Lee Thomassen's saga to obtain justice in Baltimore County Public Schools. Others may choose to skip over his detailed story.

For those who aren't familiar with mold illness, it means loss of quality of life in dozens of ways, often accompanied by loss of job, income, home, spouse or a significant other, as well as the loss of the freedom to do what a person or a family always did. For some these losses are almost beyond belief. Yet such patients often "don't look that bad." Imagine struggling to get up by 10 a.m. after a night of non-restorative sleep that began at 9 p.m., only to be too tired after emptying the dishwasher to do much more until 2 p.m. And then hear a relative talking in the next room suggesting that malingering or depression is your problem. It happens every day. We'll meet some rare physicians in *Surviving Mold* who understand enough to listen, but we'll also meet some physicians who treat mold patients with condescension and disdain. It is time to improve medical education in the field of chronic inflammatory responses syndromes for those who arrogantly deny the existence of mold illness. Malpractice insurance cariers may help physicians who won't want to take time to learn something "new" to change their ways. Litigation for failure to diagnose mold illness is here.

We'll meet as well the disability examiner and the defense attorney who attack the stricken mold patient as if his/her claim for some sort of compensation would deprive the attackers of their *own* financial security. For mold patients who've lost their vitality and self-confidence, even self-worth, trying to fend off these slashing attacks rarely leaves enough life force (or financial means) to enjoy even simple pleasures of life.

Usually the mold patient has a legitimate complaint: what just happened isn't fair. What did the nurse at Maui Memorial Hospital do to deserve having her life destroyed by her workplace? What about the child advocate from San Bernardino whose life was devastated by mold illness just by working in the musty building with wet HVAC ducts? How should she react to the list of sixteen experts the defense is planning on bringing into court to attack her and those who treat her? One day of those experts' time will cost more than what she is asking for in settlement. Do we think that a shattered life is a fair reward for helping others?

And what did the hydroponics expert from Princess Anne do to see unusual coagulation problems clot off the blood supply to part of his intestine?

I could share thousands of other stories, as everyone has a true story to tell, but they all lead to the final answer: none of these people deserve what happened to them. They just didn't know that their genetic susceptibility made them prime targets for an unremitting assault by their own innate immune systems, systems that were fully operational before they were born.

The journey to understand the immunology of mold illness has no end. I am eager to see what new insights arrive each month in the mail with the immunology journals. Today I talk glibly about TGF beta-1 and its incredible role in the aberrant multi-system functioning that identifies mold illness patients, yet it was only in April of 2008 that I could get a lab to do a commercial assay for TGF beta-1. Yet the power of this new data was such that by fall of 2008, I revised my long-standing case definition of mold illness to accommodate the additional information about illness mechanisms that TGF beta-1 brought.

At the core of why one person becomes ill from exposure to a WDB and another doesn't is understanding what is involved with HLA. Human leukocyte antigens (HLAs) reflect parts of our immune system passed down from our most primitive, one-celled ancestors. The evolutionary conservation of these defenses over ***three billion years*** is astounding. The marvelous immune weaponry our bodies use to protect us from ubiquitous foreign invaders - called antigens - has devastating personal health consequences if the inflammatory cascades set off by the (invisible to the naked eye) toxins and inflammagens found inside WDB continue unabated. The problem is that for those unfortunate people with mold susceptible HLA DR makeup (called haplotypes), the antigens response system doesn't work correctly. Call it defective antigen presentation if you will, or simply say your genetic makeup left you with a tiny glitch in the software that operates your personal protective robo-suit – think "Ironman."

[*More on HLA DR at* **http://en.wikipedia.org/wiki/HLA-DR**]

Normally, cell structures called pattern recognition receptors (the antigen detection system) will be "set off" by antigens of particular types. This antigen detection leads to the release of inflammation chemicals. Some of these chemicals are called cytokines, TGF beta-1 and split products of complement activation. Gene activation and an explosion of an ever-expanding series of cascades of inflammation responses ensue. This ever-multiplying host defense response to toxins was classically described by Dr. Lewis Thomas in the New England Journal of Medicine in 1972. His eloquent prose from observation of what just a few molecules of endotoxins do to the human host describes what we see in mold illness.

This concept of magnification of host response is opposed to the idea from toxicology that the "poison is in the dose." We now know that this idea from toxicology has little bearing on mold illness, as indeed mold illness isn't toxicological at all, but is immunological instead. Thinking in toxicological terms is somewhat simplistic as it ignores the fact that every person's innate immune system is personal and genetically coded, thus it works differently for each of us. The concept here is that the role of the poison lies in the ***initiation of the mega-multiplying response*** to a given signal. As Dr. Lewis Thomas said, "the **response of the host** makes the disease," not the antigen. Said another way, Dr. Thomas rightfully tells us that, "the reaction of sensing is the clinical disease" (NEJM 1972; 287: 553-555) and, "we are in danger from so many defense mechanisms, that we are in more danger from them than from the invaders."

And so it is in mold illness patients.

Normally the result of all this cellular conflagration is a process called antigen processing, in which antigens are engulfed (phagocytosed) by special white blood cells, called dendritic cells. Once inside the cell, the antigens are "decorated" with HLA pieces, processed further (antigen processing) and are served up as tasty morsels to special immune cells called T lymphocytes (antigen presentation). More lymphocytes are now involved, and they in turn tell antibody-forming lymphocytes, called B cells, to get to work making very specific antibodies that bind to the antigens. Normally then, the next time a <u>non-mold susceptible</u> person walks into a musty basement, his antibodies will target a given antigen and clear it out fast.

That is the way things should be. For those not adversely affected by exposure to the interior environment of water damaged buildings, that is the way it is.

But based on the occurrence of a specific group and sub-groups of 6 HLA haplotypes (out of 54 identifed by lab results in over 6000 patients, Appendix 2) found in a total of 25% of us, the antibody production process fails. These unlucky people are susceptible to water damaged buildings. This antigen presentation system can fail in so many ways (think of a computer operating system full of errors), it's a wonder that any of us make antibodies correctly!

Without antibody protection, the antigens stay around, constantly bombarding us with our own cytokines, more split complement products and more TGF beta-1. Our defenses against this self-barrage are small hormones made (mostly) in the brain called regulatory neuropeptides. The two most important neuropeptides are alpha melanocyte stimulating hormone (MSH) and vasoactive intestinal polypeptide (VIP). Eventually, the constant pounding on neuropeptides destroys the neuropeptide production mechanisms.

Now inflammation goes wild, exuberantly without parental control from MSH and VIP, resulting in high TGF beta-1, high C4a, high MMP-9, low MSH, low VIP and much more. What you now have is a shell of a person who's defenseless against new exposures and is suffering daily from inflammation. The process makes those afflicted miserable but doesn't kill them although medicine has learned that long term inflammation leads to a higher incidence of various other conditions including cardiovascular illnesses and certain brain related diseases.

If only we could patch the antigen processing system or restore the defenses - MSH and VIP! Wouldn't that be a wonderful result?

Actually, we can. I can tell you that use of VIP is a wondrous advance that can send C4a, TGF beta-1 and MMP-9 back into their proper niche. It's taken thirteen years of research to figure out the ins and outs of the Troglodytes lurking inside WDB, but now that we know what they look like and what they do, we can send them back into their dungeon.

As you would expect, cracking the treatment code depends on putting together the right "numbers" on the code at the right time. We all know that opening a lock on a high school locker requires three numbers, dialed in correct order - 3 left, 25 right, 32 left. If you dial 32 left first, the lock won't open. So it is too with mold treatments. You'll learn what to do, but if you decide to be creative by skipping steps or doing things differently from what you're supposed to do, you won't open the lock that mold puts on your life.

There are three themes I want you to follow in *Surviving Mold*:

- Mold illness is real
- Water damaged buildings contain impossibly complex 'stews' of biologically active chemicals
- Mold illness is inflammation which is caused by an immune system that has run amok.

Cast Out False Knowledge.

The key to understanding is casting out false knowledge. You can discard a lot of ideas you might've heard, such as 'mold toxins are the only factor to this illness,' or that 'spore counting is helpful.' You can toss the local pompous toxicologist into the dustbin as well if spouting 'poison = dose.' While tossing out false knowledge, add the allergists advocating shots, too. Mold illness is not an allergy. Allergy shots provide no relief here.

Let's go back to our child advocate in San Bernardino for a minute. She looks healthy until you see the list of medicines she takes. She doesn't have anything wrong, according to the standard labs her California doctors ordered. But she's living a life between her bed and a nearby chair because of her non-existent MSH and VIP, high C4a and TGF beta-1. She gets better with modern therapy yet she has dual HLA haplotypes that are both mold susceptible. She's desperately thinking about starting a family now as she's lost four years of life and the biological clock is ticking. But what will happen to her unborn child if she visits a moldy bookstore or the antique shop down the street? Do the horribly disabling inflammatory responses that will surely come after her short visit to the moldy store find their way across the placenta?

What would you tell her about starting a family? Come back to this question after you are finished reading this book.

You can't ignore your children's safety. If you know that your own genes can make your daughter ill, where will you let the child go? Don't forget the elementary school has a flat roof and the mudslides in an El Nino year cracked the foundation. The local Y was closed after the plumbing leaks and the dance studio has leaked ever since the "minor" earthquake a few years ago.

If your job is to survive mold, what will you do for your children? This book will help you figure that out.

Be Prepared For Surprises.

This book will bring surprises. Treating multiple sclerosis (MS) is an arena with surprises too. It is also an autoimmune disease. For years, I've seen people with scarring on their brain MRI scans. The scars are called "gliosis," small areas where demyelination, loss of protective myelin sheaths, has occurred. These areas of damage are called "unidentified bright objects," because anyone can see them light up on the scan, but what do they mean? If the nerve damage is worse, and physicians find lots of demyelination with neurologic symptoms, often a patient gets an MS diagnosis. I believe if the spinal fluid has unusual proteins called oligoclonal bands, together with the scars and neurologic impairment, an MS diagnosis is often confirmed. The illness is treated with expensive but effective medications that can stop the autoimmune process, but since the basis of the illness is inflammation driven by autoimmunity, no one with MS should be told a cure is coming tomorrow.

There are plenty of people I've seen who were told they had 'MS' but who didn't have oligoclonal bands. Using my protocols, I've seen treatment successes in those 'MS' patients who had been exposed to water damaged buildings. Until recently, none of the patients with oligoclonal bands responded to our therapies designed to correct aberrant innate immune responses.

Delaney Liskey is twelve years old. Her parents didn't want her to be treated with beta-seron for her MS even though she had plaques of demyelination in multiple areas. She was treated for Lyme disease with some benefit and the steroids she took stopped the developing blindness. She'd been a bright young lady, active in sports, a leader in her school, but not anymore. When she had a spinal tap done, it showed the dreaded oligoclonal bands. She saw some of the best MS specialists in Virginia, who patiently waited for the parents to agree to definitive MS treatment.

When I read her new patient packet in January 2010, I saw the neurology reports, the MRI reports, the consultant reports from the Lyme specialist and the primary care physician's comments. Pediatric MS like this wasn't common, yet all the requirements for the illness were there. I called Delaney's mother, Terri Liskey, and told her that I saw no indication that my protocols helped those with the oligoclonal bands. Terri wanted me to at least look. She would do what was right for her daughter when she was sure all other possible problems had been considered. Deep down, she wanted to be sure she had done everything. Clearly, the Lyme treatments had helped, and Guy, Delaney's father, had Lyme too; couldn't I see Delaney right away?

Terri is persuasive, to say the least. Sure enough, Delaney was a bright child with an inflammatory anvil sitting on her shoulder. The home had fairly significant mold problems. I think Guy and Terri were somewhat surprised about this. Delaney didn't have a positive visual contrast sensitivity (VCS) test—it's unusual for the VCS to be positive in a twelve year-year old—but she had lots of symptoms and exposure. She started taking cholestyramine (CSM) while the lab tests were being run. No surprise, Delaney had genetic susceptibility to mold, low MSH and high C4a. But it was her TGF beta-1 (I'll introduce you to all these acronyms and new players a bit later), that was scary - it was over 40,000. In fact, it was so high, the lab couldn't measure it.

TGF beta-1 impairs normal T-regulatory cell function, which in turn contributes to prevention of autoimmunity. There's a nice literature on TGF beta-1 in MS, but no one had ever treated MS by lowering TGF beta-1. Don't forget that exposure to WDB's often drives TGF beta-1 through the roof.

CSM is a benign medication, except it can cause some gastrointestinal problems like constipation, reflux and bloating. I've used it as first line therapy for biotoxin illnesses in thousands of patients for years. For those patients with Post-Lyme Syndrome who take CSM, they invariably become quite ill for a period of time before improving, a phenomenon associated with a rise in MMP9 and TNF that I call intensification. Delaney didn't intensify when she took CSM. Did that mean she didn't have Lyme disease? No, absence of response isn't the same as presence; perhaps she never would have the instensification anyway.

She got better. Her symptoms fell away. She became more alert, brighter, and more able to exercise. And the lesions started to clear. Hold on, said her neurologist. Let's let this play out some more. We can wait on the beta-seron.

I thought this was just a fluke, as spontaneous remission occurs in MS all the time. But it wasn't a fluke. Her TGF beta-1 fell to normal! Her C4a corrected. All her labs looked better. And then the follow-up MRI showed just about everything was back to nearly normal. In a way, I'm not convinced Terri or Guy cares about TGF beta-1, as they now have their beautiful daughter back.

In medicine, we call anecdotal findings like this one an "N = 1" study, as so many variables that could be in play weren't identified or recorded.

Great, Delaney is better; I like happy endings. But the oligoclonal bands should have disappeared. I wondered if another spinal tap was indicated, but who would justify doing that to a beautiful young lady who was doing well? Hippocrates told me to do no harm.

Still, when Ben Allen came in from Charleston, South Carolina (he was in the process of moving to New Orleans), he was disabled due to oligoclonal band-positive MS. He'd been a top government analyst, but no longer. He'd seen all the best MS docs on the East Coast, including two from Johns Hopkins from nearby Baltimore. No question about his diagnosis, even though all his problems started after he was in a moldy building. He stopped the beta-seron he was prescribed because it didn't work and the cost was too high.

Ben was VCS positive and had lots of symptoms, so he went on CSM in April 2010. He had all the inflammatory problems I'm used to seeing in his blood results, but with his oligoclonal bands, I didn't see much hope for him.

Wrong. At follow-up in June, Ben was off all his lung meds, off his Adderall, off everything. He's fine and he's looking forward to going back to work. Oh yes, his TGF beta-1 was too high, though not as high as Delaney's. Ben had been on definitive MS treatment for a long time so we don't know what his TGF beta-1 was before he was treated. So now I have an N=2 study with another MS patient coming in next week. She has oligoclonal bands too. Could there be something here?

And then there are two moldy patients with acute transverse myelitis (TM), a demyelinating illness usually placed in the same demyelinating illness category as MS. Following treatment of their mold illness, they had marked reduction of disability, one able to walk with assistance and the other lost to follow-up after getting better.

Dr. Douglas Kerr at the TM unit at Johns Hopkins has written on innate immune inflammatory problems in spinal fluid of TM patients. Could there be something here?

Then there are the chronic ulcerative colitis (CUC) patients with presence of autoimmune findings of a positive ANCA antibody. I have four mold illness patients with high TGF beta-1 and prior diagnosis of CUC, coming on after exposure. When their TGF beta-1 falls with treatment of their mold illness, the CUC disappears. Could there be something here?

We're just beginning to look at innate immune functions. So much of what I'm seeing in my patients is only recently being described in immunology journals. And under it all is the interaction of genetics, environmental exposure and inflammation. And don't forget, there is body habitus (physique), with all those hyper-mobile people having predictable inflammatory problems and the Ehlers-Danlos III people having chronic inflammatory response syndromes. Can we see what is going on there?

Can you understand why going to work these days is so exciting for me? Spending time on PubMed is now a joy. What else do we see in published literature that involves unknown sources of unusual inflammatory problems?

I guess the idea of wondering about inflammation is analogous to looking up at the Milky Way on a clear summer's night in Pocomoke, Maryland, with the only light around being the fireflies. What else is up there?

I still dream of finding answers to questions that as yet haven't been answered (or even asked, for that matter). For example, here's a mold patient with falling platelet counts. Is it just due to his chronic lymphocytic leukemia (CLL) and the 0401-3-53 haplotype, or does the mold illness cause something we don't yet understand in CLL? Does the search for answers about low platelets in the world of inflammation make any sense? Or is all we can find in CLL are studies with N = 1?

Surviving Mold is about damaged people, but it's also about enquiry. Because there's so much to learn and so many people to help.

15 Things All Mold Illness Patients Need to Know

1. Always have a plan.
2. Be careful extending your ankles in bed, as the muscle spasm, the wrenching, twisting, knotting cramps caused by capillary hypoperfusion, which in turn, is caused by inflammatory compounds seen in some CIRS-WDB
3. Never take steroid meds by mouth unless threatened with death.
4. Trust your new-onset symptoms to tell you to get out of exposure.
5. Ask for objective data from anyone who says you have Chronic Fatigue Syndrome or fibromyalgia (they don't have any). Having a positive XMRV test might not mean anything. *What do the inflammation labs show?*
6. Never do two interventions at once. If you do, you will have little hope of using scientific principles to get better. All you are doing is guessing. See why later.
7. It is a good bet your crawl space is moldy or your basement is moldy.
8. The diagnosis you had last year does not necessarily make sense this year.
9. If a health care provider says adrenal fatigue or androgen deficiency (without testing MSH), be very worried. And when they talk about reverse T3, run away.
10. Demand peer reviewed publications to support the basis of new therapies.
11. Never use cash-only labs unless you need new liners for canary cages.
12. Never trust negative air samples or those who rely on them to say "no mold."
13. Administrators are known to lie. And they do. Trust rarely.
14. Don't believe for a minute that you need virus killers, anti-depressants, fibromyalgia pills or bizarre therapies. You need a plan based on hard science.
15. Always evaluate action taken thoroughly. Never make assumptions without checking.

CHAPTER 2

How *Mold Warriors* Evolved to *Surviving Mold*

It's been five years since the book *Mold Warriors* was published, but the national and international approaches to mold illness since 2005 have evolved exponentially since then. Was *Mold Warriors* a catalyst for those changes? Perhaps indirectly in a small way. Judging by the number of depositions on behalf of patients in which I was cross-examined with the defense trying to discredit me by quoting a selected line or two from the book - I'm certain the insurance industry defense bar felt *Mold Warriors* was important. As an aside, the defense has yet to discredit me from anything written in *Mold Warriors*. They will surely try to find something they can use in *Surviving Mold*.

But the main driving force in the changing approach to mold illness has been the growth of evidence-based scientific support. Evidence that's shown how exposure to the interior environment of a water-damaged building ("WDB") causes illness. New technologies and new testing have helped to create a mountain of data such that a growing number of high level consensus panels now agree that exposure to WDBs causes an inflammatory illness. The science is so compelling that arguments against the illness now fail in court and elsewhere. However, despite the scientific reality, insurers and others are still paying attorneys to say mold couldn't hurt anyone, and, sadly there are still some judges out there who are ill-informed and so they buy into the self-serving defense palaver.

The truth is that while the substantial majority are not permanently harmed by mold - plenty of people are (about 25%) and the readily available science tells us who is, and who is not harmed. As today's mold and other pathogens in water damaged buildings evolve, more of us or our children may become victims due to our own genetic make-up.

Wrongful decisions in mold cases are contrary to the science and logic, but don't forget that judges are not as visible as the umpires in baseball playoffs who can lose their World Series credentials if they call fair balls foul and blow calls under the scrutiny of national television cameras. Although blown judicial calls can be appealed, and often are, the legal system is needlessly expensive and thus unavailable to the ordinary victim in personal injury cases.

When you hear an aggrieved participant in litigation say, "I'm taking it to the Supreme Court," just ask if that person has the cash needed to go through the legal process. An underfinanced legal battle normally means a lost legal battle- and that is usually due to procedural maneuvering. Many small claimants have learned what "papered to death" means.

If the reason a person lost a court case about mold is because a 'Frye' or 'Daubert' challenge (both relying on the Judge's own assessment of other's opinions) prevented the presentation of their evidence, just remember that truth isn't always going to be the winner in court. And what is true in a New Mexico court, on the very same facts might not be considered to be true in an Indiana court! The vagaries of a legal system that is contorted and "local" comprises only one unhelpful paradigm.

What about creating our buildings? Another example of an unhelpful system. Society must learn that construction defects, or other defects, that permit water entry into living spaces or those immediately adjacent that share the same ambient air, must be recognized and corrected *quickly*, because the link between water intrusion and human illness is now extremely well demonstrated. Face it: if something is made by man it will fail. All buildings will leak at some time.

But the process of change is slow. Knowledge about mold illness has been built and has grown in the same way a series of water droplets eventually becomes a stalagmite that grows toward the ceiling in a limestone cave. Countless patients, law suit decisions and physicians learning one by one have built the bridge that's led to today's understanding. What we know now about mold illness dwarfs what we knew in 2004. I wonder what we *will* know in six more years.

Back in 2004, there was only the hope that opinions about mold illness written by U.S. government agencies and the World Health Organization would actually mirror what we knew then to be true. Today, sitting on my desk is the U.S. Government Accountability Office (GAO) statement about mold illness that was published in September 2008. And on top of that is the July 2009 missive from the World Health Organization supporting the ideas that our group and others published since 2003. We now have clear evidence that shows that mold illness comes from exposure to the diverse mixture of compounds found in water damaged buildings (WDBs) and that the illness is based on the inflammation those compounds cause. Furthermore, it's <u>inhaling</u> those inflammagens (antigens, or foreign particles, that activate innate immune responses) that makes people sick, based on their genetic make-up.

I had four goals for *Mold Warriors*: bring information to the public, the legal system, the political process and physicians about mold illness. The order of these goals isn't random. Physicians are the slowest to realize what new information means, and slower still to apply that information to practicing medicine. The legal system often lays claim to being the slowest responding group, yet entrenched medical opinion changes glacially, if at all.

Quite frankly, leading physicians to new understanding is a slow, expensive process. One needs look no further than "Big Pharma" to see what works to alter physicians behavior. Big Pharma's success in manipulating and changing physicians' behavior through countless advertisements and visits from drug company representatives is mute testimony to its effectiveness. What would the allergist do when the mold drug rep showed up touting the benefits of EXP 3179 to lower TGF beta-1? Surely he'd want to learn and listen more about innate immunity in mold illness if the rep had a pretty smile and a nice-looking ankle, especially with low-cut socks, wouldn't he? But we have only knowledge to offer so must resort to books along with testimony when our patients need us.

For a while, the CDC promulgated the idea that mold growing indoors was just a nuisance rather than a major problem - their pronouncements about the human health effects seen in New Orleans after Hurricanes Katrina and Rita in Fall 2005 basically said, "Don't worry about it." That attitude changed 180 degrees in March 2007, when the CDC published a paper in *Applied and Environmental Medicine* that showed that levels of mycotoxins, endotoxins and beta glucans seen in water-damaged homes in the Gulf Coast were known to be associated with human illness. The papers they cited to support their conclusions had been published much earlier. Curiously, they weren't discussed in the 2004 Institute of Medicine (IOM) report commissioned by the Centers for Disease Control and Prevention (CDC) that had not found "sufficient evidence of a causal relationship with any health outcome, and had concluded there was insufficient evidence to determine an association with many health effects" and mold. Yet suddenly, in 2007 this old information was a prized bit of cutting edge scientific advancement. Am I the only one who wonders what happened behind closed doors such that papers written in the 1990's became important to the CDC in 2007 but weren't important in 2004?

Is it just my concern or does the sudden acceptance by the CDC in 2007 of old information from a decade earlier, evidence that the CDC had discounted for years, seem a bit odd? Maybe the old information was like aged 2001 vintage Primitivo wine (Salamandre Cellars, Aptos, California), better in 2010 than it was in 2003.

True to CDC form (I admit to doubting the reliability of agency pronouncements), however, one can see where the last sentence in the 2007 paper had been added. It says, "The health effects from exposure to these highly contaminated environments are still undetermined." This statement *has nothing to do with the paper*, as no recording of health effects was ever attempted, but someone, no doubt a politically responsive manager (see Chapter 19, New Orleans), must have demanded that it be added, almost certainly as a bone thrown to insurance companies faced with mold illness claims. In a very real sense we have the major taxpayer funded health agency having moved from denial that guns make loud noises to suddenly announcing that earlier test data now showed that guns not only made loud noises but shot projectiles which could penetrate people but (not to worry) health effects from being shot are still undetermined.

Systematically supplied misinformation is the kind of major hurdle those of us who know the truth about mold illness have faced. I've often wondered from where that misinformation comes, as the misinformation supply process we see today parallels what Big Tobacco put out for years about the health effects of cigarette use. As part of all the Tobacco litigation, a library was formed that is a public repository for all the shenanigans that Big Tobacco pulled off for years, complete with details on *who did what*. Indeed, I wasn't surprised to see individuals referenced in the Legacy Tobacco documents library (**http://legacy.library.ucsf.edu/**) now prominently involved with mold litigation defense. If you have the time and interest, search the Library for the names of people who might be involved in a mold defense in litigation: if you find them in the Tobacco Library, don't be surprised. Big Tobacco paid Big Medical and thousands of lawyers well for their opinions and their manipulation of the legal system. The highly paid players that gamed the U.S. health care system for years about health effects from tobacco are still alive and are still doing what they always did for their payors.

Of course, you can measure change in many ways. Educating the public and the medical community has been terribly slow if you measure our progress in days or years or in the amount of suffering and loss that mold patients have endured during that time. If pain is summed with physical and financial loss, we're now reaching stratospheric levels. But remember, it's only been a few years.

I think each of my *Mold Warriors* goals has been met, though there's yet more work to be done. *Mold Warriors* showed the world that exposure to the interior environment of water-damaged buildings hurts people. The basic feature of what was going on physiologically stemmed from inflammation from a part of the immune response called innate immunity. Let's step back a minute to ask *"What do we know about innate immunity and how long have we known it?"*

A keynote address on innate immunity and immune responses was given in 1989 by Charles A. Janeway, Jr., PhD, at the Cold Spring Harbor Symposia on Quantitative Biology in Long Island, New York. Dr. Janeway correctly predicted that recognition of antigens, engulfed into specialized immune cells, facilitated by HLA molecules, would be the fundamental basis of the host response to external stimuli. In essence, the host (you and me) response goes haywire when the innate immune system doesn't recognize potentially infectious organisms, inflammagens and toxins. Dr. Janeway said, "This immunology will integrate innate and induced immunity ... for the myriad T-cell surface molecules with triggering capabilities ... that includes a role for receptors that recognize patterns of microbial structure."

In other words, antigens determine how your body responds to external stimuli, including viruses, bacteria, fungi, and other foreign invaders. If your body's process for identifying and controlling these foreign invaders is compromised in any way, so is your health. *Antigen recognition* is a process of identifying particular structures that are conserved in evolution and remain constant in bacteria, fungi, dinoflagellates and spirochetes (sources of "my illnesses"). As will be discussed in several places in this book, when the process of antigen detection, processing and presentation fails at any stage, antibody formation cannot take place. Without antibodies there can be no removal of antigens and the resulting ongoing innate immune activation causes chronic illness. Said another way, if the antigen isn't removed by an antibody, the stimulus for initiation of inflammatory responses from antigen detection continues and the illness becomes chronic. Our entire being suffers from friendly fire from our own innate immune system.

Although I'm concerned that some physicians don't use innate immunity concepts in their daily practice and don't spend the extra *two minutes* during an interview to search for environmental clues to illness, in their defense, I must say that the information I want them to use was fleshed out in earnest only 20 years ago! Nonetheless that information is available at a moment's notice from online sources. In the Internet Age, 20 years is an eternity. Twenty years ago was 1990; who had cell phones then? Facebook? Twitter? Not even dreamed of back then.

Let me give an example of the role of the Internet in helping resolve complex medical diagnostic procedures. I was recently asked to testify in a case of wrongful death in which a child died from an unusual tick-borne illness called tularemia after having been bitten by a tick carried by a rabbit. The case went to trial twice, each time ending in a mistrial because the jury could not agree whether the illness was one that the physician would be required to know. The facts were clear: after a tick bite and removal of the tick after it was attached for at least 36 hours, the child developed fever and an enlarged lymph node (lymphadenopathy). The physician used a computer for all aspects of his group medical practice including the on-line textbook, *Harrison's Textbook of Medicine*. While under oath I accessed *Harrison's Textbook of Medicine*, typing in the key words "tick bite, fever, lymphadenopathy." That is what the patient had, right? Within two seconds, a series of references appeared, including several for tularemia. Given that the physician relied on computers for every aspect of his practice, one could ask what happened that the physician didn't use the computer to find an easy answer to an unusual presentation of an illness that was readily identifiable, especially with computer prompting.

Was the physician responsible for the child's death because he neglected to spend a few seconds to look for the answer online? The jury members couldn't agree. No one would hold that a physician must know all things about all illnesses, but the process that the physician used for his practice wasn't followed in this case and a child died of treatable illness because of absence of due diligence in thinking about what might be causing the child to deteriorate rapidly.

My point is simple. As far back as 1989, the medical "generation time" in which a physician's knowledge becomes out of date was said to be about five years. With the around-the-clock availability of computerized library resources (see the National Library of Medicine, PubMed; **http://www.ncbi.nlm.nih.gov/pubmed**), information is instantly available.

Physicians who say they don't use innate immunity in their practice actually use it every time they talk about inflammation or treat inflammatory arthritis or inflammatory lung disease. Just look at current models of childhood asthma, an illness that's exploding in incidence in the U.S. The EPA says that 21% of all new cases of asthma are due to exposure to WDB. 21%! The important concept in the genesis of the illness is based on "remodeling." Remodeling means "something" happens that the airway changes to be more reactive and more in need of drugs to assist normal breathing. Type in "transforming growth factor beta-1 (TGF beta-1)" and "remodeling" into the search field on PubMed and you'll see more than 840 academic papers discussing the role of TGF beta-1 in modeling in lung, heart, liver and kidney. The chronic inflammatory response syndrome that patients suffer when they are sickened by exposure to the interior environment of WDB includes a marked rise in TGF beta-1. Rising TGF beta-1 from mold *means* that breathing problems will follow exposure to WDB.

Of course, we now know that unusual neurologic problems and autoimmune problems also follow high TGF beta-1.

The day is coming closer when we'll look at asthma as an illness of remodeling rather than an inherited allergic illness or an ill-defined illness of modern society. Is TGF beta-1 part of an arcane arm of immunology? Type in "TGF beta-1" on PubMed and look at the papers. More than 50,000! When a physician says he hasn't heard of TGF beta-1, I have to wonder why not.

Not long after Dr. Janeway's stunningly prescient lecture, Toll (German for "Eureka") receptors were identified, then c-linked lectins, mannose receptors, dectin-1 and dectin-2 receptors, among a host of others. T-cell signaling leaped into our lexicon, primarily from the initial HIV/AIDS research; now cancer and autoimmunity researchers are looking into it intensely. Second signals of cytokine and complement now dominate discussions on mold illness and essentially all of rheumatology. Pattern receptors are linked to differential gene activation, a process that promises much in our future understanding of health and illness. Look at genomics, the study of human gene activation, a field that's now barely ten years old. How soon will practicing physicians need to know about differential gene activation?

Remember that 1989 was just 21 years ago. Since then we've witnessed an explosion of information about host responses and innate immunity that affects all of medicine. What we see in mold illness portends the next wave in immunology. If effective antigen presentation (from innate immune mechanisms) leads to antibody formation (acquired immunity), and this is indeed the case, how do we fit *defective antigen recognition* into the scheme? Mold illness will help in this emerging field of immunology, as the parallels to "over-response" to antigens due to *absence of protective antibody formation* that are seen in mold illness apply to so many other diseases, including autoimmunity and cancer.

In the end, solving defective antigen recognition holds the final cure for mold illness. Until that information is available (The time is coming nigh! T-regulatory cells anyone?), one goal of *Surviving Mold* is just that: to help you survive. Ideally, we all want to live well, but when a treatable illness is ignored or suffers from absence of due diligence from health care providers then the quality of your life is in jeopardy.

Let me introduce you to some players who hold the secrets of survival. The names may be foreign to you, but since they're the things that hold the secrets of surviving mold, meet them today and perhaps know them as friends tomorrow. I don't feel you need to be an immunology expert to read further, but I don't want you to turn away from learning more. Take the time to learn the language of mold illness and we'll try to make things as understandable as we can. No one says learning is easy, but that doesn't mean you can skip the learning process when it's your illness. Knowledge is power.

Mold Warriors was published in April 2005. By June of that year, the book no longer reflected the cutting edge of mold illness medical knowledge. Before June 2005, **C4a** was basically an unknown compound to me. Dr. Patricia Giclas of the National Jewish Medical Center in Denver had performed a series of assays on my patients for a related product of complement, C3a, with interesting diagnostic findings in patients with chronic fatiguing illnesses caused by bacterial infectious diseases, especially Lyme disease. She and her co-workers had worked with C4a extensively in the past, noting its relationship to 'Chronic Fatigue Syndrome.' I began to send her blood specimens from mold patients and the jaw-dropping results changed my approach to diagnosis and treatment of mold illness like the giant leap for mankind changed what we knew first-hand about the moon's surface. After June 2005, C4a became the inflammatory marker of greatest significance looking at innate immune responses in those with exposure to WDBs.

The new findings weren't over, either. We knew that deficiency of melanocyte stimulating hormone (**MSH**) was incredibly common in patients with mold illness and that far too often MSH didn't return to normal after therapy. Low MSH means increased susceptibility to mold illness (and others), ongoing fatigue, pain, hormone abnormalities, mood swings and much more. MSH is a hormone, called a regulatory neuropeptide, and it controls many other hormones, inflammation pathways and basic defenses against invading microbes. Without MSH, bad things happen; chronic sleep disorders with non-restful, non-restorative sleep develop and endorphin production is reduced, so chronic pain follows. Unusual germs begin to grow in strange places; the gut loses its normal defenses against large proteins and unusual antibodies to one of the key parts of gluten (called gliadin), start to appear.

The driving force behind MSH production in the brain is a compound often associated with obesity called **leptin**. If there's a surge of inflammatory responses that disrupt the normal binding of leptin to its receptor in the brain (in the hypothalamus especially), a process that results in MSH production, the body responds with more leptin release, trying to override the blockade of leptin activation of that receptor. Leptin turns on how tightly the body holds onto fatty acids. When leptin is high, one holds onto fatty acids and stores those compounds in fat, so guess what happens? Rapid weight gain and essentially, that process guarantees that weight loss with standard approaches like eating less and exercising more *will fail*. The worse the impairment of MSH activation by leptin (called leptin resistance), the higher the leptin level will climb and the worse the weight problem becomes. Let me tell you: if leptin is rising due to ongoing inflammatory responses, just forget about losing weight via exercise and cutting back on calories. Nice illness, you'd say: exposure to a WDB causes the kind of inflammation that knocks out normal leptin function in the brain; therefore that inflammation lowers MSH (which acts to assist in restorative, restful sleep AND is made with beta endorphin, a compound that controls pain perception) and raises leptin, making too many people chronically tired, in chronic pain and forever fat.

For those who saw MSH rise back into the normal range of 35-81 ng/ml, health was restored. As we'll discuss, those with a particular genetic make-up almost never saw their MSH recover. But replacing MSH wasn't possible, since there was no commercially available form of MSH approved by the FDA for use in people. What could we do to correct the loss of MSH, the master controller? Wasn't there something else we could do for mold illness? You bet.

Don't be misled by the bogus normal range of 0-40 for MSH that you will see on lab reports from LabCorp. It has always been 35-81; simply deciding that all the low MSH levels *had to be normal*, as LabCorp said they did in 2006, is no basis to change a normal range. Without knowing who is a case and who is a control, a lab may not arbitrarily change a normal range.

By fall 2005, we knew that levels of vasoactive intestinal polypeptide (**VIP**) were critical values in understanding the inflammatory responses seen in mold illness, particularly the unusual shortness of breath, especially in exercise, that this illness brings. *Mold Warriors* never even talked about VIP! By 2006, we had clear confirmation from a world of peer-reviewed literature (and data we had collected) that showed VIP plays a role similar to MSH in regulating inflammatory responses, but VIP was far more important in shortness of breath and multiple chemical sensitivity, than MSH was (see Chapter 16, VIP). VIP deficiency was just as common as MSH deficiency in those with mold illness, though it was more commonly found in people who didn't have mold illness than MSH deficiency was. Still, if VIP were available for use, then maybe we could help restore the missing *regulation of inflammation* that was typical of mold illness. And (here's the point) if we restored regulation, wouldn't that be like hitting the reset button on the innate immune problems mold illness people had?

The idea sounded good, but in the end, what happened in re-exposure to a WDB had to be prevented from causing another new round of inflammation that led to another new round of lack of regulation. Who cared if we put a "regulation of inflammation Band-Aid" on a previously sickened person if we couldn't fix the basic problem of defective antigen presentation? In other words, what ultimate good would we do if we could replace VIP, only to have the person get sick all over again when they entered another WDB? Replacing VIP or MSH doesn't solve the primary problem - how a mold illness patient's immune system becomes unable to regulate inflammatory responses. Having said that, we now know that there are a significant number of people who aren't exposed to WDB who are essentially 'cured' when they use VIP for a period of time, maybe two to six months, such that they don't have over-active inflammatory responses with re-exposure and all the associated downstream hormonal and inflammatory problems once VIP comes on the scene. Those people say good bye to chronic disabling fatigue, regain their cognitive functions, begin to breathe normally and more. I now use the "cure" word for them. Exciting? You bet.

Without adequate regulation, however, really bad things happen to our own inflammation when the initiator, the antigen, isn't removed. VIP won't help then.

Our questions became both easier to answer and then more complicated in April 2008 when measurements of plasma levels of transforming growth factor beta-1 (**TGF beta-1**) became available. At that time, there were more than 45,000 references on the importance of TGF beta-1 in the scientific literature, yet no lab offered a commercial test to measure it. Here was a compound that was a cytokine, an inflammation *producer* that could also *suppress* inflammation. It could change the basic appearance and function of cells, making them more primitive (called epithelial to mesenchymal transformation, EMT), thereby changing for the worse what the cells did. Even worse, the compound participated with other inflammation producers to create a new kind of inflammation cells called Th-17 lymphocytes. They set off an entirely different kind of inflammation than any we'd seen before. TGF beta-1 went through the roof with mold exposure.

So there you have it: fixing mold illness means correcting how your body regulates inflammation, reducing C4a and TGF beta-1, and restoring VIP. We want to stop the incredible rise of C4a invariably seen with the second or third bout of mold illness (we call that phenomenon, "sicker, quicker"), stop the TGF beta-1 from going nuts, and then we want to regulate inflammation and correct defective antigen presentation.

And once those four fixes are in, we'll want to take some time to rejoice in a sense of mission accomplished.

For the record, my use of the term "**mold illness**" needs to be clarified: it's merely an expression for a sub-category of biotoxin illness, a jargon term, a shortcut I use, one that connotes that *this illness is an acute and chronic, systemic inflammatory response syndrome acquired following exposure to the interior environment of a water-damaged building (CIRS-WDB) with resident toxigenic organisms, including, but not limited to fungi, bacteria, actinomycetes and mycobacteria as well as inflammagens such as endotoxins, beta glucans, hemolysins, proteinases, mannans, c-type lectins and possibly spirocyclic drimanes, plus volatile organic compounds (VOCs).*

There is no one element that is causative of mold illness. It is the end result of countless aspects of innate inflammation merging together.

While molds themselves may be associated with human illness, in my opinion the secondary metabolites of microbial growth and inflammagens from those microbes are of greater overall importance. Until we have a way to identify the specific interactions that each individual component of the "biochemical stew" found inside WDB can cause, we can only look at exposure to the entire interior environment as the cause of illness.

We live in the Era of Dangerous Buildings, so the likelihood that you won't ever be exposed to dangerous buildings is quite low. We're also still living in the Era of Dangerous People, so look out for enemies of truth. Be careful what you believe, as today's facts are tomorrow's fallacies. Look out for those who may have conflicts of interest. Especially look out for Internet doctors who embellish their credentials until they appear to be the second coming of Christ and Buddha all wrapped into one long PayPal account (oh yes, you will pay for those guys' empty advice, pal). You will see them for their narcissism, their self-aggrandizement and their thoroughly worthless "science." You might recognize them by their insistence that you sign forms that seemingly make you give up the right to criticize them or sue them. You can't give up these rights. If you see these forms, run out of that office, don't walk.

But if you are being billed thousands of dollars for a few minutes of phone time and receive no help, and don't recognize you are being defrauded, you aren't alone. It is common practice these days for "enemies" to complain maliciously to State Medical Boards that a physician is unfit (especially in Lyme disease!). But when certain practice parameters like fees and integrity get in the way, the State Boards often will investigate that practice. If you don't have data in your hand from the docs in question that can withstand scientific scrutiny, ask your doc simple straight questions. If the answers don't make sense, ask the State Board to help you answer some simple questions:

1. Is what he says true? Maybe the problem is just a difference of opinion.

2. Has he lied about his abilities? (Really look hard at someone who says he is credentialed by more than two organizations, especially if he has been tossed out of hospitals for failure to practice good medicine)

3. Is what he describes based on good science? Ask for the bibliographies. You will see "haruumph, haruumph" a lot on some websites. Look for the "blowhard titers."

4. Is what he is charging consistent with norms for other practitioners? I'm not suggesting that physicians shouldn't be allowed to charge for their time, but $900 per thirty minutes on a phone call is a tip-off to fraud.

5. Is he suggesting you buy drugs, antibiotics, supplements and nostrums from him since his are *so much purer* and better than others, even if they cost just a little bit (actually, a lot) more than what you can buy in a pharmacy.

6. If some doc says you have a biotoxin disease such as Lyme when all your tests are negative, but then says you have Bartonella and Babesia, too (when the IV antibiotics they sold you based on the unconfirmed diagnosis they made themselves don't help), also based on no data, run to the nearest medical body and yell fraud before you give away your gall bladder to needless use of Rocephin and your veins and gut to needless use of vancomycin.

I want you to *survive* WDB. Learn about 'mold illness' in detail; this book is a good place to start.

CHAPTER 3

Assuming Assumptions Are True: Ass² Medicine

"I never guess. It is a capital mistake to theorize before one has data. Insensibly one begins to twist facts to suit theories, instead of theories to suit facts." Sherlock Holmes.

"The diagnostic approach that assumes that previously made assumptions are correct is best described as Ass squared (Ass^2) medicine." Ritchie Shoemaker.

My practice requires that I review countless medical records of prospective patients. My practice population may be skewed in that few people would come to Pocomoke, Maryland, for medical evaluation if their own local physicians had been successful in treating them. So it's no surprise that my practice has far more new patients who *haven't* been helped by treatment than well patients coming in for routine checkups. On average, my new patients have seen just over <u>ten</u> physicians, trying to find a diagnosis and/or treatment that actually help. Usually, if you are seeing me it is because the system failed you. I can't tell you how many of those physicians' medical records show what I call *diagnosis by assumption*.

And it drives me nuts. More than anything else, the flawed diagnostic and therapeutic process that I see written in black and white has made me critical of modern day medicine in America. It makes no difference if the patient is seen in the ivory towers (some of those institutions are the *worst* for patients with chronic fatiguing illnesses!) or in medically under-served areas. The ongoing themes of sloppy medicine, lack of attention to detail and lack of adequate time to sort through complex problems drive an endless cycle of more sloppy medicine at follow-up, more lack of attention to detail and less willingness to take time to develop a broad-based perspective on complicated patients. If the problem is compounded by failure to read current literature, then education might be on the list of solutions. But the bigger problem is attitude. "Mrs. Johnson always has twenty problems; all her tests are normal and that mold stuff is a bunch of nonsense."

It is negligence in its purest form to make such idiotic statements in this day of instant access to information.

The problem of assuming anything in medicine is more than just guessing wrong at first leading to more wrong guesses and then even more wrong guesses. *Believing* that what one *wrongly guesses* to be right actually is right, hurts so many patients. Often the wrong guess somehow becomes bronzed as being right in the medical record.

All of us make guesses, though we might call them hypotheses, clinical impressions and working diagnoses, but all science is based on verification of hypotheses and all medicine is based on differential diagnosis that challenges today's diagnoses tomorrow. Instead of saying I don't believe in mold illness, a better response is to ask, "What data support this diagnosis?" If the answer isn't obvious, it becomes incumbent upon the physician to rule out the diagnosis by knowing what mold illness involves. If the physician doesn't know what inflammation from WDB can do, how can he help a mold illness patient? Guessing won't do that.

In mold illness, many times the wrong guess begins with an assumption by the physician that what he has heard from the Nay-sayers about mold illness is true: It's "implausible," or "just a bit of fuzzy mold won't hurt anyone," or "no one in my experience has ever been made sick from mold."

If the CAT scan, MRI and endoscopy are normal, not to mention the EMG, spinal tap, sleep study and more, there are physicians (# 1) who *assume* that their patients have *psychiatric* problems as an explanation for their many symptoms or they're malingerers or worse. The next physician (#2) who sees the patient reads comments in the medical record like, "diffusely positive review of systems; all studies normal, stress in life," and immediately understands the implied assumption: *the patient is a wacko*. The assumption continues: not just that the patient isn't truly ill, but physician number #2 introduces new medications and more tests (they are normal too), essentially making additional diagnosis and treatment decisions based on the original but incorrect assumption. This happens every day in American physicians' offices.

Assuming that an assumption is correct is what I called Ass^2 medicine. Once you see it in action (every day, every-where, and in who knows how many patients with chronic illness) you'll never fail to see it as the "accepted" standard of care. Oh yes, you are stressed. You are overweight, you smoke, and you don't exercise. That's the reason (pick one from the above list, take two if you want to) you are so tired and hurt and can't breathe or think or get through a night with restorative sleep.

Does anyone with a multisystem illness reading this, who hasn't had helpful interactions with physicians (perhaps multiple physicians), think I am joking?

It goes something like this: when presented with data about symptoms the physician doesn't believe, accept, understand how to put together or know about, the physician simply discards the data, believing (assuming) that there's no basis for credibility. After discarding the objective data, the physician must create an explanation in his construct of illness for what causes the patients symptoms. He cannot accept what is right in front of him; the patient has told him and he won't listen. The labs he didn't order don't tell him what is wrong and he won't read any part of the robust literature on innate immunity. He creates an assumption for what is right and then justifies his assumption with groundless action. The basis for the assumption?

I'm not going to blame the collusion of the insurance companies here. Others do. And I'm not going to blame the developing industry of throwing drugs at "fibromyalgia," all the while telling patients to eat better and lose weight, sleep better and exercise more. Sound familiar?

Meanwhile the patient has (i) leptin resistance and can't mobilize fatty acids or burn fat by direct beta oxidation; (ii) has MSH deficiency and has no chance for restorative restful sleep; and (iii) has low VEGF and low VIP and will never deliver oxygen adequately in capillary beds to actually achieve meaningful aerobic metabolism in exercise. Therefore they are wiped out for several days after trying to do something (that idea is memorialized by the CDC term "post-exertional malaise," for goodness sakes) and all the exercise did for them was to burn lean body mass, make them even more protein depleted, more tired, more unable to sleep and more unable to burn fat. The physician dutifully records that the patient didn't even try to exercise, has lost no weight and now is taking too many sleeping pills. The next stop is, you guessed it, the Ass^2 zone.

Perhaps the physician's assumption lives on as an example of the physician's clinical acumen in action. Once the created diagnosis is stated in the medical record, no one goes back to challenge the basis for the original assumption, and so the assumption of an assumption is born. And the problem is worse because no one routinely reviews old medical records for what was known, what wasn't known and what was done without supporting documentation. Like I said, I see it *all the time* in medical records.

Here's one particularly egregious example of Ass^2 medicine in a case that went to trial in Washington DC. The plaintiff ended up on a respirator in a teaching hospital ICU due to respiratory complications following mold exposure. The patient had never been on a respirator before and had never had any previous respiratory problems until mold exposure changed her life forever, yet the intern responsible for her ICU care assumed the patient had been on a respirator. I'm not sure how the intern came to that conclusion, since the patient was unconscious. Since she was on a ventilator, she wasn't talking, but there in the medical record was the comment: the patient was known to have been on a respirator before. In fact, the intern was ignorant of inflammatory pulmonary problems that occur often after exposure to a moldy building. He didn't know that mold illness can shut down respiratory function, and thus assumed that something else was the cause.

The resident learned the history of respiratory failure from the intern, and the resident repeated the story to the senior resident and then to the fellow. The attending physician, relying on what his team told him, wrote a long note in the record, saying that in a patient with severe repetitive ventilator-dependent asthma, they should consider an experimental therapy.

No one except the plaintiff's legal team (much, much later) checked the facts. There was a ton of bias in this case, because once she could communicate with the medical team, no one believed the patient. They didn't believe me, either, despite my searching every hospital in the area for the previous five years and showing that the plaintiff hadn't been admitted to any of them, much less been intubated. It's tough to prove someone's telling the truth when the chart has damaging comments in it. The defense lawyer jumped all over the statement in the chart, insisting the patient's illness was just a flare-up of a pre-existing condition. Even worse, the judge agreed. After all, I wasn't there and surely the patient herself couldn't be believed. *Case dismissed.*

Imagine reading the following series of records I saw during the last week:

> **First physician:** Mold can't make anyone sick, therefore this person who says she's ill is trying to avoid work or blackmail a landlord to get money. She's just a malingerer.
>
> **Second physician, after reading physician number 1:** Patient is a known malingerer, diagnosed in the past with factitious disorder.
>
> **Third physician, a psychiatrist:** Patient with known malingering disorder, confirmed by two physicians, has now refused anti-depressant medication, and is fixated on the delusional idea that mold exposure made her ill. Admit for treatment. DIAGNOSIS: major psychotic disorder.

These days, it's easier to correct a bad credit history than a flawed medical record.

Think I'm exaggerating? Check out this painful, but true story: Mr. Jones is living in a basement floor apartment of a water-damaged building and thinks he might be allergic to mold. Dr. Smith is the new primary care physician assigned to his case by his health plan. Dr. Smith has been up much of the night filling out the forms required for him to log onto and use the plan's new electronic medical record system. He isn't happy that he's allotted just ten minutes to meet a new patient and do the typing, too. But rules are rules. Besides, he's got more than $200,000 in medical school loans to pay off and this job pays well. His supervisor will be checking in with him later that day to monitor his compliance with institution policy and to double check the electronic medical record, the one that repeats countless prior entries with each new entry. (*Author's note*: just try to expunge wrong information from that electronic record.)

"Tell me what symptoms you have, Mr. Jones," said Dr. Smith.

"Well, Dr. Smith, I'm glad to meet you. I'll try to be brief, although I hope you'll let me tell you the entire list of my concerns. The last two physicians I saw here didn't seem willing to let me talk. For the last several years, I haven't been feeling well. My basement smells musty and I can see mold growing in my bedroom closet. I have an environmental report that shows microbial growth that shouldn't be there. I made a list of my symptoms so I could share it with you without taking too much time. On a daily basis, more likely than not, I'm tired. I don't sleep well and I awake not feeling refreshed. I ache all the time and I feel weak, too. I won't try to carry a bag of groceries up a flight of stairs without resting first. Every night I fear the muscle cramps in my calves. They're so knotting, twisting and tearing that when they come, I'm in agony. I break out in a cold sweat sitting on the bathroom floor trying to straighten out my leg. I look for blood appearing in my ankles after a night of cramps like that one because I know my calf muscles tore. And my fingers and toes contract like I'm some kind of animal. The fingers look like claws! And I can't open the fingers without pulling them back into position. I'm short of breath, almost like I have to take a series of short breaths to fill my lungs even when I'm sitting still. After just three steps, I'm winded and then my legs begin to yell at me. I cough, a dry cough, but it's all the time and my sinuses just won't let me alone. The surgery I had last year didn't work just like the surgery before then. The gastrointestinal doctor said I had irritable bowel syndrome after he scoped me from top to bottom even though my gall bladder doesn't contract right. I get diarrhea just about every night and even when I go on a cleansing fast. My brain is in a fog all the time. I don't even try to read books any more because I can't remember what I just read. I'm confused a lot, last night I put the milk away in the pantry and I couldn't find it or the cereal this morning. The cereal was in the refrigerator. Moody? Oh yeah, that driver from Delaware made me so mad yesterday when he cut me off, I could've shot him. And my wife says I'm really irritable, but you know, I usually feel sad. Some days I could eat a whole horse and other days I don't care if I eat at all. I sweat so much, a lot at night too, but I usually feel so cold. My fingers turn ice cold. And I pee all the time. Excuse me, is it OK if I get another bottle of water? My fingers get numb and my legs too after they tingle. I feel dizzy when I stand up. My taste is shot, it's almost like someone put a penny in my mouth. There have to be ants or bugs somewhere on me; I always feel they're crawling on me. Maybe that's why I get all these tremors all the time. Oh, and one more thing: I feel clumsy. It's like my feet don't quite work. Why last, week, I must have stumbled three times going up the stairs or walking downhill. And I drop things easily.

"Dr. Smith, are you there? Did I talk too much?"

"Yes, Mr. Jones, I'm here. Is there anything else about your symptoms I need to know? I've looked over your chart while you were talking and reviewed all the test results from your previous work-up. Everything was normal. Your allergist said you were fine and the occupational medicine consultant said that the levels of mold in your home couldn't possibly make anyone ill. By the way, he sent me the ACOEM opinion from 2002 that shows me that mold is just a part of everyday life and can't possibly hurt anyone. I think you might want to read this authoritative rendition of science.

"You certainly have a lot of symptoms so I can't be sure I know what's wrong with you. I see that two physicians, each of whom is well known here, say you're overly sensitive to your symptoms. Before we go too far, I want my colleague, Dr. Freud, to talk to you. Because based on what you just told me, you seem to respond to life with symptoms more than many people. No one can have that many things wrong with them and still look as healthy as you do. If you have a psychiatric problem—and I think you do—you might have what we call a somatoform disorder. Dr. Freud may be able to offer some insight into where these symptoms come from and how he can help you overcome this idea that mold is hurting you. I don't know anything about mold allergy causing all these symptoms but I don't think there's anything to that idea. Let's talk again after Dr. Freud sends me his consult."

Mr. Jones knows the mold arguments (after all, he's been through this routine three times) and he isn't about to get written off with a psychiatric diagnosis by someone without a clue.

"By the way, Dr. Smith, I know you're blowing me off. I've read the ACOEM statement. What a pitiful excuse for science: can you show me any human clinical trial that provides academic support for that report? Of course you can't, because there are none. And since you're sending me for psychiatric evaluation, do you know of any psychiatric illness that has the diversity of symptoms I have? Oh, may I look at your copy of the ACOEM piece? What is written there? It says, 'Dr. Smith, this will help you get rid of the mold wackos.'

"Perhaps you'd like to read some 'authoritative science': the 2010 Policyholders of America position paper on mold illness written by treating physicians, complete with 2000 references; the 2009 World Health Organization report; and the 2008 U.S. government GAO report, all of which support my views about mold. I'll show you the January 2007 article in the *Wall Street Journal* written by David Armstrong that shows how the ACOEM report and its bastard child, the AAAAI report of 2006, are simply publications written by defense consultants published for use by defense attorneys in mold cases. And the 2008 paper written by Dr. Craner completely exposes the fraud at ACOEM generating that paper you showed me. Which, by the way, Dr. Smith, leaves you and your OccMed buddy wide open for criticism."

"Thank you, Mr. Jones. I appreciate your concerns for my job. I'm actually quite busy these days; I won't have time to read those authoritative papers. My office will be in touch with you for the appointment with Dr. Freud. I hope you'll excuse me, Mr. Jones, but I've run out of time."

Do you think I'm making this scene up? Nope. At least Dr. Smith listened to Mr. Jones. Or did he? Did you see how quickly Dr. Smith assumed that Mr. Jones' prior work-up was thorough? And he assumed that the illness was medically unexplained, so therefore, Mr. Jones should be sent to a shrink? And don't forget his refusal to read. At Duke Medical School, as a second year med student in 1975, a grizzled in-patient physician told me one late night on Long Ward (then the in-patient medical floor), "Ritchie, if you can't get your work done in 24 hours, you may have to stay up late because the patient is counting on you. If your patient needs you, you don't sleep."

Was Dr. Smith an Ass^2 physician? You bet. And don't forget that Dr. Smith was again typing in the established medical record that mold couldn't be the cause of all those symptoms. Harrumph, must be psychiatric.

We wouldn't expect professionals in other fields to use Ass^2. Imagine you're a baseball scout charged with detecting the hitting tendencies of the big slugger for a team you might face in the World Series. Your report is incredibly detailed about how the slugger does against right-handed pitching. You see his tendencies on fastballs inside, outside high and low. You triumphantly return to the manager with your report. The manager says what a good job you did, but wonders how the slugger did against lefthanders. How about his handling of a curve ball, a slider or a cut fastball, not to mention the spike curveball from the star pitcher, Cliff Lee?

"Oh, there are such pitches? I didn't see him against a left-handed pitcher."

No, we don't see absurd examples of tunnel vision in baseball. But we do in medicine.

"Oh, there are lab tests measuring innate immunity? I never knew."

This subjective mentality in medicine reminds me of Alice in Wonderland. If she doesn't know "it," she assumes "it" doesn't exist and therefore since "it" doesn't exist, anyone who believes in "it" is silly. Ass^2 was a concept Lewis Carroll knew and understood well.

I get to see Ass^2 in published studies, too. There's no shortage of examples of *really lousy* logic that others might read and think was good science. Yesterday I read a breathless report from the rheumatology department of Robert Wood Johnson Medical School in New Jersey that asserted that the symptoms of Chronic Lyme disease are explained by psychiatric illness (American Journal of Medicine 2009; 122: 843-850). I looked to see if the authors were practicing psychiatrists, but none of the authors listed was a specialist in psychiatry, including the senior author, Leonard Sigal MD. Dr. Sigal is well known for his past attacks on the idea that patients with Lyme can remain ill after taking standard doses of antibiotics, so I expected to find evidence of bias in his paper. No surprise, the paper was full of assumptions that aren't supported by objective data. Ass^2? For sure.

I thought it odd that the entire paper - one that purports to debunk the idea that people with Post Lyme Syndrome are actually ill - was based on his subjective assessment of psychiatric symptoms, performed by the writer, another non-psychiatrist. Moreover he hadn't done any detailed assessment of pertinent laboratory findings (objective) from the patients selected from among a series of visitors to the New Brunswick, New Jersey campus. For example, the paper attacked as "non-diagnostic and unreliable" a previous study that demonstrated persistence of Lyme organisms, objectively confirmed beyond question, after treatment with antibiotics regarding Lyme; but then cited the exact same criteria from the U.S. CDC for diagnosing Lyme. The CDC says isolating the Lyme bacteria, *Borrelia burgdorferi*, was itself alone adequate to establish the diagnosis of Lyme. Naturally, a study like this one would support the idea of a Post Lyme Syndrome. Dr. Sigal didn't say that the CDC criteria were non-diagnostic, just that a paper that used CDC criteria was *not* acceptable as meeting criteria.

Double talk from Dr. Sigal aside, the occupational medicine department at Rutgers, where Dr. Sigal used to work, has raised questions about "medically uncertain symptoms (MUS)." Medically uncertain basically means there's a subjective disagreement about the cause of a given group of symptoms. And if in doubt about the source of symptoms, particularly if there were an occupational source, call in the psychiatric team for more subjective assessment. If Mr. Jones were seen in the Rutgers clinic, I suspect the diagnosis would be MUS, quickly followed by a psychiatric diagnosis and therefore (1) not mold or (2) not Lyme or (3) not CFS and most assuredly (4) *not compensable* if he tried to say the basement rental made him ill. No pertinent objective lab tests would be done. All of this palaver is about the doctor and none about objective tests to ascertain the patient's welfare since the patient after all is only you ... And you didn't go to med school.

I have no vested interest in Lyme as a diagnosis except to note that just about everyone with Post-Lyme Syndrome has two particular HLA DR haplotypes of the many that exist. Our group published this finding in 2003 and it was confirmed in 2006 by a well-regarded Lyme researcher, Dr. Allan Steere. When I see a genetic association with persistent illness, I look for and routinely find underlying problems with innate immune responses. So if I don't see any analysis of appropriate lab testing (objective) in a MUS/psychiatric source of illness study, well, the study isn't reliable. But I can see how it's difficult for a judge, trying to decide scientific issues beyond his training, assessing the weight of an institution like Robert Wood

Johnson - well, surely anything from there is top-drawer research, he might think.

In other words, any study that bases its conclusions on its own spin on interpreting symptoms (assumptions, basically) without objective measures, like lab tests, is a bunch of hot air. MUS is just such an example, and a totally subjective assessment of symptoms, especially complex multiple symptoms like those of Mr. Jones, without properly assessing innate immune inflammatory parameters, will advance neither patient care nor science nor medicine, nor the use of science in the courtroom.

There's an entire manual of psychiatric diagnoses in the latest psychiatric "Bible," the DSM-4R (R is for revised, but not yet DSM-5, which is itself coming). In that manual, you will not be able to find any basis for making a psychiatric diagnosis if the patient has an obvious untreated medical condition. Manic depressive illness in the face of thyrotoxicosis? No way. Schizophrenia following phencyclidine abuse? Nope. Attention deficit disorder associated with new onset type I diabetes? Don't be ridiculous. Thought disorder during delirium? Silly! Treat the delirium and see what you have. What about somatoform disorder when the patient has high C4a and high TGF beta-1? That diagnosis is real common in my world because invariably the inflammatory markers are ignored.

Can a psychiatric diagnosis be substantiated in the face of some other co-existing inflammatory condition? No. Now let's ask that same question, coming from a different direction: which blood tests objectively confirm a given psychiatric illness? There are no objective diagnostic tests for depression, anxiety, post–traumatic stress disorder (PTSD) and many other psychiatric illnesses.

Don't think for a minute that I believe that psychotic illnesses *can't* be diagnosed before blood tests come in, but I'll be upset if any one says Post-Lyme syndrome is psychiatric in origin when that person *hasn't looked* at the labs that show highly abnormal innate immunity parameters. Even worse, if I diagnose anxiety alone as the cause of fast heart rate and the patient has a hyperthyroid condition that I missed, wouldn't I be potentially liable for that incorrect diagnosis if the patient had something terrible happen because my anxiety medications didn't work?

Can't you just see the plaintiff's lawyer asking a physician on the stand, what did you think about the patient's thyroid studies? "Sir, I didn't order any thyroid studies; I made a psychiatric diagnosis." Can you imagine the line of questioning that would follow, all intended to show what a Bozo the Clown the allegedly negligent physician was, at least in the eyes of a jury or a judge? Of course, all the questioning was intended to show that the physician was arrogant, dismissive and biased when he assessed the data. If the doctor ignored or never ordered thyroid studies, the plaintiff's attorney would make him look foolish in the eyes of the jury.

Failure to diagnose is a common cause for litigation against primary care physicians. What about Dr. Smith's failure to diagnose Mr. Jones? His defense would begin with the evaluations in the chart from the OccMed and the allergist, both of whom diagnosed him subjectively and incorrectly using Ass^2 medicine. And wouldn't the jury like to see the engraved comments on the ACOEM report. Mold wackos indeed.

And just to finish this brief analysis, can you hear the plaintiff's lawyer ask the physician who cited Dr. Sigal's work as definitive in his defense? "Doctor, what testing did Dr. Sigal report in his paper that ruled out inflammation as a cause of the illness?" [*Editors note:* none.] Please recognize that the attorney is just getting wound up to discredit entirely the psychiatric illness idea in Post-Lyme. As soon as the physician being charged with malpractice says he didn't do proper testing based on bogus information, like Dr. Sigal's paper, the insurance carrier will decide to settle the case.

So in Post-Lyme, one is well advised to be thorough in objective physiologic assessment before blaming the patient (or blaming Lyme!) for the illness. And so one should remember that *no psychiatric illness, including depression, stress, somatization, post–traumatic stress and other similarly 'invented name' syndrome describes any physiologic condition*. Psychiatric diagnoses cannot stand alone in the face of obvious inflammatory illness.

By the way, I'm sure you can easily see the National News piece citing the prestigious medical journal that hosted Dr. Sigal's missive: "Definitive study shows Chronic Lyme is psychological in origin." Do you think for a minute that a defense attorney in a Lyme case wouldn't want to use Dr. Sigal's work as helpful in his defense? We call such publications a "made-for-defense piece." The Lyme world now has several papers similar to this one, as was the case for the mold community in the early 2000s with ACOEM Exhibit One.

As an aside, the conflicts of interest statement in Dr. Sigal's paper, now required by essentially all academic journals, says that he "does not stand to gain financially by the results reported herein." I can't dispute his disclaimer but somehow the idea just sounds hollow. Testimony in court using his conclusions might provide gain, perhaps financial and perhaps academic. I can't say his disclosure isn't accurate: I know of many who say Dr. Sigal is an honorable man who wouldn't dream of using this publication for anything other than noble ends. I can't argue with them. Maybe my cynicism about published papers like this one is influenced by the fact that such empty (and misleading) disclosures are common in the mold litigation world. Defense-testifying mold consultants, authors of treatises that said that mold could never hurt a flea, tried to bring their colleagues' similar papers into court proceedings while the colleagues (also paid by the insurers) tried to use the first fellow's paper. It was 'you scratch my back' .... In recent years such "made-for-defense" papers have been more often recognized for what they are – a half-baked solipsism cooked up for money.

The idea that symptoms alone can be analyzed isn't new. Such an analysis is totally subjective and thus slipshod at best. The problem comes in when authors use those symptoms to make massive assumptions about meaning. And then those authors assume that their assumptions are correct. Assumption of an assumption, yup, it's "ass squared" (Ass^2) medicine again. When one of my patient's rights gets stomped by an ass-squared diagnosis, I'm going to stand up for the correct diagnosis, one based on objective lab results that paved the way for appropriate treatment that produced positive results.

Let's look at another example before getting into a mold case I want you to understand. Just look at Gulf War Syndrome (GWS), an illness that's still argued about in the scientific and public literature. In 1998, Fukuda, lead author for the 1994 definition of Chronic Fatigue Syndrome, writes in the Journal of the American Medical Association (JAMA; 1998: 981-998) that chronic multisymptom illness affecting Gulf War Air Force veterans had no objective parameters to support its existence, but symptoms suggested there was something wrong that was associated with Gulf War exposure. Sounds like "Medically Uncertain Symptoms (MUS)", but did he look at innate immunity? No. The problem with this study is that the authors didn't bother asking the right (any?) questions about physiology and didn't assay for the right tests. Given the CDC's role in this paper, with Dr. William Reeves as senior author, it's not surprising that tests for a few biowarfare agents were all negative. (Dr. Reeves was the leader of the government scientist band that believes CFS is a psychiatric illness, despite his occasional protestations otherwise.) They didn't look for exposure to weaponized mycotoxins or antigenic components found in the anthrax vaccine, either of which could cause those symptoms.

Did anyone forget the tried-and-true homily that absence of proof is not proof of absence? If no one looked at physiologic responses, perhaps recording the presence of antibodies to infectious agents, then that means that if there had been defective antigen presentation, the illness *would not be detected* by antibody testing. And therefore, if the illness wasn't identified by "first layer" testing, then the illness syndromes would be medically unexplained. Talk about don't ask, don't tell!

OK, but I really want to know. Everything I know about MUS says that properly administered testing will show the way in inflammatory illnesses. Start with the easy questions: What were the MSH and VIP levels? What were the TGF beta-1 and C4a levels in this cohort?

These tests certainly were abnormal in the small GWS cohort I've seen in my practice, and my patients sure had the same kinds of medical exposure that the other GWS patients had.

What do you find if the researchers looked for symptoms only? Garbage in, garbage out. 100% subjective reporting. 100% subjective recording and analysis. MUS now simply becomes an acronym that says the authors weren't thorough. And that idea takes us to Dr. Sigal's paper.

Ahh, I don't forget my cynicism about MUS papers: The cover in litigation this MUS approach provides is huge. Can you hear the OccMed?

"Your Honor, it's a darn shame we don't know more about Mr. Jones illness, but since we don't know what's wrong with him, I agree it's an obvious MUS. Well, we do know that *something* is wrong; I just don't see that you can blame the stinking, moldy God-forsaken basement he lives in. Maybe when time has passed we will know the mold did it, but really, Your Honor, we aren't there yet."

I guess I don't see much intellectual integrity in papers that hide conflicts of interest or deliberately ignore objective lab parameters. Not too long ago, I did a PubMed search for conflicts of interest in medical journals and was surprised to find hundreds of citations. For a subject that prompts so much editorial outrage, one would think that conflicts would be less common! I'm accustomed to politicians lying about their opponents in their quest for votes and defense consultants lying in deposition, but wouldn't you think that prestigious journal editors would be attuned to flagrant abuse of conflict of interest? Nope.

One way to deal with the MUS approach is to prove prospective acquisition of illness symptoms. For example, say your grandchild has a runny nose and a viral-type illness but then he spikes a fever a few days later. Might be a secondary ear infection, right? He might need antibiotics. Sure enough, little Johnny's backed-up nose has plugged his Eustachian tube; the poor middle ear filled up with fluid as a result and the middle ear fluid is now infected. Johnny starts on penicillin, but he gets a rash a couple of days later. Your primary care doc changes the antibiotics; Johnny heals and is now just fine.

Is Johnny now allergic to penicillin? We often determine what caused the rash by waiting a while until Johnny is healed and then giving him some more penicillin. If he gets a rash, we've proved he's allergic to penicillin. That confirmed penicillin allergy might be important later in Johnny's life. We can treat the second rash, it's no big deal, but at least we know for sure about the allergy. No guessing, no assumptions.

But what if you think your elementary school in Oregon is making you sick? There's mold all over the place and lots of your fellow teachers are sick, too. You aren't going to guess. You come to Pocomoke, I treat you and you're much improved. Now we must show that later, off all medications, you don't get sick away from the school, despite ongoing exposure to the "ubiquitous fungi of the world." (In other words, it's not the fungi that are everywhere that's making you sick.) Your symptoms (subjective) don't change and your labs (objective) don't change. You were careful to include a detailed recording of everywhere you went and everything you did. Can we say that exposure to the ubiquitous fungi of the world didn't make you sick? Sure, you just did what we call a *prospective re-exposure trial*, just like we did with Johnny and penicillin. Only in this case, my patient in Oregon isn't allergic to the ubiquitous fungi of the world. Could the original exposure to the school have been the cause? Possibly, but absence of proof isn't proof of absence!

Now, here comes high risk re-exposure and the proof. You have agreed we're going to send you back into harm's way on purpose. You go back into the school Monday morning and start to get sick by noon. The rest of the day, you want to take some cholestyramine (CSM), because you know it'll help, but you can't; not yet. Tuesday morning before school, you do a blood test, looking at C4a, TGF beta-1, leptin, MMP9, von Willebrand's profile and VEGF. Labs aren't subjective, they are about you and they don't lie. By Tuesday afternoon, you really want some CSM, but again, you can't take it.

The illness is starting to get rough. We do more labs on Wednesday morning, as another blood draw fleshes out the sequential activation of innate immune elements (SAIIE) we know will occur if the school makes you ill. My God, when will Thursday come? Another day in school Hell and finally the Thursday blood draw comes. Now you're out of the school and back on CSM.

That process we just did there determines proof of causation. You suffered immensely because the mold illness was reproduced in just three days. You suffered for years, finally felt almost normal again and in just THREE DAYS, you feel like the living dead again.

If the school is the culprit, the labs will show C4a levels rising the first day, leptin rising on day 2, MMP9 on day 2-3, with VEGF rising and then crashing. TGF beta-1 changes according to your genes, with the dreaded genotype people doing the worst. The clotting factors crash, with bleeding often occurring on the third or fourth day.

Who could possibly argue with this proof?

But the defense asks Dr. Emil Bardana of Oregon Health Sciences University to comment. This case has already taken up 20 binders of documents and countless thousands of dollars. Dr. Bardana's institution is revered in his state; his opinion will have weight in court no matter what he says.

And what he says is, "stress caused the changes in labs." The plaintiff's lawyer doesn't jump into his throat on cross-examination for this completely unsupported, totally subjective, self-serving assertion and the judge says simply, "I see."

Then the mold data found in the dust of the classroom is discounted by the defense team's premier defense consultants. How can we know how the mold got into the dust from the air, they ask? There's no proof of that. *Case dismissed.*

Fortunately, such abuse occurs less often now, especially since 2006 when the EPA released its test (Environmental Relative Moldiness Index, ERMI) for fungal DNA found in dust as an indicator of what was in the air at some earlier time.

The judge in the Bardana case seemed flummoxed by how mold spores and mold DNA get into dust; in the end, the simple point that aerosolized particles do indeed have mass and do indeed act as particles that are subject to the forces of gravity, and therefore really do settle into dust, was lost on the judge. This idea that settled dust is useful in analysis of what went on in the air for a long period of time is accepted by virtually every current mold researcher. Astounding as it may seem, my patient lost her case because the Oregon judge ignored the EPA work with settled dust as the most accurate source to measure fungal DNA and determine building health.

And as it turned out, Dr. Bardana had just authored a paper (one that included the plaintiff) showing that mold illness was a psychological problem. I didn't see any psychiatrists on the list of authors, either. Though he didn't use the term MUS, he included no labs (of course) or any prospective trials, just a group of 50 patients picked out of a group using unknown selection criteria by someone with unknown conflicts of interest. The fact that my patient had so many lab abnormalities was completely ignored by Dr. Bardana in his Sigal-like leap to "nothing really wrong," and as we saw, he said some things he couldn't possibly defend on cross-examination.

And yet, I see the Bardana paper occasionally raised as a legitimate study of mold illness. The Bardana idea of "no illness from mold" quickly melts with even a gentle attack on the study's methods.

Back to Mr. Jones. If he brings suit in Portland, will his case fall prey to Dr. Bardana and his research associates like the case above?

So the big question is, how does someone trying to Survive Mold survive the assumption gauntlet? They must be prepared to show good science (that means references!) at every step of their medical evaluation. Don't let the attending physicians say, "I don't know anything about this." I had a case not long ago where a toxicologist attempted to get out of his lack of any knowledge of the immunological basis for mold illness by saying that immunology was out of his realm (no kidding). When shown the GAO report and the WHO report, each of which emphasized the immunologic aspects of mold illness, the judge agreed that the toxicologist could go home to his realm as he didn't belong here to talk about this one.

Using good science is key in this field; don't rely on blogs from blowhards from the intellectual desert or anywhere else. The hard science on mold illness is there for all to read. That science must be presented logically and briefly. *Patients have to become informed consumers and informed educators.*

Physicians who show a lack of concern about symptoms of mold illness need to hear calm logic, not shrillness, from patients who feel their case hasn't been heard. Patients must recognize when their physicians are blowing their complaints off and respectfully demand that a superior in the department or a supervisor on the team take another look. You must stand up for yourself knowledgeably, patiently and persistently.

Unfortunately, many mold illness victims are suffering from profound cognitive impairment. One reason that Dr. Smith might not have wanted to work with Mr. Jones is that some mold patients will try anyone's patience. When they are treated, they can be like any other person, but doctor/patient relationships can be blown apart as the cognitive issues of mold patients accelerate. The mold person really is changed by the illness. You can't always go back to who you were before.

People like Mr. Jones have the data needed to show Dr. Smith what he didn't want to take time to hear. Patients often tend to go on and on about their illnesses, not pausing to see if anyone is listening. No busy physician will listen very long to ranting. Mr. Jones presented a quiet, logical summary of his illness, because he knows not to trust that Dr. Smith and his predecessors would or will be either thorough or thoughtful. And in his defense, Dr. Smith wasn't provided adequate time to do the job expected of a modern physician. You can help make the best use of the little time you have by providing a brief, accurate history.

Let's go over the basics one more time. Psychiatric illness exists; it's real, it's recognizable and it's largely treatable. But psychiatric illness doesn't have the objective biomarkers seen in chronic inflammatory response syndromes. Said another way, if you see labs confirming the presence of chronic inflammatory response syndromes, it's a physiologic illness. You must treat the inflammation successfully before believing any potential psychiatric problems are the only condition. Dr. Smith didn't even think to look for a chronic inflammatory response syndrome. I don't think he was intentionally trying to hurt Mr. Jones; he just wasn't up to speed on current medicine. Shame on him.

But we must reserve greater shame, scorn even, for defense consultants who deliberately ignore the presence of inflammatory response syndromes; their actions cause undue hardship and suffering in every mold illness patient's life.

Let me share an example with you from Indiana. Here we had a clearly demonstrated mold problem, confirmed by a fungal DNA test, the Environmental Relative Mold Index (ERMI). The EPA has published multiple studies about ERMI showing that the fungal fragments and spores that *were in the air* but then settled in dust are significant. If only ERMI were available in the Dr. Bardana travesty-of-justice case discussed above!

The patient from Indiana was exposed at work, became ill and then disabled. His occupation was cognitive-based, not involving manual labor, and one that permitted multiple accommodations of changes in position, rest and frequent breaks. Yet the man couldn't work. His brain didn't work; he had multiple symptoms from multiple body systems. He applied for disability.

There were no surprises during his visit to my office. He had a series of objective metabolic problems. His levels of MSH or VIP were unmeasurable. His blood was incredibly salty, with an osmolality of 309. VEGF was dead low and TGF beta-1 was over 18,000, then one of the highest levels I've ever seen (normal is less than 2,380). C4a was three times normal...

This patient had a lot of things wrong - objectively not subjectively - all there in black and white. This case had plenty of clearly shown objective parameters, ones that could never be disputed, from high complexity facilities like LabCorp and Quest. These terribly wrong lab tests are well accepted and used all over the world and they clearly showed a chronic inflammatory response syndrome.

So when I received the letter from the disability carrier informing me they approved the claim and were beginning to send monthly checks, I wasn't surprised. But get this: the disability claims examiner told the patient that he *personally* had grave concerns about the diagnosis the patient's doctor had made (the same one I'd made). The examiner felt that the patient should have a "heart-to-heart discussion" with his personal physician and his consultant (me) about the therapies he was receiving.

Why? The insurance company asked Dr. Freud, a full professor with perfect hair from the Great University of Minds, to consult (subjectively) on the case. He noted that the patient had so many symptoms (we'll ignore the fact that he missed at least half of what was actually there), that he concluded the patient *must have* a somatoform disorder. He provided lots of psychobabble about that diagnosis, using the typical array of psycho tests to make his global diagnosis of an aberrant belief system without anything physically wrong and therefore: the patient is disabled.

Wait a minute. If the patient is a psycho and his illness is caused by an aberrant belief system, then the guy agrees that disability payments are indicated?

What?

Let's look at a couple of the bizarre aspects of Dr. Freud's opinion. He cites my lab findings as "normal," ignoring the amazingly abnormal findings like the TGF beta-1 and C4a. He ignores the verification of these labs by the attending physician! He ignores the incredibly abnormal findings on magnetic resonance spectroscopy (says the MRI "is normal") and doesn't even comment on the increased lactate and abnormal ratio of glutamate to glutamine on the MR spectroscopy report.

Come on, this guy is a full professor! The intimidating presentation of his academic credentials clearly impressed the claims examiner. But in fact, he tried to cover up his markedly deficient knowledge of inflammation and its effect on the central nervous system by ignoring its existence. Instead, he went with what he knew, relying instead on somatoform illness. In essence, his ignorance provided the diagnosis.

I wrote back to the claims examiner, applauding him for his understanding of the patient's disability. But I asked him to look at the plethora of abnormal inflammation studies; such inflammation must be corrected first before the Great Professor could conclude a psychiatric source. In fact, I felt he should have a heart-to-heart talk with the professor about the process by which he came to his conclusions.

One last point about symptoms. A standard medical history takes a patient's comments about symptoms and presents them in the medical record to other physicians. Physicians can't be expected to use patient-completed questionnaires as reliable; the doctor's own training and skill in history-taking are critical elements in sorting out what's real, important and pertinent. Having said that, if a patient knows the physician has recorded the history accurately, the physician needs to be told. If the physician has recorded the history inaccurately, the physician needs to know that, too. Once the physician's history is chiseled in stone, especially the Ass^2 aspects, it's too late to change the medical record.

Vicki Smith sent me a marked-up copy of a history recorded by a defense consultant, Dr. Phillips of the University of Pennsylvania. He was apparently hired by an insurance company to provide ammunition to be used against her in a legal proceeding. She found more than thirty errors in his recorded history, yet she wasn't able to have the record corrected.

How can patients like Vicki avoid such disgraceful but *permanent distortion* of their histories? Simple: record your interview, especially if you're in litigation or wonder if you might end up making an insurance or workman's compensation claim. Ask for a copy of the notes from their office visit, even if the notes aren't typed. Be prepared for any interview by preparing both a carefully edited but accurate summary and a thorough timeline that you send to your examining physician well ahead of the exam. Verify that the examiner has read the summary and timeline before you even let the interview begin. When a physician is handed a stack of medical records twelve inches high the night before the interview (or worse, just as he walks into the exam room), expect mistakes! If the physician has been provided adequate time to prepare for your interview and adequate information about your history, then your story can be told accurately. If you disagree with the physician's record of your case, you can certainly ask for a correction. If no correction comes, you can demand that the record be corrected.

For more information about how to correct medical records, see:

http://patients.about.com/od/yourmedicalrecords/a/howtocorrect.htm,

http://medicalrecordrights.georgetown.edu/stateguides/wa/waguide4.html,

http://www.healthtechnica.com/blogsphere/2009/08/18/when-your-medical-records-are-inaccurate/

and

http://www.cancerlynx.com/medicalrecord.html

If you aren't proactive in the cause of accuracy you will see the errors of Ass² medicine coming back to haunt you later.

CHAPTER 4 Don't Assume: Use Data to Show the Way

If our hypothetical Dr. Smith had ordered the labs that I ordered on Mr. Jones, he would have "put Mr. Jones' illness on a piece of paper." There on paper, in front of Dr. Smith, would've been the cold, hard facts confirming the presence of a chronic inflammatory response syndrome; objective and unimpeachable testimony that Mr. Jones was as sick as he said he was. His illness wasn't somatoform disorder; he didn't need a psychiatrist or anti-depressants. He needed a physician to listen to him, make the needed investigation using scientific methods, and then take an active interest in healing him.

Objective lab data demolish assumptions and then stomp on assumptions of assumptions. They also stomp on the egos of the assumers when the assumers are wrong. Those stompees often get defensive, especially when their insurance industry clients have some skin in the game or they feel that their (outmoded and incorrect) belief system is being challenged.

One would expect that the crucial role in medicine for a physician is helping the patient rather than assuaging his or her own concerns.

It's disarmingly easy to show the incredible lab abnormalities that mold illness patients have. The series of lab tests that we run routinely has changed a bit from year to year as we learn more, but the basic concepts have always been the same. We assume nothing. We test for to find irrefutable evidence of:

(i) lack of regulation of inflammation; and
(ii) lack of control of hormone function due to
(iii) absence of regulatory hypothalamic neuropeptides.

If that mouthful of medical syllables weren't enough, we're also testing for specific objective evidence of abnormalities of innate immunity as shown by inflammatory changes, especially in MMP-9, TGF beta-1 and C4a.

But that's not all. We're also using lab tests to scan for abnormalities in autoimmune functioning, looking at antibodies in the cardiolipins class as well as antibodies to gliadin, the small protein found in gluten. We are on the alert for proven abnormalities in clotting by looking at von Willebrand's profile and PAI-1, for example.

As part of a differential diagnosis, we also need to show what Mr. Jones *doesn't* have, so we also test for look:

- classic tests of inflammation like C-reactive protein and sedimentation rate
- CBC and metabolic profile
- lipid profile
- thyroid tests
- Lyme tests
- immunoglobulin tests
- and more

We anticipate *normal results in these tests* as they don't have relevance to innate immunity, but being objective demands that we are looking for *what you don't have* – just as important to differential diagnosis as *what you do have*. Only running the standard tests for biotoxin patients is like using a banana to search for missing keys in the dark. It just doesn't work as well as using a flashlight.

Each of the illnesses that I treat, including mold illness, has multiple abnormalities in innate immune findings, and there's almost always a clear delineation of genetic susceptibility using HLA DR immune response genes as the tell-tale marker.

To finish this hypothetical case of Dr. Smith and Mr. Jones, let's have Mr. Jones come to Pocomoke for evaluation. His labs were drawn and initial phase treatment initiated. The following is basically a transcript of the first phone call between Dr. Shoemaker and Mr. Jones since his visit one month ago:

> **Dr. S:** "Good morning, Mr. Jones. We have a lot to talk about today. I hope you had a chance to review the labs that we sent you. Did you read all about MSH and innate immune responses?"
>
> **Mr. J:** "Yes, I did, Dr. Shoemaker. I read through the results and I recognized some of the names from reading Chapter 4 in *Mold Warriors* and some other materials you've written. There were so many asterisks that you added to my lab results cover sheet that I may need some help and perspective looking at this many abnormal results. I get the general idea that I have lots of things wrong."
>
> **Dr. S:** "Tell me, Mr. Jones, how did you feel when you received the packet of lab results from us?"
>
> **Mr. J:** "Well, to tell you the truth, I held that nice, crisp envelope in my hand for a minute or two before opening it because I've been hoping that you'd be able to show me what's wrong *and* that you'd tell me what's wrong is treatable. So I held my breath a bit as I opened the package. At first when I saw all the abnormalities - and they were just like you said they were going to be! - I felt a sense of vindication. I knew something was wrong! I was so discouraged after Dr. Smith talked to me that I thought I didn't have any hope left. After I looked at these results, I knew I wasn't crazy. Some things are *wrong*. I also know there's nothing somatoform about this illness. This is *real*. That much I saw. But then after I had a wave of relief that

there really *was* something wrong, I looked at all of the stars beside the abnormal lab tests, and the idea of ***so much wrong*** kind of took away my joy at seeing the proof. It was like I said the world was flat and to prove it, I was going to fall off the edge."

Dr. S: "I can understand your mixed feelings. Let's go through how you've been doing since I last spoke to you in my office. Have you been taking the cholestyramine?"

Mr. J: "Yes, I am. I gather that most people don't have any issues with it and I really didn't have any trouble after the first few days. I had some heartburn at first, so I took some reflux medicine for a few days and after about a week, I took some sorbitol to ease hard stools. I added some dried apricots and cashews to my diet, and that helped, too. I only missed a dose or two, so I can't really say cholestyramine caused any real trouble. I do get some funny looks in my office, though, when I pull out my shaker and make a dose at my mid-morning break. Ken, the fellow I was telling you about from down the hall, the one who has those weird seizures no one can figure out and who's tired all the time, kept on coming over to see what I was doing. He finally asked me what the yellow drink was for. When I told him it was to fix what the building is doing to me, you could almost see him wanting to know more. Finally, I just told him that mold inflammation hurts brains, too. He might be calling you.

"You know, it's only been a few weeks, but I can remember things better now and I don't get lost any more around town. I'm at about 60% of my energy and I don't ache all day anymore. I used to pee all day long and now, I can take a three-hour car ride without stopping. My wife says I'm not as grumpy as I used to be, although when the Phillies lost to the Yankees in the World Series, I didn't talk much for two days. And that tingling is gone, just from the powder. I love it now. I was on the plane back from Charlotte the other day and I was listening to the people in the seat in front of me. The guy was just like I was before I saw you. I bet he's moldy too."

Dr. S: "Mr. Jones, you're going to see this illness just about everywhere. It's not uncommon, but so many people buy into diagnoses like the one Dr. Smith tried to pin on you. We did a survey a couple of years ago just to see how many doctors people had seen, how many different diagnoses they'd been given, and how many medicines had been tossed at them. It was incredible - more than ten for each category. But let me warn you, please: do yourself a favor, Mr. Jones and try not to help others, like Ken, until you're even better than you are now. I'm glad you're doing well, but you might do more harm than good by telling everyone what you now know. You're just getting started. You've got plenty of time to help others. Basically, as Bob Dylan sang back when I was in high school, you need to know your song well before you start singing.

"But back to you. Do you recall when you were here that I told you that I thought that we would find that you'd have that rare HLA type? Our word for HLA type is called a 'haplotype.' Because of your long arms and long legs and that extra joint flexibility that you have, it was more likely than not that you did have the HLA 11-3-52B that I worry about. Remember that we call that haplotype 'the dreaded,' for the simple reason that while this gene type isn't common (it's found in less than 1% of the population) it's associated with some of the worst grouping of symptoms and some of the worst lab abnormalities that I see. There's another gene type that's almost as bad called 4-3-53. For this dreaded genotype, there are twelve subtypes of the 04 named 0401, 0402, and 0403 all the way down to 0412. The very worst ones are the 0401's, 0402's and 0404's. If you look at your lab packet on page 11, you'll see your HLA haplotype. You have the 0401 for the 4-3-53 and the 11-3-52B. Unfortunately, you have the absolute worst combo of the immune response genes involved. You were the last person of all kinds of people to move into a moldy basement. It never should have been allowed."

Mr. J: "Dr. Shoemaker, that means my kids are guaranteed to have one of these crummy genes, right? I remember when you asked me about whether I was flexible as a teenager, that caught me off guard, but when I got home, I talked to my parents and we have several relatives who played professional basketball and my cousin played volleyball for Penn State. It's quite a hoot when we get together at family celebrations to see who has the longest arms. But also, one of my uncles had to have some surgery on his thoracic aorta. The blood vessel was actually starting to split apart, creating an aneurysm. They say he didn't have Marfan's, which raised the question about other diseases of connective tissue, I think you said Ehlers-Danlos III. Isn't that the one you said you thought I had?"

Dr. S: "Your cousin played for Penn State? She must be pretty good; they've won, what, 103 consecutive matches and two NCAA championships now?"

Mr. J: "She was pretty good alright, until she blew out her right knee playing basketball. She pulled up for a jump shot on a fast break and boom, down she went. She didn't like it when I told her she shouldn't be trying to run fast and stop fast, but come to think about it, my other uncle, her father, had the same bad knees. You know she's been down in the dumps a lot lately, too. She's got asthma now. Nobody else has it in the family. And she lives in the bottom floor of that apartment that's half below grade."

Dr. S: "Mr. Jones, after you feel a bit better, be sure that your cousin has a blood work-up. If she's got high TGF beta-1 - and I'd be surprised if she *didn't* - don't forget that TGF beta-1 really nails lungs and some immunologists think that the remodeling of lung lining cells that go on in asthmatics is just from high TGF beta-1. If she has the 11-3-52B and high TGF beta-1, then she needs to get treated. The lists of tests are all on the Internet. She can find them on **www.biotoxin.info or www.chronicneurotoxins.com.** But don't let her just take cholestyramine without a baseline series of labs. As you can see, this illness is way too complex to guess."

Mr. J: Laughs. "I know how much you don't like assumptions, Dr. Shoemaker."

Dr. S: "OK, we did talk about Ehlers-Danlos III. We talked about your genes because while you don't have all the characteristics for being Ehlers-Danlos III, you *did* have the very long fingers; you *did* have the ability to lean over and touch your hands to the floor. It's pretty amazing you could actually step through, one foot then the other, keeping your hands clasped and then bring your hands all the way up your back. You also had the ability to take your hand and put it back behind your neck, touching your long finger to just below the tip of the scapula and then you could clasp your other hand by bringing it behind your back up from the waist. Do you remember the picture on my wall of Michael Phelps that was printed in *Sports Illustrated* after he won eight gold medals? You look just like him.

"There are other gene types linked to mold illness, not just the two dreaded ones. There are specific gene types that I see present at far higher percentages in people with mold illness. Similarly, people with symptoms after they had Lyme disease and didn't get better with antibiotics will have a number of recurring genes that show up almost all the time as well. No surprise, the chronic ciguatera patients also have unique gene susceptibility as well. It's of great interest to me that essentially all of the people that I've seen with Gulf War Syndrome either had 11-3-52B or 4-3-53. Just to finish the thought, to date, the only people I've seen that had chronic fatiguing illnesses that came on after they had the Gardasil vaccine were people with the 11-3-52B haplotypes. So there's no question that the genes you have are important and they do make a difference."

Mr. J: "Well, what about my children, Dr. Shoemaker? Shouldn't they be looked at? It looks to me that they're guaranteed to have one or the other of these dreaded ones since, of my two, I'm guaranteed to be donating one to them."

Dr. S: "Right, that makes good sense. Often the first thing that a parent will think about as soon as he or she finds out that there's genetic susceptibility is the potential problem for children. I have to tell you that I'm delighted that you picked up on the genetic aspect so quickly. It's usually the Moms who think of children before Dads do.

"But Mr. Jones, getting back to the rest of your labs. If you remember when you were here, I drew a triangle on that Biotoxin Pathway diagram that I gave to you, trying to explain all that was going on. I basically told you that inflammation that developed in what we call the periphery—that means not in the brain— could result in inflammation that attacked production of the regulatory hormones, VIP, MSH and vasopressin. These are important in regulating inflammatory substances so it's almost like a battleground or a war. The inflammation is beaten back by anti-inflammatory compounds and anti-inflammatory compounds are beaten down by inflammation. In your case, the inflammatory elements won and now you are without sufficient regulatory hormones to stay and police things so the inflammation compounds are running haywire. As we'll get to in a minute, that's one reason why you have so many things wrong. We call that problem hypothalamic dysregulation, just so you know the jargon.

"Now, both your MSH and VIP are too low to measure. You'll note the LabCorp range of 0-40 of MSH isn't correct; the proper range was established for years before LabCorp changed it without any reason in September 2006. It is a always has been 35-81. *After I saw LabCorp had changed it, I called their main office in Burlington, NC, and spoke to their Chief Science Officer. He told me that they'd simply had so many low results from MSH that <u>they changed the normal range to accommodate all the low results</u>. So many physicians were ordering MSH on affected people; of course the levels were low. It's really crazy that they assumed that the low results were from normal people. I mean, not knowing who was a 'case' or who was 'a non-case' means that there's no way anyone—especially a commercial lab—can justify a lab normal range like they did. It's completely incorrect to publish normal ranges based on ill patients.*

"So, having said that, your MSH is quite low and so is your VIP. If you'll look on pages 5 and 6, you can see why I'm worried about your very low vasopressin. I don't want you to get confused because I also call vasopressin 'ADH' and 'antidiuretic hormone.' Remember that while this compound is made by the posterior pituitary and primarily acts on the kidney to hold on to free water, it also acts in the part of the hypothalamus called the suprachiasmatic nucleus, where it has a tremendous role interacting with VIP and with MSH. When you don't have these three controlling elements acting synergistically, then the result will be an inflammatory illness featuring a lack of regulatory control. When all three of these elements are abnormal, the illness is what I call "hypothalamic." Don't forget that each of these hormones has its own important regulation duties, but each acts in concert with the other two. You know, Mr. Jones, the idea of a centralized control structure isn't limited to hypothalamic hormones.

"Think for a minute about a military operation; call the regulatory structure system, the 'command and control.' If you want to render the enemy's coordination of attack and responses ineffective, simply remove control of the command. What you'll have is a bunch of chaos when everyone starts doing what they feel like doing rather than what they're supposed to be doing.

"There are lots of clichés about no command and control: inmates running the asylum; chicken with its head cut off; situation normal, all fouled up. At least in the medical world, we can show definite abnormalities in command and control."

Mr. J: "Well, Dr. Shoemaker, since I have all three regulatory hormones out of whack, does that mean that there's no hope for me?

Dr. S: "No, actually, what it shows is the incredible intensity of your illness. Just to make sure that we communicate correctly, your vasopressin level is 1.3 and while that might look normal at first glance, it's actually far too low for what it should be with your high osmolality. Over 60% of low-MSH patients have a foul-up of regulation of ADH/osmolality. What I mean by this is that vasopressin tells the kidney to hold on to free water. If there's not enough free water being kept in the body —if it's simply being peed out every thirty minutes—then the salt that remains behind in the blood will rise to a higher level. This measure of salt and water, called osmolality, basically tells us about dehydration. Because you can't stop losing free water from the kidney, you pee all the time. The salt content goes up in your blood as you lose pure water; that's why you're thirsty all the time. You get static shocks because what happens as the salt level in the blood rises, is that your sweat glands will pour as much extra salt into the sweat as they can, giving levels of chloride, for example, that are higher in mold illness patients than even what we see in cystic fibrosis."

Mr. J: "So those shocks that I get—that blue spark that jumps from my fingers to the light switch—doesn't mean my light switch is bad?"

Dr. S: "No, Mr. Jones, you've created an electromotive force, an electrical potential really, that's being discharged to ground. If you think about it, it's not just in wintertime and it's not just after you've petted your Persian cat that you get shocks. You get shocks from a drinking fountain, a car door and doorknobs all seasons of the year. Not everyone has these interesting symptoms and, to tell you the truth, I never read about this in a medical textbook. It wasn't until so many patients told me about getting static shocks that I started looking into this. Of course, that was way back in 1998. Seems like ancient history now.

Mr. J: "Hmm, before I got sick, people used to tell me I was the Energizer Bunny and now I'm just a battery. At least I'm not an abattoir."

Dr. S: "That's funny; I get it. I'm glad you enjoy a pun, one of my favorite kinds of humor. Let's move on to some good news. In the very worse cases of this chronic inflammatory response syndrome, which is what you have, there will be a suppression of the hormone ACTH for a given level of cortisol. Remember, cortisol is made by the adrenals and ACTH is made by the pituitary that tells the adrenals to get to work if not enough cortisol is present. The very worst patients will have both very low levels of cortisol *and* simultaneously low levels of ACTH. People who are early in the illness or those who will be getting better, we can pretty much profile by the high levels of ACTH and high levels of cortisol measured simultaneously. I have to tell you that when I first started seeing some of these high ACTH levels with a high cortisol at the same time, I thought I was looking at some sort of an ACTH-producing tumor because the levels were significantly different than what I normally had seen. Usually, as cortisol rises, there's a signal sent to the hypothalamus that then tells the pituitary to cut back on production of ACTH. That doesn't happen in more than 50% of the people with low MSH. That lack of regulation of ACTH production is pretty typical of all the hormone dysregulation problems that I see and talk about for this illness.

"Just to finish the point, Mr. Jones, one reason that I've used the expression 'giving cortisol to these patients is like throwing gasoline on the fire,' is that extra cortisol taken by mouth in the form of prednisone or hydrocortisone or some other substance like that can cause a new kind of negative feedback superimposed on a dysregulated system, so that the extra cortisol suppresses the ACTH, prevents it from normal functioning and not uncommonly, the feedback-controlled rise of ACTH, responding to low cortisol, becomes damaged. And the ACTH is the very last line of defense against inflammation and now it is shot. These are some of the most desperately ill people that I see and often it was the actions of well-meaning physicians using prednisone or hydrocortisone that converted the illness from being a minor problem to being disabling.

"Let's move on to some of the other important labs. The androgens, or male hormones, are as affected as the other hormone systems. Parenthetically ADH, osmolality, ACTH and cortisol will be disrupted about 60% of the time in chronic inflammatory response syndrome patients. The androgens, DHEAS, testosterone and androstenedione, will be affected about 40-50% of the time. Once again, we see really difficult problems if someone goes to their doctor for treatment of a low testosterone level and the doctor gives them some sort of androgen cream to put on their skin. The male hormone in the cream can actually suppress the body's normal drive to make testosterone, causing prolonged suppression of the production of male hormone for a long period of time. This problem is much worse in those with low VIP too, as low VIP can make the enzyme (it is called aromatase) that converts testosterone to estrone and estradiol, female hormones, go absolutely nuts. The more testosterone the patients use, the worse their testosterone deficiency becomes!

"You know, when I first started working with people with chronic fatiguing illnesses, often patients who were seeing alternative providers, it bothered me a lot to see how many were taking DHEA as a magical remedy. I can remember back in the 1980's, when DHEA was first announced as an anti-aging drug and as a panacea for so many illnesses. It is not. It's an androgen that's eventually processed to make testosterone. But now I measure DHEA routinely and it's not uncommon to find low levels. All I can say is, if you're going to use the drug, you should at least show that there's some sort of deficiency in it.

"Moving on a bit, the normal measures of inflammation, like C-reactive protein and sedimentation rate are *invariably not a problem* in patients with chronic inflammatory response syndromes. Yours are completely normal.

"The 'new kid on the block,' TGF beta-1, has risen to the top of my interest chart. TGF beta-1 is released in increased amounts in people who have connective tissue problems like you do. The research going on in Marfan's syndrome, primarily led by Dr. Hal Dietz at Hopkins, has now resulted in a clinical trial of a drug that's designed to block TGF beta-1 effects in tissues. The idea is to prevent aneurysm formation in Marfan's patients by using the drug losartan. Until recently, Dr. Dietz didn't measure TGF beta-1 in blood like I do, but his group published a paper in August 2009 showing the same things I'd found the previous year - that high TGF beta-1 in blood was a problem in connective tissue disorders. As an aside, since April 2008, we have collected TGF beta-1 data on more than 3,000 patients, including samples we had saved from our "library" in the freezer. High TGF beta-1 is far worse in people with either of the dreaded genotypes, but especially in 11-3-52B. TGF beta-1 serves as a rapidly changing marker to help us understand what goes on with inflammation in mold patients. It remains one of the most important determining factors in other inflammatory illnesses (such as Lyme disease) of who gets better and who stays ill. Interestingly, some of the highest TGF beta-1 levels are found in ciguatera patients."

Mr. J: "So that's why you asked me about getting sick from eating fish?"

Dr. S: "Well sure, but differential diagnosis is what I do every day and I had to be able to sort out if a confounding biotoxin - a different one if you will - could be confusing the issue or at least making your pre-existing mold illness worse. This concept really kills me about the Chronic Fatigue Syndrome field. If you've got, say, a virus that inserts itself into your DNA (like XMRV, for example) and does bad things to your inflammation, does that help you fight inflammation induced by mold exposure? No, of course not. Now mold illness *and* a virus are your problem. If they'd just recognize the abnormal physiology of a chronic systemic inflammatory illness, they'd be so much better served trying to help their patients.

"You know Mr. Jones, with your shortness of breath, it should be no surprise to you that your TGF beta-1 is elevated, because this compound has really adverse effects on how lungs function. If you do a PubMed search from the National Library of Medicine and simply type in 'TGF beta-1 and remodeling,' for example, after you look at the results, you'll see what I'm talking about. There's a large literature on how TGF beta-1 *transforms* the lung by *remodeling* it. Maybe that's where it got its name? TGF beta-1 can cause changes in lungs that are the same ones that people say are a result of asthma. Is asthma basically a TGF beta-1 process? Certainly it's true that's the case for many patients. We know that TGF beta-1 will stimulate normal cells lining bronchial tubes to change, almost de-differentiating from smooth, soft, feathery lining cells into thick, tough fibrin forming fibroblasts. It's the activity of fibroblasts, resulting from transformed lung cells, that causes the lung fibrosis that we find in so many people who are waiting for lung transplants."

Mr. J: "Dr. Shoemaker, I don't want to interrupt, but are you telling me that I'm going to need a lung transplant because my TGF beta-1 is so high, because I'm an 11-3-52B haplotype?"

Dr. S: "No, Mr. Jones. I'm not saying that. What I am saying is that fixing TGF beta-1 now becomes of extreme importance to preserve your lung function and to prevent any more deterioration. But if you had idiopathic pulmonary fibrosis and the pulmonary lab said you needed a lung transplant, fix the TGF beta-1 first, as the literature is very secure about that illness and TGF beta-1.

"You should remember that the First Responders to the 9/11 attacks have a huge increase in lung transplants. I've not seen that many of the 9/11 First Responders, but all of them have high TGF beta-1.

"Moving right along, the MMP-9 test is useful because when we think about inflammation proteins - well, remember the cytokines I told you about? - what we're looking at are proteins that effect three different areas of the body: the cells that make them, called the autocrine effect; cells that are nearby where they are made which is the paracrine effect; or they can move into the bloodstream, where we can measure those affecting different sites, which we call the endocrine effect. Your MMP-9 is quite high, which tells us about the sum of cytokine effects rather than just one simple part of it. We know that MMP-9 can go up with lung disease as well as with blood vessel problems. It also affects nerve, muscle, lung, brain and joints.

"Take a look at your lipid phenotype, if you will. I use the very old-fashioned Fredrickson phenotyping to see whether a cholesterol problem is driven by a genetic problem of over-production of LDL (Type II) or an insulin problem (Type IV), for example. Your phenotype is normal. But what's not is your PAI-1—it's too high. PAI-1, together with MMP-9, helps deliver oxidized LDL out of blood, across several different membranes, through a basement membrane, into the repository for LDL cholesterol in the areas outside of blood vessels (sub-intimal space) where blockages begin. As an aside, it drives me bonkers every time I see an ad for a cholesterol drug on TV that implies that cholesterol is being deposited *inside* blood vessels when actually it's deposited *outside* blood vessels and pushes its way in. I guess I'm not very good at marketing because people look at those developing obstructions, become alarmed and pay for expensive cholesterol drugs that they may not need.

"We looked at some tests to show what you *don't* have as well, especially with so many symptoms and fatigue. You've had blood count and metabolic profiles done before. They're still normal and that's reassuring, but you and I know that biotoxin illness almost never creates such a problem. There are exceptions, of course, like abnormal liver functions that appear when your bile stops moving. We also look at a test that measures activity in the small cells that line the gallbladder ducts. This enzyme, called GGTP, is the most important one looking at the reduction of flow of bile in the small bile ducts that we see so frequently in illnesses like yours. High GGTP can help answer questions why given people have problems with cholestasis, for example."

Mr. J: "Do I have that one, Dr. Shoemaker?"

Dr. S: "You're quick on the draw today, Mr. Jones. You don't have that."

Mr. J: "Well, you know, I am feeling better. I can joke; make some word games. Before, I used to forget what I read - and make a pun? Forget it. I forgot the words I was punning on.

And before, my God, it was just last month, there was no way I could process a question that made sense that fast. Oh sure, that didn't stop me from asking questions. My wife used to give me a hard time because she said I'd ask another question even before I heard the answer to the first one. I couldn't process the answer and I forgot the question I asked, so I thought *she* was going nuts."

Dr. S: "Yes, that's really a pain to deal with from our point of view. More than one physician has complained to me that mold patients really should just shut up and listen. No questions about it, you guys are the most difficult and demanding patients anyone could imagine. But look at you now after you're fixed. Now, about your nasal culture ..."

Mr. J: "You're not going to do another one of those, are you?"

Dr. S: "No, Mr. Jones, one is all we need to start your healing but the fact that you had a normal nose culture is a tremendous benefit. For those that have unusual organisms that are resistant to many classes of antibiotics, their case takes longer to heal. As I told you before I did your culture, patients aren't going to get much better from cholestyramine alone until the unusual germs are eradicated from the deep recesses in their nose.

"Over 80% of the patients I have seen with low MSH will have a multiply antibiotic resistant coag neg staph (MARCoNS) growing in biofilm in the deepest parts of their nose. We call them commensals to help others who use that jargon term understand what the bacteria do while living at our expense. Less than 2% of people I have seen with normal MSH will have the multiply antibiotic resistant staphs. Don't forget that these organisms aren't invading you and they aren't like their big bad brothers, Staph aureus. They just sit on mucus membranes, silently sending their hemolysin proteins into your blood stream, turning on cytokine attacks 24/7. If you have one of these previously benign staphs, you won't get better until they are eradicated.

"These bacteria live in a protected environment, surrounded by the polysaccharide matrix that is the biofilm. Biofilm has a great technical name too: slime. No kidding. Inside the slime dome, or igloo, something magical starts to happen to the individual bacteria. After living all their lives as free swimming single organisms (planktonic forms) they start doing different, specialized tasks, but not *everything* like they had to before, but almost as though they have become a *multi-celled* organism. Incredible. Imagine these bacteria having pot-luck lunch, with each bringing his favorite dish to the party. No two do the same kind of cooking and after the meal; they just sit around trading DNA like some sort of baseball card swap meet. Differential gene activation is the key to what makes these organisms hard to kill with antibiotics, as the different chunks of DNA they share can code for antibiotic resistance. It isn't a surprise that resistance to methicillin is so common; these organisms may indeed be the reservoir for the methicillin resistance that we see in the scary methicillin resistant staph aureus (MRSA).

"Since the biofilm protects these germs from topical sprays, saline solutions, dry air, moist air and oral antibiotics, how do we kill off these limpet-like surface dwellers? Good luck! Just ask the cardiac surgeon how he clears the bacteria from an artificial heart valve; or the orthopedist who has to remove the foreign hardware from your hip or the dialysis doc who has to clear your peritoneal catheter. The answer comes from Dennis Katz at Hopkinton Drug (Massachusetts) who developed a nasal spray with EDTA and antibiotics in it. With the EDTA, the biofilm is dissolved, the antibiotics can access the inner sanctum of the biofilm and the germs are killed. Our ability to make you well rests on a number of factors and one of them is ridding you of any biofilm harboring these MARCoNS.

"Now as to the measures of VEGF and erythropoietin, they tell us about oxygen delivery in the capillary beds. What I mean by that is that if there's reduction of blood flow in the small blood vessels, called capillaries, all because of some increase in cytokines or TGF beta-1 or MMP-9 or C4a, or in any combination, we'll see the capillary beds responding with a message that says to a gene controlling factor called hypoxia inducible factor, HIF: 'Hey, we're not getting enough oxygen, we have hypoxia, do something. Now!' In response to HIF activation, both VEGF and erythropoietin are poured out. Extra blood flow follows (see Chapter 6, Follow That Oxygen). With high TGF beta-1, however, these two HIF-controlled compounds are often significantly out of whack. In fact, low VEGF is one of the most common abnormalities that I see in CIRS (chronic inflammatory response syndrome) caused by exposure to the interior environment of WDB.

"As to cancer-I have to tell you, Mr. Jones, that a lot of people ask me about cancer and mold exposure because some of the mold toxins are known to cause cancer. I just don't see very much cancer in my groups. In fact, our levels of breast cancer in women, colon cancer in men, and cancer in all comers is less in the mold patient then it is in the normal population *if VEGF is low*. If VEGF is high, then there's no difference between mold patients and non-mold patients. Something important about VEGF is going on. By the way, Mr. Jones, if you felt like reading more about mold, VEGF and cancer, you may want to look at a book about Dr. Judah Folkman. I think it is called *Dr. Folkman's War*. He was working in Boston looking at cancer cells in tissue culture and wanted to find a way to stop cells from growing rapidly and moving. In an unusual, serendipitous "experiment," for some reason, a tissue culture agar plate was left uncovered overnight in his lab and it was exposed to mold growing nearby. Sounds like Alexander Fleming and penicillin, doesn't it? Well, in this case, what Dr. Folkman's research staff saw was that the lining cells, the endothelial cells exposed to the mold stopped growing normally. What a breakthrough that was. They eventually found the substance, called fumagillin, named for the *Aspergillus fumigatus* that made it. It was the first agent ever used in what is now called anti-angiogenesis. Now VEGF blockers are key elements in treatment of many kinds of cancers. With your low VEGF, it's functionally as if you already have a VEGF blocker on board.

"Now let me warn you about the hard one: gluten. If we find you have antigliadin antibodies, but not celiac disease, your life will really change. You will need to eliminate all gluten from your diet for about 3-6 months. I'll back up: the autoimmune tests that we have measure anti-cardiolipins and antigliadins. Because of your unusual joint symptoms and the fact that someone thought you might have lupus, I ordered some additional tests as well. Your ANA, the most commonly ordered autoimmune test, is normal, as is your lupus panel. So you don't have lupus, even though at one time you did have that false positive ANA. Of interest, though, is that the cardiolipins, especially the IgM class and the antigliadins of the IgG class, are commonly found to be positive, that means present, in people with CIRS at a level that's astoundingly high compared to control groups. You have *both of these* and that's a real problem. We will talk about a no-gluten diet for you because until we have this auto-immune problem under control (driven by TGF beta-1), the gluten that you eat will turn into gliadin, then the gliadin antibodies will make the inflammation process worse for you. We'll get to that a bit later.

"I also want to tell you that you do have the anti-actin antibodies. These used to be called the smooth muscle antibodies. If they were present, we worried about liver disease. However, it's now simply showing us what's going wrong with cells called regulatory T-cells that should control autoimmunity but that just start being fouled up by high

TGF beta-1. So TGF beta-1's overall role in rheumato
is ***giant***. My prediction is that someday we'll have a b
new text book of rheumatology, which will talk about autoimmunity and TGF beta-1, because most textbooks don't account for that and quite frankly because they don't, they're just not right. Given your high TGF beta-1, body habitus and other autoimmunity, it's no surprise that you have this problem of positive anti-actin antibodies, too.

"Your C3a and C4a results just came in the other day. I hope you received them in the most recent mailing we sent you. C3a tells us indirectly about bacterial membranes present in blood. The good news is you don't have anything there, but your C4a is over 20,000. Any test result over 2,800 is of concern; the average for people with CIRS starts at 10,000. People with more serious illness and repetitive exposures will have an almost exponential increase in their C4a with re-exposure such that when people are sick the second time, C4a peaks out usually at around 20,000; the third time will bring 50,000; the fourth time will bring 180,000 and the fifth time too high to measure. I'll talk a lot about C4a later, but basically you need to know that as it is now, with your level over 20,000, you're in a high-risk category and you really must stay away from any kind of water-damaged buildings. You have about 5 - 10 minutes to get out of a building once you know that it's moldy or else you'll start suffering the slings and arrows from C4a. (*Editor's note*: the minimum time needed to activate the enzyme that makes C4a is about five minutes.) I don't want to be an alarmist but I'm just telling you, for example, if the building ERMI is over (–1), which itself is a low number, and your C4a is over 20,000 and you're in a WDB, you'll get sick while someone else will not.

"There are some other labs to pay attention to. You don't have Lyme disease. Your thyroid is nice and normal. Your iron studies are fine, including the iron storage protein called ferritin. I sure do see a lot of people with elevated ferritin levels, including a really disproportionate increase in people with hemachromatosis. Can't tell you why. The A1C hemoglobin we do to make sure that diabetes isn't lurking in the wings and the IgE test tells us that the illness that you have from mold exposure is *not* allergy or acquired immunity: it's simply due to innate immune problems. "I want you to look at your von Willebrand's profile. Like me (see Chapter 5, Blood in the Sink), you'll start carrying around medicine with you when you travel because you're at high risk for developing some rather impressive nose bleeds. What you have is called *acquired von Willebrand's syndrome*, which is a bleeding disorder caused in part by the effect of high C4a on how the individual pieces of von Willebrand's antigen, called monomers, aren't able to be linked together, monomer to monomer, to form what we call multimers. This polymerization process is one that's needed to help secure the integrity of a clot in a blood vessel, especially your nose. High C4a disrupts that. That abnormality does correct with treatment so that it's possible you won't bleed to death tomorrow from re-exposure to your moldy basement.

"The last test we did for you is the genomics assay, the PAX gene tube. What we have done with this tube is essentially preserve evidence of gene activity in white blood cells in your blood. We can block the process of destruction of the messenger (mRNA) made when a part (a gene) of the DNA is turned on. This mRNA tells us about gene activity, such as when DNA becomes activated in the gene, out comes mRNA. By preserving the mRNA that would otherwise be rapidly destroyed in whole blood samples, measuring mRNA we can find out what genes are active. There's a computer program called Rosetta Reader that reads about 35,000 genes of inflammation. It's an incredible advance to be able to look at your illness at a given time and see which genes are active. The beauty of this test is that there's a big difference between normal people (controls) and those with biotoxin illnesses (cases). Also, there may be significant differences among distinct biotoxin illnesses that otherwise look pretty much the same as far as symptoms, VCS and labs."

Mr. J: "I've got to ask, Dr. Shoemaker, what did my Rosetta Reader show? I don't see it here. I mean, basically, it sounds like you're telling me there's a fingerprint of mold illness in our genes. If that's true, couldn't you just go into a building, draw one tube of blood and figure out who has mold illness?"
Dr. S: "I thoroughly enjoy seeing your brain come back alive again, Mr. Jones. That's precisely what we think we'll be able to do when we get the genomics assays finished. The time isn't 10 years away, either. It's now.

"We have to then look at what proteins are present in your blood at the same time and what we're doing is combining what scientists are calling genomics and proteomics.
By putting both together at the same time, we create a fingerprint of your illness. We already know a lot about differential gene activation seen in mold illness and Post-Lyme and ciguatera and CFS, for example; we've got to do a lot of work on control groups and the effect of day-to-day changes, and we're doing that. In science, the skeptic must be heard. When someone says these results aren't significant, well, we've got to show an answer. Basically though, in the face of obvious abnormalities that are controlled and are part of a well-designed study, skepticism is inappropriate."

Mr. J: "Well, Dr. Shoemaker, does that mean that I can go back to everyone in my apartment complex who might also be breathing some of that air and do this genomics test and come up with a unique profile that would tell me whether they're moldy are not?"

Dr. S: "That's what it looks like now. We've got to raise a lot of cash to do the work; it's not cheap. If the money is there, we've got the samples stored to give us the answers."

Mr. J: "Holy cow."

Dr. S: "Mr. Jones, we've got to talk about the treatment that's your next step..."

Mr. J: Oh and one last question. I remember reading about CSM in Desperation Medicine. How does it work?

Dr. S: CSM isn't absorbed by our body and it won't react with anything in our body outside of the inside of the gastrointestinal tract. It acts like glue to absorb a large list of compounds that we fear, like DDT, DDE, PCBs and dioxin. If you take it with food, the food becomes glued, so that is why CSM must be given on an empty stomach. It can cause a lot of abdominal distress, like constipation, bloating and belching, but it works well enough and fast enough that literally over 8000 people have put up with the side effects to feel better. Of course, most people don't have side effects from the drug whatsoever.

The CSM / liver partnership is important. While the innate immune system and its creation, acquired immunity, seek out dangers by patrolling the blood stream, the other safeguard is our liver. It processes the poisons that are carried to it by the blood and has two ways of doing this. When it can, it neutralizes them so they pose no danger and they are then excreted. When it cannot, it does manage to divert them down the trash chute – the bile ducts. These ducts lead to the intestine and most harmful compounds would then be excreted in the normal way. Not so the biotoxins. Without CSM, the negatively charged toxins, we call most of them ionophores, simply get reabsorbed further down in the gut and rejoin the party.

As fate would have it, just about all the toxins we have to treat are very small, with a group of oxygen, nitrogen or sulfur atoms that form a ring in three dimensions in the real world. These large atoms share electrons such that the ring develops a net negative charge. The radius of the ring just coincidentally matches the size and shape of the side chains of CSM that have a net positive charge. When CSM gets together with toxins in the gut, the bear trap-like positive charge traps the toxins and they aren't reabsorbed. Down the toilet they go.

As time goes by, with CSM and with removal from exposure, the deficit seen in visual contrast testing (VCS) disappears. Here is the time where people often stop CSM but if they are exposed to any of the 25% of buildings in the U.S. that are moldy, CSM must be re-started.

CHAPTER 5 Blood in the Sink

I'm trying to will my nose to stop bleeding. Nothing else has worked. In desperation, I've turned to mind over matter - if I lie still enough and concentrate hard enough, maybe the bleeding will magically cease and I can sleep.

Not tonight. For the fourth time tonight, I can feel the blood welling up in my left maxillary sinus. This kind of bleeding feels different from a runny nose when you have a cold; I've come to dread what follows when the bleeding comes. After all the surgeries I had in 2004, how can anything be left up there to bleed? I've got just enough time to jump out of bed before either waking my bride or bleeding all over the fresh sheets.

I bolt for the bathroom sink, paper towels from the bedside table catching most of the bright red blood drops that drip painlessly through the packing the surgeon put in this morning. One-thirty a.m. turns to 2. Although I've tried everything I know how to do as a physician to make it stop, my left nostril again has gushed blood. Whether I stand or sit doesn't matter, but with direct pressure and after lots of soaked towels, the bleeding finally stops.

The last bleeding just stopped an hour ago. That time it took 30 minutes of constant pressure, pleading and prayer.

I didn't worry much about my first nasal hemorrhage many months ago. Back when the bleeding began, I was pretty sure the nosebleeds were a hoot. I probably just blew my nose too hard. The bleeding stops soon enough, what's the big deal? Just for fun, I stuck my finger in the washbasin full of my blood in the hospital room (why wasn't it clotting?) and wrote "REDRUM" on the mirror just like Jack

Nicholson did in *The Shining*.

Good joke, right? REDRUM, murder spelled backwards, right? No one else laughed.

And then the surgeries started and failed once; started and failed again, and then three times. The bleeding isn't very funny anymore.

Back then, the bleeding brought horror, but now the bleeds bring pure terror. I've already died once, just not long enough, because somehow they were able to revive me.

During the past few months, I've learned how to figure out when the left maxillary sinus bleed is coming. (You learn fast when bleeding to death.) I didn't learn about that feeling in medical school. No professor taught me about mold bleeds and no textbook had my problem listed in the index. What *is* in the textbook about bleeds like these is just plain wrong. For example, if someone asks me about my blood pressure one more time, I'm going to bleed *on* them. And no, I don't pick my nose all day long, either.

Maybe the past four surgeries have clued me in, but I know now when the bleeding is coming. Once blooded, twice wary; three times just makes me weary.

For a moment, I think of all those moldy babies in Cleveland with nosebleeds like mine and lung bleeds of their own and how the CDC wouldn't agree that their mold exposure was making them bleed. If only the doctors caring for those desperately ill infants then in 1999 had done the testing I now know in 2010 to be so easy to do, maybe those babies would be healthy today. Oh, I've read a lot about the manipulation of science from some people in NIOSH: in a deposition in a civil case on which I was the plaintiff's expert, one consultant for the defense even seemed proud about his role in making the Cleveland physicians back off from their findings that mold exposure had caused these kids' terrible illness. His seemingly boastful statements were just such a disgrace to medicine. Tonight I've got to get over it: Complaining about *manipulation of science and overt negligence* by Federal health agencies won't help me stop bleeding.

I remember a patient I admitted to Duke Hospital years ago as a second-year medical student. It was his 45th admission for bleeding from his "classic" hemophilia; it was my first. I had no idea then of the dread he must have felt when one or two of his joints quietly filled with blood, without reason or control. The hospital was his only hope. There, we would give him some clotting factors, especially Factor VIII, and hope the bleeding didn't start in the brain. It could; the patient knew that, too. These days, managing hemophilia is quite different; patients can take charge of their illness as an outpatient using the miracles of genetically engineered clotting factors. Hemophilia rarely results in mucosal surface bleeds like mine. I don't have hemophilia, so those clotting factors wouldn't help me. I just bleed when I've been exposed to moldy, water-damaged buildings.

I got this way because I am suspicious and curious. Early in my mold practice I made it a point to visit the places my patients told me had make them sick – their homes, their places of employment. I wanted to see for myself and so did proper Sherlock work – work which involved multiple exposures to dangerous WDBs. Only now, I'm sicker, quicker and bloodier, too. It used to be that only massive exposures made me hemorrhage. Now just about any exposure brings about my own red tide. (*Author's note: red tides are occurring worldwide in ocean waters near almost every coastal land mass. Caused by blooms of toxin-forming dinoflagellates, these red tides kill fish and cause acute illness in some people, with chronic illness occurring rarely. See Discovery Channel, Planet Green: Toxic Files, 12/12/09. True red tide has nothing to do with nasal hemorrhage – except, that is, for the color.*)

Clots taste different than fresh blood. Talk about perverse horror, if I swallow hard against a clot, maybe I'll dislodge it and wind up dripping over the sink again - don't want that! Still, I've got this huge anterior and posterior pack stuffed into every crevice my nose has, and for what? To stop bleeding.

Music comes to my sleep-deprived mind from out of nowhere: *Blood on the Tracks; Bleed on Me; My Heart Bleeds for You; Red River Blues*. None of those tunes help me rest.

My sister, Ellen, is so musical. One of her comedy routines sung from the piano goes something like this:

"One of my favorite, sensitive cowboy songs is Blood on the Saddle.

It goes like this:

There was blood on the saddle, and blood on the ground,
And a great big puddle of blood all around.

The cowboy lay in it, all covered with gore,
And he won't be ridin' his bronco no more.

Oh, pity the cowboy, all bloody and red,
for his bronco fell on him, and mashed in his head."

Ellen's song didn't help, either.

I don't think I've spilled blood on the floor or all over the cabinets tonight, not like the first time I ended up in the hospital. Still, I feel like I'm back in the same Stephen King novel as before, where some dark force hiding under my bed or behind the curtains maniacally turns on a spigot in the front of my brain that spews blood until the dark force is done playing cruel games with my sleep and my psyche.

The doctors have looked at just about everything, particularly my coagulation system; I showed them my past blood studies where the von Willebrand's factors (call them vWF) change after I'm exposed to a water-damaged building. Even though vWF falls low, I don't always bleed. And because my bleeding tests with funny names like INR and PTT don't show anything abnormal, they say I don't need clotting factors given intravenously.

Why not give them a try? Nope, shouldn't do that, they say.

Besides, who knows what kind of unusual problems the clotting factors, isolated from large numbers of blood donors, might bring? Everyone says you can't get HIV, hepatitis or Babesia from transfusions, but I see the case reports on PubMed, the NIH medical library, where people did. And don't forget about hepatitis: if we know about Hepatitis A, B, C, D, E, G and some others, who's to say that some heretofore unknown Hepatitis "Q" isn't stuck to chunks of the cryoprecipitated clotting factors, and I'll get Hep Q in a few months? So they pack my nose with gauze and send me home, telling me to lie absolutely still. They smile bravely at my worried wife and gloss over my gaze that silently, surreptitiously asks, "Are you sure this is enough to keep me alive?"

No one seems to be able to get to the bottom of the problem: I'm a mold bleeder. And a few others are, too. Nowadays, the vWF profile is one of the first tests I do when someone comes in with unexplained, significant nasal hemorrhage. Looking for mold exposure sure shortens the diagnostic process in those people.

I'm afraid to close my eyes: the dark ceiling is like a movie theater that doesn't change, and bizarre images flash in front of the blank screen behind my eyelids. I see a warm light, soft and comforting. That one makes me sit up quickly. Get thee behind me! I don't want *that light* in front of me. I saw that one a few years ago in the ICU when nothing, including multiple surgeries and catheter-guided plugging of bleeding blood vessels, stopped my hemorrhage. I can remember the absolute horror on the face of the confident radiologist who came to my bedside after he'd fixed my bleeding when the surgeons couldn't; as the blood began to ooze down onto my left cheek even after he confidently said that embolization would likely replace surgery in cases like mine. Yes, I had the light that night, only that light at the end of the tunnel was a warm, beautiful sky blue.

Oh, my God, that color was Carolina blue! We used to chant, "Go to Hell, Carolina, Go to Hell" during football games when I was an undergrad at Duke. Now, here I was, literally going towards a warm, nourishing Carolina blue! Surely, I was going to Hell.

(*Author's note: Dr. Ken Hudnell, as dedicated a supporter of the University of North Carolina, Chapel Hill, as any graduate from that school could possibly be, assured me that the warm Carolina blue was clear evidence that I was indeed going to Heaven, as not only was the Devil wearing Duke blue, but God had made the sky Carolina blue.*)

In the end, Dr. Charles Schaeffer had done intricate, life-saving surgery on tiny blood vessels that ran between my brain and my face, fixing all the bleeders in 2004. Those vessels couldn't be bleeding now, and the packing should've stopped everything else.

Earlier today, I asked him for some Factor VII, the cure-all for unknown bleeding. No, let's wait, he said. The pack will take care of the problem.

Time for Plan B. I didn't ask Dr. S if he'd object if I took some DDAVP, but I bought some anyway. If I had a vWF problem, and the pack didn't work, like now, the artificial antidiuretic hormone (ADH) would stop the bleeding quickly. I said fine, I'll try it; down the hatch!

The bleeding stopped in one hour.

DDAVP causes blood vessel walls to release the heavy, polymerized von Willebrand's proteins, almost like an emergency tourniquet.

Now, DDAVP is a drug I use often in my practice, as low MSH is associated with abnormal (usually too low) production of the real ADH in more than 60% of patients. We know that DDAVP is safe enough: we use it all the time, including in pediatrics, as it helps stop little boys from wetting the bed. Sure DDAVP has side effects— fluid retention and weight gain, but those problems aren't that big a deal compared to seeing the Carolina blue light again...

Since it worked, I thought that maybe now I might be able to sleep, but no. How long will I need to take DDAVP? Would my bride still like me if the DDAVP made me balloon to look like the Pillsbury doughboy? How can I stop the process that starts the bleeding? Where could I live if I bled a unit or two after just ten minutes in a moldy restaurant in an airport? Was my life going to consist of a bubble of my home and office, but no place else?

Or was the benefit of DDAVP just a coincidence and the monster was just tormenting me more by letting me think I could stop the bleeding chemically?

And then I remembered Don Malloy, who offered his support and concern earlier today when he stopped by, not hiding his look of horror after getting a good look at my nasal packs. "Physician, heal thyself." Thanks, Don. Was that the solution, to heal myself? If not only myself, then wouldn't my learning help others?

So I took DDAVP for thirty days, with only a few side effects. When could I stop and still be safe?

Fortunately, the maniac only returns now when I know I've overstepped my safety bubble by venturing inside water-damaged buildings. I don't go anywhere without DDAVP anymore. And don't ask me to go into the moldy courthouse on Constitution Avenue in Washington DC! (*Note:* See Chapter 21, A Tourists Guide to Moldy Washington DC.)

What does it all mean?

We've got the lemons, where's the lemonade? Meet MASP2!

Now when I get a mold hit and start to bleed—and this scenario has been reprised a number of times - I can take DDAVP, hoping that it'll fix the hemorrhage almost immediately. So far, it's worked every time. But what if it doesn't? I mean, I'm getting *worse, not better* with all these innocuous exposures to water-damaged buildings.

To be blunt, all this bleeding is pretty scary. Sure, I've got an antidote that's worked to keep me alive, but since "sicker, quicker," is my greatest concern for people made ill once from mold exposure, what can I learn about the interaction of clotting system problems and inflammation associated with activation of innate immune responses from mold exposure?

In the end, Surviving Mold isn't just an ideal: it is my Grail.

Let's go back to Francis Bacon, father of inductive reasoning, and assemble everything we know about a problem and start the inductive-thought process. The following information wasn't known in 2004 and only some of it was known in 2006. The big breakthrough in 2007 came when I started to learn about pattern detection receptors, including lectin receptors and the role of glycoproteins (proteins with sugar molecules stuck on them) in mold illness physiology. (*Author's note:* look out, research jargon ahead! Skip the next section if molecular biology has little interest for you.)

Maybe it would help you understand how mold exposure makes me bleed if I told you a bit about clotting mechanisms:

> Making blood stop flowing involves multiple, interacting pieces of a puzzle. Clotting factors can be deficient; I told you about factor VIII deficiency, classical hemophilia. There are others, such as deficiency of intrinsic factors
>
> Factor IX, XI and XII, are also on the list. Another arm of clotting factors, the so-called extrinsic factors, is involved too. The complexity of control of one protein activating another is a theme in human biology; this idea is well demonstrated in the complexity of clotting pathways, each integrated with many other systems, including inflammation. We can't disregard the role of thrombin activation from prothrombin, leading to clotting. That process has all kinds of interacting mechanisms, with anti-thrombin and thrombin all in the textbook.
>
> Platelets are involved too; they act as the final plug in a hole in a blood vessel if the body has laid the proper beds down alongside broken blood vessels for platelets to snuggle in. If you don't have platelets, look out, here come more acronyms for funny names for bleeding disorders, (ITP and TTP, for example) but platelet disorders rarely result in making people

exsanguinate from nosebleeds.

But what about those beds the platelets lie in? Here's where we need our good friends, the von Willebrand's (vWF) multimers, basically polymers of monomers. In a way, the bleeding problem I acquire is straightforward. No monomers, no clotting. There are several elements to think about in vWF illness. Please don't let the names bother your thinking about the bleeding problems. Factor VIII is an acute phase reactant, and levels will fall with most kinds of innate immune inflammation, especially mold illness. Ristocetin-associated cofactor is a dominant player interacting with Factor VIII and vWF monomers, as is the von Willebrand's antigen itself. When I order a vWF panel, I get four results: Factor VIII, ristocetin associated cofactor, vWF antigen and a measure of the high-molecular weight forms of von Willebrand's factor (the multimer). People who have abnormally reduced amounts of these compounds from birth have von Willebrand's syndrome. People who *acquire* these abnormalities, like me, don't have classical von Willebrand's syndrome. Acquired von Willebrand's syndrome can be associated with cancerous diseases of the white blood cell lines; but in the absence of hematologic malignancies, acquired von Willebrand's is incredibly rare, as shown by a review of the medical literature. A recent paper in the New England Journal of Medicine, "A Bloody Mystery" (Cuker A, et al, NEJM 2009; 361: 1887-1894), provides some useful graphics and text that clarifies the vWF ideas.

Two studies from Italy, each among the world's largest, reported a paltry 20 cases of acquired von Willebrand's syndrome in more than 100,000 patients analyzed. Until Sept. 14, 2009, the peer-reviewed medical literature didn't include exposure to water-damaged buildings as a reported cause of von Willebrand's-related bleeding episodes. After that date, however, acquired von Willebrand's syndrome certainly is in the peer reviewed literature on CIRS-WDB.

At the international conference dedicated to WDB - *Healthy Buildings IX* - held in Syracuse, NY in a conference presentation (accepted following peer review) on 815 mold illness patients and 132 controls, we reported 22 cases of acquired von Willebrand's syndrome. Let me see, 22 cases in 815 mold patients is about one in forty. Twenty cases in 100,000 is about one in 5,000. One in forty compared to one in 5,000 is a 125-fold enrichment of illness in mold patients compared to the standard population.

The lab testing for von Willebrand's isn't any different now compared to five years ago: what *is* different is a focus on a particular population of people with a chronic inflammatory responses syndrome. By looking at those people who are likely to have the problem, case finding becomes much more efficient. Don't forget the old medical school saw about Willie Sutton, the American bank robber from the early 1900s. When asked why he robbed banks, Sutton, a man of few words, simply said, "That's where the money is."

Back to the biology. What drives down the ristocetin-associated cofactors and the multimers of vWF? Curiously, as Francis Bacon would notice,

The higher the C4a, the higher the risk of having acquired von Willebrand's syndrome. Could there be a role of C4a in (1) dissolving multimers, or (2) preventing the release of sequestered multimers or (3) preventing the polymerization of monomers in the first place? Each of those possible mechanisms could answer the pathogenesis question.

If C4a were the problem, then wouldn't that explain all the observed worsening of bleeding occurring with sicker, quicker?

We know that if C4a levels are normal, then there almost never will be any problems with low levels of vWF and certainly no acquired von Willebrand's syndrome.

Observations and Therapy

We know that in our work with patients undergoing repetitive exposure trials, using a standard research design abbreviated as ABB`AB, we've been able to show that there's a time course of change in objective physiologic parameters following exposure to water-damaged buildings.

> Following re-exposure of patients previously treated with cholestyramine (CSM) we see from the results of daily blood draws that levels of C4a, a split product of activation of complement often called an anaphylatoxin, rise within four hours. Leptin levels will rise in approximately 24-48 hours, reflecting the increased cytokine effect damaging the receptor for leptin in the MSH production pathway. MMP9 will rise in approximately 48-72 hours, reflecting the time course for cytokine effect to stimulate gene activation of the MMP9 production pathway. VEGF levels initially rise rapidly, followed by a significant drop. Clotting factors also change rapidly with unprotected re-exposure, with Factor VIII levels falling concomitantly as C4a rises and then recover to normal after 48-72 hours. Von Willebrand's factor and ristocetin-associated factor are maintained for 24-48 hours, falling dramatically to a nadir in 72-96 hours. This change is often associated with unexplained bleeding if the multimer levels also drop, usually either nosebleeds (epistaxis) or lung hemorrhage (hemoptysis).

- While laboratory tests show that not all patients exhibit all the inflammatory changes described above, <u>no control patients show any of the lab changes</u> over a four-day course of daily blood draws.
- Furthermore, <u>without</u> another exposure to WDB, no changes in any lab parameters performed daily over four days will occur in patients with prior mold illness.
- <u>With</u> repeat exposure to a WDB, those people found

to show bleeding problems before in the repetitive exposure trial simply have another bleed, though a bit worse than before.

A second course of therapy with CSM will return the abnormal lab values back towards values seen following initial therapy.

Furthermore, we know that injections of low-dose erythropoietin (epo) can safely reduce levels of C4a and raise levels of ristocetin-associated factor, each occurring exactly at the time just before the bleeding stops.

So if we can stop the bleeding safely with DDAVP, what do we care why and how? The answer is key to *Surviving Mold*. Survival means one must absolutely defeat high C4a. Epo does the job to stop C4a over-production but what are we going to do about the treacherous manufacturer of C4a, an enzyme that's capable of turning itself on, becoming the auto-activator, the gift that keeps on giving C4a? If you think like I do, the self-activating production of C4a is like the robot machines from the future, the Terminators, in that they keep on re-starting, no matter what the heroine did to them. As you read this chapter, you don't know how we're going to defeat the C4a-Terminator (see Chapter 16, VIP), but we will. We need to be able to reprogram our defense system – our incredibly complex, incredibly interactive and incredibly powerful innate immunity.

We need to turn off the friendly fire.

So what do we have? Exposure to WDB results in a rise in C4a, occurring faster in those with prior illness than in those with a first-time illness. C4a is produced when either the classical pathway or the mannose binding lectin pathway of complement is activated. That means that C4 must be activated for any C4a to be made. There are a lot of possible substances that activate C4; indeed, Dr. Patsy Giclas says that

C4 is like a protein with a big sign on it that says, "Activate me!"

Let's take a closer look at C4 since its activation is a basic mechanism that controls so much of what makes surviving mold exposure so tough. Beware, as the medical jargon gets thicker here. C4 basically is what we call a "pro-molecule" - like a loaded pistol in your pocket it doesn't hurt anyone unless it is triggered. However, once your amazing immunity turn on C4, splitting it into active pieces which include C4a, all kinds of bad things are guaranteed to happen if C4 activation isn't turned off. The C4a initiates production of C3a, and then sequentially turns on the rest of the complement cascade so bad things will happen in other arenas. The important aspect for me and others with these syndromes is the activation of the enzyme that makes C4a appear. We want to stop that enzyme, as it is the driving force behind the constant high readings of C4a. And if there is hemorrhage, *we must **STOP** abnormal ongoing activation of C4.*

What is the "complement cascade?" Complement involves three separate pathways: classical, alternative and the mannose binding lectin (MBL) pathway. For my purposes, MBL is where the action is. MBL, like all lectin receptors, is used by the body to recognize specific foreign invaders, antigens that have a particular kind of shape and structure. These receptors are turned on by ficolins and glycoproteins, especially those that have the sugars mannose and N-acetylglucosamine. Next, when MBL is turned on, now the heavyweight enzymes, MBL-associated serine proteinases (MASP), come into play. As it turns out, in this incredibly complicated system passed on by our ancestors (the ones who survived long enough to pass on their genes), there are several MASPs, with MASP-2 being the key enzyme activating C4 and there thereby releasing C4a.

Basically, if you knock out MASP-2, the whole system that creates more C4a grinds to a halt. Hmmm, sounds like stopping MASP-2 is a good idea if you want to avoid the terrible price that high C4a imposes when our innate immune response system hurts us.

Make that price a bit steeper, if you will, when you factor in the concept of auto-activation. Turning on MASP2 once causes more MASP2 to be regenerated. Talk about the snowball of a positive feedback system rolling down the ski-slope! C4a itself only lasts a few seconds before it's broken down, yet just about all chronic mold patients have C4a that's way too high. How can this be? Because MASP2 keeps on making more C4a even as it makes more of itself.

Our innate immunity and auto-activating immunity is the gift which never stops giving until we die. It creates an amplifying biological cascade.

In a few words: Why does the bleeding start and stop?

Remember the concept of sugar proteins, our friends the glycoproteins. If we're to link mold exposure to lectin receptors and MASP2 activation, with a resultant increased C4a, we should find that glycoproteins are involved with vWF and erythropoietin. We do. Further, we'd like to find glycoproteins involved with the changes in clotting factors that accompany re-exposure of previously ill patients. We do.

Basically: (a) high C4a is produced by MASP2 as part of the innate immune response to foreign particles and proteins, the "antigens"; (b) antigens are found in WDB; (c) C4a blocks multimer formation; (d) no multimers, no clotting; and (e) no clotting, so we get Stephen King's bloody terror.

When we block the high C4a with epo, C4a can no longer cause disruption of multimer formation. OK, that enables the normal process of polymerization of monomers of vWF to proceed and clotting can start. Add in DDAVP and we stimulate release of pre-formed multimers from their reservoir alongside edges of blood vessel walls. Net result: no more stimulation to bleed from absence of monomers (thanks, epo) and no more reduction of pre-formed multimers (thanks, DDAVP), effectively stopping the existing bleeding.

And the final result: now when mold bleeder patients are re-exposed, they're no longer destined to bleed again and again. In the perfect world of *Surviving Mold*, we'll also be able to stop MASP2, prevent the hyper-reactivity that otherwise would result from exposure to the moldy airport restaurant, thereby avoiding the need to even carry DDAVP for protection or the need to have access to epo. That time is just now arriving (Again, see Chapter 16, VIP).

My life and that of so many others is changed by DDAVP, a pill that costs about a nickel. Stephen King's monster is chased out from its darkened lair under my bed, vanquished by the light of science. To be safe, however, I don't leave my safety bottle of DDAVP pills at home when I travel.

As time has passed, this understanding about the coagulation problems seen in CIRS-WDB patients has been expanded to finally reach a unified medical end. What mold illness involves is the entire group of chronic inflammation pathways, not just coagulation. These pathways are all linked, so cytokines, MMP9, complement, coagulation elements, T regulatory cells, autoimmunity, VEGF, MSH, VIP and TGF beta-1 are all interacting simultaneously, each multiplying what the other compounds do, like a grand symphony reaching a crescendo in Beethoven's Ninth.

To paraphrase TS Eliot, thanks to some unusual nosebleeds, we've reached the end of this stage of all our explorations into mold illness. At the end of all our exploration is our first survival goal. We'll know where we started for the first time. But as you know, biology never yields all her secrets, and we're just now starting out in our explorations again.

Bleeding, then nearly dying provided me with a personal goal of survival. Now no other mold illness patient need go through visions of warm, blue lights at night after simply blowing his nose in a moldy building.

CHAPTER 6 Follow That Oxygen!

The point of this chapter is simple: without adequate oxygen, cells end up starved for energy.

When cells are starved for oxygen and energy, they don't work properly. When cells don't work right and the cause of inadequate oxygen delivery is chronic, systemic inflammation, people develop chronic symptoms, with fatigue, cognitive problems, joint symptoms and respiratory compromise leading the list. When the systemic inflammation is cleared, oxygen delivery is restored and chronically ill patients become the functionally living again.

Chronic fatiguing illnesses, including Chronic Fatigue Syndrome (CFS) aren't always due to systemic inflammatory response syndromes, but in my experience over the past thirteen years and with well over 8100 patients, for every patient with a "non-inflammation" source of CFS-like illness, there are 100 patients with inflammatory sources of illness. Mold illness leads the list of causes of inflammatory sources of chronic fatiguing illnesses by a huge factor, perhaps because WDBs provide invisible traps or ambushes where the unknowing and unprepared are injured. When one challenges the unsupported diagnoses so often made in America, as I must, especially those of fibromyalgia and CFS, what one finds is typical Ass^2 medicine. Glorified guesses at best. These conditions have no diagnostic blood tests and for their "existence" rely on consensus statements by self-identified experts; the names chosen give no clue about the physiology of the illnesses or how they begin and change over time. (*Author's note:* When there's no defining physiology for an illness, watch out for what panels of 'experts' decide is true! Always ask about conflict of interest in such 'experts;' you'll find it more often than you might want to believe.)

Perhaps the world of chronic fatiguing illnesses will change. In October 2009 it was reported that 68 of 101 patients with chronic fatigue syndrome (CFS) in the US were infected with a novel gamma retrovirus, xenotropic murine leukemia virus-related virus (XMRV). But then in a January 2010 study in the UK, 186 CFS patients were examined (**http://www.ncbi.nlm.nih.gov/pmc/articles/PMC2795199**) and none of them harbored XMRV. Maybe we'll find that the newly described gamma-retroviruses (XMRV) really do have something to do with CFS. Maybe not, as now the existence of a different virus that is similar to the XMRV is getting a lot of new press. But the bottom line says, "Is there a viral source for the bulk of CFS cases?" At the moment it seems questionable. Indeed two such studies using objective methods producing such diametrically opposed results an ocean apart might make one wonder if the subjective diagnosis of "CFS" in and of itself presents a problem. Perhaps those who study "it" may be misidentifying it? Perhaps there are many patients that have been told they have CFS – a disease without known physiology – who are actually ill from another, objectively confirmed, cause.

Still, in the end, maintenance of cellular function with adequate supplies of oxygen and energy is the underlying force that separates the chronically well from the chronically ill.

If you read the jargon in the CFS literature about oxygen delivery, the underlying physiology might appear to be so complicated that no one can describe it. Just look at the terms: post-exertional malaise (my favorite overblown term); delayed recovery from normal activity; reduced VO2 max; decreased anaerobic threshold. These terms don't tell us much about how things work or don't work. All these different names come from issues that stem from a simple problem: capillary hypoperfusion. Think physiology. Capillary hypoperfusion means reduced blood flow in the smallest, most abundant blood vessels (and the tissues in the body they serve), which means a lack of normal oxygen delivery.

Heaven forbid that someone at a CFS conference asks the speaker to discuss the mechanisms that *lead* to post-exertional malaise. Or even more discouraging, ask something like, "Please discuss (1) the prospective documentation of abnormal physiology that's been reproducibly observed in multiple venues (2) and confirms the existence of CFS or fibromyalgia."

You'll hear an answer, but not one worth writing down. Neither will you hear much discussion about innate immune responses or about mold or other biotoxins. As for prospective acquisition of illness (how people got these illnesses in the first place): Forget it.

That's too bad, because what's been named fibromyalgia and CFS by a consensus panel - a manufactured symptom-based diagnosis - is all about innate immune responses, and mold illness is first in line as far as possible causes. It's true. We've done enough retrospective and prospective studies that identify the abnormalities in mold illness; it's pretty routine these days.

But anyone with a chronic fatiguing illness, including everyone chronically ill from exposures to water-damaged buildings, should learn the elegant idea behind the term "capillary hypoperfusion." It's an elegant idea because of its "pure simplicity" and "purity of concept." Once you understand what happens as a result of the inadequate delivery of oxygen, fluids and nutrients into capillary beds and then into the cells those capillaries nourish, you can see what needs to be done to correct the problem. It sounds so simple: restore normal perfusion and then watch as the devastating fatigue and all of its associated symptoms fade away. If the illness interventions *don't* correct capillary hypoperfusion, with all attendant causes, the patient won't get better.

Why do I feel so strongly that *you* need to learn about capillary hypoperfusion? Because by the time you're done reading this chapter, you'll know why so many "CFS" patients are actually chronic biotoxin illness patients, with mold illness being a clear number one cause.

What I want you to know might appear a bit complex at first, but I'll take a shot at explaining the mechanisms of normal blood flow in the tiniest blood vessels. The abnormal physiology underlying just about every aspect of multisystem illness that plagues chronically fatigued patients is based on what capillary hypoperfusion does to *how our cells burn sugar*. The abnormal, inadequate delivery of oxygen results in wasteful burning of limited cellular resources, including "make-up" energy sources like fat and protein - especially protein.

Let's go through this. Each of our cells can store a limited amount of sugar in a storage compound called glycogen. This substance is the classic complex carbohydrate. When our cells need sugar for fuel, a signal is sent to enzymes that regulate glycogen: release some sugar. If there is glycogen, out comes the sugar, the sugar is broken down in the cytoplasm of the cells to release two dollars worth of energy molecules (called ATP), plus breakdown products called lactate and pyruvate. Lactate and lactic acid are markers for accumulation of sugar breakdown without oxygen. If there's enough oxygen delivered to the cell, the breakdown products are sent to the energy powerhouse of the cell, the mitochondria, where the magic of energy generation takes place. Out of just two little pieces of sugar comes 36 dollars worth of ATP. But if there isn't enough oxygen, the mitochondria might as well not even exist, because no 36 dollar bonus pay will occur. But the poor cell doesn't care where the cellular energy dollars come from, it says I still need to go up this hill, run this street, think this thought, I need energy/cash. So the cell's ATM machine, glycogen, says OK, here's some more glucose, don't waste it. But then the cell goes ahead and wastes it anyway.

So if not enough oxygen is around, is the illness a mitochondrial problem? No way, though you'll hear that a lot in the CFS world.

So, if there isn't enough oxygen, the cell only gets about 5% of the total available energy from sugar that it should. The other 95% is wasted, with rising lactic acid a good measure of how little oxygen is being delivered. The amount of glycogen in any cell is limited. If you burn up what small supplies of glycogen you have, you're in trouble. It takes at least ***two days*** to replenish the glycogen, a time in which people will have delayed recovery from normal activity. That means they are dead exhausted from just trying to be normal for a day. The impenetrable jargon term, "post-exertional malaise," that tries to tell us about how exhausted people are when they do too much, used by the CDC and others, to describe the results of inefficient glycogen burn, makes me wonder if anyone in the CDC really wants to help those with chronic fatiguing illness understand why they must pace themselves. But then, maybe the CDC boys don't know, since they kept trying to say in years gone by that CFS was psychological and mold illness doesn't exist.

Let's make the fatiguing illness worse. You are having the rare good day. You have glycogen, you'll feel OK (relatively). So you'll try to catch up on what you didn't do the last week when you felt crummy. But when you burn up your glycogen, you'll be exhausted for two days. *Over and over again*. Only now, you continue to push yourself because you're sick and tired of being sick and tired. Despite no glycogen, you tell the cells to work anyway. No oxygen and no glycogen mean trouble. And trouble in the cell means burning fat reserves (unless your leptin is too high, making fatty acids unavailable). If there isn't adequate fat around to burn, your body quickly turns to protein: give me some amino acids that are turned into sugar fast! Protein agrees to sacrifice itself, knowing the amount of protein we have is genetically programmed to be replenished when the protein wasting (burning) is over.

Except the CFS guy never stops burning protein. And then someone says, "exercise more and eat less." Just about everyone with chronic fatiguing illness hears that **idiotic advice**. Here the cell is starving for glycogen and the patient is told to eat *less*? The body is burning protein like crazy just to walk up the stairs and the patient is told to exercise *more*? Wrong, wrong, wrong.

So what happens now? As the lean body mass is wasted— "going catabolic" is the medical term— the body holds onto all the fatty acids it can. The patient is losing protein weight and gaining fat weight as a result of capillary hypoperfusion. What a nice illness: patients are chronically fatigued from inflammation, now they're also protein wasted, adding to the chronic fatigue, and they have extra fat on board, which adds to the exercise intolerance, adding more fat and protein loss. It's an endless cycle.

Then we order a pulmonary stress test that shows the oxygen we need isn't being delivered (that's the low VO2 max) and the pulmonologist says the patient is deconditioned and simply needs to exercise more.

Any questions on how the chronically fatigued patient gets lots of wrong advice about what is wrong and what to do about it?

Therapies that make a difference to those with chronic fatiguing illnesses must correct (i) the abnormal use of energy at a cellular level, (ii) the protein loss, and (iii) the fat storage. Each of these therapies starts with oxygen delivery. We want all cells to have the sugar (glucose) they need to produce energy. We want to increase oxygen delivery to cells so that glucose can be burned efficiently inside the cell for energy.

If your lawn mower runs out of energy and the engine is "dead," you add more fuel, often gasoline. And if the fire in your wood stove dies down and the fire is dead, you add more fuel, usually wood. But if a cell runs out of energy, it doesn't die like the lawn mower engine or the fire in the stove; the cell lives, but at a tremendous cost: it must burn something besides sugar from glycogen for fuel.

Let's make the analogy a bit more complex. If there's adequate gas in the lawn mower, but the air filter is so clogged that no oxygen makes it through the carburetor, the mower won't start easily (if at all) and it won't run right, either. But don't toss the lawn mower aside in frustration— just clean the air filter.

The wood stove is no different: if the stove doesn't provide enough draft, and not enough oxygen is delivered to the coals, the stove will smoke too much or won't burn at all. Keeping a wood stove burning properly can be a delicate art, and you have to balance all sorts of variables. If you want a nice fire in your stove, be sure to have the right sequence of fuel sizes to burn as well as the right balance of fuel to oxygen.

Simply stated, you need both fuel and oxygen, each mixed with the other properly, to mow lawns and stay warm in winter.

In many ways, those patients with chronic fatiguing illnesses, especially those with mold illness, are ill because their cellular machinery doesn't get the right amount of oxygen delivered in the right amounts at the right time and in the right sequence because of ongoing capillary hypoperfusion. Reduced oxygen causes the cells to burn fuel inefficiently.

If only treating these patients were as easy as buying a new air filter for the lawn mower or correcting a faulty air intake channel in the old wood stove.

It's easy to understand almost intuitively why some chronically fatigued people have some "good" days among a string of "bad" ones: if they have some energy on a good day, most try to catch up on work they've postponed for so long because they felt bad, and they overdo it. The result: they're wiped out for two to three days. The CFS case definition calls this delayed recovery "post-exertional malaise." Malaise? No, they're just out of fuel. They've burned up their glucose and their glycogen stores are zilch, and it takes a few days to replenish them. Once they restore glycogen, they enjoy a good day. So now it's time to do the grocery shopping or maybe spend thirty minutes in the garden. Now what happens? Their glycogen is soon gone again and they're down for a couple of days until their glycogen stores are replenished and they have a good day again, only to repeat the cycle and become exhausted again.

When will CFS folks learn that a crash-and-burn cycle doesn't work?

Until people who are chronically fatigued learn how to pace themselves, their life is this vicious cycle which continues until they finally say, "Forget it. I give up." Some do.

You can't measure body glycogen stores by looking at someone. Ask a chronically fatigued patient how many times they've asked a physician for help, only to hear, "Well, you don't look sick. All you need is a little exercise; and to lose some weight." And then they charge hefty fees for such empty-headed, patronizing and worthless opinions.

It's especially egregious that some physicians, hired by defense attorneys in mold cases, espouse profound opinions that the person didn't look sick and therefore, the mold didn't hurt them. No one could have so many symptoms and look that healthy, they say. No kidding, these omnipotent physicians basically tell us, under oath, that they can sniff out disease just by looking at people! It's a credit to the restraint of the chronic fatigue community that we don't have an epidemic of strangled physicians!

Imagine how a mold patient feels after the tenth time he's told that he couldn't be ill, even though the physician didn't do any objective tests and didn't look at anything physiologically. Did anyone look to see if the air filter was clean before they said the lawn mower was too cold or too stressed by heat? Why didn't anyone look at capillary hypoperfusion in a patient with chronic fatigue? It isn't hard to do.

I can't tell you how commonly I hear mold patients tell me, "My doctor said he couldn't find anything wrong with me, so it must be in my head." Let me see; the patient is ill, but the doc can't figure out the problem—not to mention the solution—so therefore, the physician concludes the patient has a psychiatric problem. The horror of diagnosis by assumption follows when someone believes those assumptions. Ass^2 redux.

I see that 'medical thinking' every day when I review prospective patients' records. When patients ask the doctor to do something unusual, like measure capillary hypoperfusion, VO2 max, anaerobic threshold and innate immune responses, sometimes the physician feels his autonomy is enormously threatened. Ask a mold victim how many docs have been angered by their requests and don't be surprised that the answer is more than one.

A word to my fellow physicians: a chronically ill patient asking you for help isn't presenting a challenge to your integrity and isn't trying to waste your time. When the patient is still ill despite everything that's been done so far, don't construe the plea for help as professionally insulting if you're asked to do a bit more. And please, never make assumptions about absence of illness if you haven't looked at innate immune responses. Your world of imagined psychiatric illness crumbles when you start measuring items like C4a, VEGF, TGF beta-1, MMP9, VIP and MSH.

Maybe docs should read through *Change: Principles of Problem Formation and Problem Resolution*, published in 1974, by Drs. Paul Watzlawick, John Weakland and Richard Fisch, published by the WW Norton Company. They talk about correcting the stated problem as one requiring "first order change." A simple example is a door with loose hinges. A spouse says, "Please tighten the loose hinges on the door." The other spouse fixes the door and the problem ends. But if the first spouse instead says, "I've asked you for three weeks to fix the door and you ignore me completely," the other might respond, "I'll fix the door when I want to. And I don't want to because all I get from you is nag, nag, nag"- that's an example of "second order change" at work. The door still needs to be fixed, but the argument becomes centered on the *attitude* expressed by the demand for first order change. Nothing good will come from arguing about the reaction to suggested change if the source of the request for change isn't corrected.

When the mold illness patient says, "I'm tired," he's requesting a first order solution. If the doc replies, "Well, you look pretty good, so it must be stress or something that I can't measure," that's an invitation for a second order change battle. And it happens all the time. Even worse, when a mold illness patient says a building is making him sick, he's asking for two first order changes: fix him *and* fix the building. If the building owner hires a mold defense outfit whose expert parrots, "Acquisition of illness by humans is implausible from exposure to those levels of mycotoxins in the air," the second order change is obvious. Instead of arguing about what caused the illness, the first order of business is to show there *is* an illness. Yet look how quickly the argument in court swings away from illness to the exposure, understanding, of course, that levels of mycotoxins have almost nothing to do with mold illness.

I see that scenario played out all the time.

Let's think about the glycogen-depleted patient again. First order change (good) is to replenish the depleted glycogen reserves (with fuels and sugar derived from what you eat or drink), which are delivered to tissues by normally perfusing capillaries. Second order change (bad) is to patch the sugar problem by breaking down fat or protein and converting it into sugar without fixing the capillary hypoperfusion. The second order change here continues the problem of capillary hypoperfusion and now *compounds* the problem by adding protein wasting to the equation.

I don't think Dr. Watzlawick and his co-authors thought they'd be cited in a paragraph about glycogen physiology when they wrote their classic psychology treatise!

What goes on when the available reserve of sugar plummets? Second order change won't stop this disaster in the making! If the human body has an adequate supply of fat inside fat cells, one way to generate an emergency fuel source for glycogen-depleted cells is to tell fat cells to immediately release fatty acids from fat into the bloodstream. The fatty acids can be used for fuel too; in fact fatty acids give us almost three times as much energy as glucose.

Endurance athletes, such as runners and cyclists, know how it feels when their sugar stores are at rock bottom; they hit a "wall." After they've hit the "wall" from sugar depletion, the energy surge they get from mobilizing fat and burning fatty acids feels terrific - they call it a "second wind." Most endurance athletes would rather not run out of glucose, as they usually don't have a lot of extra fat on them and fatty acid mobilization won't work for very long. During the Boston Marathon and the Tour de France, watch how often the athletes drink sugar-enriched fluids to avoid dehydration and glycogen and fat depletion.

What do cells do when they run out of fat? Or, what happens if the body won't permit burning fat, say if you're thin and don't have any extra, or if you've got extra fat and excessively high leptin (those who are leptin resistant)?

Leptin controls how fatty acids mobilize from fat cells. If leptin is high, a giant brake sign is flashed in front of all the fat mobilizers—they just can't work and fat stays put. All the demand for fatty acids when glycogen reserves are gone goes for naught.

Fat, fat, everywhere and not a fatty acid to burn.

The cost of running on empty on glycogen and then burning up fat reserves is often collapse. Remember the runners sprawled on the track at the end of the long race, too exhausted to get up? Here comes the medical team, hustling to start an IV containing sugar and salt water, thus saving cells on the verge of dying.

Some athletes can avoid collapse by reaching for their third source of fuel— protein. A little chemistry tells us why: Proteins are chains of amino acids and some of those amino acids, especially alanine and glutamine, can be rapidly broken down into glucose, releasing lactic acid into the spaces between cells. The sugar made from protein is the very same glucose we make from starches; it isn't any different from the glucose we normally store in glycogen. But this emergency use of protein burning comes at a steep price. When an athlete burns protein, he has to get rid of the extra, unwanted lactic acid, as it will cause harmful and painful spasm in muscles and blood vessels if it stays around too long. If there's an adequate blood supply to the muscle areas, then the body removes the extra nitrogen and the muscle happily burns protein for a while. But if other waste products, made when sugar stores are exhausted, pile up as well, they shut off the flow of blood in tiny capillaries by diverting the blood away from the sources of waste products.

What happens next is predictable: muscles go into spasm and agonizing cramps come on. Just watch a football game in summer or an indoor basketball tournament in winter; when you see the star player clutching his calf muscle, unable to walk, remember capillary hypoperfusion. He's becoming glycogen depleted, trying to mobilize fat beyond his capabilities. He's burning protein for fuel and suffering from reduced removal of lactic acid, a waste product of sugar metabolism. Rest, fluids and sugar will permit him to recover enough to play the fourth quarter and win the game. He doesn't have a metabolic reason for cramping other than massive overuse; his physiology isn't damaged by inflammatory compounds that won't go away. His illness is first order change.

He isn't a mold illness patient; that person suffers from second order change.

But what the athlete experiences after sudden, forceful activity is the same thing, magnified many times over as what mold patients often experience: the same series of physiologic events that mold patients endure after trying to vacuum the house or plant some two-quart azaleas in spring. Capillary hypoperfusion is the unifying theme. Without correction, reduced oxygen delivery results in protein burning.

Patients with mold illness are invariably low in MSH. The stimulus for MSH production comes from increasing leptin, released from fat cells in response to rising levels of fatty acids in the blood. Rising MSH does all kind of good things for us immunologically, but it also stimulates release in the brain of melatonin (provides restorative sleep) and endorphin (provides control of pain stimuli). Rising leptin and rising MSH makes people feel good. Just go back to a pleasant Thanksgiving as an example. After the turkey, buttery mashed potatoes, fatty gravy, something fried and ice cream, what happens? A nap on the couch and no pain. Thank you, leptin from fatty acids and MSH, for easing our perception of pain and for giving us a nice nap. Not the low MSH patient, however. He doesn't get the nice comfy snooze after the Thanksgiving binge. His MSH won't permit such a pleasant afternoon. Chronic pain and chronic sleep disruption are part of his life.

Back to first order change: When the non-mold patient replenishes his glycogen stores, using the extra sugar to replenish any depleted protein stores and whatever else is added as fat, he's now ready to spend his energy efficiently and have a good day.

But not for the mold patient who doesn't have MSH - his signal from rising leptin doesn't turn on extra MSH production, so he doesn't get the endorphin surge. In fact, his pain can become worse! Without MSH, leptin levels rise to try to push MSH production, but to no avail. The leptin rise worsens leptin resistance, which, in turn, worsens protein burning.

Let's go back to the protein burning. I think you can see just how many things that can go wrong and do go wrong when reduced oxygen delivery occurs. People who burn protein chronically are in a whole lot of trouble —they simply won't feel good. Our protein stores are genetically programmed and we need them intact for normal functioning. Ongoing protein wasting - call it a state of negative nitrogen balance - has a long history in medical research. Often seen in surgical patients who can't eat, negative nitrogen balance became treatable when researchers found that putting a patient back into positive nitrogen balance is the single most important factor in recovery. Intravenous amino acids, fatty acids, vitamins and high doses of sugar save the patient from in-hospital disasters like infections, delayed wound healing, blood clots and more. These patients aren't the same as mold illness patients but both groups are burning protein like crazy.

In its extreme stages, protein wasting is obvious: two examples are malnourished children and end-stage cancer patients. Cancer is particularly horrific, as blood vessels are forced to feed the ravenous, sugar-consuming tumors at the expense of their own body stores of fat and protein. But protein wasting in less extreme examples isn't necessarily obvious in a quick bedside once-over. A tip-off that patients are burning protein instead of mobilizing glucose from glycogen often comes from the unusual symptoms that mold illness patients invariably have.

Listen to the patient: they'll tell you what's wrong.

More than 60% of mold patients have ongoing muscle cramps, and I'm not talking about minor problems that need short-term stretching. Unusual cramping is quite common, including seemingly bizarre muscle contractions at the "end of the blood vessel line," like in fingers and toes. These uncontrollable finger and toe muscle contractions can cause a "claw contraction," a form of spasm in which the digits have to be physically straightened. When I ask mold patients about the claw-hand or claw-foot, essentially validating their unusual symptoms, most look at me with relief, because they understand that I know what's going on in their illness and thus, their bodies.

"My fingers make a Dr. Spock," one person told me, spontaneously assuming the split finger "V," Star Trek fans all remember. Only his V wasn't voluntary, it just happens. When his capillary hypoperfusion—the source of such unusual contraction of the intrinsic muscles of his hands—was corrected, the V disappeared.

Over 40% of mold patients have such unusual contractions. You won't find that symptom listed in a medical textbook under capillary hypoperfusion. If the physician asks about these unusual contractions, but doesn't factor capillary hypoperfusion into the diagnostic equation, look out; predictably, a prescription for restless legs or nocturnal leg cramps will follow. Yikes! Ask your physician to seek measurements of capillary hypoperfusion.

MEASURING OXYGEN DELIVERY

What does your physician do to check if oxygen delivery is satisfactory? How about measuring the total oxygen in a blood sample taken from an artery? Putting a needle into an artery to do a blood gas assessment not only inflicts unnecessary pain but the result doesn't tell us anything about oxygen delivery in the tiniest blood vessels, the capillaries. The problem in mold patients isn't oxygen delivery to large blood vessels. So what can we know, not infer, not speculate, not assume? Let's not work the Ass^2 side of the street, let's be objective and use science not dialogue, testing not prejudice. Here we go.

VO2 max

Measuring what's going on in blood vessels barely large enough to see requires a systemic approach. One of the best tools to use is a pulmonary stress test—put the patient on a stationary bicycle in a pulmonary lab. Measuring the expired air during exercise tells us when the patient can no longer deliver enough oxygen to tissues (the anaerobic threshold). This number is the "VO2 max."

Normally, VO2 max rises with physical conditioning and falls with illness and aging. VO2 max falls dramatically with capillary hypoperfusion. Measuring VO2 max is commonly done to see how bad a heart failure patient's cardiac reserve is. The *AMA Guides to Evaluation of Disability and Impairment* lists reduced VO2 max as a way for physicians to assign a disability rating. Most mold patients will have a VO2 max so low they qualify for a Class III (really low VO2 max) or Class IV cardiac disability (terribly low VO2 max). When you treat capillary hypoperfusion, VO2 max improves.

VEGF

When VO2 max is found to be low, levels of vascular endothelial growth factor (VEGF) will usually be low. VEGF is a growth factor, and increased levels lead to increases in blood flow, which can cause new blood vessel growth to supply rapidly growing tumors. Remembering my comments about the anti-angiogenesis agent, fumagillin and prevention of oxygen delivery, high VEGF increases oxygen delivery; low VEGF decreases it. Did fumagillin lower VEGF (yes) and could the common occurrence of low VEGF in mold patients be due to fumagillin or substances like it? You bet.

Don't forget that Dr. Folkman was father of the anti-angiogenesis movement in medicine, one that heralds great advances in treatment of oxygen-guzzling cancer cells.

His work all began with mold. So many mold patients have low VEGF and actually have a *reduced cancer rate* if their VEGF is low. High VEGF - well, that's a different story, especially if the cancer has a

VEGF receptor, meaning that VEGF will drive both the growth of the cancer and its nourishment. (*Editor's note*: many mycotoxins are known carcinogens, cancer producers, but in spite of that fact, the reduced rate of breast, colon, pancreatic and other cancers in Dr. Shoemaker's roster of patients is clearly demonstrated. The incidence of cancer in his group does not mean that exposure to WDB does or does not result in cancer or indeed has any association with it.) To ease hypoxia inducible factor (HIF); selective genes respond to HIF by increasing VEGF. Low oxygen, then, turns on these genes, but the normal gene regulation disappears in mold illness. VEGF should go up when HIF is released, but in mold illness and all other biotoxin-caused illnesses, there's a secondary failure of gene regulation and VEGF falls rather than rises when it's most needed. TGF beta-1 may have something to do with the turn down of normal VEGF, but as yet the answer isn't clear.

It gets more complicated: the other genes turned on by HIF produce erythropoietin (a lot more on that later) and transforming growth factor beta (TGF beta—sorry about all the acronyms). If there's just a simple lack of gene activity from HIF in mold illness, then erythropoietin levels will fall, just like VEGF. But nope, that doesn't happen. And if VEGF is low, then we'd expect TGF beta to be low, too. Nope, that doesn't happen either.

Having said all that, when VEGF is increased due to therapy (see Appendix 1, "What Do I Do?") all kinds of good things happen; VO2 max goes up, symptoms go down and exercise tolerance improves. From now on, every time you hear that a therapy improves exercise tolerance of those with CFS, ask if the study authors controlled for VEGF (did they measure it before and after the therapy); if they didn't, their conclusions are at best worthless, and at worst, misleading.

VIP

Vasoactive intestinal polypeptide is a neuropeptide first found years ago in the gut. It regulates blood flow to a certain extent, hence its name. VIP researchers, including Dr. Mario Delgado and his colleagues, figured out much later that VIP regulates inflammatory responses and pulmonary artery response to exercise. We know now that VIP plays a critical role as a regulator of hypothalamic responses to inputs from retina and nose. Whether VIP deficiency leads to multiple chemical sensitivity remains to be seen, but it certainly could be a factor in that unusual syndrome. As far as oxygen goes, we need to look at VIP's effect on blood vessels in pulmonary artery circulation.

Remember that blood is pumped out of the left ventricle in increased amounts during times of extra demand, like exercise. The extra blood that gets to the left side of the heart comes from the right side of the heart (thank you, William Harvey, 1620). But in order for extra blood to get out of the right ventricle and into the left ventricle, it must face the gauntlet of pressure changes in the pulmonary artery and then the small blood vessels in the lung that eventually drain into veins that go to the left side of the heart. What happens if pressure in the pulmonary artery falls? More blood is pumped into the pulmonary artery. That's the normal response to exercise. Makes sense, doesn't it?

But what if the pressure in the pulmonary artery (PA) either *doesn't fall* or even *rises* during exercise? That would mean less blood is pumped out of the left ventricle during exercise and that would mean people would become short of breath. OK, let's look: how often do patients with mold illness become short of breath during exercise? Research says 75% of patients experience that symptom. Does PA pressure and shortness of breath correlate with VIP deficiency? Almost. How about when you add VEGF deficiency to VIP deficiency? Bingo—almost 100% of patients with both deficiencies get short of breath when they exercise.

Stress echocardiography

If VIP is low, how can we measure pulmonary artery pressure during exercise? Well, we could do a heart catheterization, in which we put a catheter into the femoral artery in the leg, make sure the patient stays deathly still so we can measure all the pressures and blood flows that we need to, then have the patient get up from the cath table and boogie on a treadmill for awhile. On second thought, let's not.

How about we float a Swan-Ganz catheter, or another one of those specialized devices that measure pressure in pulmonary capillaries when the patient is supine and immobilized, and then have the patient operate a bicycle?

It's true that *invasive* studies would mean we could really be sure about our measurements, but *non-invasive* studies can tell us indirectly what the pulmonary artery pressures are, and the best of those is stress echocardiography. Echo can measure the velocity of flow across the tricuspid valve and that measure can be correlated with the pulmonary artery pressure. This indirect measure can tell us rapidly, safely and inexpensively what VIP is (or is not) doing to pulmonary pressure.

VIP increases the levels of a compound called cyclic AMP (cAMP, mono or one phosphate) inside the cell. This compound was one of the first to be described as a "second messenger" within the cell. It's an energy molecule, derived from the better known ATP (tri, or three phosphates) and ADP (di or two phosphates). Cyclic AMP decides the response to a given input. Not enough cAMP and you have a cellular phone situation like that we see in rural areas: no signal, no communication. No VIP, no communication from cAMP.

Sequential stress echocardiography studies almost always show that pulmonary artery pressure doesn't fall during exercise in those with low VIP. When VIP is corrected and the echo is re-done, the pulmonary artery pressure falls during exercise, exercise tolerance increases and not surprisingly, cognitive issues are reduced. On the other hand, if VIP doesn't respond to interventions, the patient will continue to suffer and pulmonary artery pressures will remain dysregulated.

Magnetic resonance spectroscopy (MRS)

Perhaps the biggest breakthrough in understanding the relationship between mold exposure and deficits in executive cognitive functions comes from magnetic resonance spectroscopy of frontal lobes and hippocampi (left and right). Sure enough, the brain's chemical structure, as determined by its magnetic spectrum, tells us a lot about how inflammation affects blood flow. In October 2003, I realized MRS could perform a non-invasive chemical analysis of the brain and wrote about it, but my dream of actually doing it didn't come true until July 2006. Progressive MRI opened a new facility in nearby Salisbury, Md.; they were looking for referrals, and they did MRS.

What a change in my practice. Now, with consistent head positioning giving reproducible results, I could see the differences between cases and controls regarding multiple different chemicals in crucial areas of the brain during information acquisition (frontal lobes) and information storage (hippocampi). We don't see any abnormalities in white matter, cell intrusion, chemical neurotransmitters, or the structural/glial elements of the brain. These normal findings help with differential diagnosis: the illness isn't due to a lot of conditions like Alzheimer's, tumors, psychiatric conditions, MS and the like. We do see an almost uniform increase of lactate (showing that oxygen delivery is reduced) and then a resultant reduction in the ratio of glutamate (excitatory) to glutamine (inhibitory). We call this ratio the "G/G."

What does this mean? Imagine a room with a nice warm glow from overhead ceiling lights when each one is working well. All is fine: lactate and G/G are normal. Now shut off the lights (turn off the oxygen). How much can work can you get done in the room? In the same way if you don't have enough oxygen delivered to capillaries of the frontal lobes, how well will you concentrate? How often will you be confused? And how about the hippocampus, the seat of memory? Will there be a problem with recent memory? Yes, quite a bit.

The best part of finding the abnormalities in central nervous system metabolism is that when the source of the capillary hypoperfusion that causes the increased lactate and the reduced G/G is fixed, the brain returns to normal metabolism and cognitive issues abate. The underlying brain function hasn't been destroyed. The lights were simply dimmed not destroyed.

As we'll see in other chapters, maintaining normal brain metabolism means avoiding all further mold exposure. Let me say that again: *Surviving mold means avoiding all further mold exposure.*

Low dose erythropoietin

Who'd have thought that an injectable drug used to treat anemia would hold the key to treating cognitive issues in mold illness? Actually, life-changing effects of injections are the norm these days: anti-TNF drugs have revolutionized treatment of psoriasis and rheumatoid arthritis. Interferon has changed the way we treat hepatitis C, and don't even get me started on injectable steroids and human growth hormone in athletes.

A simple literature search would've shown epo's tremendous ability to correct capillary hypoperfusion and stabilize C4a. In other words, epo should correct most of the refractory illness caused by exposure to water-damaged buildings after the patient has been removed and toxin aspects have been corrected. And it does.

The molecular biology of what epo does is a wish list to prevent consequences of innate immune responses. Yet in May 2007 and then again in November 2007, the Food and Drug Administration (FDA) placed a so-called "Black Box" warning on the use of epo. Epo had been shown to increase the risk of heart attack, stroke and blood clots in patients on renal dialysis when their hemoglobin rose above 12. It was also noted that patients with head and neck cancer did worse with epo than without. None of my patients who have received epo have been on dialysis and none have had head and neck cancer, and epo helps a lot. The scary data used by the FDA were from much older studies; there was nothing new to worry about regarding epo yet the Black Box is pretty intimidating to anyone thinking about a new treatment. We use lots of informed consent before using epo and to date have routinely see excellent results and nothing in the way of complications. We monitor for problems with blood counts, clotting, blood pressure and more. I guess the idea is similar to those who work with high voltage power lines. We receive immense benefit from things that only they can do yet the "things" involved can bite you. OK, only use epo if you know how to be careful. What else is new in medicine?

The benefit from epo comes from (i) opening capillary beds in the brain; (ii) lowering C4a; and (iii) fixing refractory cognitive symptoms. Don't use epo if you haven't any experience with it, but if you know the drug, it'll help your patients immensely.

Final thoughts

Capillaries are the smallest blood vessels in the body, but they're the most abundant. Looking at large blood vessels, as Big Pharma does in an attempt to prevent heart attack and stroke, has nothing to do with where the action is in innate immune responses. Capillaries are subjected to an incredible number of influences by cytokines, complement and adhesion molecules, among others. Reducing blood flow in capillaries reduces delivery of oxygen and nutrients that cells need. In the end, fixing capillary blood flow is mandatory to fix the symptoms that we see in biotoxin-associated illnesses, especially mold illness, (the bogus diagnosis of) fibromyalgia and Chronic Fatigue Syndrome.

Looking ahead, the future for research in this field is bright.

CHAPTER 7

Shannon's Story: Courage and Perseverance

by Laura Mark, M.D.

INTRODUCTION

Life is what happens while you're making other plans! Over the years I never would've predicted how often that phrase would echo in my mind as I learned what mold exposure did to my family, me and my career. As I sit here today, the phrase is now my mantra. What more do I need to say? I listen to a simple, gentle reminder of the tenuousness and fragility of life; the frequent need to reassess; and now, as I regroup after I regain my health, I will once again venture forward with confidence as one of the curves of life redefines me. I am a physician, a mother and a person. I care about those around me and if that caring means I take back seat to seeing others do better, well, I enjoy seeing life that way.

Giving to others comes from my roots, my parents and my heritage. Caring is what I do.

What I will tell you is my true story. I am alive and surviving in Williamsburg, Virginia. I will not fail to win this battle to survive and thrive. Yes, Dr. Shoemaker has helped me, but I have helped myself. I am an informed advocate. I teach physicians, administrators and all who will listen. I am not invincible; mold hurts me in an ever increasingly severe pattern and I am one of the very few who can't take VIP. That is a setback. The good news is that I benefit from meds that knock out the profound explosion of C4a that I suffer within five minutes of exposure to a WDB.

I will win this battle, not because of meds, but because of who I am.

Let me begin by telling you about myself and my family. While my detail is cut short because I can't write 5000 pages here, I could. My hope is that you will see how easily, rapidly, and unpredictably the health of any individual (or couple, or family, or school, or community) can be devastated by an unfortunate combination of genetics and exposure to teeny-weeny molecules that we always find inside WDB. But only inside WDB. Amazing, really. If you are never exposed to a WDB, you can't really know what I mean. But your family members might be exposed. Ultimately, people like me and my family may be able to gain control over these tiny molecules. That's what this book is about.

First and foremost, I'm a mother. My four children age from 20-29. When this saga began, they were infants (or maybe even fetuses and that really scary idea gives me no rest, ever). I had no warning that my family would be hurt by wet buildings when my children were young. During those infant and toddler years, I finished college, med school (Georgetown) and became a board-certified psychiatrist after completing a psychiatric residency at Maine Medical Center in Portland. My four-year payback assignment for the National Health Service Corps was in New Bedford, Mass., and my time there was extended to 10 years. I loved my work and always believed I learned at least as much from my patients as I hoped I was able to offer them in return.

In 1995, because of an economic downturn and family matters, we relocated to Virginia where, with lightning speed, our health, that means my entire family, crashed. I became ill, but I wasn't willing to be a patient.

My children experienced their share of the usual childhood illnesses, with occasional deviations from the norm. At first I didn't think too much of the oddities. Some were cute: like who was double-jointed, who could bend their thumbs down to their forearms, who could curl their tongues into the shape of a clover leaf or reach their tongues up into their noses, who never got a temperature above 94 degrees, who had to wear hats and dark glasses all the time even indoors? We developed cute nicknames like "aqualung boy" and "the green goober kids" and "the stinky poop kids" and "little miss fuss-budget." Doesn't every family have its own little stories to tell? And when I look back on those days, how could I have missed the obvious?

As I look back on the clues, now obvious to all *Surviving Mold* readers (but let me tell you, NOT TO a lot of physicians who need to learn), I have to say I wonder how life would have been different if I had known then what I know now. I mean, imagine if a pediatrician had said in 1989 that Shannon had a connective tissue defect due to high TGF beta-1 and that she was therefore going to be hammered by going into the basement of the University of North Carolina's Medical School. *Now* I know that I never should have let her go.

I think about that at night when everyone else is sleeping. It isn't guilt, but it is a feeling so powerful that I can't let another Mom go through what I have endured, at least not if I can educate her doctor. And Mom, too.

Shannon had so many bewildering symptoms; they made no sense. Her pediatricians came up with a bewildering array of medical diagnoses. No, no one said high TGF beta-1. Eventually, I realized that the medical community would not give my family the answers we needed. Over several years, over thirty of this country's most highly trained physicians took her medical history, examined her, and performed extensive (and expensive!) diagnostic tests. By the expression of disbelief on many of those doctors' faces, they clearly weren't convinced that Shannon could really be experiencing *all* of those symptoms, which didn't fit together anyway. Besides, she looked so "good" on the outside. I knew they'd pegged her as a hypochondriac, somatizer, Munchausen's patient, or worse yet, had decided that we were a "folie-a-deux" tag team! For those of you not familiar with these French medical terms, they basically mean she's faking it and I'm encouraging her.

Dr. Shoemaker talks about Ass^2 medicine. I've seen the assumption raised to the 100th power (Ass^{100}).

It was heartbreaking to watch Shannon wither. That withering image comes closest to the memory I have of my daughter back then, and those days, I was fearful that she'd wither and *die*. Dying of a known illness is one thing; psychologically people can cope better when they know the enemy. But with Shannon, no one could tell us what to cope against.

So let me tell you all about Shannon.

PART ONE – SHANNON'S STORY

Shannon was born in Maine and spent her first years living in an old farmhouse that her Dad and I were remodeling. We moved to Maine so I could do my psychiatric residency in Portland. We would embrace a "back-to-the-earth lifestyle." Those years were hectic but wonderful. New career, new family, new life; we had everything. Money never meant much to me.

The first 13 years of her life reads like a storybook. Her Dad and I used to call her "the girl with the Midas touch." She excelled at everything she did. She was a superb student, studied classical piano from age four, was a phenomenal artist, and a talented soccer player. She was admired and respected by her peers and teachers, developed a strong social conscience and planned to pursue a career in medicine.

The first inkling of a health problem occurred when she was 13. For over a year she complained frequently of feeling hot, asking me to take her temperature, which was always about too low at 94 degrees. She appeared to be a little flushed, but nothing else seemed wrong. And nothing bad happened after we didn't do anything except observe. Her Dad and I joked, "She'll do well at the Sarah Bernhardt School of Theatrics." Did we patronize her? No way. She was doing just fine.

One spring night, though, things were different when she came into our bedroom saying she didn't feel good, that her belly hurt, and that she felt hot. I knew she was acutely ill. Even a psychiatric resident can have medical judgment. Appendicitis?

Yes, she did. But the reason I'm telling you this story is that I hope you and your doctor learn the importance of **listening** and **believing**.

Two days after the appendectomy, her surgeon told us that her appendicitis was caused by a small carcinoid tumor growing in her appendix. The tumor was benign, thank God, and was "localized and had not reached the outermost layer or broken through the wall of the intestine," meaning, no spread to other areas of the gut was likely. Carcinoid can be a malignant killer, especially if it responds to VEGF, but it is otherwise very slow-growing. It produces bursts of hormones that cause episodes of flushing, wheezing and rapid heart rate. Our little Sarah Bernhardt did have a "biological basis" for her sensations and complaints.

Now the plot thickens. Two weeks after the appendicitis, in the midst of soccer season, she came down with mononucleosis. What kind of coincidence was that? Her spleen blew up like a basketball and she had to stop playing soccer for the next six weeks and "take it easy." Her health never quite seemed the same after that. She never quit pushing toward her dreams even though she just wasn't right. Meanwhile the inflammatory response to mono was making her now susceptible to organisms inside water-damaged buildings, ones that should never have bothered her.

At 15, in high school in Virginia, the illness begins to blossom. Stress? She had it - her parents' separation, moving to a new state with a new school and friends, and then her father's death. But these stressors *were not* the explanations for her increasingly diverse and frightening array of symptoms, though I heard that idea a lot (*Editor's note*: Ass^2 medicine in action).

I sure have learned that diagnoses from health professionals, like "it must be stress," are often an easy answer for health providers to give to a patient who walks into their office with many seemingly unrelated symptoms, a fairly unremarkable physical exam and a routine set of results for a routine set of lab tests.

Did anyone ask if Shannon was moldy in Virginia? Nope.

Let me back up. In psychiatry, I spend a lot of time with each of my patients, a lot more than the physicians who let me down about Shannon, get to spend on average with their individual patients. So, in fairness to the health professionals, per patient experience has become increasingly controlled by the clock. We know that an experienced primary care doc on average takes less than 30 seconds to make a diagnosis for an acute problem when seeing someone he knows. Throw in newness of patient to doctor, throw in a complicated chronic illness and the time demand rockets up. Medical history taking means time taking. When someone is cognitively impaired, and just about all moldy patients have executive cognitive problems, it's difficult to evaluate symptoms that wax and wane, don't seem to make sense or fit under one umbrella and that cause patients to become very irritable, angry, and despondent over time. We have to face it: these patients are often not pleasant to be with (GOMER means "get out of my emergency room") and commonly become frustrating to treat.

From my take on mold illness, the problem is really more than just time for an office visit or patience to put up with a shattered brain. It is education. It now takes me about two minutes to see that I am NOT working with a biotoxin patient. If the health provider knew what I now know, if they actually had some basic knowledge about chronic systemic inflammatory diseases, I am sure they wouldn't find it as easy to "tune out," focus on one symptom, hand out a prescription, and maneuver the patient out the door.

But back to Shannon at age 15. As her headaches increased in frequency and intensity she went to neurologists over and over again. Yes, they all agreed she had migraines (*Editor's note*: she didn't). When excellent migraine meds didn't work, here came the referral to a psychiatrist. One particular period of headaches was accompanied by nausea and vomiting, abdominal pain, sore throat and swollen lymph nodes. She was so desperately ill she was hospitalized. She was admitted to treat dehydration, but laboratory tests surprisingly showed markedly elevated liver function tests. Her liver was so swollen that it extended all the way down into her pelvis.

You won't believe this, but she was told that she had acute mononucleosis *again*. Granted, as physicians, we continue to revise our understanding of viral illnesses, but this diagnosis was just too hollow for words. Still, what this saga has taught me, as a physician and as a mother observing my daughter's decline over the years, is to *never assume* that a "real problem" doesn't exist simply because we can't pinpoint a precise cause.

Shannon managed to get on with her life. She excelled in academics, sports, and the arts. She graduated at the top of her high school class, was voted as "the student most likely to succeed" by her classmates, was co-captain of the high school state championship soccer team, and was awarded prizes for her artwork. Even with her gradually increasing fatigue, frequent headaches, and decreased stamina, she made everything look easy. Too easy. Others thought she was fine when she wasn't. "She looks good to me. Can't be that sick." Although she had hopes of pursuing a career in medicine, her unpredictable health asked the question – "could she could manage the rigors of medical school". By now the recurring bouts with mononucleosis (*Editor's note*: any one else feeling skeptical?) were occurring on a two-year cycle at ages 13, 15, and 17. Off to the University of Virginia as an undergraduate student in the sciences. Guess what happens at age 19? After several weeks of increasing headaches, fatigue, sore throat, abdominal pain, and nausea, somebody says mono again. Let's not forget that giant liver.

By then I was convinced there was something drastically wrong with her immune system. Here was the list: an intense reaction to the DPT vaccine, four mild cases of chicken pox, a non-malignant endocrine tumor on her appendix, and now, recurring episodes of the Epstein Barr (EB) virus. As Dr. Shoemaker's toxicology colleague, Dr. Mark Poli, might say, "All of this is just plain wrong."

I consulted with a long-time family friend, an internist at the University of Virginia Medical Center. His team indeed felt that Shannon was experiencing a recurring reactivation of the EB virus. Their recommendation was to treat with an antiviral drug at her next outbreak. At least we had a plan.

She completed her undergraduate studies in two-and-a-half years despite her weakness and failing strength. By then it was apparent that med school wasn't in her future. She decided to take a break from academics, accepting a one-year position as a counselor at a residential program in Scotland for young adults with autism and developmental disabilities. She loved her work, but fatigue, headaches, and bout number five of apparent EBV required her to return home after 10 months. No one asked about the moldy buildings. She spent the next few months essentially bed-ridden, but as her energy picked up again, so too did her ambition and momentum.

What to do? She knew she wanted to help people, and she loved gardening. She needed flexibility and control over her work schedule; occupational therapy seemed like a good choice. Specifically, she wanted to provide horticulture therapy to young adults with disabilities. After discovering that the University of North Carolina at Chapel Hill had a top-notch occupational therapy masters program, she applied and was quickly accepted. She moved to Chapel Hill, taking a job at the UNC Autism Research Residential Program. Her masters program started in August 2005, and she was among a group of 18 students attending classes in the *basement* of that UNC Medical School building.

My God, I didn't know to tell her to stay out of basement classrooms.

Shannon was getting worse. She described a series of increasingly confusing and varied symptoms. While trying to minimize their effects so as not to worry me, she was also working desperately to complete her program. By then she'd met a young man (Jonathan) completing his agriculture degree at the local college. As fate works, it is the connection to her friend that led us later to Dr. Shoemaker.

I can't help wonder still, 'what if'? Yet Shannon is a treasure for all those who benefit from knowing her and there is no reality in 'what if.'

The summer of 2005, we were seated in the office of a renowned

University of North Carolina School of Medicine rheumatologist who opined to now 24-year-old Shannon, "You have fibromyalgia and you'll have to learn to live with it. I'd recommend you see a psychiatrist. I have a colleague who uses biofeedback. Maybe you could learn to control your stress and symptoms better with that technique."

I recall that moment distinctly. I was screaming in my head, "The hell she'll have to learn to live with it!" He then proceeded to proudly (and pompously!) describe a research study he'd just completed for publication concerning "fibromyalgia and the histrionic personality."

Infuriating! Of course he believed that fibromyalgia patients have histrionic personalities, I remember thinking. By the time the average patient reached his level of specialty, they'd already been sick, frustrated, and doctor-fatigued for years. They'd surely been disregarded, patronized, or simply not understood, while being billed for extensive and expensive tests that "showed nothing wrong." To be informed, once again, by that physician to "learn better how to handle stress" would've been grounds for just about any self-respecting patient to have a well-deserved "histrionic" outburst. Like screaming at the jerk! I mean, come on, he'd not even read the medical records I'd sent ahead of time, didn't really listen to what my daughter was telling him, focused on the history of her father's death and disregarded the rest of her history, and made outrageous assumptions about her diagnosis and prognosis! By the way, the exclamation marks used in this paragraph do not *begin* to reflect the degree of rage I was experiencing by that time in our search for answers. (*Editor's note*: Dr. Mark's frustrations and experiences are ones essentially every mold patient comes to share.)

Desperate, we followed the advice. Seeing a psychiatrist turned out to be a wonderful experience for Shannon. He had an open and positive attitude, listening with compassion while she vented her frustrations about being a young adult with a chronic illness that defied diagnosis and a very uncertain future. Most importantly, however, he reassured her that, "It's **not** all in your head." When

Shannon was hooked up to the biofeedback equipment, she displayed an unexpected, severe reaction to the testing. The technician hadn't seen such a reaction in all her years of using the procedure. Shannon had an over-reactive and poorly regulated nervous system, which explained for us a number of Shannon's symptoms. So, actually, a lot of it *was*, in fact, in her head *and in the rest of her nervous system.*

In early December 2005, Shannon's illness flared. She called me, crying from her severe pelvic pain. I jumped into my car and drove to meet her in the Duke University Medical Center ER. Who wouldn't be certain that Shannon's symptoms of flushing, rapid heart rate, and severe diarrhea meant that her carcinoid tumor was back?

But at the Duke ER, she was examined and sent home. A few labs were normal. "See your doctor tomorrow." No. Unacceptable. I phoned my sister at Hopkins, herself a Duke graduate. She paged a former colleague in the GI department who saw Shannon right away in the ER. He found a pelvic mass, confirmed by an abdominal CT scan. But what should be done about the pelvic mass? I still feel that if the pain had continued, she would have been admitted, but because she felt better, home she went, with instructions to follow-up with her gynecologist the next day.

It was during that ER visit that I finally told Shannon about her previous diagnosis of the carcinoid in her appendix. She had never been told previously about that carcinoid tumor. My husband and I had been advised, "We got it all," and the follow-up testing she had every six months for a few years was consistently negative. So we figured it was best not to tell her until adulthood or on a need-to-know basis.

During the long ride back from Durham to Williamsburg, we both were worried and quiet. What does a pelvic mass mean? Once you think of cancer, any kind of cancer, not much else enters your mind. Shannon didn't say anything about her not having been told when she was younger about the carcinoid, but we've known each other so closely that talk wasn't necessary. Carcinoid is a killer when it comes back and there aren't many therapies to use. But deep down I knew that it would be odd if a pelvic mass was carcinoid. Yet everything about Shannon's case was odd!

Shannon's gynecologist saw her first thing the next morning. After a very thorough physical exam what she found wasn't a mass at all, just a flipped uterus. We almost kissed her! Still, I had to wonder, how did Mr. Duke's Hospital mistake a flipped uterus for a pelvic mass? Had somebody been in a hurry?

Next was an urgent evaluation by another local contact, an endocrinologist. Again, her history led to additional testing. No carcinoid was found, but the endocrinologist simply stated that finding carcinoid tumors was nearly impossible when they are present but small. Meanwhile I was imagining a carcinoid making her liver huge, her colon obstructed, her lungs too tight to let air move and her heart moving as a quivering mass.

Nope.

Fact defeats fear. I can breathe again.

What should we do? Treat her based on symptoms? If all the blood and urine tests were normal, I wasn't too keen on just treating empirically. In one hand was the prescription for injections of a hormone designed to suppress the carcinoid. In the other was my over-worked phone. Shannon and I agreed that we needed a clearer picture of the situation. With only a couple of weeks left in her first semester at UNC, Shannon was determined to complete her coursework, and agreed to go to Hopkins if needed for a quick "tune-up" during her winter break.

It was cold in Baltimore in early January 2006; my sister arranged for Shannon to be admitted by a GI oncology surgeon. And he was terrific, pursuing with clear and focused determination the goal of finding that carcinoid tumor and, I was sure, removing it, thus ridding her of its chaotic effects. Shannon's body, however, was equally determined not to give up any secrets. Tests one through ten were normal. More tests. Still normal. More tests? She had them all. From blood and urine to ultrasound, X-ray, magnetic resonance imaging and PET scans, the answers were all the same: no evidence of carcinoid anywhere.

Each day I waited; the nurses knew I was a physician, yet they understood me as Mom. To help me keep track of our direction (and always asking, "What are we all missing?"); I drew a representation of Shannon's body. On one side of the page, I drew arrows to which organ system appeared to producing her symptoms; on the other side, I drew arrows to which organ system may be implicated from a slightly abnormal test result. I was increasingly confused by the extent of multiple organ system involvement.

Shannon was evaluated by experts in multiple medical specialties. She got used to the daily 2 week drill: Here was the consultant in his crisp long white coat, arriving after morning rounds, smiling and pleasant, with a series of residents and students crowding into the small room behind him. No kidding, Shannon answered questions asked by attending physicians from *surgery, hematology/oncology, neurology, endocrinology, internal medicine, neuro-ophthalmology, dermatology, anesthesiology, intensive care, rheumatology, infectious disease, allergy and immunology, gastroenterology, gynecology, and finally, public health and community medicine*. And then the med students came in followed by the interns and residents, with an occasional fellow thrown in. Shannon was what trainees in tertiary care institutions call a "Fascinoma."

She was not a Fascinoma: she was my daughter.

No one asked about mold (oh yes, especially including the allergist). No one asked about low MSH. No one asked about impaired antigen presentation. No one asked about Chronic Fatigue Syndrome. For all their combined expertise, and with all their combined research grants, I just didn't think any of them grasped the entirety of what was going on with Shannon.

What became increasingly obvious during that hospitalization was the difficulty we physicians have in seeing the forest for the trees, the number of detours we can explore with our array of medical tools, how unpredictably and severely the human body can react to assaults to its integrity, and how sometimes, the answer is closer than we think. To get to that final point of reflection, however, a few more pieces to the puzzle were waiting in the still-darkened recesses of human physiology, each burning brightly even while kept under wraps, and each as obvious as winter's driven snow if only someone would look.

I remember those next months after Hopkins as nothing much more than an endless succession of days that Shannon spent in bed. I'm glad that she didn't have carcinoid. That much we now knew for sure. But clearly, the assault her body had been through at Hopkins took a lot out of her. And clearly she wasn't getting better despite everything that had been done for/to her. She moved back home for several months, stating at the outset, "I just need to take a break from it all." She was plagued by intense itching, aggravated by showers and toweling herself off. With their relentless presence, the torturing effects of diarrhea, fatigue, headaches, shortness of breath, flushing, dizziness, brain fog, and nausea persisted.

But by late spring, (who knows why) Shannon was beginning to feel some semblance of what had become her more limited "normalcy." With her spirits rebounding, she made plans to return to NC and get back into the swing of her previous life. She rejoined her boyfriend, returned to school and her job, while continuing to experience ongoing but less intense symptoms. Sometime that summer, in a phone conversation, she off-handedly mentioned that, "Jonathan has noticed that the mole on my shoulder has tripled in size and gotten a little darker over the past few weeks."

A short time later she returned to Williamsburg for a visit. While eating breakfast one morning, I asked her about the mole. She shrugged, saying it had shrunk back to normal size and color. I took a peek at it. She was right; it looked like a regular mole ... but not one inch away from it was a **black spot** no more noticeable than the period at the end of this sentence. "How long has that been there?" I asked. Peering at it over her shoulder she answered, "I think a couple of months." What was barely evident to my naked eye was transformed into a mini-volcano erupting from the surface of her skin under the magnification of a 100X lens.

I quickly arranged for her to see a local dermatologist where she states, "My Mom wants this taken off." She returned home a short time later saying, "The doctor said you have nothing to worry about, but she took it off, just like you wanted." She drove off the following morning, back to her life in North Carolina.

Then came the phone call a few days later. "Mom, the doctor said you were right. It's malignant melanoma. You can call her for more information. I have to go back for a larger excision to make sure they got it all." We both burst into tears. All the fear about one kind of cancer from a few months ago was gone; but who was ready for this? Although her surgery was successful and Shannon was then apparently cancer-free she was still ill. What did I need to do next to stay on top of her health? I'd gotten pretty desperate.

Look again at the huge variety of symptoms Shannon, in her own words, described:

- within a few minutes of eating, my body gets hot and my face and neck get bright red (and they did); then my heart starts to pound like it's coming out of my chest (and her pulse would increase up to 130-140 beats per minute)
- after I talk for a few minutes, I get out of breath
- I can't walk up a flight of stairs anymore without feeling like I'll faint
- I can't see in the dark anymore; when I go outside Jonathan has to hold my hand or I'll trip; I can't drive at night
- My memory has gotten really bad – I have to cram for a test the night before and read all my notes right before the test, but the next day I don't remember any of the material
- I see more blue coloration out of one eye and more green

out of the other; when I look at a flat wall, I don't see a smooth surface and the coloring isn't consistent

- I feel like I'm going to faint now almost every time I stand up
- I have to pee so often that I hardly drink anymore
- When I press over my stomach I can make the flushing happen
- Light really bothers my eyes, I can't even watch TV anymore or go to the movies because moving my eyes back and forth gives me headaches and vertigo
- All of my joints hurt and sometimes my arms are so weak I can't lift them up to brush my hair
- I'm having diarrhea 20-30 times a day and I feel nauseated a lot
- I feel so jittery and tremulous when I wake up in the morning and my pulse is 120; it's not anxiety but that's how it feels
- I wake up drenched with sweat in the middle of the night a lot
- My breasts have been swollen and really sore for a long time and my nipples are always erect; it hurts unbelievably to have sex
- I get headaches all the time and nothing makes them go away – they're not migraines and Imitrex does nothing to them except make me feel worse
- It takes me hours to fall asleep even when I'm exhausted, then I can't get out of bed in the morning because everything hurts

- My hands and feet stay ice cold
- My life consists of sleeping until classes, cramming to take a test, having forgotten it all by the next day, and sleeping some more
- Jonathan says I walk around with my shoulders hunched, mouth open, flaring my nostrils with each breath; I just don't seem to get enough oxygen; he also says I make funny gulping noises with each breath.
- I got lost coming home yesterday, I couldn't figure out which way was north and I got on and off the ramps going in opposite directions a few times until I figured it out (*Author's note:* at which point the poor thing burst into tears, and so did I!)
- If life doesn't get any better than this, what's the point?! Every time I think I'll do at least OK for a while, I get wiped out again ... Will I ever feel normal again?!!!!!

Who doesn't hope for hope? By September 2006, my daughter said, "Mom, please read this book. Jonathan's Dad gave it to me and I just can't focus on it right now." That book, *Desperation Medicine***,** written by Ritchie Shoemaker MD, turned out to be our miracle. And part of the miracle was a serendipitous connection – Jonathan's Dad was a colleague of Dr. Shoemaker's research partner, Dr. Ken Hudnell...which is how that book finally got into my hands.

When Shannon asked me to read *Desperation Medicine*. I did, and wow, ***the light bulb turned on*!!!** Finally, things made sense.

My world was soon turned upside down both personally and professionally. As I read each chapter, the fog of "no answers" surrounding Shannon's health began to lift. With a growing sense of excitement, an understanding of the fundamental *hows* and *whys* for her poor health began to take shape. From that text emerged a clear scientific framework from which I would *reconfigure my perspective of medical diagnosis and treatment*. Just imagine my excitement! I could improve not only the physical health of my family and my patients, but also the emotional aspects of our lives. Why? Because with this new knowledge, we became empowered. With this knowledge we could then *take action*. We learned that previously misunderstood, disregarded, and "untreatable" symptoms had a physiologic basis that could be demonstrated with a variety of specific and readily available diagnostic tests; tests that were already being widely used in our medical and research communities.

One of the advantages to having grown up in a multi-generational family of physicians is that there's no shortage of people with lots of medical insight who care about Shannon (and me). I'm fortunate to have a father who, as an 80-year-old physician and a generalist in every sense of the word, continues to know vastly more medicine that I could ever hope to know. He's taught himself, and challenges me frequently, to "think outside the box." His insights about "how physicians can very quickly get trapped into thinking and proceeding along one path of assumptions, and thereby miss other possible explanations for problems and solutions" have been especially revealing during this process. Family members have been there when needed to shore up our rapidly diminishing stores of financial and emotional security - I can only wish other patients had such support. I'm also fortunate to have an incredibly talented sister who is an anesthesiologist at Johns Hopkins, who's responded with lightning speed and expertise to help on a moment's notice.

PART TWO: The Path of Discovery

September 2006. Desperation! How well that single word captured Shannon's present and God forbid, her future. Like so many other parents whose children are burdened with a serious or lethal illness they neither deserved nor had any control over, desperation could completely overwhelm my soul during the few brief moments of stillness I allowed myself. Grief bubbled up when I imagined the future she'd never have. Anguish, frustration, anger, fear, isolation ... these were emotions that could cripple me in the blink of an eye, creep into my consciousness when I least expected it and, I had to remind myself frequently, could also destroy any hope we had of finding answers.

While she slept most of that weekend, I was quickly absorbed by the mystery and fascinated by the significance of the discoveries. With growing excitement, I felt "the earth move under my feet." It all fit!!! It was beginning to make sense!!! The history of exposures, the progression of symptoms, the multisystem involvement with its dizzying arrays of expression, the obvious derangements of her immune system ... the pieces of the puzzle seemed to be falling into place.

Bright and early Monday morning, filled with renewed vigor and a restored belief in the miracles of modern medicine, I called Dr. Ritchie Shoemaker. I reviewed as concisely as I could Shannon's history and asked if he could review her medical records, which by then were voluminous. (A note of advice: ALWAYS obtain copies of your medical records as you go along; more on this later.) With his affirmative response echoing in my mind, I rushed off to get those records into the overnight mail. Within a couple of days, he called, saying, "She's pretty sick; get her up here as quick as you can."

I didn't have time to read *Mold Warriors* before the hurry-up visit even though his office staff asked me to read as much as I could before the visit. Maybe I should have postponed our trip until I did because as it turned out what he told us was right there in black and white, acting like the flare in the sky that illuminated the dark recesses of physiology we hadn't seen. He would put Shannon's illness on a piece of paper, lab results really, objective findings that showed in just a few pages what was wrong with Shannon, something that nearly 1000 pages of records from Hopkins, Duke and others hadn't.

We arrived at Dr. Shoemaker's office in October 2006. Dr. Shoemaker proceeded to get a more detailed description of her progression of symptoms. Next came their correlation with her history of various exposures. Her particular list of inflammatory triggers included living and working (or attending school) in buildings that had water-damaged interiors and air quality problems, EBV infection(s), and tick bites. Dr. Shoemaker then proceeded to teach us about his biotoxin pathway model. What we both noticed during that session and others that followed, was that we had difficulty understanding much of what he was trying to explain. This was partly because his level of grasping the molecular complexities was beyond what we could comprehend, even with a healthy brain. I read his Chapter 4 (explains molecular mechanisms of illness) in *Mold Warriors* but it just didn't sink in. The other reason, though, was that our brains were fuzzy from inflammation, not to mention being irritated by the fluorescent lights. Being a physician, I was particularly embarrassed by the difficulty I had in fully understanding what he was saying. Returning home from our visits, Shannon and I went through a series of "debriefings" in our attempt to remember what he'd said. A tape recorder helped; reading helped more. It came as a relief when, three months later, after I'd become a patient of Dr. Shoemaker's and received my own lab results back, that I began to understand why I had been struggling so much.

From that first meeting with Dr. Shoemaker our lives changed dramatically. We found out that Shannon had to avoid potential triggers for an inflammatory response. That idea was critical to her recovery. And so I took a crash course about how she could "live life in a bubble." By "bubble" Dr. Shoemaker meant that she could not be exposed to the indoors of *any potentially water-damaged building*. The idea of Shannon moving to Arizona and living in tent (not a moldy one!) for a few years to reset her innate immune (incredibly heightened) reactivity and to recover some strength was overwhelmingly tempting. Instead she went to live with a close friend near us in Williamsburg whose house we considered to be relatively safe. During her six months in the safe house, each of the rest of the family members were tested by blood and VCS. What a surprise; the illness with a genetic basis actually held true. We all had similar laboratory abnormalities, again explaining the vague but multiple complaints the rest of us had been voicing (or not). It quickly became clear that unless we all made dramatic changes to our lifestyles, particularly in the realm of exposures, we were headed towards the same disastrous health consequences as Shannon had. As my son so poignantly commented one day, "Shannon took the bullet for the rest of us."

Where to begin and determining what we needed to do became the next priority. Shannon was clearly in no condition to be able to do anything for herself. She spent her days in bed, in the dark, and away from noise and light. Her brain was a mess and her batteries were dead. With Dr. Shoemaker's initial referrals, I contacted multiple informed experts – remediators, neurotoxicologists, researchers, physicians, and other mold victims who had already been through this process. What I quickly discovered was that there was a huge lack of correct information, a huge amount of misinformation, a huge and *intentional lack of cooperation from government agencies* charged with protecting the health and legal rights of its citizens, and a huge amount of cover-up of the truth of mold illness.

Because our whole family was clearly affected, we needed have a safe place to live. For us that meant purchasing another house that could be remediated and made safe. It meant throwing out, selling, or donating most of our possessions and starting over. And it meant huge expenditures - we were hemorrhaging money but at least I hoped the hemorrhaging of our lives would be stopped. It breaks my heart to imagine the incomprehensible numbers of fellow citizens who are similarly affected and don't have the options we had. That's why it's so important to understand the issues and build, remediate, diagnose, and treat correctly the first time. It took several months for us to be able to get into the safe home. The struggle was worth it. We could begin to get our lives back.

Shannon's absolute dedication to the practice of "living in a bubble" and following Dr. Shoemaker's recommendations regarding treatment and other lifestyle changes has been the proof that his perseverance and his research are paying off. After 2 years of avoidance and treatment, Shannon's inflammatory markers returned to normal, her hypothalamic-pituitary axis somewhat normalized, her autonomic nervous system function dramatically improved and the quality of her life became worth continuing. There are still a number of health issues she needs to "fine tune" and I don't expect her ever to get back to a "normal" life, but the joy I now feel when I am with her and the "living" I see she has been able to get back to is truly impressive.

The critical importance of maintaining hyper-vigilance regarding potential triggers became crystal clear when several months ago she was bitten by a tick – her inflammatory response was massive and down she went ... hard. Headaches, severe fatigue, night sweats, flushing, shortness of breath, joint and muscle aches, and brain fog all returned with a vengeance. Luckily, again, the right treatment in the nick of time seems to have gotten her over that hump. Until the researchers can figure out how to turn off the genetic switch that results in the inflammation cascade for so many folks, avoidance, hyper-vigilance, and rapid treatment after suspected exposure will remain essential components of our lives.

Unfortunately, the additional struggles Shannon and so many others with truly disabling illnesses from moldy buildings endure are a disgrace. Just look at what our overburdened, uninformed, antiquated, immoral, unethical and downright inhumane social security, disability, and legal systems *don't do*. They don't serve those who need help.

It is beyond my comprehension that the government of this great country has allowed a system designed to assist those less fortunate to evolve into a system of denial, abuse, and neglect of its citizens. The system is clearly designed to cause attrition of huge numbers of potential beneficiaries, as determined and supposedly protected by law. The unwieldy process of applying for entitlement benefits, so confusing and frustrating as to generally require legal assistance, seems to me to be a legal kick-back scheme of attorneys supporting attorneys. The arrogance of "judges" and their so-called "medical experts" who clearly have no knowledge of, or interest in these relatively common conditions and who are clearly exploiting the system is appalling. After a financially and emotionally exhaustive 3 year process of applying for benefits to which she was clearly entitled, it took phone calls and letters to our Senator's office to finally get anything accomplished.

The "Pledge of Allegiance" is running through my mind ... where is the "liberty and justice for all" in this picture?

One final example of injustice, and this one illustrates prevalence (the 20-25% number shows up here as well!) of mold illness. Shannon's master's degree occupational therapy class, held primarily in the moldy UNC-Chapel Hill medical school building, was comprised of 18 students. Several of the students began to complain of the musty smell and respiratory problems. There were signs throughout the building to leave the air conditioners on at all times due to moisture problems. Complaints were made, a petition was developed, and supervisors were approached. As is typical in many work settings, the students were advised to "just get through it." Face it: the supervisors didn't want to make waves. One student developed severe headaches which were accompanied by total loss of vision for several hours at a time. A second student developed acute onset of "rheumatoid arthritis;" her symptoms subsided the following semester when she no longer had classes in that wet basement classroom. A third student developed severe and permanent hearing loss in both ears that required use of hearing aids. Then there was Shannon – that's 4 severe responders out of 18. And that's not counting those students who may not have recognized or reported changes in their health as related to those circumstances.

I would have predicted that this cohort of sickened people would be of great interest to researchers from UNC. No, the research project didn't happen.

Similarly, it wasn't until my own family's experiences occurred that I began to better understand the connections between building health and human disease. As I looked back in time at my family's pattern of symptoms the pieces began to fall into place. I suspect if we could peek into the futures of that entire class of students, we may find as well that other students were "sensitized" by those exposures and may have gone on to develop more widespread and prominent symptoms and illnesses.

A very significant epilogue to the UNC story was revealed to Shannon after her health demise had reached crisis proportion and she was forced to drop out of the program. One of her treating physicians, whose own office was housed in that building, shared stories with her about numerous colleagues from that building who developed unexplained pulmonary and other health problems, had to work from their homes, developed cancers ... an epidemiologic cluster of unexplained illnesses, occurring right in the middle of a prominent medical school setting. One would think that would set off bells in some high-up administrator's head, right?! And trigger a rapid course of investigation, remediation, compensation and PREVENTION, right?!

As I am now looking at the research opportunities this incredibly common illness presents, I can't stop thinking what a perfect opportunity for UNC the moldy med school building presented. Researchers wouldn't have to go far to study the effects of a highly toxic water-damaged interior environment on a group of medical clinicians and support staff. After all, the target population would be highly motivated to work with the investigators to find answers. Add to that the fact that the study would be a legitimate medical expense for those with university health insurance that would cover the huge number and variety of diagnostic tests that surely would be ordered as part of the differential diagnosis of their exposure-related illness. Many people identified as ill would be required to establish a safe place to live and work to preserve their health for their futures. And all would, I expect, wish to have family members tested to establish genetic susceptibility, then-current baselines or multisystem health or illness, and, again, develop a plan for PREVENTION.

Well, guess what? This series of events never took place. Surprised? Oh, we had the proof of prior awareness of the University of the problems (maybe the sign that closed off the door into the basement classroom that said, "Biohazard, Do Not Enter," was a clue to inside knowledge). Nope, no study, just a series of attempts to gloss over extensive problems present over a number of years, each that were addressed with grossly inadequate measures. My attempts to talk with top UNC officials by phone resulted in letters from UNC attorneys. And so it went. Again. And again. And again, again! No justice for Shannon or anyone else. More futures knowingly and unknowingly demolished. Admit responsibility? Where was the Tiger Woods-like apology to all those students who had been cheated out of their lives?

This process of stonewalling real illness is nothing less than criminal.

Action must be taken. Now.

Mind, body, and spirit are too precious to waste.

CHAPTER 8 The Belperrons: A Family Adrift in a Sea of Ignorance

All that was standing between me and Susan Belperron was a stack of medical records.

For several years now, her family had been seen by physicians from all over Massachusetts, yet she, Marcel, Pascal and Andre were still sick. Pascal first became ill in 2002, followed by Andre, Marcel and Susan herself. No practitioner really helped, though physicians like Dr. Py tried. The Belperrons never stopped seeking answers.

When they called for an appointment with me in May 2006, it was clear that they had enough of "no answers." Imagine that. I could see their mounting frustration and hopelessness, after I finished personally reviewing the nearly 5,000 pages of medical records Susan had sent. I counted the number of times in those 5,000 pages that the collective medical minds from some of Massachusetts' finest academic institutions mentioned mold: one. That was it. 1. Once. Uno.

So many physicians had multiple opportunities to correct their oversights, including neurologists, environmental docs (OccMeds) and attendings. And the one mention of mold (by an allergist) led nowhere. Pathetic. Actually, this case is worse than merely pathetic. The failure here to diagnose an entire family, despite a series of repeat visits, with ample opportunity to correct errors of arrogance and omission was incredibly well documented. No one even considered mold illness.

Ass^2 to the max.

None of these highly trained specialists even considered the potential for WDB-acquired illness, and that cost the Belperrons financial loss *and* adverse health effects. These physicians' clear neglect in their duty to be thorough and inclusive in assessing the Belperron family meant that the family suffered great and unnecessary harm.

I don't think the Belperrons would say Boston can claim any academic interest in the inflammatory illness that's so well defined by exposure to water damaged buildings. It's a pity, really, because if Boston's wonderful scientific minds would simply open their eyes, they'd see the exciting interaction of genetics, environment and inflammation that plays such an awesome role in 21st Century illnesses. But I guess it should be no surprise to me, given that more than a few medical people from Boston and its suburbs make a very nice living *selling* naysayer medical opinions testifying for the defense in mold exposure cases. I can think of a long list of Occupational Medicine practitioners in the Northeast who don't actually understand (*Author's note*: or refuse to acknowledge) the expanding medical and scientific literature about indoor molds and their fellow dwellers in WDB.

Don't get me wrong, we've made some progress: the current head of pediatric rheumatology at Children's Hospital is now more than willing to order C4a and TGF beta-1 for Marcel; and a professor in Clinical and Biochemical Genetics at Mass General, Dr. Marsha Browning, has taken an active interest in the family's cases. But the road at Mass General hasn't been smooth (*see Pierre Belperron's letter to Peter Savin, MD, CEO at Mass General at the end of this chapter*).

One of the docs who participated in this family's care, which resulted in no benefit, no insight and a whole basket load of unnecessary medical bills, heads an occupational and environmental health training program in Boston! Sheez. Only one of the highly anointed and appointed docs even knew enough about mold to ask about mold exposure.

(*Editor's note*: If you are a new resident in Boston learning about

Occupational and Environmental Medicine and the Director of your program *has no absolutely no clue* about environmental illness caused by exposure to the interior of water-damaged buildings (as was the case for the Belperrons), and therefore, has no possible means to discuss diagnosis and treatment of such an incredibly common illness, what will you likely learn from such an instructor?).

As Susan once wondered aloud, "How can so many smart people - famous people at that - *be so absolutely stupid and then so arrogant* about their ignorance at the same time?"

A good North Carolina born boy like me would say, "Shame on ya'll."

I look at the gallery of professional people who failed their duty to the Belperrons over and over again —it's a disgraceful collection of an "all-star cast" that includes the power of Harvard, Children's Hospital, Boston this and Boston that. Add to that list Senators, Congressmen, Coast Guard officials, Merchant Marine Academy officials, University Vice-Presidents and a lot more, too. The Belperrons have all of the names. Should a medical board be apprised of the problems the Belperron's have suffered? Probably, but I'd like to see mandatory continuing education for the Boston Brahmin docs to make that referral unnecessary.

Who could've known back in 2006 that the Belperrons would become the model for what's wrong with American health care for those with biotoxin illness? On the other hand, Susan Belperron is also a fantastic example of what's right with American Moms who don't allow health, education and government bureaucracies to ignore their family's needs.

Biotoxins changed Susan's life. What she and her family had to endure should never have to be repeated.

Susan might weigh 140 pounds after Thanksgiving dinner, distributed on a wiry, athletic frame. Gentle, musical and soft-spoken, Susan is refined and sophisticated. Don't make her mad, though. Not telling the truth about her children is the best way to see Susan bristle.

Most of us would fear the female dog standing just three feet from our carotid arteries if the dog were a heavily muscled, snarling 140-pound Rottweiler with saliva dripping down her lips, ready to defend her pups. Yet plenty of physicians, politicians, administrators and lawyers tried to flick Susan away, as if she weren't worthy of their time and care. Were it not for a seemingly never-ending series of professionals who underestimated Susan and didn't help "her pups," this chapter wouldn't have been written.

Potential adversaries take note: do not underestimate Ms. Belperron. Physicians, educators, administrators and politicians: You're expected to be nice, polite, competent and responsive. Don't make sudden, aggressive moves. You are being watched as aggressively as if you were a known puppy killer. And don't say really stupid things, like, "Well, we think the school is perfectly safe," when she has the fungal DNA analysis that says it isn't. And don't say that you've never heard of fungal DNA testing; Susan has and she knows more about it than you do. Oh, and don't lie about not having the report, either.

All that was necessary to turn this situation around was some simple human compassion, and open minds with understanding and a willingness to seek solutions to seemingly difficult problems. While the Belperrons have found some good physicians along the way, what they have largely seen is Ass^2 medicine at its worst. Those shortcomings weren't limited to the healthcare community; the lack of compassion, understanding and willingness to seek solutions also extended to other institutions, including education and government.

As I look back on this saga, I'm hopeful that the heroes in the end will be the sons, Andre, Pascal and Marcel, because each of these enormously talented young people will have a gift to share with others, a gift that only comes from surviving adversity. The gift comes from sacrifice on their parts, of course, but especially sacrifice from Pierre and Susan. The mark of evolutionary fitness of an individual like Susan is her offspring's accomplishments. I guess it's possible that the boys could become cynical and disillusioned, each burned emotionally by their losses and their hurtful experiences.

But I don't see destructive tendencies in their future; having survived, despite a society that's illogical and assumptive, I see the intensity of their life experiences leading them to outstanding individual achievement. They're each musically talented, extremely bright and pleasant to the girls' eyes. My staff tells me Pascal could be the next Bachelor on TV with 24 sweet young ladies wanting to be his bride. (Sorry, Andre, Pascal has better hair).

But in order for Andre to become a landscape architect, for Pascal to become an entrepreneur and for Marcel to work his way up to become the President of the USA, they'll have to survive their incredibly heightened sensitivity to environmental exposures. That's where I come in.

I should mention Dad and husband Pierre Belperron, who works 75 hours a week—I'm not kidding—to keep the family safe, clothed and fed under a roof that doesn't leak. Well, maybe they're not so safe in the local middle school, the high school, Drexel University and the University of Massachusetts in Amherst. If you want a registry of ERMI scores in a college, ask Pierre what we know about buildings at UMass. UMass won't tell you just how many buildings they have that are thoroughly unsafe for those with sensitivities to indoor microbial growth, but Susan and Pierre can because they have the ERMI scores. And Andre and Pascal have rocketing C4a levels, acquired following exposure. The scenario isn't pretty.

Susan's been busy seeking help for her boys, so now just about every politician in Massachusetts has those Amherst ERMI scores as well.

Finding out what was wrong with his family drove this mild-mannered father, born in Algiers to a French father conscripted to fight in the Algerian Civil War and an American mother, to refuse to knuckle under to arrogance, deception and outright fabrication from agencies and hospitals. When you read his letter to the CEO of Massachusetts General Hospital, you'll know what I'm talking about. This man is a musician, an instructor for the "In Control Crash Prevention Training" program and a service manager for Saab. He's gentle and helpful. He's also Susan's rock.

Note to those who'd hurt Pierre's family by their desultory attention to public responsibility: he doesn't give up when he's right. And he's right this time.

He wasn't happy about what he heard from Susan's visit to Harvard Teaching Hospitals.

Perhaps the worst moment of all was the one that took place in a specialized 'environmental disease' clinic operated by Children's Hospital, a Harvard Teaching Hospital, when a highly regarded toxicologist and specialist in environmentally triggered illnesses (such as Lyme disease) refused to listen to his patient. In fact, he simply "waved away" her suggestion that she and her children might be suffering from mold-linked toxins and inflammagens.

"He was in a hurry and he was obviously irritated," says Susan Belperron today, remembering the despair she felt when Dr. Alan Woolf out-of-hand rejected her contention that bioaerosols generated by microbes in water-damaged buildings might have made her family ill.

"He had the full weight, the full authority of Harvard behind him," she says today, "and he basically just laughed in my face. He was almost scornful, and I know I'll never forget the dismissive, patronizing look on his face when he told me, 'Mrs. Belperron, I can assure you of one thing with complete certainty: mold does not make people sick!' "

He was dead wrong then and still is now. His attitude to Susan wasn't "just a bad day."

Those who know say that having cancer changes your life forever; maybe knowing mold will change Dr. Woolf's life. Susan doesn't agree.

Naturally, she never went back to that physician. Susan's phone call to my office, made not long after her lone Dr. Woolf encounter, wasn't unusual and didn't foreshadow the Belperrons' role as pioneers in public accountability.

"My three sons are sick, perhaps from a tick-borne illness, and so are a bunch of kids in our neighborhood," she told me. "The Health Department thinks we're a Lyme cluster but says that no one can help us."

So what else is new? Lyme misdiagnosis is rampant with under-diagnosis and over-diagnosis daily events across the U.S. Usually both sides are wrong since residual inflammatory illness can't be treated by calling the illness "fibromyalgia" or "antibiotic deficiency," either. My stance - that each is way off base in thought and therapy - doesn't win me a lot of friends on either side, despite the publications our group has and the verification of what we say coming from both sides of the argument.

Then Susan said, "We live in Ipswich and I think there's some kind of toxin in the water here. No one at Harvard's teaching hospitals will even listen to me, and I'm just tired of being pushed around and sloughed off. Please help."

I didn't know much about Ipswich, so off to the maps I went. The Ipswich River runs into Plum Island Sound and then into the Ipswich Bay. No, this isn't the same Plum Island, site of the top-secret chemical/biological warfare research labs of the Long Island Sound. The water from the Ipswich River is used for drinking water by several upstream towns, so that in summer, the Ipswich River nearly dries up further upriver, creating stagnant pools that could support harmful algal blooms, primarily cyanobacteria. (Blooms of blue green algae-like *Microcystis* and *Cylindrospermopsis* are another public health problem too often ignored by our elected officials and agencies). But the town of Ipswich has twice-daily tides of up to nine feet, so fresh-water dwelling, toxin-forming cyanobacteria aren't a likely problem for the Belperrons. But then I look on the NOAA Harmful Algal Blooms website and *bingo*, there are plenty of blooms of brevetoxin-forming dinoflagellates that poison the previously famous "Ipswich clams." I wouldn't eat them now. So then I looked on the Ipswich Riverkeeper's website, which breathlessly reports on the current flood conditions.

Hhhhmmm. Does anyone else reading this book think that flooding can cause human health effects from microbial growth that can be mistaken for Lyme disease?

"Ms. Belperron, is your home located near the Ipswich River?"

"Yes."

"Does flood water ever enter your home?" I ask.

"Rarely, there's been a small amount of water in the basement when the water table rises, but not from the river," she replies. "Our home has massively thick stone walls with a lot of 'sweating' (condensation) on the indoor side."

"Do you see visible microbial growth?"

"Yes."

"Did the visible microbial growth precede your sons' illness?"

"Yes," she answers, "but so did Andre's Lyme disease, confirmed by Western blot testing."

"Are the other members of the Lyme cluster also affected by flood conditions of the Ipswich River?"

"Yes."

"Do their homes have microbial growth that you can see?"

"I haven't been in all of them, but yes, from what I've seen."

A history of exposure to the interior environment of water-damaged buildings, with time of exposure preceding illness, doesn't necessarily mean that a chronic inflammatory response syndrome either exists or is caused by moldy conditions. But without such a history, I don't find mold patients. Just being outside or digging in compost or sitting in a stack of wet leaves in fall won't do it.

Mind you, the conversation above took about 30 seconds during our first phone call. No one else had even bothered to ask Susan about her home and the time course of the illness. Why did I ask about moisture and mold? Simple: Sutton's Law, which we were constantly reminded of in med school. Like Willie Sutton, who said he robbed banks because that's where the money was, I was looking for where the action is. Medically uncertain illness (*Author's note*: does the commentary about Dr. Sigal and Lyme in Chapter 3 take a new meaning here)? Think innate immunity and mold first, and often, you won't be wrong.

Here's the cover letter that the Belperrons sent to me in mid-May 2006:

> Dear Dr. Shoemaker,
>
> Our family has been living with illness that we've not been able to resolve for a number of years. Having read about you, and having read Desperation Medicine, we're hoping that you might be the one physician, following many, who actually gets to the bottom of our illness.
>
> Our odyssey started with our middle son Pascal (now 16) about four years ago. He suddenly started having severe, intense and frequent motor tics, so much so that it seemed he was having seizures. Physicians diagnosed motor tics and prognosticated a temporary condition. Later, Pascal added vocal tics, and as time passed, physicians seemed poised to let it pass as 'chronic multiple motor tic' or Tourette's if he showed any hint of compulsion. That supposition changed when a neighborhood child started to exhibit the same symptoms, then another boy nearby.
>
> Soon there were enough patients with tics that the Environmental Disease Clinic at Children's Hospital wanted to take a look. They

looked and passed the baton to the state DPH, which made a show of studying the situation, but were not diligent in keeping track even of records sent to them, and they did nothing but write a report two years later that said the numbers involved were not 'statistically significant.'

Seven has now grown to thirty (that we know of), including four in our household, with other families also having multiple sufferers. While the early sufferers all had prominent visible motor tics, later, there were manifestations that were more subtle in terms of other sorts of neuromuscular unintentional movement manifestations. What were consistent were many of the other symptoms.

We've received mostly non-diagnoses. Some physicians have been willing (not necessarily ours) to lay the cause of all of this at Lyme disease. We don't dismiss this out of hand, because although Lyme is so prevalent here, the manifestation of the illness is not exactly classic for Lyme. So we've looked into all sorts of co-infections or combinations of biological pathogens and toxic exposures, but so far to no avail in us or any of the others. An interesting finding, though, is that those who thought they had Lyme (or their Dr. thought so) and who took antibiotics, improved and sometimes become symptom free, but would relapse within weeks of stopping antibiotic therapy. Although non-specific use of antibiotics is far from our nature, we tried this and got some relief, though the success has been mixed.

Because the size of the group has grown, and we've been aided by a health-care consulting firm, Boston Market Strategies, the MDPH is now launching another investigation headed by Dr. Bela Matyas. BMS has also drafted a survey that our town will disseminate so that we can identify as many others as possible, and we do hope that at some point, a proper epidemiological study will be performed. Throughout this process, we've encouraged the group to present a big picture that might reveal the cause of the illness, while never giving up on the thought that one physician might find in one of us the cause, which could help everyone.

Ipswich is a lovely old New England town on the Massachusetts coast about 35 miles north of Boston. Its population is diverse ethnically and economically and numbers about 14,000. Behind the four-mile barrier island that forms our beach are tidal marshes, which place a buffer between the populated portion of town and the ocean. The Ipswich River flows prominently through the middle of town and then through the marshes on its way to the sea. Many of those afflicted have been recreational users of the water here. Virtually everyone boats and uses the beach, and most spend some time in or around the river. In studying the river, we did learn that upstream there was once methyl mercury contamination, but no one has tested positive for this. Also, the clamming flats, which have been closed for years, reopened in recent years, before the first person got sick, and we always wondered if there was something locked in the sediment that was suddenly released (we don't eat clams). In reading your book, it was hard to ignore the similarities between your area and ours, and while some of the specific neurotoxin-producing algae may not come this far north, perhaps there's another organism doing something similar. Susan has also heard of stripers with lesions, and she's willing to query fishermen directly when they return to the shores here in a few weeks.

Another characteristic of this area that's suspect is the insect population. We're overrun with ticks of many varieties (and the Lyme and co-infections they carry), plus rampant salt and fresh water mosquitoes, and greenhead flies. The deer herd is out of control. We've had issues in the past with the chemicals used to control mosquitoes. Until about 15 years ago, Malathion was sprayed. We noted that our son Pascal had respiratory distress on the nights they sprayed. The town stopped that spraying but adopted marsh spraying with Bti. Last summer out of concern for Eastern equine encephalitis they sprayed with Anvil, and it was right after our two sons drove behind the spray truck (not knowing what it was, they'd never seen that before), that Andre got ill. Coincidence? Perhaps.

> Perhaps you wonder how Susan developed her Rottweiler-like Dr. Jekyll from her benign Ms. Hyde. And perhaps you wonder why she's bringing her family all the way to Pocomoke again next week to start a trial of vasoactive intestinal polypeptide (VIP).

(*Editor's note:* after this section was written, the Belperron family, though not Marcel, as he's not yet eighteen, has had a remarkable improvement, including resolution of motor tics, with VIP therapy. There are so many threads to this story; maybe we should start at the beginning.)

A Nightmare in Ipswich

Their lives changed forever on a mild summer evening in 2002. That's when the nightmare began.

During the four years that followed the onset of their family health disaster, Pierre and Susan Belperron would watch in horror, as one by one, their three children fell prey to a disabling malady that periodically left them wracked with agonizing pain, crippled by nervous tics and covered with hideous, pustular sores that oozed viscous green fluid, leaching through their skin, with the thick goo mixed with their non-clotting blood.

Had Marcel, Pascal and Andre Belperron been poisoned by some unknown toxin in their small-town environment? Were they the victims of a mysterious genetic disorder that left them utterly exhausted for days on end in their home in Ipswich, Massachusetts, while their pain-ripped muscles twitched violently? Their parents looked on, unable to stop what no physician could diagnose. They'd take the boys to the emergency room when things like this happened, but doctors never found what was wrong.

During the third year of what Susan Belperron now refers to as "our desperate odyssey," she became ill as well. Weakened by severe joint and muscle pain and dazed by short-term memory and mental focusing problems that left her staring into space, she lived with a secret terror: What if she became so enervated by this mysterious ailment that she could no longer care for her children?

August 29, 2002: Ipswich, Massachusetts

They were in the middle of supper – vegetables from a local farm and roast chicken– and they were enjoying themselves. And why not? More than any other time of the day, the Belperrons loved suppertime, that bright, magical hour when they could all sit down together at the cherry-plank table in the kitchen and concentrate on the two activities they loved most: eating good food and vigorously debating the pennant prospects of their beloved Boston Red Sox.

The Belperron boys were on the verge of a new school year. André was starting high school, Pascal was entering seventh grade and Marcel was embarking on first grade. Talk turned to bus schedules, soccer practice, violin lessons and teacher assignments. Did everyone have cleats that fit? Did anyone need a haircut before school started? It was a few minutes past six o'clock now, and the shadows were already deepening on the front lawn of the old, stone house in the historic seaport town of Ipswich. Gazing through the dining room windows, Pierre Belperron realized all over again how privileged he felt to reside in this antique Yankee homestead, and how blessed he was to share it with the laughter-loving Susan, and with these boisterous, healthy children.

"Pascal, what *is* it?" asked Susan, a tinge of alarm in her voice.

Pierre blinked. Hearing the note of concern in Susan's voice, he swung his gaze back to the table.

"Pascal, are you al*right*?"

Pierre looked across the room at his middle child and was stunned by what he saw. Pascal's brown eyes had begun to roll in his head and the muscles around his mouth and chin were jumping and twitching. Pierre watched a flood of spastic tics ripple across his son's face.

"Are you making those faces on purpose?" he asked his son.

Pascal was shaking his head, no, but his facial muscles were still jerking and quivering, one after another, as if an endless stream of electrical waves were rolling across his distorted features.

"What's the matter with him, Pierre? What's wrong?" asked Susan, concerned.

"Is Pascal sick? What's *wrong* with him?" asked little Marcel, who was barely out of kindergarten.

Pascal frowned. "I'm OK. But I'm not doing this; I can't help it."

"It's okay, son. It's okay," said his father. Pierre was immensely relieved to see that the child's eyes returned to their normal position and that the frantic facial tics had begun to recede. Then just moments later, without warning, the tics started again.

After sitting with Pascal and observing him for the evening, Pierre and Susan moved him into his bedroom. It was nearly two AM before the still-frightened parents settled down to get a few hours of sleep themselves. At this point, only one thing seemed clear: in the morning, Pascal would not be going to school with his brothers. Instead, his fretful mom would take him to a local clinic for an emergency consultation with the Belperrons' family physician, Dr. Frank McDermott.

As she drifted off to fitful sleep, Susan Belperron didn't imagine that her family's "desperate odyssey" was about to begin.

From "A Chronology of Tics:
The Health Journal of Pascal Belperron"

[The following excerpt and others that follow are quoted directly from a journal kept by Pierre and Susan during their three sons' illness.]

September 3: Saw Dr. McDermott. Blood tests done to check for strep or Lyme infections, and to check thyroid. All tests negative. Diagnosis was neurological tic, and Pascal was referred to a neurologist.

September 6: Pascal was seen by neurologist Dr. Muriello. She prescribed re-do on strep and Lyme tests, and to hold off on getting any of the alpha-blocking medication.

We first noted tics at the end of August, just prior to the beginning of school. Pascal's entire face wrinkled with spasm. The next symptom noted was that his eyes would roll back into his head, in conjunction with other involuntary facial movements. At their peak, Pascal's face would be ticking 25-30 times per minute. Pascal's inability to focus visually in school before another spasm would hit him tired him out a lot. At this point, Pascal's skin was looking very gray and he had dark circles under his eyes.

September 12: Tests for Lyme and strep came back negative. Because final cause of tics was not yet determined, no decision made on medicating him.

September 18: Pascal seen by Dr. Moskowitz. He felt that more extensive neurological diagnostics were necessary to make an evaluation. Based on his examination of Pascal, remedy Bos Runa was administered without any effect. Dr. Muriello deemed Pascal's condition just a tic, and didn't see the need for anymore testing. She suggested medicating him with clonidine. We requested the opinion of another neurologist. Dr. McDermott said he would try to reach someone but nothing ever came of this.

In an effort to try to prevent a worsening of his condition, Pascal stopped playing sports and the violin. September 20 Pascal started to tic in his back and shoulders. By evening his arms and his legs were also ticking. He noted that when trying to focus a camera, he was unable to.

September 22-23: Pascal was very tired; he stayed in bed. His body now has some part always in motion. School became difficult. Leaving for treatments and doctors' appointments, plus difficulty reading due to eye tics all added to Pascal's exhaustion, which made school even harder. As he got further behind, the fatigue and stress made schoolwork even more difficult, and he was withdrawn from school.

The irony here is that while Pascal was withdrawn from school, he still went to the school itself almost every day for non-academic activities: orchestra, tech-ed and even lunch. Little did the Belperrons know at that time that they were sending him into a mold-riddled environment that would cause ongoing damage to Pascal. It was years later when a diagnosis was finally confirmed.

"Never Forget: We Have Each *Other*!"

During the years that followed Pascal's attack, Pierre Belperron and his family would be tested to "the absolute limit." Again and again, as the three boys grew increasingly ill, the 42-year-old cellist and Saab service manager would struggle against encroaching despair, while his three sons and his beloved spouse struggled in a twilight world of neurotoxin-triggered sickness that left them helplessly immobile for weeks at a time.

He often told himself that it seemed ironic that they'd all been so healthy (and so happy!) up until the evening when 13-year-old Pascal was felled by the bizarre onset of tics and muscle spasms that soon prevented him from playing sports or the violin.

Until that night on their quiet side street in Ipswich, Pierre, Susan and the three boys had enjoyed an almost idyllic existence.

Because both parents were graduates of the renowned Boston Conservatory, where Susan had studied the flute as Pierre was perfecting his skills on the cello, their home had always been a place of music, a gracious and refined family setting where the strains of Beethoven and the sonorous harmonies of Brahms could be heard at all hours.

While Pierre held down a highly demanding job at Charles River Saab, 40 miles from Ipswich, Susan spent her days looking after the boys, volunteering at the schools and community supported farms, tending to her own beautiful gardens and enjoying the beach and Ipswich River. Theirs had been a deeply fulfilling life for nearly 15 years, as they chauffeured their growing sons from one soccer or basketball game to the next, or sat in delighted appreciation through violin recitals, where each of their children drew rave reviews from teachers and friends.

But then the nightmare began. Remembering their endless trips to consult with more than 30 local physicians, along with the numerous lab tests and MRI scans that each child was required to endure, Pierre shakes his head in profound amazement. "Quite honestly, I don't know how we survived it," he'll tell you with a groan of remembered aggravation. "First Pascal began to develop his tics and muscle cramps, and they quickly got so bad that he had to quit school. Later he was covered with these strange blisters full of pus. We had to wrap him in sheets for a while; they were so irritating that he couldn't stand to put on clothes. The smell was so bad we had to keep the windows open."

"But that was only the beginning. Not long after Pascal's first attack, Andre's face began to swell up terribly. He developed vision problems to the point where he could barely read. He'd been an all-star goalie on his soccer team at school, as well as an all-star pole vaulter and he had to quit doing both. We took him to doctor after doctor and they all told us the same thing: 'Oh, it's just a virus, that's all. Don't worry; it'll go away soon.'

"But it *didn't* go away."

Andre stayed sick, and now the youngest of the three brothers, first-grader Marcel, was also beginning to develop difficulties. It started with reading problems, which eventually were tied to neurological vision problems, and the symptoms grew from there. Once again, the distraught parents started making appointments with neurologists and toxicologists and chiropractors (including an acupuncturist) in the Boston area.

Susan Belperron says she'll never forget "the pain and the torture" that went on month after month; as the exhausted couple fought to help their children survive their mysterious malady. "It was just a blur of anguish," she'll tell you today. "I remember one terrible afternoon, when I was sick myself, and I had to take Andre to a new doctor. I got lost on the way; it was an unfamiliar address, and I couldn't concentrate, I couldn't focus. I was just weeping.

"The worst part was not knowing what was wrong with us. The uncertainty, the misery of going from one doctor to the next, and not one of them could help us. Not one! I don't know how we survived. But I guess the turning point came one afternoon, after all three boys had become terribly ill. Pierre and I were talking on the phone and all at once I had a thought, and I told my husband, "Look at the good side. We have each other. And all at once, I felt my spirits lift. And I said to myself: We're going to make it! I don't know how, but we *will*!"

From The Health Journal of Andre Belperron

In late August 2005, André began to suffer some lethargy, loss of appetite and a stiff neck. The lethargy was noted even by his high-school soccer coach. The stiff neck was attributed to something that happened during soccer. At the beginning of the school term, André felt somewhat sick and had a fever. Very quickly he developed other symptoms, including loss of appetite, extreme fatigue, pains in his head, neck, eyes and back. Then he developed tingling in his left arm and both lower legs and feet. Visits to the primary care group suggested possible strep throat, a virus that was making the rounds, or a small chance of Lyme disease.

André was referred to Dr. Joseph Py for examination on September 13. Dr. Py felt that André likely had Lyme disease, despite the absence of a recognized or identified tick bite. Diagnostic blood testing was postponed, since it would likely yield a negative result if not enough time had elapsed since infection. A course of antibiotics was prescribed in case André did have Lyme or a coinfection, but there was little concrete evidence in clinical evaluation to confirm the diagnosis. We hesitated to do anything further with Dr. Py because he's not in our insurance network.

Symptoms continued to develop. A mild fever would come and go, along with an upset stomach, sore throat, photosensitivity, poor memory, a foul taste in his mouth and odors emanating from his body. We took this information back to the primary group to see if anyone within the insurance network could help us. There was a referral to infectious disease specialist Dr. Steven Keenholtz. Because this physician antagonized and was rude to André during his examination on September 15, André walked out. However, Dr. Keenholtz had apparently seen enough to suggest an acute problem and notified the primary care group that André should go to the emergency room. He was seen at the Beverly Hospital ER later that day by Dr. Debra Hillier. She performed a spinal fluid tap, which led her to diagnose viral meningitis (*Editor's note*: this diagnosis was never confirmed). She felt that André should be feeling better within a week or so, given the length of time he'd already been ill.

But André didn't get better. Having received a recommendation to go to Tufts-New England Medical Center (NEMC), we tried to see a physician in Infectious Disease, but couldn't because of André's age. Instead, we were referred to Dr. Herman Meisner. André saw him September 21. Dr. Meisner was dismissive of our concerns, thought that André just had viral meningitis and would be well very soon. When we mentioned our concern about Lyme, his response was "no tick, no rash, no Lyme." When we informed him about our son Pascal and his problems, and our concern that there might be a link, Dr. Meisner took on a very defensive posture, said he knew all about our Pascal and his circumstances. (How, we wondered?) He made it clear that he didn't want to have anything further to do with his situation or André's. His advice was to go back to Children's Hospital and call the Department of Public Health.

André hasn't been to school since the first day of class. He has tutors and tries to do a little academic work each day. He's still lethargic; sleeps an enormous amount; has a very poor appetite; swelling and transient splotches and welts— pustules really—on his face, chest and back; chest pain that impacts breathing; difficulty lifting his arms; persisting pain in his back, neck and head; trouble enunciating his words; cognitive problems; and difficulty with memory. He's been in for an EKG, EEG and X-Rays per Dr. Jackson, as a preventive measure.

Understand that three months ago, André stood on the grounds of the U.S. Naval Academy in Annapolis, Maryland, possessing all the academic accomplishments (GPA 4.0, math SAT 780, full honors/ AP course study) and athletic prowess (three varsity teams per year including all-state selection in track and all-league selection in soccer) required to be an outstanding candidate for admission to that institution. Knowing he cannot possibly fulfill the physical demands of his candidacy, he knows not only that he will not head to Annapolis next fall, but that he's in danger of not being able to fulfill his high-school requirements so that he can go to college at all next year.

Evaluation before Pocomoke:
beginning to understand what's wrong

Pascal

When the Belperrons arrived in July 2006, mold wasn't a major part of their vocabulary and certainly not a major part of their health concerns. What they know now far exceeds what anyone else in Ipswich—or Massachusetts, for that matter—knows about human health effects caused by exposure to WDB.

Pascal's medical records before 2002 numbered less than 20 pages. After 2002, there are several thousand. He's told he has facial tics, with a referral to a neurologist made early on. Labs that one would expect to be done are normal. No strep, normal thyroid, Lyme negative, neuronal antibodies negative, blood culture is negative. MRI of brain is negative. Antibiotics don't help when they're given for an unrelated skin infection. Pascal refused clonidine, Neurontin and Orap because of their side effects, and the revelation that they were offered only to hide the symptoms, not to remedy anything. His illness now includes "involuntary motor tic involving significant facial grimacing and eye deviation, becoming more frequent as the day goes on and he becomes fatigued. He's also developing twitching in his arms and legs." His illness was not thought to be due to social problems or Tourette's syndrome.

In January 2004, he's sent to another neurologist at Children's Hospital in Boston (a resident sees the patient and does the basic work then presents the case to the attending Pediatric Neurologist), who says he might develop classic Gilles de la Tourette syndrome if he progresses to have sonic tics. The neurologist notes the concept that the Belperrons are "boating people," but he says *there's no toxin* exposure. (*Author's note*: this is a fairly outrageous assumption. There is no basis in any process of medicine or any process of basic science to say "no toxin exposure." The use of VCS testing and labs for dinoflagellate toxin exposure, ones that would be implicated by exposure to estuarine toxins, were well established in the medical literature at that time. To say "no toxin exposure," without any data and using only an unsupported assumption was a sad error here. *If the proper testing had been done* here, the answer would have been obvious, though the identification of toxin exposure would still be incomplete.) No diagnosis is made and a referral to Dr. Woolf is made in the Environmental Clinic.

In the interim, an acupuncturist saw Pascal, noting that in Chinese medicine, twitching that moves from place to place is described as "internal wind." Pascal improves significantly on this treatment, but the absence of a sustained response following treatment, a vital clue, is overlooked. It's the ongoing relapse that prompts the family to seek allopathic care again in January 2004. In retrospect, the acupuncturist was so close to knowing the illness, as the innate immune inflammatory process, once activated in biotoxin illness, is an internal zephyr that will "raise the dead" and make the living act like they're dead.

Pascal was just the index case—the whole Belperron case file is a disgrace to the ideal of American medicine.

They completed multiple forms and went for a first visit to the

Environmental Clinic. The Belperrons are asked if environmental testing had been done in the school or home, or if there were environmental hazards. Questions are directed to obtaining information on pesticides, cleaners, solvents, furniture refinishing, fumigation, home water supply, use of cigarettes, lead paint and more, but nothing about mold. Imagine how much time could've been saved if the Environmental Clinic had simply asked if there'd been exposure to water-damaged buildings. But that didn't happen.

The river was a focus of the environmental evaluation, and they considered heavy metals, thorium and contamination from a nearby gravel manufacturing site. (*Author's note*: to worry about thorium (*rare illness*) and not mold (*common illness),* is illogical, at best). The issue of a neurologic motor disorder was raised, "with the State Department of Environmental Hygiene and Protection - **confidentially only** - to alert them to this cluster and to see if they have any knowledge of similar cases from the Ipswich watershed area."

By April 2004, the Clinic had nothing to offer other than a referral to neurologists and a psychiatrist. Susan Belperron's concerns that Andre was getting much worse were recorded but nothing was done. A meeting with Suzanne Condon of the Massachusetts Department of Public Health was arranged.

Another attending neurologist ruled out West Nile Virus, coincidentally with the same resident doing the history as he had at the prior Children's Hospital attending neurologist.

Neuropsychological testing was done in June 2004 (five hours of testing cost $5,000). Pascal had clear evidence of problems with executive cognitive functions such as working memory and sequencing, with decreased assimilation of new knowledge. Couldn't someone have read the neurocognitive testing paper on *Pfiesteria,* published in "The Lancet" in 1998, one that would've helped to identify a toxin as the cause of the illness?

The neurology group sees Pascal in September 2004. They don't seem too happy with the family's use of alternative providers for care. The same resident physician who'd seen Pascal twice before recorded his parents' concern about the possibility of a Lyme disease diagnosis. Again, he's not treated but is told to return in another six months.

The next month, Pascal has a tick bite, but no rash. His temperature goes to 104; at Children's Hospital Emergency Room, they prescribe doxycycline. The physician orders a careful search for tick-borne illnesses, but all tests are normal. An Infectious Disease consultant felt that the illness was due to a virus and suggested stopping all medications. Pascal passed out on two occasions after the ER visit, so more searches were begun. Cardiology is consulted; Holter monitor and an echocardiogram are normal. An EKG was unremarkable (no prolonged QT noted).

For the first time, Pascal had a coagulation study performed that showed a prolonged partial thromboplastin time (PTT). No von Willebrand's profile was done and the rest of the family wasn't tested. (*Editor's note*: all the Belperrons have prolonged PTT, with Marcel and Andre having the lupus anticoagulant).

Rheumatology gets involved in November 2004. At least this physician noted that Pascal had missed the seventh grade due to the tics, but said the illness was viral. The attending ordered complement studies, but not C4a. (She was so close to finding the answer; if she had just ordered the right complement test!) Now his anticardiolipin titer is turning positive, though no TGF beta-1 was done. She suggested that the Environmental Institute of Autoimmune Disorders at NIH be consulted. His Lyme titers were still negative, confirmed at least six times previously.

Here we've got to ask the simple question: if the problem now needs an autoimmune work-up at NIH, wouldn't it make sense to at least ask if there were exposures to water-damaged buildings, a condition that is well-documented to be associated with abnormal functioning in autoimmunity?

The Children's Hospital Primary Care team is consulted. This physician says that no one else in the family is ill (oops, the collective Keystone Kops are wrong again; Andre is affected; now Marcel is getting ill, too, as is Susan). This physician wonders about allergy and makes a referral.

The allergist sees Pascal in March 2005, nearly three years after the illness began. *He notes the presence of mold in the basement in their home.* Pascal has some mild allergy to trees but not much else. There's nothing further noted about mold.

Now the parents are getting desperate. Next up are Lyme labs paid for with "cash donations." I looked for the PT Barnum testing results but didn't find that one.

By December 2005, the family is back with the *same* resident in neurology, with the same psychiatric diagnoses now figuring in the diagnosis, but a referral for Pascal and (now) his brothers back to Dr. Woolf is recommended. Round and round they go on the referral merry-go-round. Only the ride isn't very merry.

The Environmental Clinic decided that a "full conference" was indicated to review this case. The conclusion is that, "we need to review the environment in more detail." At that time, the U.S. EPA had released its ERMI testing results, which were available commercially. **For $300, the conference of the Environmental Clinic could've obtained the same results I did several months later: massive amounts of toxigenic and inflammagenic filamentous molds throughout the home and schools.** (*Editor's note*: ERMI won't quantify the total burden of fungi in an environment; it will tell you qualitatively which fungi are in the area sampled.) I'm aware that in 2006, there were a few physicians who didn't agree that exposure to the interior environment of water-damaged buildings could make people sick, even though the big CDC report from the Institute of Medicine was published in 2004, and the much more comprehensive report from the EPA/University of Connecticut came out later that same year.

Just look at what resulted from physicians' attitudes (which basically could be called *believing, but having no data, there was no illness*) did to Pascal, Andre, Marcel and Susan.

Because a cluster of students in the area (at least five others) were ill, the Public Health authorities were brought in. They worried about radon in the basement but didn't bother with mold there. As my later conversations with this agency showed, there wasn't much interest in pursuing the possibility that there was a public health issue.

Andre

Andre's case showed just why the Belperron family was so focused on Lyme. As his illness began to emerge in 2004 and into 2005, and his older brother Pascal deteriorated, Mom and younger brother Marcel are showing signs of an unusual illness. A Lyme test on Andre sent to Igenex in October 2005 was positive. I am not going to argue with others about Igenex testing. I don't use the test; others do.

One month earlier, in September 2005, Andre has an unusual illness, with fever headaches, nausea, and stiff neck. He's evaluated at Northeast Hospital Emergency Room and a lumbar puncture is performed. The spinal tap shows no initial abnormalities. When the spinal fluid comes up positive for an unidentified enterovirus (a notoriously false positive test) and is negative for Lyme, they call his illness viral. Still, no one can fault this diagnosis. Antibiotics are prescribed. Let's not believe that non-specific, polyclonal assays for viral antigens are truly accurate in spinal fluid. But who can argue with a lab, right? But if the illness is confirmed by a lab to be viral, what are the antibiotics going to do? Logically, if the docs believed the enterovirus test, then no antibiotics would be used.

Two weeks after the spinal tap, Andre is seen by a pediatric infectious disease doc. No one reading his note would know that what the doc said that made Andre so angry that he walked out of the office. Another infectious disease consultant pondered the illness carefully, deciding that a virus was at fault.

So the backdrop was set for Mom and Dad: two kids with diagnoses of "I don't know, maybe a virus" and another with "a virus, definitely not Lyme." Now a physician who listens to them and has earned their trust says he has Lyme. IV antibiotics are prescribed followed by a series of other antibiotics.

Meanwhile, back at the MRI machine, Andre is found to have a small, ill-defined mass in the left side of the brain. As time passed, this lesion is shown to be a stable structure in the left hippocampus, but now, Pascal has a fever and is in the Emergency Room with presumed Lyme disease.

If Susan weren't now feeling so bad herself, she might've been able to deal with her concern about her children better, but the brain mass was an evil blow. And then, Andre is sent to evaluate a prolonged QT interval on his EKG. He didn't have the problem. (*Author's note:* this same scenario would be repeated in 2009 as Mass General worried about an EKG problem the boys didn't have. In case someone wonders about the value of reviewing old medical records compulsively: here it is).

Because of the concern about the brain lesion, the Lyme, the headaches and the general absence of a diagnosis for anyone in the family, the same Children's Hospital neurologist who was seeing Pascal performed a spinal tap *again* on Andre in December 2005. The Lyme work-up on the fluid was negative, as was the evaluation for cancer cells. Andre is also referred to Dr. Woolf; imagine the parents' dismay.

Once again, the Environmental Clinic doesn't even record any data on the home or school environmental conditions or ask about mold.

Another evaluation at the Lahey Clinic in January of 2006 ends with the prophetic words, "*A neurotoxin seems unlikely* to explain Andre's difficulties." We may wonder, what was the basis for this broad unreferenced dismissal of the role of neurotoxins? Does the term, "seems unlikely," fit into a rigorous differential diagnosis process? Or is it Ass^2? Because it was dead wrong. I previously called the combined work-up of the Massachusetts and Boston physicians pathetic. This part is worse.

In March 2006, Andre has neurocognitive testing. He does "well," with functioning noted "well above average." No one talked about Andre's tics and pustules and cognitive decline. When the neuropsychologist reviewed the results with André, oblivious to his cognitive state before his illness, she happily reported that he was "college material" and that his math aptitude was in the 60th percentile. Remember, this young man took the SAT once (he was too sick when he was scheduled later on), scored a 780 on the math portion and knew which questions he got wrong. When the neuropsych gave him the "good news," André asked, "But what happened to the rest of me?" Sixtieth percentile might be fine, but not for someone who was 99th percentile a few months before the illness began.

What would you do if everything you've been told was basically wrong and everything you'd done for your two oldest children simply resulted in their getting worse? And then, Marcel was ill, too.

Marcel

After Pascal saw the same resident neurologist at Children's, in 2004, so did Marcel for problems with focusing and scanning. He'd also developed asthma and teachers were telling him he was lazy. Following his bout with mononucleosis, strange discolorations began to appear on his body. No testing was done, and a simple follow-up in six months was requested. No mention was made of the positive ANA with a speckled pattern from 2003. No mention was made of the Lyme Western blot from 2003 showing multiple bands.

Marcel's tics were becoming just as common as Pascal's, with Andre not far behind. Marcel's story is yet to come.

Susan

In the four years I've known Susan, I've always wondered about her favorite piece of chicken. When I was growing up, my mother always wanted the back so her three kids would get a leg, thigh or piece of white meat. It took me a while growing up to figure out that her favorite piece wasn't the back, but I'm sure if given the choice, Susan would request the chicken back so that Marcel or Andre or Pascal would have a meatier piece to eat.

Susan wondered about neurotoxins, despite the clear decision from Lahey Clinic that *neurotoxins had nothing to do with the illness that affected the entire family*. Yet when she's seen with an extremely similar syndrome as her sons, the Lahey physician changes his mind and sends her to a series of agencies and specialists. The results of such investigations are no different for Susan than they were for her children.

She was seemingly "well" until January 2006. Prior to this, she'd suffered a life-long battle with various respiratory problems - asthma, pneumonia on multiple occasions, and other more common pulmonary complaints that somehow didn't respond to normal courses of treatment. She had the near-sudden onset of multiple health symptoms from multiple body systems. Cognitive effects, abdominal effects, joint symptoms, pain, chronic fatigue, clumsiness - you name it, she had it. The PT Barnum Lyme test came back, is called "positive," and antibiotics begin. (*Author's note:* there was nothing to support anything that could be called a positive Lyme test here.) No wonder she and the entire family become fixated on Lyme as the diagnosis for all their ills. *Nope.* Maybe the time will come that people stop looking at antibody testing in Lyme and instead look at the genomics and proteomics testing that we now know actually shows us the way. But 2006 was light years ago in our knowledge of what goes on in illnesses like the Belperrons'.

Susan was seen by Dr. Rose Goldman, MD, MPH, Chief of Occupational Medicine at Cambridge City Hospital, another Harvard Teaching Hospital, in February 2006. Nowhere in the three-page note was there a remote consideration of exposure to mold. Her discussion of tics basically amounts to a dismissal of any organic grounds: Susan is labeled as a faker. The only bright spot in this visit was the surprise appearance of Dr. Michael Lappi, first introduced to the Belperrons at the Environmental Clinic at Children's Hospital. He sat in on some of the evaluations as he was studying Public Health at Harvard while on leave from the U.S. Navy. When speaking with Susan privately after Dr. Goldman had left, he advised Susan that he'd seen her body tics when she'd brought the boys to Children's, some months before she was even aware this was happening. Then he spoke to her about her contention that she thought there might be some mycoplasma or other microbe co-infection along with a tick-borne illness. While he wouldn't divulge things he said he knew from his work with the Navy, he told Susan that she was on the right track and to keep pushing. If there was one thing Susan knew to do, it was to keep pushing.

Let it not be lost on the reader that Dr. Goldman is at **Harvard** and is the Chief of Occupational Medicine. Who could argue with her? Simple: anyone with data. Dr. Goldman didn't have any.

But not so fast. Here we see the basic problem with the entire field of litigation with respect to mold illness. Let those with tons of data argue with God-like anointed Chiefs with no data and who will a jury believe? The God-like Chief has an advantage over "lowly" practicing physicians, where image is everything and data is just too much (scientific) information. The song-and-dance scene in "Chicago," where Richard Gere taps a razzle-dazzle to win the case for his client isn't a bad analogy to use for the Dr. Chief's of the world in court in mold cases, even now.

Let me say that I don't see Dr. Goldman's name involved with mold illness testimony, but I do see a few other Occupational Medicine docs from Harvard-associated hospitals plying their litigation trade, endorsed indirectly by the Harvard name. At least, and to her credit, Dr. Goldman said she didn't have a clue what was wrong: "I can offer no specific environmental etiology (cause) to her symptoms." The others testify under oath that exposure to moldy homes and schools won't cause illness.

As a rule, OccMeds are well advised (you heard it elsewhere before I say it here) to start telling the truth when under oath that they have no knowledge of any aspect of diagnosing and treating mold illness.

And we know that such expert testimony otherwise is based on fantasy masquerading as science and has no basis in any fact.

It would've been so simple for Dr. Goldman to ask for an ERMI test. Imagine the months of suffering she could've prevented.

All her children were affected, and then Susan starts with tics. What would you do? Susan started to search the Internet. Who deals with issues like this? I guess that's where I come in.

The Belperrons are typical patients in my practice, though most don't have the detailed documentation of "care" they received from Boston medicine.

July 20, 2006 Pocomoke, Maryland

Doing new patient work-ups for entire families is grueling. No one else is scheduled for the all-day affair. Patients are asked to bring things to do to occupy themselves while they are "in-between" parts of the evaluation. In this case, the Belperrons were ill before they left Ipswich, and after a nine-hour drive, perhaps a bit antsy to do something other than sit around and answer countless questions, take VCS tests, have a cup of blood drawn and then to listen to opinions on four new patients. But overall, the day went well.

VCS was diffusely positive, though not in Pascal. Symptoms rosters were typical of those seen in biotoxin illness patients. Magnetic resonance spectroscopy of the brain showed the expected reduction of oxygen delivery (capillary hypoperfusion; See *Follow That Oxygen*), and abnormalities in neuronal function with reduction of the ratio of glutamate to glutamine.

The genetic testing results weren't surprising, nor the results of inflammatory markers. C4a was sky high all around in the Belperron's lab results.

Because of the epidemiologic consistency of all the Belperrons with thousands of other cases, their clearly defined potential for exposure as shown by the presence of visible mold and absence of any known confounders, including Lyme that could do all these things, the family was started on CSM with nearly immediate improvement in all four.

After they arrived home, ERMI testing was first. They all took CSM and symptomatic improvement was noted rapidly. That one ray of hope didn't take the sting out of the ERMI result of more than 14, a dreadfully high number. Their home had massive contamination with *Aspergillus penicilloides, Aspergillus versicolor, Stachybotrys, Trichoderma* and *Wallemia and* they hadn't even sampled the basement.

Finally, the house was on trial. Verdict? Guilty of causation of illness. The sense of relief that the Belperrons had now that a diagnosis was confirmed lasted about a minute. This was the family home that Pierre had remodeled with his bare hands; imagine the flood of emotions this news brought.

How would you feel if you had to leave your family home on short notice, leaving everything behind? All because of an enemy that had been invisible before, but that you could see everywhere now. And once the Belperrons could see what the mold and all its microbe fellow travelers could do without any consideration of mercy, there was an intimidating aspect to their lives that hadn't existed before. They hadn't known what hurt them so badly for four years. Now they did, and it felt like they were surrounded by mold castles.

In terms that Margaret Mead might use, survival from the elements was now their basic level of societal organization. Shelter from the storm had to be shelter without the pathogens.

The questions cascaded like Niagara Falls: where can we go? What about the kids' schools? And then the questions became more poignant, as the family had to decide what to take and what could *not* come with them because it was or might be contaminated. What about sports equipment and awards? Musical instruments and prized sheet music? Every family faced with a mold evacuation asks the same general questions but no one has answers that can heal the fear of more illness. Beds? *Trash.* Bedding? *Trash.* Most everything else? *Trash.* Everything was discarded—furniture, linens, bedding, appliances, electronics. Sheet music, CDs, books, toys, and every prized possession. Only family photos and official documents are saved, and they, double wrapped in plastic, were hidden away in the garage attic, along with Pierre's childhood cello.

Susan and Pierre quickly found a safe place to live. "Buy new" is a brutal policy when no new money is coming in and the bills for everything start piling up fast. However, when faced with the choice of health over belongings, choosing health was easy.

By early 2007, the Belperrons may have thought that their worst nightmares were all behind them. Some certainly were. They finally had a medical diagnosis and treatment. They'd sacrificed their home of twenty years to move into a safe haven. All were feeling somewhat better, and then reality set in. With the knowledge they'd amassed, they had to figure out how to live with a mold illness in the larger world.

What they didn't realize was that the hurdles that lay ahead were even higher than the ones they'd already passed.

Andre was feeling better, though he was still not quite right. He'd enrolled as a freshman at Drexel University in Philadelphia, where he became worse early on. The Belperrons had calculated the risks from WDB that André would face at school. André withdrew, trying to regain his strength, and in the interim worked on an accommodation plan with Drexel for his return later in 2007. He also tried to take classes at the local community college. André was quickly divining the protocols of living with mold, and he realized that both campuses of North Shore Community College had water damage that made him sick. After countless hours of dealing with the Disability Services Office and various Deans and administrators at Drexel, it was clear they didn't want André. His disability was inconvenient for them, or worse, might open a can of worms they could never close. Beyond agreeing to ERMI test his dorm room, no other substantial accommodations were offered, and André was told he could simply wear a respirator, even in the cafeteria.

André then conferred with the University of Massachusetts, where he'd been previously accepted, and their Disability Services Office assured him that they could provide all the necessary accommodations. André entered UMass in January 2008.

Meanwhile, Pascal and Marcel had gone back to school after being out of their home for six weeks and out of school for nearly a month. They were feeling great.

In retrospect, the Belperrons realized that they should've known better. A few weeks after their return to school, Susan walked into Marcel's elementary school, and within minutes she was experiencing a recurrence of symptoms. A few days before she'd watched—their new home bordered the back of the school—as a March torrent had flooded the area behind the school. This certainly wasn't the first time. *Oooops, water damaged building*. She withdrew Marcel from the Winthrop Elementary School and he'd never return. Pascal, meanwhile, was a student at the Ipswich Middle High School, and he came home after the same storm and said that there'd been dumpsters outside the school and that carpet and building materials were being thrown out of the second floor. *Sigh: Water-damaged building.* Pascal was withdrawn from that school as well.

They should've never let Pascal enroll in the high school, either. Though opened only in 2000, it was a disaster from its first days. Pipes froze and burst, roofs leaked and waste water holding tanks backed up. To this day, in 2010, more shoddy construction is being revealed, this time due to a Nor'easter that tore away a portion of the roof flashing, only to reveal that significant sections of rain and ice barrier were never installed. It was the sight of the dumpsters and the water-damaged materials that jolted the Belperrons to take decisive action.

Marcel's elementary school, while not the perpetual disaster of Pascal's school, was old (Susan had gone to school there), flat topped, and at the foot of a hill. They later recalled that when Marcel was younger, he got sick every week by Friday, and had to recover over the weekend, but then he'd be better just in time to start school on Monday, and was always better during the summer than when he returned to school each fall. It was attributed to allergies, and Marcel ended up taking Thursdays off since Fridays were deemed more important. Allergies, huh? Pascal and Marcel finished out the remaining months of school at home.

That summer, the Belperrons tried in earnest to get an educational plan —a 504 plan—in place for each son under the ADA. Pascal's plan went together fairly easily. The administrator was the Assistant Principal with whom they had a good relationship. Remember, this is a small town and they had a first-name relationship with virtually everyone in the school system. Pascal was to stay at home, have a marginal amount of in-home teaching by a tutor, and the balance of his classes he could take on-line. He was a senior in high school, and this seemed to be a reasonable way to finish out his time. In fact, although he was an excellent student, he didn't do well when confined to remote instruction and its inherent isolation, but he kept his gaze forward, to bigger and better things.

For Marcel, things were much more complicated right off the bat. They first visited the Middle School principal, Cheryl Forster-Cahill, and other staff in May 2007, trying to establish protocols to have Marcel taught in their home. Though there were lots of meetings, the Ipswich Public Schools didn't have Marcel's education plan in place that year until November. Later, a school department employee who was in on the internal meetings at the school told the Belperrons that the accommodations offered to Marcel by his elementary school drew outrage (from whom?) because they were establishing a precedent. When the Belperrons tried to have a 504 Accommodation Plan drawn— they learned this was the only instrument to protect the civil rights of their sons—they found the school system protecting its interest by leaving the 504 plan vague and without any substance or detail (see Chapter 20, Fighting For Our Lives, by Lee Thomassen).

Finally, a meeting was convened, including lawyers, to get this plan written. After more than five hours, with Susan on speakerphone because she couldn't enter the school, the best the Belperrons could achieve was to have one substantial sentence added to the 504 document: "Marcel cannot enter the Ipswich Middle School to access his education." A five-page document the Belperrons had drafted had been reduced to less than one, and none of the tangential agreements that were discussed ever materialized.

Ultimately, all the Belperrons, including Marcel, were treated with contempt by the school system: they were lied to; they failed to provide Marcel with an appropriate education; and they made developing the protocols as arduous as possible, seemingly to drive them from the system. The most egregious hoodwinking came when, after an extensive discussion with Dr. Shoemaker, Middle School principal Ms. Forster-Cahill asked to have an ERMI test done to disprove the notion that the school was moldy. She had the test run by Mycometrics, which meant that Dr. Shoemaker had a copy as soon as the test was done (and therefore the Belperrons had it, too), but she refused to give the results to the Belperrons. Although she had the test results by early November 2007, it wasn't until the Belperrons filed a "freedom of information" demand in late December that the received the results. We could see why. One of the rooms tested had astronomically high mold loads, with a composite of more than 60. When confronted regarding these results, Superintendent Rick Korb responded by saying, "Mrs. Belperron is reading the results differently" and went on to hide behind the fact that there are no Federal or state guidelines regarding mold. The Belperrons were getting a taste of schooling with mold and living with mold cover-up.

It happens every day. School systems often put teachers, students and staff needlessly at risk. So what if 25% of attendees are going to suffer physically and intellectually.

Pascal got some hard lessons in 2008. Having been a sailor and motor boater from an early age - he and André bought a Boston Whaler when Pascal was 13 - and having worked at the local boat yard about as long, Pascal dreamed of a life at sea. He applied to the U.S. Merchant Marine Academy and received not one, but three congressional nominations. He passed the physical fitness test with no problem. He was accepted by the Academy!

The joy was short-lived, though, when the Department of Defense Medical Evaluation Review Board (DoDMERB) rejected him. In their eyes, his medical history made him too much of a risk. This wasn't completely unexpected, and Pascal moved to Plan B: the Massachusetts Maritime Academy, which was not as rigorous, but he'd receive the same basic training and licensing. Being responsible, Pascal notified the MMA of his disability early on. Essentially, he advised them that they'd need to ERMI test a dorm, academic buildings used by marine engineering majors (this is a small school, so it was only one or two buildings), and their ship. If the ERMI tests failed, Pascal understood that he couldn't attend, and he wasn't about to ask them to fix anything. In June, he was summoned to the campus to meet with the Dean and campus physician. They tried to talk him out of attending the MMA for fear the Coast Guard wouldn't grant licensure because of his condition, and without a CG license he couldn't graduate, unless he switched majors to a degree not requiring a license. Pascal was sent home, crushed, to think about his decision. Then, a letter came that rendered a decision for him. The Dean sent back his deposit, and advised Pascal that his acceptance at MMA had been rescinded.

Rescinded? Do you think the MMA was afraid of something, afraid enough to violate someone's civil rights? The Belperrons, though not litigious, were mad enough that they contacted a civil rights attorney. When they did, they were told that while Pascal's rights had absolutely been violated under the provisions of the ADA, that bringing suit against the MMA was futile, because the MMA would be defended by the Attorney General of the Commonwealth of Massachusetts, whose unlimited funds would allow them to defer and delay until the Belperrons tired or had given all their savings, in vain, to an attorney in the hopes of prosecuting this fight. As a last resort, and with his entire future turned on its head, Pascal contacted the University of Massachusetts and took a place there. Pascal was dealt a cruel hand, and learned about living with mold.

With André and Pascal both at UMass, it would've been delightful to write that UMass had delivered as promised on the accommodations. It was on this supposition that Pascal had decided to follow his brother there. But after sharing a bedroom for 18 years, it's not like they needed to be spending more time together. Some mistakes in protocol were made in André's first semester. The Belperrons kept calm and stayed the course, believing that the Disability Services Office had André's best interest at heart and the competence to correct their mistakes going forward. Those were pretty generous assumptions. André's second semester was handled worse than his first, and as a result, he was exposed to very moldy buildings and got quite ill. Susan and Pierre were at the Amherst campus, a hundred miles from Ipswich, the next day, and somehow managed to put together a meeting in which about a dozen faculty and administrators gathered to discuss André as well as Pascal's forthcoming arrival. They were told that everything necessary would be done to test buildings, have contingencies, notify André in the event of water damage and so on.

The sad truth is that UMass did very little to help André and Pascal. ERMI testing is a double-edged sword. It tells you where you might be safe, but it also serves as an interdiction to entry. In the case of UMass, it was more the latter. André is a Landscape Architecture (LA) major. He cannot enter the LA building because it's suffered too much severe water damage. André and Pascal live in a new dorm complex, which so far tests OK. If there's a water leak, though, where will they go?

Last year, a sewage failure and the resulting mold "explosion" in the one cafeteria on campus that had a normal ERMI prior to this failure resulted in the Belperron boys having no food service, and now they cook all their own meals. During the spring semester of 2010, none of the classrooms that Pascal had scheduled had safe ERMI levels, so he's had to attend each and every course via Skype. Ask Pascal what he thinks about paying full tuition and fees, and then having to attend through the portal of a lousy video camera and barely audible audio. Most ironically, neither Belperron can get any health services or blood testing at UMass. Why? The Health Services Building has severe flood damage and astronomic ERMI values!

During their matriculation, some accommodations have been achieved, but at a great expense of time—mostly Susan's—and emotional energy. André and Pascal have endured contemptuous behavior of the staff, administrators (including one Dean whose offer of help was to assist them in transferring out of UMass) and even the Trustees, who, to a person, refused to even acknowledge receipt of a letter that their parents had written. The Belperrons have been subjected to yelling, lies, circular pass-offs, being hung-up on and even threats. There have been some brave souls who have stepped up to help them, including Dr. Elizabeth Brabec in the Landscape Architecture Department, who has advocated some for André, and John Dubach, CIO and Special Assistant to the Chancellor, who talked about helping early on.

In an effort to further the cause of their own children, plus champion their concerns regarding water damaged buildings and their effects on health, Susan and Pierre started contacting local, state and federal agencies for help. They've spoken to or corresponded with the Ipswich Board of Selectmen; the Ipswich Public School Committee; the Massachusetts Department of Public Health; Executive Office for Education; Center for Disease Control; the Department of Education's Office on Disability, Office Against Discrimination and Department of Higher Education; the Attorney General; the Federal Department of Education Office for Civil Rights; and the list goes on. In most cases, we aren't talking about intake level staff here - Susan persisted in getting her message to the highest levels at every office. When frustrated at these bureaucracies, they also contacted Massachusetts Representative Brad Hill; State Senator Bruce Tarr (both of whom at least took the time to come to their home and lobby locally on their behalf); Congressman John Tierney (no interest in helping); Senator Ted Kennedy; Senator John Kerry (whose staffers at times seem to want to help, but can't seem to get anything done); and newly elected junior Senator Scott Brown. What the Belperrons learned is that many of these agencies don't really want to confront school systems, preferring to make inquiries, and then finding in favor of the school system. Other agencies cited that they have no jurisdiction or purview. One kindhearted state agency head told Susan that she was right, but he felt "powerless and unable to help" her.

Meanwhile, work was going on locally, trying to bring awareness about mold into the schools. After sending detailed information to the Ipswich School Committee, including the ERMI test results in 2007, Pierre was asked to testify at a School Committee meeting. Pierre raced to the meeting after work, the evening commute taking nearly 90 minutes. As he started to address the school committee, his phone started to vibrate. It was Susan (remember, she can't go into those school buildings), which Pierre found to be strange, for she knew what he was doing and was watching him on local access TV. In fact, no one goes to committee meetings anymore in Ipswich; they just watch the proceedings at home on TV.

When Pierre got home, Susan was enraged. So was Pierre when he found out why: when he started to testify, the TV feed went dead, and came back as soon as he was done speaking. Coincidence? Even more galling has been the lack of any follow through from the school committee on mold in the school and health effects.

A year later, Pierre was at a wedding where he ran into a fellow musician. That musician told him something that blew his mind. She told Pierre that she'd been playing in the pit at the high school musical production, but had taken the evening off to attend the wedding. Not to worry, she told the director, she'd engaged a substitute. Later, the orchestra director called her, concerned about who the sub was. "It's not Mr. Belperron, is it? We can't have him here at the school," the director said. Pierre wasn't the sub, but that comment shocked Pierre and explained something he hadn't previously understood. He'd been adjunct faculty at the Ipswich High School for years, and his were always the best cello students in the system, and for years he'd always had a student of his be the principle cellist in the regional youth orchestra. As a result, he had lots of students, and usually a waiting list. Over the past couple of years, since the Belperrons started making noise about mold in the schools, no new students came his way, and last year when he graduated three seniors—again, the best three at the school—there were no more students to take their place, and he was out of a teaching job at the school. At one time, he thought his disappearing student pool was coincidence, but with this revelation from his friend, Pierre was also learning another facet of living with mold: the world wants you to keep your mouth shut.

Treatment advances come to Ipswich

Meanwhile, CSM alone was not adequate to correct all the illness symptoms. No one should ever assume that it will. As time passed, the boys and Susan used all the treatments I had. The biofilm-forming organisms were eradicated. The abnormal cytokine responses corrected. Dehydration was fixed, and C3a was normalized. The approval by the Ethics Committee (Institution Review Board, IRB) for the erythropoietin study provided a ray of new hope. (Author's note: human research should be approved by IRB agencies.)

We already knew there were genetic linkages that the Belperrons' had that weren't common. They all had "stuff" in their urine which was due to a markedly increased 24-hour urine excretion of uric acid. They all had unusual clotting profiles but only one (Marcel) had the lupus anticoagulant, though he doesn't have lupus. They all had somewhat unusual EKG, with a QT interval of around 0.4 seconds. They don't have the long QT interval syndrome but why they all have that problem isn't clear. Susan, Andre and Pascal all had a *marked elevation* of pulmonary artery pressure with exercise.

They came down to Pocomoke for the two-week erythropoietin trial. What they received for their troubles was dramatic improvement. Their abnormal MR spectroscopy scans resolved to normal within two weeks. Their symptoms disappeared as did the tics. Andre's pustular acneiform eruptions, ones that no antibiotic or Accutane had helped, suddenly improved. And of course, as we now know, the C4a elevation melted away. There were no side effects of the drug and no adverse events.

Erythropoietin is expensive so the time they enjoyed freedom, released from the tiger cages of relentless and merciless inflammatory attacks, was limited. They were not able to obtain insurance coverage for the only drug that stopped the illness cold.

The illness slowly returned, as we had not turned off MASP2. Erythropoietin can't do that. What would be next?

In January of 2010, the VIP clinical trial received IRB approval. Here came the Belperrons. If the VIP stopped MASP2 (it corrects so many things, see Chapter 16), couldn't they have a life again? Marcel was too young for the trial as he was for the erythropoietin trial. The results of this trial were simply spectacular. The best finding was the correction of the pulmonary artery pressure but seeing the few tics of Pascal again disappear was nice, too. The drug was tolerated well and there were no side effects or adverse effects.

Is it tempting to say forget it? Not for the Belperrons

There's been a certain futility in the Belperron's efforts. They have something of an education in place for each son, but it's certainly not what each of them deserves. Even though they're now my patients, they have continued to knock at the doors of the hallowed halls of Boston medicine, trying to get local care for their ongoing health problems, and while they've found some sensible and empathetic souls, mostly they've been handed more bizarre and inept care and fights with their insurance carrier. They've learned that bureaucrats seldom have a sincere desire to help, and often times disappear, only to be replaced by another who's equally clueless and inept. There was one, Drew O'Brien, a head staffer at John Kerry's Boston office, who seemed to understand. Once he listened to their story (only after Pierre and Susan camped out in his office until he'd meet with them), he saw the implications: their plight involved a convergence of health, education, civil rights, public accommodation and health insurance. While he was insightful, he unfortunately has failed to deliver any assistance to the Belperrons other than another push to get on the endlessly revolving bureaucratic merry-go-round.

You might think this would drive them to despair and to give up. Remember that Rottweiler analogy? These guys don't give up. And, much as they kept going through physicians until they found their way to Pocomoke, they'll keep hounding local, state and federal bureaucrats and elected officials until someone steps up to protect their sons' rights and works to bring accountability in the Ipswich and UMass systems with regards to the condition of their buildings and the effects those buildings have on the students who occupy them.

Living with mold has cast an omnipresent shadow over the Belperrons' lives. There were some obvious changes in behavior that they conceded: most friends' homes are off limits. They go to only one movie theater. There are few stores that don't make them sick. There are only a few buildings on the UMass Amherst campus confirmed to have safe ERMI levels. Then there are the less ordinary limitations. Voting now requires an absentee ballot. Pascal contested a traffic violation, but decided not to pursue the matter when he saw the water-damaged ceiling tiles in the courthouse (See Chapter 21; and Pascal, stay out of the Federal courthouse in DC!). What happens when jury duty comes around? There are protracted "negotiations" with their medical insurance provider, because they can't find care in Massachusetts, and their insurance doesn't want to pay for their care in Maryland. At one point, an insurance case worker advised them that Pierre ought to abandon his employer of 21 years and get a job elsewhere that would have insurance more to his liking. One physician, in talking to Susan about their plight, suggested they should move to Maryland.

These insults and deep wounds have taken their toll in other, less direct ways. With the inflammatory damage they've suffered, they also may be more susceptible to other biotoxins. This was first noted in Marcel, who had severe onset of symptoms on a couple of occasions after spending a day at the beach, the correlation being too strong to ignore. Others noticed it, too, and Susan discovered that there had been significant algal blooms in their area. (*Editor's note*: although the local expert in the toxin-forming *Alexandrium* blooms is on the faculty at UMass, she didn't respond when Dr. Shoemaker emailed a request for a discussion.) So this family, which once spent virtually every summer day at the beach or in their boats, now never goes to the beach, except perhaps for a walk, and they're selling their boats.

Lastly, they've learned to be ostracized. Some people and institutions treat them like criminals. Others, being a bit more polite, just keep their distance, or walk away when they start to talk about water-damaged buildings.

So what should the Belperrons do about mold illness and life in the 21st Century in Massachusetts? The Office on Disability wanted a letter from me as its basis for intervention:

> Dr. Myra Berloff
>
> Director of the Office on Disability
>
> State of Massachusetts
>
> 12/15/09
>
> Dear Dr. Berloff,
>
> I'm writing to you at the request of the Belperron family of Ipswich, Massachusetts. They are seeking your assistance in obtaining accommodations for their two sons, Andre, the oldest, and Pascal, the middle son, such that each will be able to continue their studies at the University of Massachusetts in Amherst. I'm hopeful that the approach you will take to solving the problems these two young men face will provide a precedent so that when Marcel, the youngest son, now being home-schooled, will also be able to attend U Mass.
>
> The problem that each of these three young men face is that they become ill rapidly if they are exposed to biologically produced toxins and inflammagens, substances invariably found within the interior environment of buildings with water damage and

microbial growth. We know a great deal about the physiology of their illness, based on an ever-expanding array of studies done on animals and people alike, which unveil the various aspects of inflammation seen in the Belperrons' illnesses. The current mold literature is vast, with studies published all over the world, each year bringing a near-exponential increase in knowledge. In fact, trying to stay current with the basic science is a full-time task. We know a great deal more about the physiology of their illness based on a database of more than 5,900 patients seen, diagnosed and treated at my office beginning in 1998. Our group has published multiple peer-reviewed academic papers, each expanding the information base that supports clinical decision-making for the Belperrons and others.

The application of the established basic science that supports the documented medical interventions that make a difference clinically remains dependent on two simple concepts:

(1.) We must prevent re-exposure. In the end, all the new knowledge about basic science and all the old knowledge that we have about the physiology regarding what happens hyperacutely in re-exposure, beginning about five minutes after exposure in sensitized individuals, falls to no importance if the patient is "hit." Every hit counts; every hit adds to the inflammatory burden leading to what we call, "sicker, quicker." The physiology here is clear, with an enhanced rate of production of C4a and TGF beta-1 to higher levels faster with each subsequent exposure. Our request for you then, is to help the Belperrons avoid these "hits."

(2.) The second aspect of our requested accommodations is (i) to provide adequate facilities to document illness as it occurs; conversely, (ii) documentation of absence of illness as avoidance of re-exposure is confirmed; and hopefully we won't need this aspect, (iii) facilitation of early therapy if illness is reacquired.

We know that the lab testing involved requires kinds of blood drawing tubes found in every diagnostic lab in the U.S., and the tests require adherence to some simple instructions. The tubes for some tests must be chilled; for others, the tubes must be spun within five minutes of draw, and for others, an additive must be placed into the tube within two minutes. Therefore, the specimen processing is more complex than simply drawing a CBC and a metabolic profile. We provide instructions, written in English, with specimen requirements, lab codes, diagnostic codes and specimen handling instructions and yet, the likelihood of error in specimen processing at distant labs remains distressingly high. We've seen this problem, one so easily avoided, on several occasions in the Belperrons.

Let's add one more layer to the blood drawing problem: if the site is located in a water-damaged building, as it is in Student Health at UMass, then Andre and Pascal would have to travel more than 70 miles to access proper (safe) lab facilities. After much ado, a Quest lab 50 miles away agreed to draw the blood, but only on one occasion, as the lab failed to do their work properly and wouldn't agree to correct their errors.

One might ask, couldn't the Belperron boys simply react to the onset of symptoms by leaving the exposure? *No*. We know that symptoms are actually a late developing aspect to the illness, in some instances taking two to three days to appear following exposure. Couldn't they just use the on-line VCS tests from **www.chronicneurotoxins.com** or **www.biotoxin.info**? No, the VCS test won't change for a day or so. Given that the boys need to be able to confirm their innate immune responses are going into overdrive following re-exposure on very short notice, access to labs that can do the proper testing <u>must be available</u> on a day's notice.

Returning to the main problem, that being assessment of a building index of contamination, *assessed before* the boys occupy a school classroom or residential space; some simple accommodations are indicated. Given that the Belperrons will be asked to attend classes, meetings and gatherings in buildings, it's not unreasonable that a building history be developed (preferably using multiple sources of data, especially including work orders generated from maintenance records of requests for intervention for moisture-related events, including changing ceiling tiles, re-painting and much more), with such information made available on an Internet site, accessible before either of the Belperrons were asked to report for class. The model I'm suggesting was one instituted for another of my patients who attended the University of North Carolina. As it turned out, there were 28 buildings identified at that campus that had water intrusion, either ongoing or by history, without confirmation of safety using fungal DNA testing.

While this next suggestion seems straightforward, notifying the Belperrons of new water intrusion events, similar to an "Amber Alert," should be routine. A posting on a public website would be simple enough. Instead of watching the outdoor Weather channel, the Belperrons could simply access the "Indoor Weather Channel." It would follow that documentation of action taken, with an assessment of efficacy of that action, would also become public knowledge. As an aside, a parallel example of what I'm proposing is obtained from observing the approach of several Mid-Atlantic States to the potential problem of exposure to toxin-forming dinoflagellates. Maryland decided to post on their DNR website the location and risk from exposure to blooms of dinoflagellates, while Delaware did not, even though the same environmental conditions affect both states. The results? Maryland has seen a plummeting health problem from such dinoflagellates, while Delaware, choosing to keep constituents in the dark, suffers an ongoing human health problem. You might want to review the Discovery Health Channel program on *Planet Green, Toxic Files* that first aired December 12, 2009, in which one of my patients from Delaware is interviewed in depth. I'll suggest that Massachusetts wouldn't look "good" in national media eyes for withholding pertinent health data as Delaware did. One can only imagine the headlines: "U Mass fails to inform straight A students at risk for serious illness caused by their own classrooms" would have on staff and student recruitment, alumni donations and the like.

One would have to be naïve to think such a result *wouldn't* occur if UMass used the Delaware model of public information.

The environmental testing in items 1 and 2 above should be the standard for assessment: the environmental relative Mold index (ERMI), as has been used extensively by UMass to date, per my request. My concern is that UMass hasn't acted consistently to protect its students, including the Belperrons, from exposure to high ERMI areas.

I need to be clear with you: I'm not a mycologist. I'm a treating physician. Any environmental data should be discussed with an expert mycologist, including Dr. Lin at Mycometrics in New Jersey. Failure to provide the results of such consultation as public information about health risks wouldn't be either logical or ethical, in my opinion. I'm aware that failure to disclose inside knowledge is regarded as an actionable offense in stock trading, financial transactions, sports deals and political interactions, to name a few. Transparency and openness are not buzzwords: they're the standard of expected behavior.

Again, to perhaps re-state an obvious point, one must expect that evaluation of action taken, with publication, will be the standard of both medical and building intervention. If UMass "fixes" a water-damaged building, let us all see what the ERMI was before and after intervention. If human illness recurs, let us have another ERMI assessment. If the building is cleaned properly after remediation, illness will not recur. If the building is not remediated and then cleaned properly, illness *will* recur. The demonstration of illness is not based on symptoms alone; objective parameters are readily available to all observers to assess the physiologic effects of re-exposure.

The next point is one I suspect you've already noted. If the building has a history of water intrusion, be it last night, last year or in 1930, and there's no evidence of remediation (even assuming that any remediation was effective), then all students at risk, based on prior history, including but not limited to the Belperrons, *may not attend classes in that building.* Once UMass knows what's wrong with the Belperrons—and based on my notes of prior communications, they do— then they may not act as if the potential for other students to be affected is a new idea.

As the science of building health assessment proceeds at an exponential pace compared to just three years ago, I reserve the right to amend these recommendations to upgrade building assessments in the future, based on peer-reviewed publications.

Let us turn to human health, as I think the essential elements of building health, public accountability and disclosure have been addressed. We have biochemical markers that reflect predictable changes caused by short-term exposure—i.e., less than 30 minutes—that show changes in blood measured 240 minutes later. These changes, published in peer-reviewed venues, give us the ability to determine if the health effects were caused by exposure. We should access these data routinely, as discussed in item (2) on page 2 of this letter. The procedure to assess a building of unknown safety is to simply test before and after blood results. The testing involved, C4a and TGF beta-1, are stable; they won't change over time.

I don't think I need to emphasize that the blood drawing facility itself must (i) be without evidence of an elevated ERMI; (ii) must not have evidence of water intrusion; (iii) and must not have visible microbial growth or musty smells. Such documentation may actually be an advertising point for this increasingly "Green Building" world, in that UMass could promote the safety of their structures as a positive aspect of attendance at UMass, as opposed to nearby institutions with really old buildings, for example.

In all the discussions I've had to date with UMass—and there have been many—I've found no single individual who was either told to learn about inflammatory illness caused by exposure to the interior environment of WDBs or who had thorough knowledge of these two patients. I've felt this absence was illogical at best and bizarrely negligent at worst. Does your office expect staff members to be informed of special needs students in state and federally funded institutions? UMass needs a person to become an informed liaison with the Belperrons.

Returning to ERMI, we know that levels between negative 1 and 2 are problematic for the Belperrons in residential areas. Such correlation is incomplete regarding classrooms. If elevated levels of a few fungal species, but not enough to absolutely drive the ERMI away from acceptability, are identified, then we must be concerned about marginal exposures. A reasonable "hedge" against such marginal exposures is to permit the use of portable HEPA filters by both Belperrons in such classrooms.

We must also deal with the issue of cross contamination, as materials that are porous, and especially cellulose-based materials, are carriers of the minute fragments of fungi, bacteria and other microbes that create toxins and inflammagens. Simply stated: the Belperrons may not handle products that have been taken from a known moldy environment. This proscription may seem easy at first, but probably is the most difficult to enforce.

Given the extensive work that's been done to date to ensure that the dormitory accommodations for each of the Belperrons have been guaranteed, and that guarantee has been substantiated, we must consider the possibility that their current "sanctuary" could become violated. Here my concerns are less strident. I'd suggest that there be another site identified that could be used as an emergency relocation site for the Belperrons. If their current site could be verified to be safe, then the back-up site wouldn't be necessary. Having said that, I always save my computer work on two un-linked locations. The concept applies to the living situations for the Belperrons, too.

This next item may apply more to Marcel than to Andre or Pascal. One cannot separate the maturing process of education from the maturing process of social interaction with peers. If the education provided is isolated, such as an on-line or Skype experience would provide, the valuable aspects of peer-to-peer interactions are lost. Such a loss would be a significant setback and should be avoided.

I have a long series of interactions with the Belperron family. They have suffered an egregious series of mis-diagnoses and possibly negligent care from some of Massachusetts' finest physicians and institutions. There's no question about their diagnosis, and now we must focus on long-term management, which means accommodating their environmental illness.

Without accommodations, they are condemned to become ill and likely drop out of school. With accommodations, we have two candidates for magna cum laude.

I suggest you choose accommodation as outlined above."

Sincerely,
Ritchie C. Shoemaker, MD

(*Editor's note*: after the Office of Disability received this letter, they rejected it, demanding that the letter be submitted in a bullet format. After the re-formatted letter was sent, no action was taken. The idea of putting things into bullet format to make it easier to ignore is now a standing joke around Dr. Shoemaker's office.)

Given the problems that this family has suffered, one institution should stand out as a beacon of integrity: Massachusetts General Hospital. Here was another Boston bastion of medical excellence. But such excellence was spotty at best.

Here is Pierre, finally having had enough, writes to Mass General:

Susan and Pierre Belperron
Ipswich, MA 01938

February 20, 2010

Dr. Peter Slavin,
President, Massachusetts General Hospital
55 Fruit Street
Boston, MA 02114

Dear Dr. Slavin,

We are writing to you to recount the circumstances surrounding the medical care our sons have received at Massachusetts General Hospital over the past year. Our sons, André, Pascal and Marcel share a congenital autoimmune inflammatory syndrome that's triggered by exposure to microbial growth in water damaged buildings and other biological neurotoxins. They're being successfully treated for this out of state. However, circumstances arise where we can't throw them in the car and drive 500 miles at every turn.

Last winter, Marcel had a significant return of his symptoms (pains throughout his body, tics, exceptional fatigue, weakness, head pain, etc.) plus a number of new symptoms, including blurred vision, uncontrolled movements/tics in his jaw, a sensation of pressure inside his head and inability to control his eyebrows and lids. After conferring with the physician in Maryland, seeing our PCP and an ophthalmologist, and when Marcel deteriorated significantly on a weekend, we went to the ER at MGH. While nothing was found upon exam and testing, we were counseled by the nice staff that Marcel's chronic condition might be best examined by MGH, which sounded good to us, since we didn't relish having this treated so far away, and with insurance complications as a result. Plus, they said an advantage at MGH was that pediatrics was small and the specialists collegial, which would help with a condition that traversed specialties. Unfortunately, rather than help us solve our medical issues locally, after a year of dealing with MGH, we now have two nightmares: the medical one our kids came with, and then the one created by MGH.

Marcel was seen for a follow-up to his ER visit by neurologist Dr. Schliefer. He was later seen by Dr. Sims, and was later referred to Dr. Browning (and these physicians sent Marcel to a number of other MGH specialists) in genetics, since Marcel shared a condition with his brothers and mother. Dr. Browning seemed sympathetic to Marcel's situation and indeed wanted to see his brothers as well. She also suggested that since Marcel's illness transcended disciplines that the Coordinated Care Clinic would be valuable to our family, and we were eager to have such help. Unfortunately, none of that promise came to fruition, and worse, miscues at MGH added to our considerable burden.

The following are incidents that we have found to be unacceptable, perhaps reprehensible, in our dealing with MGH.

MRI on ER visit states "NEW FOCAL NEUROLOGICAL DEFICIT ASSESS MASS/LESION, BRAIN TUMOR." Marcel doesn't have a mass/lesion/ tumor, never did. His brother, André, does.

- Physicians' notes are LOADED with errors (in as much as we can get copies). If we were to recount just a portion of the errors, it would fill page upon page. Some of the problems stem from the obvious "lifting" of notes from one visit to the next, so that errors are perpetuated. There is also extensive innuendo, alluding to Marcel's demeanor, the veracity of his symptoms, and considerable attribution of physical symptoms to mental disorder.

- Neurology associates seemed bent on attributing Marcel's physical problems to psychiatric disorder. Marcel did work for an extended time with a psychologist outside of MGH, Dr. Greenberg, in 2009 on biofeedback to help him work with his pain. When we asked Dr. Greenberg, who had extensive one-on-one sessions with Marcel over a long period of time, if he thought Marcel had psychiatric problems, he thought the notion was ridiculous.

- Marcel was scheduled to see Dr. Mischa Pless on a referral from our PCP. When we arrived, we were told the appointment had been canceled. We had not been contacted. The desk staff stated they'd left a message on our home phone. Mrs. Belperron advised them we have no home phone, and showed them my cell phone history—no call missed or message left. Then, because I was angry, we were ushered to another office. There I was counseled to walk over to Mass Eye and Ear and see if we could get seen there.

- On December 1, 2009, we brought Marcel to see Dr. Browning for a follow-up on an October visit, only to be told he did not have an appointment.

- Dr. Browning first frightened Marcel with a diagnosis of Wilson's disease, based on low urine copper. When she called for a 24-hour urine test, I (Mr. Belperron) drove to MGH to get the jug. There were no instructions. Having done this before, I asked for them. When I returned the next day and tried to leave the jug at the lab, it was not accepted because there were no physician's orders to go with the jug. I left and went to work. After calling Dr. Browning's staff, I had to drive back in to MGH to leave the jug. We ended up there five times that week for the various tests and appointments. Then we waited and were never apprised of the results, which were negative on Wilson's—until we went in for another appointment a month later.

- Dr. Browning arranged a sleep test for Mrs. Belperron and Marcel in December, because both have had ongoing sleep issues. We have yet to find out the results of those tests, and we have asked. It's been two months!

- During a visit with Dr. Schliefer and Dr. Sims, when we were advised that Dr. Sims had contradicted Dr. Browning's recommendation that Marcel be placed in the Coordinated Care Clinic. When Mr. Belperron tried to address Dr. Sims, she kept interrupting me, over and over. When I pleaded with her to let me finish my question, she smugly bellowed that she didn't need to, that she already knew what I was going to say. This was perhaps the rudest, most obnoxious and condescending affront I have been subjected to. Though blessed with virtually infinite and unflappable patience and calm, I was enraged and left the exam room. Mrs. Belperron noted that Dr. Sims had a grin upon my departure, clearly humored that she could thus rattle me in front of my wife and son. Shame on her, and you, too, if you find that behavior acceptable.

- Marcel was sent for a chest X-ray by a pulmonologist in December, where nothing was found, but Dr. Browning, back in October, felt some reason to send Marcel to see this physician, because he was good with medical puzzles. He didn't solve our puzzle, we never heard what the result of the x-ray was, and had added more incorrect information to the medical record.

- Marcel, Pascal and André saw Dr. Browning on December 23, 2010. She seemed especially harried that day. During that visit, she declared that she had discovered that Marcel and Pascal had Long QT, that Pascal had hypertension, and that all other concerns had to be put on hold until these matters were addressed. This one visit led to a flurry of remarkable mistakes, miscues and unacceptable behavior.

  - During the December 23 appointment, Pascal was handed a cup for a urine sample. He noted that it was labeled with his brother's name.

  - Dr. Browning, at the end of the appointment, turned to Mr. Belperron and stated that there was more to talk about and asked me to call her. This made sense, since there were new diagnoses, and the kids were told to abandon all physical activity. When I called the next day, I was told she was on vacation and would not be back for a couple of weeks. I left a message. She did not return my call upon her return. I emailed her. She eventually responded, and asked when I would be available for her to call me. I gave her a three-hour window the next day. After waiting the three hours, plus an extra hour, I gave up and emailed her that I was leaving. She did call me that night, but made no apology. When I started to state my concerns, she said that it sounded like a long conversation and she didn't have time. Her suggestion: make and appointment and come into the office to talk.

- With enormous difficulty, Pascal was scheduled to see Dr. Ellinor for his supposed LQTS and Dr. Traum for his supposed hypertension. Dr. Traum's diagnosis? Pascal's BP was checked with the wrong-sized cuff! At MGH! PCP also found this. Dr. Ellinor stated Pascal did *not* have LQTS, but indicated since there had been a diagnosis, to be certain, he should have further testing. To add injury to insult, because of a change in our BCBS to the "Option 3," we now faced some $890 in testing fees from Dr. Traum's visit. Because of the wrong cuff!!!

- Marcel was sent to see Dr. Rosales to check his LQTS. She stated Marcel did not have it at that time. However, we confronted her staff with the statement on the medical record that Dr. Browning conferred with Dr. Rosales to arrive at the original diagnosis. Dr. Rosales' staff denies this and states Dr. Browning conferred with another cardiologist. Dr. Rosales, in this visit and reading the EKG from the visit with Dr. Browning, declared that there was no LQTS there. Who are we to believe?

- Dr. Ellinor believes that Marcel and Pascal should have ECG's to confirm an absence of any cardio problems, and to further prove the negative. If they should, don't you think MGH should do it, since it is their misdiagnosis? No, we're told we should have our PCP arrange this, or the nephrologists (for LQTS?), which makes no sense. MGH should not only do the procedure, they should eat the expense. There's no reason we should pay MGH, or BCBS pay Anna Jacques Hospital if referred there by our PCP, because of an MGH snafu.

- When Marcel was scheduled to see Dr. Rosales at the North Shore Children's satellite, Mrs. Belperron asked about the condition of the building pursuant to water damage and explained why. The receptionist was indignant that I would suggest that their facility could have such a problem; it was a hospital, after all! When we arrived, I was aghast. Right above the receptionist's head, the ceiling tiles were completely water stained. I know this site isn't necessarily your purview, but we did encounter a surly attitude every time we tried to find the condition of a building in advance of an appointment at MGH.

- Note that when we tried to voice our concerns through channels of due diligence, and dealt with the Office of Patient Advocacy, we were treated rudely, and when we sought to speak with a supervisor, instead, we got a letter. Rather than the letter expressing any advocacy on behalf of the patients, it was an attempt to cover the back-side of the hospital, full of citations of chapter and verse, and a retelling (inaccurate at that) of our encounters over the LQTS. Denise Flaherty, who signed the letter, may advocate for someone, but certainly not for us.

Dr. Slavin, we can appreciate that as you read this letter, you may not find our indignation justified or the errors cited egregious. The details are too vast in number to enumerate, yet this outline loses much of its bite because of the lack of detail. We feel that a few things must happen:

- There are items now in Marcel's and Pascal's medical history (we don't have all of the records, so there may be more problems once we finally get all records) that could affect their ability to get health insurance, life insurance, employment, licensure, etc. in the future. All mention of LQTS in Pascal and Marcel, the assertions that Marcel's issues have a significant psychological component, and Pascal having hypertension should be expunged. They never had these conditions.

- A number of physicians' notes make mention of depression. Marcel does not have it, yet one doctor mentions it, and then more pick it up in their notes. (*Editor's note:* classic Ass^2.)

- There are dozens of other factual errors throughout, regarding immunizations, prescription drugs, afflictions of others in the family—including notations of other family members having a variety of inaccurate conditions, as well as inaccuracies in notes from Mass Eye and Ear.

We know that there are mechanisms for addenda to be attached to the medical records when there is a disagreement, but this family has been put through enough. Life is a struggle without spending my weekends writing a letter like this, let alone chasing down and citing the mistakes of others in a hospital record and hoping that our request for addenda is granted. We've had some helpful personnel at MGH insist that we go to whatever lengths necessary to have these mistakes expunged completely. You need to have someone go through those records and delete these, and all other mistakes. If need be, have someone sit with us and we'll help them correct the records. Then you need to be sure that if there are any charges for services post-December 23, 2009, that we are not held accountable for them.

There's nothing you can do to undo the fear that Pascal and Marcel endured after they were told that their LQTS had two symptoms, fainting and death, or the weeks they spent not playing sports or participating in simple pleasures like cross-country skiing and ice skating, things they'd normally be reveling in this time of year. It's up to you, Sir, to make these right.

Finally, there is the medical issue that brought us to your doorstep a year ago. We'd hoped that MGH would help our sons so that we would not have to go all the way to Maryland for treatment and monitoring of their condition. Seems we were wrong. If you want MGH to have another crack at it, I would ask that you, yourself, call Dr. Ritchie Shoemaker for an explanation of his protocols for treating chronic inflammation due to neurotoxin exposures, and perhaps corroboration that we're sane people trying to get our kids well. I am sure he'd take your call. If you're uninterested or determined not to help us medically going forward, I would request that you put this in a statement. We're not looking to damn anyone, but hope that if you declare your inability to treat our sons, perhaps, if nothing else, we can get BCBS to cover the out-of-state physician.

In a discussion today, BCBS stated that there ought to be some face time allotted to us with someone in your staff to review all of this. Thank you for taking the time to review our concerns, and we look forward to your reply.

Sincerely,

Susan and Pierre Belperron

Epilogue

As I write this in July 2010 dramatic changes are ongoing in the lives of the Belperrons. VIP has worked beautifully (Susan finished a rapid 12-mile run several days ago, leaving Pierre in the mid-summer dust), but the cost remains high. One would think that health insurance companies would welcome provision of coverage for a drug that stopped the hemorrhage of dollar bills on needless medical interventions.

Despite the unsupported diagnosis of prolonged QT syndrome, that diagnosis is starkly present in the medical record. Ass^2 medicine? You bet. Marcel was denied a work permit because of the non-existence of the diagnosis, only being allowed to work when his new physician intervened to affirm that the diagnosis was wrong. More lessons from the School of Hard Knocks for the precocious Marcel.

Andre and Pascal too continue to suffer from their own "paper injuries" from Massachusetts General Hospital, an institution that has yet to acknowledge their errors.

Does any reader think that UMass will have new helpful ideas about the need for accommodations?

To answer the vitally important question about Marcel attending the moldy high school in Ipswich, the Belperrons moved away to southern New Hampshire. What difference forty miles can make! All the parents wanted was a safe school with a staff willing to recognize that Marcel had been injured by his years of mold illness and he won't always be 100%. Can you imagine the reduction of expectations of public institutions of the Belperrons when their main concern is a normal education process for Marcel?

Yet, by the Belperrons leaving the Massachusetts battlefield, they effectively give the remaining adversaries the right to declare victory. I can just imagine them saying, "Thank goodness the Belperrons are gone."

But what was the victory? What did the school authorities and Massachusetts' authorities actually win? The moral victory that comes from empty promises from empty suits said without any sincerity? Did they lose any sleep over their lies and deceptions?

What do the Belperrons take away from this endless Kafkaesque journey? If Pierre can work enough to pay for VIP (See Chapter 16: VIP), and without the drain of all the battles for truth he might be able to do that, he will have his family back. Susan can return to being the gentle mother, leaving the Rottweiler costume behind. Marcel can grow; Andre and Pascal will survive and grow too despite the conditions of their college education. Pierre might even have time to play the cello again for his own peace.

But should anyone again attempt to deceive the Belperrons like the Ipswich authorities did, the Rottweiler will return, only this time experienced, savvy and cold-blooded.

The family has learned the lessons though that don't come from books. Pierre quoted a phrase that may be appropriate here: "Nulla tenaci invia est via." For the tenacious, no road is impassable.

That idea fits the Belperrons, for sure, but it also applies to all Mold Survivors. As you read about Jerry Koppel, Liana Jones, Lee Thomassen, Dr. Laura Mark, Erin Culpepper and so many others, the courage to adapt to a new life and not be trampled by it requires courage, self-worth and tenacity. Not only is no road impassable but nothing is impossible!

CHAPTER 9 Jerry Koppel: The EPA Cost Him His Hypothalamus

Few will have the greatness to bend history itself; but each of us can work to change a small portion of events, and in the total of all those acts will be written the history of this generation.

— Robert Kennedy, during a speech he made to the young people of South Africa on their Day of Affirmation in 1966

We judge the fitness of an organism not by its success but by the success of its offspring.

— Paraphrasing Charles Darwin, 1859

The Koppel's story

In 1968, Bobby Kennedy challenged the nation's youth to get involved in government and change "the system" for the better. One of the tens of thousands who were inspired to do just that was Jerry Koppel, who was working to complete his Master of Science research in natural resource management. Koppel's response: to dedicate himself to improving U.S. environmental policy. [His real name has not been used as he fears repercussions from his powerful former employer – the US government.]

During what turned out to be a 25-year quest, Koppel worked for four different federal agencies, a Congressional committee, and a national nonprofit environmental organization. He loved his work and believed he was helping to improve the nation's policies, just as Bobby Kennedy had urged.

In interviews in early 2008, Jerry spoke positively about his life-long professional commitment. "Over the years, I was privileged to play a role in many important environmental efforts: helping communities organize for the first Earth Day, developing national environmental education programs, funding environmental research, promoting national land use and coastal zone planning, conducting national wilderness studies mandated by Congress, and helping to draft and enact legislation to permanently protect America's natural treasures by designating them as new National Parks, National Wilderness, and Wild and Scenic Rivers," he said.

But let's go back to 1991, when the U.S. Environmental Protection Agency hired him to work at their D.C. national headquarters building on M Street to help improve EPA environmental regulations. Two years later, the fit jogger, hiker and wilderness backpacker began to experience flu-like symptoms. The "flu" didn't go away: the fatigue, achiness, sore throat and low-grade fevers were worsening rather than clearing up. Even more symptoms appeared out of the blue.

After months of these symptoms and repeated visits to his physician, standard medical tests had provided no answers. At this point, his doctor raised the specter of the misnamed "yuppie flu." "I can't find anything else causing your symptoms," his physician admitted. "This yuppie flu is caused by stress, so I'll refer you to a psychologist to talk about stress reduction."

That explanation didn't make sense at all to Jerry Koppel, whose life was exactly what he wanted it to be. (*Editor's note*: Chronic Fatigue Syndrome, cynically called the yuppie flu in the early 1990's, was first described in 1985 by Drs. Dan Peterson and Paul Cheney, each of whom remains involved in unveiling the puzzle of this as yet unexplained illness. Recent breakthroughs in viral identification from Dr. Judy Mikovits and her colleagues at the Whittemore-Peterson Institute have raised the issue of causation from a retro virus, XMRV but other recent work in the UK questions this. CFS has nothing to do with stress.)

"I was happily married to the love of my life, my soul mate; our lives were filled with the joy of our delightful five-year-old son; and I was enjoying my EPA job - in fact, the EPA gave me awards for outstanding performance each of my first two years there," he says now.

"So what, this doctor is suggesting that I was stressed from too much happiness in my life?

"That was ludicrous," Jerry scoffs, just remembering that deluded diagnosis. His wife Loretta agrees. "When Jerry told me his doctor suggested it was stress, I couldn't believe it; it would've been laughable if it weren't for how serious the illness had become on a daily basis. I told Jerry we needed to find a different, better doctor fast."

As a young woman, Loretta had been introduced to Jerry by a co-worker in Washington. They married in 1984; son Jack was born in 1988.

Loretta and Jerry knew even then that stress wasn't the reason behind his health problems. The question remained, what was, and what kind of doctor could find the real reason? Koppel would later learn that his 1993 physician - like way too many doctors then and sadly, now - was seriously misinformed about the nature and causes of "yuppie flu," and he couldn't have been more wrong in suggesting it was caused by "stress."

By December 1993, Jerry's flu-like symptoms were joined by migraine-type headaches, dizzy spells, and extreme sensitivity to light and sound. He went to a neurologist, but blood tests and repeated MRIs of his brain didn't show any abnormalities. The doctor restricted Jerry's diet to eliminate possible migraine "triggers," and prescribed a series of migraine medicines. The result: they found that Jerry's diet didn't cause the headaches, and the medicine they'd prescribed didn't stop the headaches, but caused severe adverse side effects.

In early 1994, he also began having respiratory symptoms - coughing, shortness of breath, sore throat, swollen neck lymph glands and hoarseness. The headaches, fatigue and achiness were worsening and the light sensitivity and dizziness were persistent. Innumerable medical tests turned up nothing, although lots of horrid possible causes were eliminated, which was reassuring and worthwhile. But the answer to the question of what was causing Jerry's symptoms remained just out of reach. But as Jerry became Sherlock Holmes, eliminating all logical possibilities, the impossible loomed as probable.

"It was troubling and mystifying because Jerry was becoming more and more ill, and four different doctors still hadn't been able to point to a cause that made sense," says Loretta, recalling that period. The neurologist was puzzled, too, so he suggested that Jerry keep a daily diary of what symptoms he was having, when, and what he was doing at the time.

Thirty days later, Jerry went back to the physician to go over the diary. "I was stunned," recalls Jerry, still astonished all these years later. "The diary revealed that my symptoms were directly tied to being in the EPA Waterside Mall building. I wasn't having symptoms anywhere else."

The diary showed clearly that his symptoms would start after Jerry entered the building and get worse the longer he remained there, but when he left the building for the day, they'd improve. The neurologist suggested something in the workplace might be causing Jerry's symptoms, so he advised him to continue an hourly diary that might help pinpoint what exactly it was in the building that was causing the illness. At the same time, he referred Jerry to a pulmonologist, looking for help understanding Jerry's respiratory symptoms.

"Back then," remembers Loretta, "we'd noticed that after leaving work, by the time he was home, he'd already started to feel better.

And on weekends, when he was away from the EPA office, he had no symptoms at all - either at home or at any other location. Until the neurologist pointed to the EPA building as the possible culprit, it simply had not occurred to us that Jerry's illness might be caused by the EPA's headquarters." Loretta and Jerry would soon learn that this symptom pattern was characteristic of an illness dubbed back then "Sick Building Syndrome."

Was it really possible that something in the EPA building was the cause? Jerry and Loretta found that difficult to believe - it was the EPA, for heaven's sake. "It defies logic that the very Federal agency that's supposed to protect our air and water - and protect people from toxic exposures - has a building that's making its employees sick," says Loretta.

The pulmonologist ran numerous tests, concluding that it was not allergies or pre-existing lung disease that caused Jerry's respiratory problems. He was intrigued about the possibility that something in the EPA workplace was causing them, though, so he requested that Jerry do 30 days of "pulmonary peak flow measurements" throughout each day, at home, at work and everywhere else Jerry went. Peak flow readings are a standard means of measuring whether lung function significantly decreased from time to time during a day's activities. Such monitoring is used by many physicians, including the rare occupational medicine physician, as a validated method to document occupationally-related respiratory disease.

During those 30 days of pulmonary testing, Jerry continued to work in the EPA building each day as usual. The pulmonary testing results were dramatic: Jerry had normal lung function - and no symptoms - in all other locations, including his home, but shortly after entering the M Street facility, his symptoms would begin to progressively worsen and his lung function would steadily decrease during the day by as much as 40 percent. His cough now didn't disappear with removal from exposure: it was so severe at times that it disrupted meetings, and he had to take periodic "breathing breaks" to go outside, hoping "fresh air" would help the coughing and shortness of breath to improve. His pulmonologist said the results proved that something in the building's indoor air was causing Jerry's respiratory problems, and he diagnosed Jerry as having occupational asthma and occupationally caused reactive airway disease.

Loretta remembers that time well. "On the one hand, it was a relief to finally know the source of Jerry's illness. On the other hand, you're wondering: What now? Where do we go from here? How do we find out exactly what's causing the illness? Should Jerry immediately stop working in the building?"

When Jerry discussed his physicians' conclusions with his supervisor, she admitted for the first time that the building had a "history as a sick building" that had made other employees ill. She suggested Jerry go to the EPA Health Unit and report his doctors' suspicion that the building was making him sick.

"I'd never heard of illness caused by a 'sick building,' " recalls Jerry. "And this was the first time in my three years at the EPA - and the first time since I became ill - that anyone at the EPA had mentioned the building as a cause of illness. Why hadn't my supervisor mentioned it much earlier on in my illness?"

He immediately made an appointment at the EPA Employee Health Unit, reported his symptoms, and asked for their help in conducting indoor air quality studies to identify what was causing them.

Jerry was surprised and perplexed by their response - no, they would not offer any help other than noting his symptoms in his file. In fact, the Health Unit physician would neither diagnose him nor comment on what might be causing Jerry's symptoms or whether any other employees were sick because of the building. They wouldn't provide any help to Jerry's doctors, either, and his supervisor would have to contact EPA's Health and Safety Division if Jerry wanted the next step to be taken, namely indoor air quality testing, or if his physicians wanted indoor air quality information about the building.

Jerry promptly reported this response to his supervisor, and asked her to contact EPA's Health and Safety Division to formally request indoor air quality studies to identify the source of the problem and solve it, before things got worse. At that point, he assumed that EPA would quickly take action, the problems would be solved, his health would return to normal, and he could continue enjoying his EPA work. No one suggested he be transferred.

About that same time, Jerry was contacted by an EPA career employee who introduced himself as the "Health and Safety Officer" of the EPA professional employee's union, and as such, served as the employees' representative on the EPA's Safety and Occupational Health Inspection Team, which conducted periodic inspections of the building and investigated employee indoor air quality complaints. Jerry hadn't even been aware there was a professional employees' union or an "inspection team." The union's Health and Safety official then proceeded to lay out facts that left him dumbfounded:

- poor indoor air quality in the building had been causing many employees to become ill since 1988 - six long years.

- Jerry's illness symptoms were typical of what indoor air quality experts referred to as "sick building syndrome." In fact, an EPA indoor air quality study found that more than 1,900 (38% of the total) EPA employees in the building reported "sick building syndrome" symptoms, similar to those Jerry was experiencing

- the EPA's former Health Unit director, an occupational medicine specialist, had concluded the building presented a major public health problem for its employees, and he'd resigned his position when the agency failed to take corrective action to promptly cleanup the indoor air and protect employees' health

- indoor air experts had documented that the building had seriously deficient ventilation that allowed the indoor air to become contaminated with high levels of toxic chemical fumes from thousands of computers, printers, copying machines, formaldehyde-laden office furniture, carpeting, and even vehicle exhaust fumes from the parking garage under the building. The building had never been designed to be an office building housing EPA's 5000 employees, so the fresh air ventilation system was undersized and couldn't possibly deliver enough outdoor air into the building to flush out the chemical contaminants

- the building had continual water leaks with ongoing mold contamination problems in all building sections. The heating and ventilation system had been contaminated with mold, but the EPA failed to stop the water leaks or properly clean up the mold contamination

- the EPA had an "alternative workspace" location outside the building for employees who were made ill by the building, but the agency purposely made it difficult to qualify for being assigned there. Instead, sick employees were often told their illnesses must have been due to pre-existing allergies, and were either made to keep working in the sick building or they were pressured to resign or retire.

Despite the fact that the EPA management was hindering Jerry from getting to the bottom of the problem causing his illnesses, he remained reluctant to conclude that they were denying there was a problem and/or trying to cover it up. Jerry believes it "was ironic, to say the least, to learn that the EPA - the nation's environmental leader—had a 'sick building' that was making its employees ill." Loretta pipes in, "Initially, it was hard to believe that the EPA was failing to immediately try to solve the problems or move sick employees out of the building. It was even harder to believe that EPA officials were engaged in cover-up and denying there even was a problem."

But soon, the union Health and Safety Officer's assertions were proven correct. In fact, that union official was one of the "unsung heroes" of the EPA's sick building saga. Jerry learned that the union official himself had been made ill by the building, and subsequently courageously exposed the EPA cover-up and tirelessly advocated for, and protected the rights of employees who'd been made ill by the

EPA's building.

Meanwhile, concerned about Jerry's rapidly deteriorating health, his pulmonologist wrote the EPA requesting they at least temporarily assign Jerry to workspace outside the headquarters building until air quality studies could pinpoint the source of his illness and the indoor air contaminants could be cleaned up in the headquarters building. The pulmonologist urged prompt action to protect Jerry's health from further deterioration.

Nope. Jerry was dismayed when his supervisor initially agreed to temporarily let him work outside the building, but then reversed her decision and ordered him to continue to report for duty in the building as usual. "She said if she assigned me to a different building, then she'd have to do the same for other staff who said they had illness problems in the building," remembers Jerry. (*Editor's note*: how many times have we heard that?)

In addition, the EPA's Health and Safety Division refused to initiate air quality studies of Jerry's work areas. Instead, they hired a consultant who, ignoring all the facts and the many other employees with similar illness symptoms, merely walked through the area and concluded that he saw nothing that could be causing Jerry's symptoms. He rejected the medical report from Jerry's physician, and instead suggested Jerry's home was causing his illness, or it was caused by "pre-existing" medical conditions. Outrageous, no? The fact that Jerry's symptoms cleared up when he was home, the pulmonary function tests proved he had normal lung function while in his home, and his physicians had clearly documented that there were no pre-existing allergies or pulmonary problems at the root of Jerry's illness was ignored.

The EPA officials informed Jerry that in order to qualify to work in a different building, his physicians had to "identify the specific air contaminants" that were making him ill and write detailed medical reports of each of Jerry's symptoms "proving those specific contaminants were the cause." (*Editor's note*: though we still see such sleazy tactics from "**two-bit**" defense lawyers in mold cases, such absurdly impossible demands for "specific causation" i.e. deciding that only *one compound* out of a mixture of similar acting compounds is the culprit, remained in force until the 2008 US GAO report was released, supported by the 2009 WHO report on moldy environments. Specific cassation was a standard defense ploy for years as it put impossible demands on plaintiffs.).

"At that time, EPA's national clean air program office was announcing across the nation that indoor air pollution and 'sick buildings' were big problems nationally that adversely affected people's health," recalls Jerry now. "Yet another part of the EPA - the ironically titled 'Division of Health and Safety,' whose mandate supposedly was to protect employee health - was denying that their own headquarters had indoor air problems that were making employees ill. The ironies were astounding."

Now Jerry and Loretta were no longer skeptical of allegations of an EPA cover-up: it was clear EPA officials were obstructing Jerry's doctors' efforts to protect his health and claim his rights to a safe workplace. Even worse, EPA officials were dismissing employee air quality complaints as figments of their imaginations, hysteria, or alleging the employees had "pre-existing allergies". Moreover they were refusing to conduct objective air quality studies.

"This was a classic 'Catch 22' situation they were setting up," notes Jerry, still angry today about the agency's duplicity. "The only way to identify specific air contaminants causing my illness was for the EPA to conduct air quality tests and provide my physicians with data from previous indoor air studies in the building. Yet EPA refused to do so, making it impossible for my physicians to identify specific air contaminants! My physicians were incredulous, but determined to support my case."

During this time, while Jerry and the doctors battled the bureaucracy, he'd continued doing his job, working in the building every day as his supervisor had ordered. The trouble was, his symptoms were becoming more severe and complex. The symptoms no longer cleared up when he was away from the building, and he was becoming chemically sensitive, reacting to common everyday materials and places that would in turn dramatically increase his already severe fatigue, headaches, dizziness, and respiratory symptoms. He was having neurocognitive problems, too - memory loss, difficulty concentrating, and problems organizing his thoughts.

"I know now that continuing to work in the building overwhelmed my body and led to permanent disability," says Jerry. "If I'd only known at the time that it would've been better to just quit my job and protect my health, I certainly would have done so. But I simply had no idea it would lead to permanent disability."

Despite the EPA's refusal to provide indoor air quality studies, Jerry's occupational medicine specialist and his pulmonologist submitted more than 40 pages of medical reports supporting their diagnoses and their recommendation that Jerry's health would deteriorate further unless he was promptly assigned to work in another building. He'd been diagnosed with (i) occupational asthma; (ii) reactive airways disease; and (iii) headaches induced by toxic exposures in the building. That was the short list; he also now had chronic fatigue syndrome, dysautonomia, and neurocognitive impairments, as well as other contributors to his overall disabling condition that were not yet diagnosed.

The EPA's response? Can we all guess? Don't forget that some of the same EPA officials back then are *still on the Federal payroll* now in 2010, not counting the other placeholders who have come along saying the same kinds of things to make the current EPA approach to mold a national disgrace. Lip service? You bet. Action? Forget it.

EPA managers rejected his physicians' reports due to the alleged failure to identify specific air contaminants causing the illness, neatly springing their "Catch 22" trap. His supervisor again ordered him to continue to work inside the headquarters building.

"It was clear that if I didn't report for work inside the building, they were planning to fire me," remembers Jerry. "The union official warned me that the EPA had done this to other employees who had health problems due to the contaminants in the building. My physicians were now clear: they told me that if I continued working in the sick headquarters building, my health would worsen because medications couldn't stop the illness."

"Get out now or die" was the command from Wyatt Earp as he was policing Dodge City in the Old West, not from the 1994 EPA, Washington DC.

When his supervisor rejected a final plea to allow him to work in a different EPA building, Jerry reluctantly hired an attorney to represent him. That attorney's intervention finally got the EPA to agree to temporarily allow Jerry to work in another EPA building. In what was obviously a retaliatory action, his EPA supervisor's next stripped him of his major job responsibilities and instead assigned him duties previously performed by junior-level staff.

"It was devastating to Jerry," remembers Loretta, "for a person with his credentials, experience and abilities to be essentially demoted because of an illness caused by EPA's own building. He'd been such an effective part of the EPA system, a team player and a leader, and suddenly, all those things were taken from him." She notes with disgust, "The next blow was, they suddenly gave him a lower performance rating, while in each previous year he was recognized for outstanding performance."

The news got worse: unfortunately, by this time, the continued exposures to contaminated indoor air had overwhelmed Jerry Koppel's body. His daily symptoms were severe and now permanent. If he went into any other building except his home, the symptoms became immediately even more severe. He'd exhausted all his sick leave attempting to recover, so in 1995, too ill to continue reporting to work, he had no alternative but to retire. He reluctantly handed in his paperwork, and hoped that with some rest, and without the daily battle against the bureaucracy, his health might improve.

Jerry Koppel was now chronically ill and physically very weak, with daily dizziness and headaches, severe fatigue and digestive problems, painful joints and muscles all over his body, extreme light sensitivity, brain fog, and he was unable to sleep more than a few hours at a time. To maximize the potential for recovering his health, the Koppels went all-out to make the house into a "safe haven," installing an expensive whole-house HEPA and activated-charcoal filtration system designed to filter chemicals and particulates. Jerry had become so severely chemically sensitive that he, his wife, and son had to throw out all their normal clothing and buy organic cotton clothes, which were all Jerry could tolerate.

After his retirement, he spent most of the time in bed, unable to venture outside his home without becoming so severely ill that he'd be bedridden and in even more acute pain for weeks. At times he was too weak and dizzy to safely walk by himself from one room to another. Medications only caused severe adverse effects, with no relief from his daily illness symptoms. He was no longer able to participate in normal life: there was no socializing outside the home with his wife and son, no participation in his son's school or extracurricular events and activities, no travel or vacations. He was often too ill to even care for himself. When family and friends came to visit, they had to change into organic cotton clothing the Koppels kept on hand for visitors.

"Naturally, my illness turned our lives upside down," said Jerry with characteristic understatement. "My wife was a full-time working professional, who now essentially had to become a single parent and breadwinner for the family, while also taking care of an invalid husband. It was an incredible burden. And believe me, during the years when my condition was so bad that I was in constant bad pain and unable to care for myself, too often I was not a joy to be around. So I feel very blessed to have had such a loving supportive mate who's stuck by me all these years, and a son who was a constant source of joy and love."

For her part, Loretta remembers those years with characteristic empathy. "It was devastating for Jerry to not be well enough to get out of bed. Before he'd become ill, he'd been such an energetic and active person," remembers Loretta. "Of course, it was stressful for me to suddenly have to shoulder all the family responsibilities while working full time and taking care of my son and a bedridden spouse. To suddenly lose Jerry's income only added to the stress. But we were fortunate that I had a good job that was capable of supporting us. I'd come close to quitting that job when my son was born; if I'd quit then, we would've had a real financial nightmare when Jerry's illness forced him to resign."

Despite the difficulties, however, the Koppels never let it get them down for long.

"We don't feel sorry for ourselves," Jerry Koppel says today. "You don't have to look far to find people with more severe challenges in their lives. So we've tried throughout this to count our blessings each day rather than remain focused on what we lost." Loretta adds, "We look back on the past 15 year ordeal, and are so thankful for having survived it so well. We attribute that to several things: prior to the illness, we had a strong, loving relationship with each other and with our son Jack; most of the time, we maintained a positive attitude of being grateful each day for all the good things in our lives; we had the savings to pay for cutting-edge experimental medical treatment; and of course, a bunch of superb doctors who we refer to as our heroes."

The EPA's cover-up and their failure to promptly heal their "sick building" devastated the lives of many EPA employees besides Jerry Koppel. There were single professionals who lost their salaries and had to move in with their parents because they could no longer afford their mortgages or care for themselves, married employees whose marriages fell apart from the additional stresses of job loss and chronically ill health, and support staff who didn't earn a lot to begin with and who were forced to go on welfare after being too ill to keep working; naturally, they were unable to afford to seek medical care from specialists. "Compared to many, I was very fortunate,"

Jerry Koppel insists.

Due to privacy laws, it's impossible to find out exactly how many EPA employees were forced to quit in those days due to disabling illness caused by the EPA Headquarters, but from word-of-mouth reports, Jerry estimates that at least several hundred employee's careers were ended and their lives turned upside down by the disabling illnesses they suffered. And it was all caused by that one contaminated government building. It won't be a surprise then to find out that Washington DC has a raft of moldy buildings, where employees struggle for their rights to a safe cubicle (See "A Moldy Tourists Guide to Where ***Not to Go*** in DC").

Forced to retire early, Jerry wasn't qualified for full retirement benefits. As a federal employee, his other recourse was to file for federal Workers' Compensation. [Federal law bars federal employees from filing lawsuits against their agencies for damages except in cases of discrimination.][1] Facing high medical bills not covered by insurance, and the loss of most of his income, Jerry asked his doctors to prepare medical reports for a Workers' Compensation claim. By this time, Jerry was diagnosed with a long list of disabling medical conditions: occupational asthma, reactive airways disease, chronic fatigue syndrome, dysautonomia, orthostatic hypotension, migraine headaches, chronic dizziness and ataxia, photosensitivity, neurocognitive impairment, and multiple chemical sensitivity.

His Workers' Compensation claim added a final insult to his previous injury. As they'd done with numerous other EPA employees forced to retire after the building made them too ill to work, the EPA opposed Jerry's claim, asserting there was no evidence that the building caused his illness. The Department of Labor's Office of Workers' Compensation Programs rejected Jerry's claim with a familiar reason and that old "Catch 22": Jerry's physicians had failed to identify the specific air contaminants causing his specific illnesses. As a final

[1] Workers' compensation: (colloquially known as workers' comp in North America) is a form of insurance that provides compensation medical care for employees who are injured in the course of employment, in exchange for mandatory relinquishment of the employee's right to sue his or her employer for the tort of negligence. (From Wikipedia, the free encyclopedia)

insult, the federal Office of Workers' Compensation told Jerry that his supervisor had denied that the building made him ill.

How could his physicians identify <u>specific</u> occupational causes of Jerry's illnesses, when the EPA denied the building was "sick" and refused to provide the necessary indoor air quality data from indoor air testing? (*Editor's note*: this absurd demand to find one weapon in a firing squad that killed the immune response, i.e. specific causation, is now destroyed by the US GAO (2008) report and the WHO report (2009), each of which condemned the idea.)

Disgusted with these absurd bureaucratic games, Jerry was determined not to have his rights denied. He pointed out that Department of Labor regulations required that agency to ask the EPA for all relevant data. The Department of Labor did so, but when the EPA didn't respond, the Department of Labor (DOL) again rejected Jerry's claim, telling him he could appeal the decision if his physicians could provide proof that the building caused Jerry's illness and disability. (*Editor's note*: the DOL finally recognized mold exposure as a disabling condition in 2009 in Haefner v. FAA, see Appendix 6.)

He was no quitter. In 1998, in the midst of Jerry's severe daily illness, a lawyer friend helped Jerry file a Freedom of Information (FOI) request that the EPA provide all indoor air quality studies on the EPA's Headquarters building, including all data on ventilation, indoor air chemicals, water leaks and mold contamination. The EPA's response: they'd provide the data, but Jerry would have to pay $18,000 for "copying and search time."

Jerry's lawyer friend told him this was illegal, bureaucratic nonsense - another of the EPA's attempts to make him go away and keep the truth hidden. So Jerry filed an appeal with the EPA General Counsel's office, pointing out that by law, he was exempt from any such charges because he needed the documents to file his Workers' Compensation claim, and the EPA was legally obligated to provide them free of charge.

In the meantime, Jerry continued his quest to find medical treatments that would restore his health or at least improve it. "We were very fortunate that we had some savings, and with Loretta's income, we were able to afford to pay for medical treatments our insurance wouldn't cover," he said.

Despite that, by 1997, his condition had steadily deteriorated, and his primary care physician referred Jerry to a leading expert on Chronic Fatigue Syndrome (the same Paul Cheney, MD mentioned earlier). The specialist confirmed the CFS diagnosis as being caused by indoor air contaminants in the EPA building, and performed exercise ergometry with gas analysis on Jerry—a test recognized even by the Social Security Administration as being the gold standard for determining the degree of disability.

Jerry failed the test miserably, collapsing physically a few minutes into the test. The specialist's conclusion: Jerry was fully disabled and unable to work; he had the abnormally low cellular cortisol levels typical of many CFS patients, and stress rapidly caused his cellular functioning to become anaerobic, which would result in physical collapses.

The specialist began Jerry on a series of what then was the most promising experimental treatments for CFS symptoms. Some treatments helped somewhat, but by 1998, Jerry's symptoms worsened and he began experiencing severe orthostatic hypotension problems and seizure-like collapses. Confined to bed 24 hours a day and unable to care for himself at all, he was getting IV saline infusions via home nursing care, which helped stabilize the orthostatic hypotension.

Orthostatic hypotension is a part of autonomic nervous system dysregulation ["dysautonomia"] - both dysautonomia and orthostatic hypotension are common in CFS or CFIDS[2] patients. Orthostatic hypotension, also known as Neurally-Mediated Hypotension (NMH), is a miscommunication between the brain and the heart that can cause lightheadedness, dizziness, weakness, fatigue, nausea, mental

[2] CFS is also known as Chronic Fatigue and Immune Dysfunction Syndrome. (CFIDS)

confusion, headaches, muscle aches and profuse sweating. Instead of sending signals to speed up the heart, the brain tells the heart to slow down and the vessels in the arms and legs to dilate. Blood pools in the extremities and not enough stays in the brain. Without enough blood circulating throughout the brain, providing oxygen (among many other things), the patient faints or merely feels very sick. The autonomic nervous system plays a key role in regulating many of the body's basic functions—such as heart rate, blood pressure, digestion, sleep. During these seizure-like collapses, Jerry was only semi-conscious for a period, during which he was barely able to speak.

"It was very scary, especially when they first started happening, and we had no idea what it was or how to care for him," remembers Loretta. "He was barely conscious, barely breathing, with his eyes rolled back into his head. After a few minutes, he'd become a little more conscious but not responsive when you talked to him. This could last 10-15 minutes. As he started to recover more, he could hear us speaking to him, and was able to mumble 'yes' or 'no' to questions. As the seizure-like collapse ended, he got alternating chills and sweats, his muscles cramped, and he'd be left so weak after he recovered, that he'd have to be helped to bed.

"The first time, we'd been outside in hot, humid weather, and he suddenly felt very weak; as we started to go inside the house, he collapsed in our carport. I thought it might have been sun stroke. My son and I weren't strong enough to lift him to carry inside the house, so we tried to care for him while he was lying on the carport slab in the shade. We brought him water and ice to keep cool him down and hydrate him; then blankets after he started chills. His breathing was so shallow, I kept reminding him to keep breathing, to talk to me, to tell me what was going on," she remembers. "As he became more conscious, I suggested we should call an ambulance, but he wasn't having classic heart attack symptoms, and we were afraid that with his chemical sensitivity a trip to the hospital could do more harm than good. Eventually, he felt better and with our help, he was able to get inside."

Jerry's son, Jack was 10 years old when he first saw his father collapse mysteriously in front of him; he'd watch many of his father's collapses in the ensuing years as he grew into a young man.

"I was just six years old when my dad became so severely ill from the toxins in the EPA building. Prior to his EPA illness, he and I had a lot of fun together doing physically active things like scouting, playing ball, and swimming. When he became disabled by the illness, I suddenly no longer had a father who could do those kinds of things with me. I worried a lot about him, as he kept getting sicker and weaker," said Jack.

"But the first time he collapsed in front of me was really scary," he remembers. "As I think about it now, it was too much for a six year old to understand. I do now and that understanding is part of who I am. That started a long period of Dad being even sicker and weaker than before. His doctors explained the collapses to my mom and me, and how to care for him when the collapses occurred, so when they happened again (they did a lot), it was a lot less stressful than that first time. My mom and I became an effective care team for him.

"Of course, always in the background was the worry about whether he'd get better or continue to get worse. But one of the hardest things initially for me as a boy was that suddenly my father couldn't participate in my life away from home. I missed that a lot."

But then Jack reflects further. "Looking on the positive side, one thing I learned over those years is to have a positive attitude and to be thankful for all the good things in our life. Lots of people have much more difficult lives than do we, so like my parents, I focus on what's good in my life."

In 2002, Jerry's seizure-like collapses were under control enough for him to try experimental hyperbaric oxygen therapy (HBOT) under the direction of a neurologist in Naples, Florida. The neurologist's clinical research showed that in some patients, HBOT improved CFS symptoms, as well as neurological problems like dizziness, headaches and neurocognitive difficulties.

Jerry was too ill, weak, and chemically sensitive to fly anywhere. So once a year between 2002 and 2005, Loretta and Jack would load up the family minivan for the two-day drive to the clinic in Florida, where Jerry would get HBOT treatments twice a day for two weeks. The trip was exhausting for Jerry even though he'd sleep on the trip in a makeshift bed in the backseat; it was at least as tiring for Loretta, who had to do all the planning and driving.

The trips required weeks of advance planning; Loretta even had to go there in advance on scouting trips to find a hotel that Jerry could tolerate—one that didn't have mold problems (a real challenge in southern Florida), and that didn't use chemical room cleaners and fresheners or new furniture or carpets laden with chemical fumes that would cause Jerry to collapse. Heavy-duty air cleaners had to be shipped to the qualifying hotel ahead of time, and the rooms had to be rented two days in advance of their arrival so the air cleaners had a chance to clean the air in the rooms prior to Jerry's arrival. Special non-chemical cleaners were shipped ahead for the housekeeping staff to use in prepping the room. Jerry's organic cotton and wool bedding was carried in the car; as soon as they arrived, Loretta and Jack would make up the beds with Jerry's chemical-free bedding. Sometimes, they arrived to find a problem in the hotel room, which would lead to an exhausting late-night search for a last-minute replacement. The next day, the hyperbaric oxygen therapy began.

Although the HBOT therapy helped, the trips themselves back and forth "were exhausting," says Loretta today. "We traveled with so much, it was like moving."

Several years later, when Loretta's job prevented her from staying in Florida for the two weeks of Jerry's treatment, Jerry's "Aunt Rosie" volunteered to take care of him during the treatments. "Aunt Rosie is another of our blessings," says Loretta. "Without her, Jerry never would have been able to get those treatments that helped him."

The HBOT treatments improved—but did not cure—Jerry's headaches, dizziness and light sensitivity. But improvements in CFS symptoms of fatigue and aching muscles and joints were only temporary, and after 3-4 months, previous levels of those symptoms returned. And of course, the exertion of the trips alone sometimes caused a relapse and left Jerry exhausted for weeks or months. The neurologist and Jerry finally concluded that for the time being, HBOT had done as much good as it could do.

"They may not have 'cured' me, but HBOT played an important role in stopping my downwardly-spiraling condition, and significantly improved some of my neurological symptoms," says Jerry today. "Of course, insurance wouldn't pay for any of these treatments, and the costs of lodging were very high. I felt very fortunate that we had the savings to pay for all of that."

Let's return to 1998, and Jerry's Freedom of Information Act (FOIA) request for EPA to provide his doctors with all indoor air quality data and studies on the EPA's headquarters building. Remember the EPA said it would cost Jerry $18,000 to get the requested documents? After Jerry's lawyer friend threatened to file a FOIA lawsuit against EPA, the EPA finally threw in the towel and dropped their demand for $18,000 in fees. By 2004, Jerry had obtained more than 500 pages of studies on the EPA headquarters building's indoor air quality problems.

The illuminating documents proved that Jerry's work areas and the building as a whole had seriously deficient fresh air ventilation; Jerry's workspaces and the building generally were contaminated with 13 species of toxic molds and bacteria and 27 species of allergenic molds and bacteria; the building's HVAC system was contaminated with mold, which the EPA had failed to professionally remediate; and Jerry had suffered high-dose occupational exposures for three years to at least 23 different chemical toxins in the indoor air.

Was it any wonder the EPA had fought so hard for so many years to prevent Jerry and his physicians from getting this data?

Armed with this new information, Jerry's lead physician—an

occupational medicine specialist, and another hero of this saga—prepared a 50-page medical report with hundreds of pages of appendices that documented the specific contaminants that caused Jerry's disabling medical problems. Jerry submitted this as part of the appeal of the denial of his Workers' Compensation claim.

The U.S. Department of Labor ignored the new medical evidence, and 12 years after his original claim, Jerry and Loretta are still fighting the denial.

"We'll fight them to the end," said Loretta, who's rapidly approaching retirement.

"We're going to need the money they owe us in order to continue to afford the medical care that's kept Jerry alive this long—it's cost us an average of $20,000 a year for medical treatments not covered by insurance, and we won't be able to afford that if I retire when I'm supposed to."

In 2005, Jerry's occupational medicine specialist suggested Jerry consult with Dr. Shoemaker because of his pioneering work with biotoxin-injured patients. "I was in really bad condition when I went to my first appointment with Dr. Shoemaker," remembers Jerry now. "It was a real revelation to learn from him that my EPA workspaces definitely exposed me to mold toxins that were the cause of my worst illness symptoms. Dr. Shoemaker explained that I had a genotype, he called it the dreaded genotype, that made it difficult for my immune system to recover from exposure to mold toxins; my son had inherited that same susceptibility to being made seriously ill when exposed to biotoxins; and the mold toxins had been in my body doing damage for so long that my hypothalamus now functioned minimally, and unless that could be corrected, I was unlikely to be able to resume a more normal, less disabled, life."

But even now, several years later, he remembers how wonderful it felt when, for the first time in more than 10 years, his condition began to stabilize after months of treatment with Welchol, one of the drugs we use to substitute for cholestyramine in some patients.

"Although HBOT had helped me a lot, when I first saw Dr. Shoemaker, I still had really bad daily symptoms that severely restricted how long I could be upright and prevented *any* physical exertion," recalled Jerry. "After Dr. Shoemaker put me on Welchol for several months, I began to improve. There were no longer the severe day by day swings between being bedridden in severe pain and being able to get out of bed, take a shower and sit upright for a while.

"The Welchol didn't eliminate the daily symptoms of fatigue, aching body, headaches, weakness, orthostatic hypotension, light sensitivity, chemical sensitivity, the dysautonomia, and so on," says Jerry. "But it reduced the intensity of those daily symptoms—and believe me, any reduction in the severity of the symptoms is significant as far as I'm concerned. And it's fantastic to now have my daily condition stabilized—more constant, instead of wildly swinging day-to-day from 'bad' to even worse."

Jerry's symptoms rapidly get worse if he goes out of the house and into other buildings or if he's exposed to low-level chemical fumes found in everyday articles like clothing, building materials and furnishings, vehicle exhaust, and so on. And if he's upright too long, he still experiences hypotension symptoms.

"If I overexert, I have the seizure-like collapses, but they're not provoked by minimal activity like they used to be, and when they do occur, they're not as severe as before," he says. "This is such a dramatic improvement compared to my condition over the previous 12 years! By giving me this improvement, Dr. Shoemaker has given us an amazing gift that's added tremendously to my quality of family life. He gave back to me and my family a part of our lives that had been missing for more than 12 years. He's truly one of our heroes."

Despite that improvement, Jerry remains extremely disabled and is unable to resume anything approaching a "normal" life, but stabilizing his condition nonetheless made it possible for him to occasionally participate in some important family experiences that were impossible before. For example, for years, he hadn't been able to attend social or school events with Loretta and Jack without a severe set-back in his condition, resulting in being confined to bed afterward for weeks or months.

"I still can't go to such events regularly or often," admits Jerry now. "But now I can occasionally share a very special event with my family, knowing that afterwards, I'll have worse symptoms, but it won't be as severe as before. So I got to do some things I never thought I would: I got to see my son graduate high school, I got to see him receive a national journalism award, and I watched him start his first day of college."

Jerry Koppel doesn't mention that after each of those events, he was exhausted, but his wife remembers it well. "We joke that when he's had an exposure or gone to an event, he feels like a truck hit him, so now the running joke is, was it a semi or just a little Toyota?"

Loretta credits Dr. Shoemaker with giving her husband a better energy level. "He sleeps better and doesn't run out of energy as quickly. He gave him a little—but important-- piece of his life back."

Jerry counts those "little pieces" differently. "What a blessing—and what joy added to my life by this exceptional, caring doctor! It made my heart soar to be able to see my son graduate from high school and start his college life," exults Jerry today. "Dr. Shoe has a very special place in our hearts."

Jerry's son Jack nods his head in agreement. "It meant a lot to me that our family was participating together again at special events like my graduation and going off to college. For so many years growing up, when Dad was too ill to participate, he always said he would 'be there in spirit', but it sure was wonderful to have him there in person. Thank you, for that, Dr. Shoe."

Loretta reports that overall, Jerry's been pretty even-keeled about all of this, and she believes her husband's positive attitude has been a large part of his being able to deal with this as well as he has.

For his part, Jerry Koppel still counts his blessings instead of listing his losses. He remains optimistic despite the long-term difficulties he's suffered and the unresolved Workers' Compensation claim.

"As bad as things have been, I've never lost sight of the fact that I have many blessings in my life," he says philosophically. "A loving wife who stuck by me all these years and a loving son who brought daily joy to my life. Being fortunate enough to have savings to pay for my medical care all these years. Exceptional physicians who dedicate their lives to trying to find cures for those of us with disabling chronic illnesses like CFIDS and MCS, and who played key roles in stopping the rapid downward spiral of my medical condition and became aggressive advocates for my case. Dr. Shoemaker and my other key physicians have been a major blessing in our lives."

The Man with No Hypothalamus

By Ritchie Shoemaker, M.D.

In 2010, it's easy to recognize a patient who has Jerry's illness. Others with that syndrome don't come to Pocomoke more often than once a month. I have no idea how many Jerrys there are; most couldn't even dream of traveling the way he had to. Yet in the group of seriously ill patients I see daily who are acutely, and then chronically, sickened by exposure to water-damaged buildings, what Jerry taught me is put to good use consistently. Once you've recognized one Jerry, it's not hard to recognize others.

How can the casual bystander recognize Jerry? Just look for the pale, undersized person, lying on the floor in a darkened corner, surrounded by water bottles, his eyes covered by super-dark glasses. Don't talk too loudly or you'll make him sick. And don't let him arise too quickly, or he'll be right back down on the ground within seconds. Don't "shine his eyes" with your little doctor flashlight, either, or you'll put him into a near-seizure.

You'll also recognize Jerrys by the many wrong diagnoses they've been given: Chronic Fatigue Syndrome, neurally mediated hypotension, stress, malingering, and who can ignore the ubiquitous (but ultimately nauseating diagnosis) fibromyalgia. Nope, none of those pseudo-diagnoses are correct. Jerry and his biochemical clones can't possibly work, yet no one with the authority to authorize disability payments for them will come forward with a check.

Why not? Just listen to the crapola spewing from the payors' medical consultants: No one could be that ill. We don't agree that smells can make people ill, and besides, his EEG is negative. He doesn't have seizures. He must be faking. And no one has published a case series that others can cite.

Now when I see another "Jerry," it is a quick path to the use of DDAVP to correct volume depletion, omega-3 to lower MMP-9, epo for high C4a and VIP to restore quality of life. If the next Jerry has enhanced chemical sensitivity, I am never surprised.

Just think about the research study one would need to do to publish the results of all these hypothalamic deficient patients. A case series of Jerrys? Face it: no busy primary care doctor wants a case series of Jerry-illness victims to become regular patients, even if the physician wanted to write the definitive academic paper. Why? MDs don't want to help Jerry because he takes time to interview, he can't tolerate the meds we always use in others, he doesn't get better and his litany of complaints drives the doctor nuts. Besides, all his blood tests are normal. Wrong!

Jerry visited me at just the right time. If he'd come in before 2005, when I finally figured out that depressed levels of vasoactive intestinal polypeptide (VIP) were a critical missing link involved in peripheral inflammation, which was part of chronic mold illness, MSH deficiency and blunted response of vasopressin to salty blood, I would've missed his diagnosis. But now, simply ordering these tests makes life easy for those of us struggling to show exactly what's wrong with Jerry and others like him.

Basically, Jerry was living without normal levels of hypothalamic regulatory hormones. Without VIP, he didn't have much chance at a normal life, but his lack of VIP wasn't the only thing wrong with his hypothalamus. In each of the hypothalamus' main regulatory centers, the coordinating, regulating and controlling elements are gone. What does Jerry have left to deal with the inflammatory, hormonal and neural activities of daily life?

Nothing.

Now, when I see the water bottle supply like Jerry had (he drank three 32-oz. bottles in a 90-minute interview, yet was still incredibly dehydrated), I know the patient is dehydrated. Forget finding normal levels of vasopressin. The light and smell sensitivity essentially guarantees VIP won't be detected by a LabCorp assay. How about the chronic fatigue? You can bet MSH will be too low.

Put the three hormones together and you've got the triad of hypothalamic abnormalities that Jerry taught me to look for. "No hypothalamus" might be a bit of an exaggeration, but the Jerrys of the world are walking examples of what central nervous system regulation of inflammation and hormone function is all about.

Think about it: since these illnesses aren't in the textbook, if you're the busy physician who has to guess why these people must avoid looking at fluorescent light (to then avoid having near-seizures), wouldn't you be tempted to assume psychiatric illness? What they have isn't a seizure. And why can't they stand up without their blood pressure going into the tank? And who orders that MSH test anyway? I'm tired of hearing physicians say about MSH, "I've never heard of it, so therefore it must not mean anything,"

Jerry's case was the first that made the answers crystal clear to me. When you review his history, you can practically read between the lines when the mold murdered his hypothalamic regulatory apparatus. At first, he got better when he left the building he worked in each day, but when his MSH production pathway died, he stayed ill even when he was away from the building. When he couldn't handle postural changes, I knew his vasopressin had departed. And when the smells and light bothered him, his VIP was gone. Seems simple now, but getting to know Jerry was the path that led to a vast array of people like Jerry that no one had defined.

No one knew how to document the biochemistry of his illness then. We do now. No one knew to stop the innate immune cascade that was the dagger through the stalk of the hypothalamus. We do now. No one knew how to correct his huge deficit of intravascular volume. We do now.

What happened to Jerry is an embarrassment and an outrage. He asked for help from the Environmental Protection Agency, the governmental organization responsible for "protecting human health and the environment." None came. He asked for readily available accommodations. None were granted. He asked for compassion and basic human decency, only to find that the EPA managers' collective ego wouldn't permit any logical approach to resolving his health claims. Is it too much to say that his hypothalamus was murdered by the EPA itself? Anyone believe in avenging angels?

In many ways, Jerry gave much of his most productive life to help many others. I don't think he meant to be so noble, but we shouldn't waste his sacrifice by ignoring the lessons from his loss. And while we're at it, you'd think the least we could do is to help him a bit. Perhaps we can; we're looking at the era of VIP replacement. And you bet it works! (*Editor's note*: VIP replacement has provided Jerry with another step back to normality.)

I guess admitting guilt isn't in the cards for the EPA: twelve long years later, they're still denying Jerry's Workers' Compensation claims.

Meanwhile, let's revisit the Koppels. Genetically their son Jack was as likely as Jerry to be as affected by mold, mildew or any biotoxin, so Jerry and Loretta were proactive, and they drummed into the youngster from a very young age that he had to make sure he'd never be exposed to anything that would set off an inflammatory reaction - if he did, his genes would ensure that an illness much like Jerry's would quickly take away his good health.

While in high school, Jack flourished, based on his intellect and his work ethic, but ever since he'd been exposed to a moldy classroom, he hadn't been right. With Jerry's dreaded, inherited HLA DR, he suffered from low MSH and too-high C4a, among other things, which means that if and when he ran into mold or another toxin, his immune system wasn't likely to banish it easily, if at all.

Jerry knew what was wrong with Jack and what had to be done, yet who'd believe that Jack was affected by mold except someone who knew what innate immune abnormalities really meant over the long haul? Did anyone think that Jack, winner of one of the most prestigious college scholarships based on merit, was a hypothalamic wreck? Yet Jerry knew Jack's future better than any Tarot reader could spell out: the answer for Jack was spelled out in horrific detail in his abnormal lab tests. Jerry could see that what happened to him would happen to his son if he didn't act to change the future.

What would you do if you were Jerry? We've all seen a hundred science fiction movies in which a small change in the past leads to large alterations in the future. If Jack went to college and did nothing exceptional to protect himself, how long would it take before he was lying in the corner with the lights off, surrounded by water bottles and hand rails to protect him from a fall that was likely to occur when he stood up too quickly? Would Jack's future be determined by his HLA 4-3-53 (the 0401, no less, the worst of the worst), his life scarred by thoughts of what he could have become?

Jerry did what any parent should: he protected Jack.

When Jack came home from college his freshman year a bit tired, Jerry was worried; Jack didn't think much of it.

After a few question-and-answer sessions, though, Jerry found that Jack was often tired after he'd been doing laundry in his apartment's laundry room, and he'd also noticed a "musty smell." Jerry's heart sank. Hadn't they always stressed to him that a situation just like this could make him as sick as his father? A few phone calls and quite a large number of explanations later, Jerry had his answer: the laundry room in Jack' dorm was infested with mold. Each time Jack entered, the mold spores entered his nose and caused the very reaction Jerry and Loretta had long feared.

Jerry took charge. Even before Jack came to Pocomoke to become my patient, Jerry found out what the labs showed, and then he started to fix the first two elements of treatment. One, get out of harm's way! Then attack whatever persistent toxins are left over from exposure. Jack took Welchol after Jerry insisted they clean up the mold in the laundry room in his apartment. His "first-time" illness status meant Jack improved quickly. But we know that should he be re-exposed to another moldy environment, it wouldn't be long before he became "sicker, quicker." His C4a needed to be monitored before and after exposures.

Case in point: Jack goes to California on a summer internship program to teach and mentor low-income high school students, helping them prepare and apply for college. He develops a bizarre upper airway problem. One doctor says it's an allergy, another says it's simple anxiety. Jerry asks for environmental testing of Jack's apartment because he thought it looked moldy. Testing came back "negative," though nothing close to a decent indoor air survey was done. Meanwhile, Jack's getting worse. Jerry brings him back to my office in Pocomoke, Md., where VCS testing is normal and all appears normal except for some muscle spasms in the upper airway.

But all is *not* normal: his C4a comes back 9004, after having been in the low 3000s. Jack was getting hit even when his exposure was so low as to not bump his VCS into the abnormal range—sicker, quicker, indeed!

Can you imagine how it felt to be Jack's family, wondering what was happening to their only child? I spoke mostly with Jerry, but his wife Loretta was now responsible for two hypothalamic-injured patients. She had seen Jerry's former successful life fall into chaos. They'd put up with incredible health demands and somehow found the strength to fight against all comers.

By now, Jack was just beginning an ambitious public service career, one that would take him to countless exposures that would appear innocuous, but they wouldn't be. Jack could choose to live life in a bubble, connected to the outside world by vapor-locked corridors and computer wires, like a modern day Howard Hughes, or he could risk that the next airplane ride or the next hotel room would be the toxic insult that would turn his life into one like his Dad's.

Which option would you want your child to choose?

Jack took a one-year leave of absence to work on Barack Obama's political campaign. Signing on as a field organizer, he rose quickly in the volunteer ranks. When one ponders the basis for the Obama campaign's enormous appeal to young people in this country, some will focus on the candidate. Others, like me, will focus on the candidates' campaign staff. I don't know which factor takes priority, since without the candidate, there's no campaign and without the campaign staff, there's no candidate.

Time will tell if Jack made a mistake trying to have a normal life. We know that new therapies will help his hypothalamus, because change in medical therapy is no different from change in a political campaign: the people who choose medicine and politics to make the world a better place will demand change for the better. As opposed to politics, however, change in medicine isn't subject to the faddish opinions of talking heads, power brokers and 30-second sound bites. Change in medicine demands the strength of conviction even in the face of criticism and fatigue.

I look at Jerry as a model for others to emulate. I told him I wouldn't rest until he could run up a flight of stairs and feel good at the first landing. I may not reach that goal for Jerry, but because of him, I predict that, with precautions, Jack will lead a normal life.

Where do we put the Jerrys of this world in the evolutionary scheme of things? Love of a Father for his Son seems an obvious theme. Then, add to that idea the concept of fitness. We don't determine the evolutionary success of an individual by his or her accomplishments.

Instead, fitness is determined by the success of our offspring. In this regard, defending the children is a marker for a successful life.

Take the idea one step further and put Jerry in charge of a school district facing a budget crunch; they have no money for building maintenance. That scenario fits many if not most school districts these days. Where do you put your energy if you're the one responsible for defending the children? Do we spend money to maintain and secure the envelope of a building so that 25% of them won't be endangered by the predictable environmental consequences of water intrusion? Or do we spend the limited funds on programs that are deemed to be politically correct for bringing up our children?

I'll take a few more Jerrys on my team and a few less politicians any day.

Update, Fall 2010: Jerry and Loretta Koppel are still waiting for Jerry's Workers' Compensation claim to be settled. Jerry's health remains limited, but he still has that positive outlook. Loretta still works fulltime as a governmental policy analyst, but she plans to retire in 2011. Jack Koppel followed Barack Obama's campaign in its final push toward the Democratic Presidential nomination. His health is good and he promises that he remains wary of any "weird smells" or odd health problems.

Jack has continued to have an active and rewarding life that's leading him toward a career in public service. So far, he's managed to stay pretty healthy by avoiding any significant exposures to moldy environments, while traveling across the nation with the Obama Campaign in 2008, returning to finish his college undergraduate degree, doing a 2010 summer study project in Europe, and traveling to professional conferences. That Jack has so far avoided major mold exposures is the result of a lot of hard work by an unusual but effective strategy he developed with his dad's help. They both joke that Jerry has become Jack's "executive assistant"--a euphemism for the proactive team approach they have developed to minimize the potential for Jack's life becoming compromised by mold exposures. They developed and refined this strategy during the 2008 presidential campaign when Jack's work for the Obama Campaign took him to numerous states beginning in January 2008 and ending with the President's election in November 2008. Jerry's role: to use the phone and internet to find mold-free housing options for Jack whether he's traveling the U.S., renting an apartment at college, or traveling abroad--a very challenging task indeed. But so far it's worked extremely well. And Jack recently was informed that he was selected as an "International Scholar"--an award for nationally outstanding young people who want to dedicate their careers to public service in government or the non-profit sector to help improve the lives of others. And he's just received invitations to speak at international conferences in Denmark and Canada.

His career opportunities in governmental relations are broad.

The Love of Father and the Love of a Mother for a Son may bring a Nobel Peace prize to the Koppels or maybe just a Senate seat. Jerry might not be there to see his son on the awards platform in Stockholm, truth be known, but I wouldn't bet against Loretta and Jerry, smiling at each other as Jack goes on his journey, one that will be awesome, filled with love and respect and understanding. Sure, Jerry is a Mold Survivor, but it is the family that provides survival lessons for all of us.

CHAPTER 10 The Oliff Syndrome: Erin Culpepper Regains Her Life

The gelding's name was Skyes The Limit and Erin Culpepper would never forget that sunny afternoon when he nailed down three straight blue ribbons, en route to achieving a "clean sweep."

A champion!

Was there any thrill quite like hearing the crowd's spontaneous applause each time Erin tapped her heels into the horse's flanks, feeling him respond to her "Horse Whisperer" touch as she sent him soaring through another heart-stopping leap?

Months later, when she was so sick with her chronic mold illness that she couldn't even crawl out of bed, this highly accomplished professional rider would remember the hours of her sweetest victory. Joints aching and muscles on fire, she'd lie on the sofa, dreaming of that bright, breezy afternoon in April of 2005, when she and Skyes The Limit had thrilled their audience with a dazzling display of equestrian brilliance.

"Erin, you did it! You did it! Congratulations!" In memory, the bay's proud owner was once again hugging the expert rider and then turning to embrace the victorious steed's chestnut-hued neck. In memory, the show judge was still smiling benevolently from his elevated perch above the ring, and the big crowd of horse-lovers at the Crystal M Arena was still cheering.

Erin Culpepper had been 27 years old on that unforgettable afternoon at Doswell, Virginia, on the outskirts of Richmond. In that hour of splendid victory, her future as a professional show-rider and equestrian coach seemed wide open, full of exciting promise.

Blessed with enormous natural talent and gracefulness that came from in-born athleticism, the blue-eyed and adventure-loving Erin seemed ready to explode onto the world of equestrian competition. There were murmurs of the Olympic trials, yet she didn't do the kind of riding that would bring an invitation.

Skyes The Limit!

But all of that had vanished now. Those bright and shining hours were gone without a trace. Sadly, hope was vanishing too.

By March of 2006, she struggled to get through each day. Oh sure, there were a few good days, days when she felt like she could do the laundry and vacuum the living room rug, but if she even did that work slowly, she would be exhausted for two days afterwards. More often, she could barely stand for long enough to make a sandwich, cutting short the preparation to avoid the sharp shooting pains stabbing her if she stood too long. The fatigue she could fight through; she still had riding students she had to teach on weekends, but the day time job, one she trained years for, was gone. Helping young children overcome disability was too much work for her. Buying groceries became a task when she found herself clinging to the side of her shopping cart at the local Suffolk Farm Fresh where she fought desperately to find the strength required to wheel the grocery cart out to the car.

Erin didn't want to live like this. And deep down, some days Erin wasn't sure she wanted to live. Maybe her expectations had been too high. Maybe the fall from her personal ideal life was too much.

Once she bounded up steps like she was made of air. Now every joint hurt, causing stiffness that lasted all morning. What a loss, as she had always had been so loose-jointed that doing splits and putting her foot behind her head while sitting on the floor was part of what she expected her body to be able to do. It seemed like everything was wrong with her now. She still could fold her fingers over each other and pull her thumbs back to her forearms, but so what.

And as she remembered her PI years (Pre-Illness), back when she was able to ride 20 miles at a stretch through the rolling, wooded hills of her native northern Virginia, the nearly disabled graduate of both William and Mary College (B.A., English) and the University of Virginia (M.A., Education) was unable to remember phone numbers or focus her wandering mind for more than a few seconds at a time. Frequently dizzy and struggling with blurred, hazy vision, she now spent most of her life twisted and curled up, mostly on her back, but only for a few brief moments of sleep before she awoke with blaster muscle cramps if she dared straighten out her ankles. She was lost in a twilight world, full of an inflammation and pure exhaustion that had no name, full of an illness that was stealing her life for no reason.

Badly frightened by her own growing fantasies of "finding a way out," she feared becoming a wheelchair-bound invalid, sitting all day long on the back ward of a "permanent care" facility, snowed by a heavy load of medication for pain, cramping and depression.

A terrifying prospect? Undoubtedly. But it didn't happen to Erin Culpepper.

Instead, after receiving a chance referral from a friend, she limped into my reception room in Pocomoke on a cool, drizzling afternoon in May of 2006. For more than two hours, she told me the story of her agonizing odyssey through the labyrinthine hall of Ass^2 mirrors that is the contemporary U.S. medical system. Still the story that emerged wasn't some fatal disease that jumped out of the bushes dragging her down from her saddle on the 16-hand gelding. It was just mold.

Battling a Cold, All Winter Long

When Erin Oliff was a sixth-grader, she achieved a distinction that seemed highly unusual for an 11-year-old. She became famous for being sick.

"It was really pretty bizarre," she told me during a detailed review of her lifelong medical history. "During the sixth grade in Fairfax, I caught a cold that lasted the entire winter. No matter what I tried, I couldn't seem to shake it. All I did was cough and sneeze, for months at a time.

"After that, I'd come down with the same cold at the start of *every* winter, and it always lasted until spring. It got so bad that I finally went to the hospital to be evaluated, and they found all these nodules on my vocal cords. Then I started having these weird infections on my skin. Next came several attacks of bronchitis, which brought on a hacking cough. Really, it was almost comical; at the high school I attended, my classmates started referring to my constant illness as 'The Oliff Syndrome.' I could still laugh about my constant illnesses back in those days, because they weren't life-threatening."

Raised in the northern part of the state by upper-middle class parents – her father is a successful patent attorney who launched his own firm - Erin managed to shake off her illnesses long enough to gain acceptance to William and Mary, a highly regarded liberal arts college located in tradition-rich Williamsburg, Va. But no sooner had she moved into her 100-year-old, ivy-covered dorm than her health woes began again.

This time, however, her "winter-long cold" was no joking matter. By January of her freshman year, her health had deteriorated to the point that she suddenly passed out in a restroom in the English Department.

Panicked at discovering an unconscious student on the bathroom floor, the English professor who found Erin ran straight back to her office and dialed 911. The ambulance rushed the slowly reviving student to a hospital, where she was given a wide-ranging battery of lab tests aimed at uncovering the source of her fainting spell.

Strangely enough, however, the blood work and other biochemical assays failed to pinpoint the cause of her sudden collapse, and her deteriorating health remained an enigma. During the next four years, as her college career unfolded at William and Mary, she fell sick again and again. But now a host of additional symptoms had emerged to torment her, along with that all-too-predictable "winter-long cold." Increasingly, she struggled with muscle aches and joint pain. Again and again, she awoke from sleep in the middle of the night to find herself drenched in cold, clammy sweat.

Besieged by abdominal cramps and "weird shooting pains, like being stuck with pins everywhere," she also experienced unnerving periods of mental confusion, along with a growing deterioration of her short-term memory. Once a gifted student who found academic assignments easy to manage, she now struggled daily to keep up with her painfully difficult schoolwork.

After receiving her undergraduate degree in English and Education in 2001, Erin would achieve another dream by placing into a graduate-school slot in the University of Virginia education program. There she studied both Learning Disabilities and Early Childhood Special Education and soon decided that she wanted to work among economically disadvantaged, "high-risk" students after landing her master's degree and her teaching certificate.

At first, her career plans seemed to be perfectly on track. Hired to run a regional pre-school program for at-risk kids operated by Head Start in Fredericksburg, Va., Erin thrived on the challenges. Her assignment: design classroom exercises and work with teachers so that her struggling preschoolers could get the very most out of each hour they spent in the program.

"The job was very high-stress, very draining at times," Erin would tell me later, "but it was also very rewarding. We were making a real difference in the lives of these at-risk children, and you could see them learning and growing right in front of your own eyes."

It was an exhilarating time. In spite of occasional health problems during this period, she held her own. She kept on with her weekend work too, at her family's longtime farm in central Virginia teaching riding and "coaching kids" on the fine art of competitive riding.

"I had a fulltime job *and* a part-time job," she told me with a wry grin, "and the demands on my time and energy were really fierce.

"But I hung in there. Participating in horse shows and giving riding lessons was important to me, because our family had spent our weekends and summers on that farm, ever since I was in first grade. For me, participating in equestrian events – like the 'clean sweep' hunter show I won that day at the Crystal M – was a way to maintain a link with my past.

"Looking back, I don't know how I managed to keep going. But I did, at least for a while. And then I started getting really sick. And when that happened, both of the worlds I was living in just fell completely apart."

Getting "Unseated" In a Way You Didn't Expect

By late 2005, Erin was hovering on the edge of exhaustion—she was a pale, trembling young woman whose days as a superbly conditioned athlete were far behind her. During a particularly harrowing visit to her treating physician that year, she had bluntly announced: "Let's face it. I either have cancer or I have an auto-immune disorder and I really need to figure out which one it is."

The doctor nodded and called for more tests. In May of that year, several of those tests appeared to confirm the doctor's long-held suspicion that Erin was suffering from chronic Lyme disease. Soon she was taking huge doses of antibiotics, day in and day out. And yet her condition failed to improve. Weak and helpless, she lay in her darkened room all day long wondering: *If I've got Lyme, a bacterial disease, why aren't the antibiotics helping? How many times do I have to kill these spirochetes anyway? (Editor's note*: Most patients with Post-Lyme syndrome will have an illness that is readily defined as a chronic inflammatory response syndrome. Correcting the inflammatory problems is mandatory after an initial round of antibiotics is completed.)

Her world was sinking fast. "It's a very strange feeling," she would tell me later, "when you realize that you can't even lift a bale of hay. Until I got sick, I'd spent most of my spare time around horses, lifting heavy buckets and picking up heavy saddles. When you ride show horses, you have to be able to control a 1,100-pound animal, and you've got to be pretty strong. So when things got really bad for me, I felt like a different person.

"I didn't even know who I was anymore."

For the once indefatigable Erin Culpepper, this transformation – from fearless professional athlete to invalid – would prove to be the most disheartening part of the illness. Hadn't there been times when she was prepared to risk life and limb for the sake of a blue ribbon?

Make no mistake: jumping show horses over obstacles can be a hazardous occupation, as the entire world had learned back in May of 1995, when famed actor and horseman Christopher Reeve broke his neck during a jumping accident.

Reeve's tragic fall had occurred during a jumping competition that took place only a few miles up the road from the Crystal M Arena, the scene of Erin's "clean sweep" victory aboard Skyes The Limit. In a split second, the famous *Superman* star's movie career ended, and he began what would be a valorous, nine-year struggle that finally ended with his death from an infection on October 10, 2004.

With the celebrity actor's dreadful fate in mind, Erin had always been cautious in the ring. Yet she understood that there are ultimately no guarantees; Reeve had been wearing a safety helmet during his massive spinal cord injury, after all, and he was far from being a reckless competitor. "Unfortunately, you can't always control the way you land after a fall," Erin would reflect later. "You can learn the art of falling, and after you've fallen a couple of thousand times, you know what you need to be doing in the air. You learn how to hit the ground and roll, how to absorb the shock.

"But every once in a while, you'll get unseated in a way you didn't expect, and then there's not much you can do."

Utter Despair . . . And Then a Gleam of Hope

By late winter 2006, Erin had been unseated from both of her burgeoning careers. Sick and enervated, she no longer could manage education programs for Head Start, and she no longer competed in horse shows. Instead, she spent most of her life in a shuttered bedroom, motionless and miserable. "I just lay there," she recalls today, "struggling with a great amount of pain. My joints were killing me, and I felt like my whole body was inflamed."

Yet on the weekends, she pushed herself out to the barn and prepared fifteen horses for the weekends' teaching lessons and academies.

"I had heating pads all over me, and I was taking huge amounts of ibuprofen. But nothing seemed to help. And whenever I made the mistake of leaving that bedroom, I paid dearly for it. Forget going to the grocery store."

In spite of her devastating health problems, however, Erin was blessed with one advantage that would prove to be decisive in the struggle ahead. To this day, she's convinced that her marriage (to real estate attorney Michael Culpepper) in August 2003 actually saved her life. Why? It was simple. As her illness deepened and her strength waned during the years immediately after their marriage, Michael stepped in frequently to coordinate her endless doctor visits and lab tests in the Suffolk-Virginia Beach area. At the same time, he urged her not to give up hope.

"He stood beside me, and he took me to all my doctors' appointments, and he absolutely refused to let me quit," she says today. "I think Michael is the real hero of this story, and I'll never be able to thank him enough for all he has done."

Even with her husband's help and support, however, Erin knew her health was continuing to decline. By April 2006, in fact, she and her doctors were completely out of answers. For one thing, the Lyme diagnosis now seemed highly dubious, since the massive doses of antibiotics hadn't made her feel any better. Looking back, she says that she probably reached the low point one spring afternoon, during a visit to her treating physician. After hemming and hawing for a while, the doctor finally announced: "Erin, you must be depressed, that's all. Quite frankly, I think *depression* is what's causing all your health problems."

Erin could hardly believe her ears. After all the medications and all the laboratory tests... after all the earlier discussions of Lyme disease and taking antibiotics for months at a time, now the doctor was trying to blame her illness on *depression*?

"I told her, 'You don't understand,'" Erin recalls today. "I'm depressed, all right – but it's because I'm really, really sick, and no one can help me!

"That was a truly terrible moment, that afternoon when she tried to diagnose me with depression and start me on antidepressants. At that moment I felt absolutely helpless. I couldn't see how my situation was ever going to improve, and I didn't want to live with my illness anymore.

"I just didn't know what to do. I was totally desperate, and if someone had told me to jump off a cliff or hang upside down in order to get well, I would have tried it."

Strangely enough, however, the misdiagnosis of Lyme disease would prove to be a key step on the long road to her eventual recovery. "A cousin of a friend from Fairfax, Virginia had struggled for a long time with Lyme disease," she recalls, "and had then been treated successfully for it at a clinic in Pocomoke, Maryland just up the Bay Bridge Tunnel. She told me her story and urged me to make the trip there. I had my doubts, because I'd been through so much disappointment and disillusionment with doctors in the past.

"You know, at first I told myself: 'Okay, so this is another physician with yet another approach to Lyme disease. Do I really care? Do I really believe it's even *possible* for me to get well again?' I struggled with myself like that for a few days, and then I finally managed to overcome my doubts. And so I made the drive over to the [Maryland] Eastern Shore, where his clinic is located. But I didn't expect much, really. At that point, all I wanted him to do was to find out – once and for all – whether or not I actually had Lyme disease.

"Imagine my surprise, when he started by taking my lifelong medical history, then ordered extensive lab tests and tested my vision for 'contrast sensitivity.' And when I asked him why he was testing my *eyesight* in order to try and uncover what was wrong, he replied in a way that nearly knocked me out of my chair."

Erin, let's think about that drafty old farmhouse where you spent so much time as a kid. And that ancient red barn where they kept the horses, and that 100-year-old dormitory at William and Mary, and that damp, cracked basement where you worked at Head Start...

As we have now seen, Erin, those buildings were all water-damaged. I think you may have been struggling with a chronic inflammatory illness caused by exposure to biotoxins and inflammagens in the interior environment of those buildings all these years, and nobody ever knew it.

First Step: Remove the Toxins

The differential diagnosis in Erin Culpepper wasn't too large. Sure she had a number of individual symptoms (28 to be exact) and each of those could be caused by many things, but ***what single illness or group of illnesses could account for all the symptoms, the lab abnormalities she had and the VCS?***

VCS testing is disarmingly simple. Patients who can see well enough (better than 20:50 vision) are asked to identify the direction of orientation of sinusoidal wave patterns (lines) placed with varying intensity of gray against a gray background. Sure enough, in biotoxin patients there is a predictable decline in the neurologic function of contrast (the ability to see an edge), that is almost never seen in control patients. Finding a VCS deficit, like Erin had, didn't mean that mold or Lyme or ciguatera was making her ill. But some illness that had a biotoxin property to it sure was. (*Editor's note*: confounding occupational exposures to metal fumes, metal dusts, solvents and hydrocarbons were ruled out during her first visit to Pocomoke.) As Erin's illness unveiled itself in the vast number of abnormal measures of innate immune responses, she looked for proof of exposure by doing ERMI testing. She found it.

Why did she get so ill? More than anything else, this high-spirited horse-lover had simply been the victim of a genetic inheritance of immune response genes (HLA DR) that made her as vulnerable to poisons from a water-damaged building as a human being can become (*Editor's note*: at most **0.03 percent of the general population** shares her level of vulnerability), she'd also experienced the misfortune of having grown up in a world that was rife with exposure to the interior environment of WDBs.

How ironic it now seemed to Erin to discover, at age 28, that the "Oliff Syndrome" her friends had joked about in high school was *real* – but that its true name was actually "chronic inflammatory response syndrome."

It was a painful realization but now that the correct diagnosis had been made through Visual Contrast testing and follow-up blood tests that showed the healing response to treatment, Erin's true recovery could finally begin.

The first step, as always, was cholestyramine (CSM), which was long ago approved by the FDA as an effective cholesterol-lowering medication. CSM is inert, an anion-binding positively charged resin that adds nothing but does subtract certain negatively charged foreign substances from the gut of sufferers.

Within a few weeks of beginning her daily CSM regimen, Erin would discover that her symptoms stopped getting worse. CSM perhaps was helping but it wasn't the whole answer. But she wasn't well-not yet. One by one the biomarkers for her illness returned to normal yet she still was ill, worse when she went into moldy apartments, stores and homes. Because of her illness-related inability to use oxygen efficiently at the capillary level (in order to burn glucose), she was still chronically starved for energy.

When would she be cured? Maybe never; but certainly not when her C4a was so high. And her VIP was so low. And her capillary hypoperfusion had to be fixed.

We had answers for those problems. Just take things one step at a time.

As you have already read, the problem that Erin had, euphemistically called "post exertional malaise," simply was inadequate oxygen delivery to tissues during and after exercise. One way to defeat this inflammation-caused oxygen starvation is to use exercises that prevent going beyond the "anaerobic threshold" in exercise to re-train muscle beds to extract oxygen more efficiently. The program starts at very low level activity (exercise bike or treadmill indoors) done the same way every day for an increasing time period as tolerated for up to 15 minutes. After the first 15 minute period has been slowly reached, then floor exercises such as sit-ups, crunches and leg lifts can begin. Slowly increase the time of doing the same thing every day after the 15 minutes of bike or treadmill until another 15 minutes of floor work is tolerated. Then add free weights to work the upper extremities, steadily increasing the exact same work-out until a third 15 minute span is tolerated. Now, with stretching in between each of the 15-minute sets, return to the first 15 minutes and jack up the exercise intensity. Write down what you do so you always do the same things the same every day. No days off.

Erin loved this idea, ideally suited to her approach to life. "You mean I can *take control* of my own illness by anaerobic threshold exercises? I am on it. How far can I go?"

You should see her exercise records. Hundreds of sit-ups (thousands per week, I'm not kidding); crunches and more. Every day. She took another leap up the ladder to recovery here.

Still C4a didn't fall. We had completed a clinical trial (approved by all the necessary regulatory groups including an IRB) that proved that low dose erythropoietin lowered C4a and improved capillary perfusion, especially in the brain. Erin was ready to go for the shots, despite the possible risk of some clotting and blood pressure changes (which never showed up ever in her) and the cost. By taking Procrit on a regular daily schedule and also engaging in the anaerobic exercises, patients struggling with cytokine-induced oxygen starvation and the inability to metabolize glucose efficiently can gradually rebuild their strength and endurance over time.

Her recovery matched the correction of excessive levels of lactate seen in the brain using MR spectroscopy. High lactate means low oxygen delivery. The erythropoietin worked well but she was still so susceptible to minor exposures. Nothing I had done fixed that increasing reactivity. After discovering mold in her condo in 2003, she and Mike moved to a small apartment temporarily. Mike kept looking for safe houses, but finding an ERMI of (-1) because of her incredible rate of rise of C4a with just 10 minutes of exposure in a coastal area of Virginia is tough. But not impossible.

Her recovery, which is now well advanced, illustrates the key medical insight to emerge from my decade (and more!) of studying biotoxin illnesses such as those caused by mold: *Based on recent breakthroughs in brain-imaging technology, practicing clinicians can now accurately measure the metabolic disturbances caused by immune-system inflammation – and these brain-measures can then be used to monitor therapies based on appropriate medications (and anaerobic exercise) that will eliminate metabolic abnormalities and thus correct symptoms.*

Still Erin wasn't there yet. She politely asked one day what would happen if she became pregnant. She was living in a safe house. She had learned all about mold exposures and knew within ***two minutes*** when she had to get out of a new building. There were no guarantees in mold illness and pregnancy, like we can just about give to the expectant Mom with rheumatoid arthritis (RA). The RA Mom-to-be always feels better; not so the mold-illness-Mom. And besides, all the meds she needed were really all posing possible risks for their use in pregnancy. My track record of women carrying pregnancies after conception in moldy areas wasn't good. And a lot of the kids born to moldy Moms just didn't think well, though some did just fine. But she wasn't herself yet, so honestly I didn't think pregnancy was a good idea.

Still, there was a robust literature on the *benefits* of epo in pregnancy (I gave her a bibliography), so maybe she could continue the one drug that had done her more good than any other even while pregnant.

Coincidentally, one of the blood tests I drew that day was TGF beta-1, the new king figure of cellular immunity and the key player in the many bizarre autoimmune, neurologic, respiratory and cognitive impairments that accompany mold illness. Sure enough, her level was way too high, over 15,000. Was this the missing link?

Cambridge Biomedical was the first lab to agree to run TGF beta-1 for me. I begged LabCorp and Quest, but without a track record of demand, they were justifiably reticent to invest in a new test. "But it isn't a new test. Just look at PubMed! There are 45,000 references to TGF beta-1." Al Correia, Vice President at Cambridge did look at PubMed and saw what I saw.

"We can set this up for you as a research assay. If it turns out the way you think, then we can offer it to the major commercial labs." In April 2008, Al got the job done. By late 2008, both LabCorp and Quest took the tests on which meant that my patients and those of docs across the country could access insurance coverage for this spectacular advance in diagnosis. And they did.

Erin looked at me when I told her that we now had to fix TGF beta-1 by using losartan, a blood pressure pill that had a break down product (called EXP 3179 by Merck) that lowered TGF beta-1 without correcting high blood pressure. I excitedly had sent Merck a request to do a study on high TGF beta-1 and EXP 3179, but they expressed no interest, as losartan was going off patent. But losartan could drop her blood pressure too much, so we couldn't work on TGF beta-1 that way.

"Well, we could try VIP. It is allowed to be compounded so maybe the benefit I have seen in so many others would be repeated in you."

I think the active word I should use here is "Bingo." By using low dose epo and VIP and staying away from re-exposure, Erin was now in control of her inflammatory responses for the first time in years. We had a winner and now Erin was on a Mom quest. No, she doesn't just sit back and passively enjoy having a life. Not Erin. She has had so many hurdles to leap over, and now she has arrived. Not all Moms who bring children into this world will know how to protect their children from mold exposure. Erin does. Mike too. Given Erin's genes, the child is going to have one or the other of the dreaded genotypes. Yet that doesn't mean a life sentence to an existence like Erin had for so many years, not when she knows as much as she does now!

Climbing Back on the Horse

Ask Erin Culpepper to describe her life today, and she'll share her excitement as she tells you about her recent return to the equestrian world.

"I'm back in the saddle," she says with a bright, ringing laugh, "and I've also returned to teaching children how to ride, only now it is kids with disabilities. Therapeutic riding, as they call it, is a wonderful idea where getting to know the outside of the horse is good for the inside of the person. "I can make the difference that I have always wanted."

"I really don't think I can find the words to describe the positive turn my life has taken, after I finally got an accurate diagnosis for my chronic inflammatory response syndrome illness and then got started on appropriate treatment. My life isn't easy, and there are still some days when I feel stiff and sore, with aching muscles and joints. And those days are difficult, because they remind me of the 'living death' I endured for so long, all through the years of my undiagnosed illness. Really, I shudder to think where I'd be right now, if I hadn't taken my friend's suggestion and driven over to Pocomoke that day."

She pauses for a moment, reflecting, and then her blue eyes light up with youthful energy and hope. "I'm taking it day by day," says Erin Culpepper. "I'm working on my exercises, and taking my medication, and watching what I eat. And I'm feeling better all the time. I really do feel like I'm on the road to fully recovering my health.

"Hey, I'm back on my *horse* – and I'm ready to ride!"

Post Script: a Word about Hypermobility

I mentioned that Erin was incredibly flexible, with long arms and hypermobility. Her HLA DR of 11-3-52B and 4-3-53 wasn't a surprise. It happens all the time. For years I had noted the occurrence of an increased risk of profound fatiguing illness in people like Erin, with wingspan greater than height. The literature is robust that hypermobility is associated with Chronic Fatigue Syndrome. Why?

It turns out that the production of TGF beta-1 is incredibly increased in people with the body habitus (shape) that Erin has (See Chapter 4).

So if there is hypermobility, a genetic trait that is associated with high TGF beta-1 and that person has a linkage to HLA DR 11-3-52B and HLA DR 4-3-53, wouldn't it make sense to monitor kids like Erin who have Oliff Syndrome for elevated TGF beta-1? Just imagine the benefit we would see in reduction in learning disability, chronic fatiguing illness and chronic autoimmune problems associated with high TGF beta-1.

And while we are at it, let's start looking at high TGF beta-1 in cases of asthma. Do a PubMed search for TGF beta-1 and remodeling, a vital component involved in the pathogenesis of asthma and you will see changes in cells of lung, heart, liver and kidney lighting up in study after study. Add the burden of remodeling, caused by TGF beta-1 producing fibrosis in lung, liver and kidney and you will see my concern.

A simple example might suffice: a local physician called wanting to know if I had information about nephrogenic systemic sclerosis, a condition that is increased in people with pre-existing renal disease who get a contrast dye called gadolinium that is used in radiology procedures. Sure enough, the two papers that are published on diagnosis of this condition both talk about high TGF beta-1.

So what is Erin's case telling us? Hypermobility, genetics, environmental exposures and illness are all connected. What a surprise. Our genomics studies should be done soon that will tell us what is the molecular basis of gene activation for mold illness, sorted by genotype.

Post post script:

September 9, 2010. The phone call came today. Erin is now three weeks pregnant. Joyously, we all agreed to hold our collective breaths for the next eight months or so.

CHAPTER 11 Benomyl: Mankind's Worst Self-Imposed Environmental Disaster?

"Genetic force is the most awesome power Nature can wield."

– Jeff Goldblum in *Jurassic Park*

Fungi eat everything first.

Here's a simple question: would you rather eat strawberries with softened fruit and black spots or perfectly shaped berries with no blemishes? Or put another way: would you eat strawberries that already have been digested by the world's ubiquitous fungi? Because the spots and defects on the fruit you throw into the compost rather than into your cereal bowl are the results of fungi digesting food. Give fungi a chance to digest something that has energy, whether it's a strawberry, a cantaloupe, the drywall in your closet or the wood framing holding up your roof, and they'll eat it.

No one wants fungi eating their fruit before they do. And you don't want to eat the fungi themselves, either, although the fuzzy little critters probably wouldn't hurt you if they end up in your personal sewage treatment system. Now put yourself in the strawberry farmer's position: He knows that his berries must be free from fungal rot if he's going to make any money. So how does he stop fungi? Since World War II, the answer has been simple: use fungicides. Better living through chemistry. (*Editor's note*: In fairness, besides chemical use, modern advances in seed production, planting, harvest, storage and transport of crops have enhanced the profitability of agriculture.)

The fungicides' cost to the farmer is offset by the enhanced income he receives from his beautiful, spotless, non-mushy strawberries. Just look at the fruits and vegetables section of today's mega-grocery stores: we've grown up in the chemical era of agriculture, and we expect to be able to buy gorgeous berries, nearly year-round. Do you remember your parents telling you when you were a child to wash your fruit and vegetables first, because "you shouldn't eat the chemicals on them?"

So fungus-controlling pesticides became widely used. Everyone's happy. The farmer can sell his crop, the middleman can take a big chunk of the profit and we're all happy to buy nutritious foods (that we wash) at affordable prices.

Except one thing—not all the fungicided fungi die. The *survivors* are resistant to the chemical, just like Staph germs can become resistant to antibiotics when all of them aren't killed by those antibiotics. Only the resistant survive in the face of constant selection pressure from chemical use. It's a basic concept of Evolution: natural selection shifts the abilities of populations of survivors over time.

Face it: when we selectively kill particular organisms, we have no idea what will be the outcome years after the killing is over. In biology, we know that Nature finds a way.

In the case of the particular mutagenic fungicides I'll be discussing, the natural selection hasn't been so natural and the time lapse for changes in populations has been compressed from eons to fractions of decades—just one reason I think these mutagenic fungicides are the source of the biggest agricultural disaster in the world.

Are the resistant survivors the same as the dead fungi? Of course not. Do we know what monsters we'll create in 2016 from the fungicides we use now? No, we don't, but by using chemical selection, we should always be reminded that monsters might emerge tomorrow as a result of what we use today.

Yet we often blindly trust regulatory governmental agencies to label a fungicide as safe for use. It's scary but true that too often those agencies *don't have a clue* about the long-term consequences of using these fungicides on ecological balances. It's also true that people who work for these agencies are being influenced by the mega-companies they're supposed to regulate. Once the agencies have stamped the product as safe, you can forget ever having any post-marketing recall done —say, after allegations of safety problems have been made. If the government says safe, well, that really means *safe from blame*. Their stamp of approval provides an umbrella of protection from liability as well, though recent court decisions may alter the enormous protection from legal claims provided to manufacturers by some of our Federal agencies.

This chapter's point is simple: we've introduced microscopic monsters into multiple ecosystems by using a breakthrough chemical for crop protection, the same one that everyone used in the late 20th century. DuPont sold more Benlate than all other fungicides *combined*; it became the all-time leading seller in the fungicide market. But there are consequences to the widespread use of any chemical and these are directly related to what the chemical does, what happens when the chemical breaks down (what do the "degradation daughters" do?) and what other unknown effects the chemical eventually has. If the chemical is a "pollutant," we're creating a plethora of polluted sites.

Four Important Questions.

(1) Have we acted as a powerful force of natural selection by using one of the most mutagenic agents ever seen?

(2) Have we unleashed resistant microbes that make toxins and thus can cause profound human health effects?

(3) Do these toxic organisms inhabit our croplands, estuaries and lakes, our offices and factories, our homes?

(4) is the toxicity and spread of the organisms related to appearance of anew large segment of DNA, a chunk of genes, if you will, that spreads from one kind of one-celled creature to another?

The answer is "yes" to all four questions.

In this chapter, I want you to see why using the systemic fungicide benomyl[1] started out as the agricultural "best idea" ever, and why in the end, it became a disaster. Based on the latest research, this agricultural disaster also may be a human health disaster (1) as well. (*Editor's note:* noted references are listed in Appendix 4, with pronunciations listed in Appendix 5.) I'll tell you about the resistant strains of fungi that emerge in agricultural products after benomyl applications, but I want you to think about what else mutagenic agents like benomyl could be doing to the world's microbes in years to come. Think about gene-altering benomyl, and its effects on the cell-division processes of all growing organisms.

Why did we need systemic fungicides?

If you were a commercial fruit grower in the 1950's, the final result of uncontrolled fungal growth in your farm had a more common name: bankruptcy. In the years before benomyl, you were like the owner of the plant from Outer Space in "Little Shop of Horrors" that wouldn't stop eating and growing—your only choice to keep the ravenous fungi at bay was to continue to dust, spray or drench your crops with a surface-applied fungus-killer. If the costs of buying the fungicides weren't bad enough, imagine how much it cost to apply the chemicals *again* every time it rained and then again at every stage of plant growth. Imagine the difficulties you'd face if you wanted to grow wheat, corn or soybeans or sell any plant product, because fungi ate it all—seeds, sprouting seedlings, immature plants, fruiting bodies, ripe crops, harvested crops sitting in storage and processed crops. The bottom line costs of agriculture in the chemical era kept rising, but with the use of fungicides, at least *there was* a bottom line.

Just look at what pre-chemical era farmers had to do to keep fungi at bay after food crops were harvested: peaches, pears and apples were picked when they were too green to eat and then the fruit was stored in coolers. Eventually, it would fully ripen on the shelf with questionable quality. "Shelf-life" referred to how much time would elapse before you'd see fungal rot, and "perishable" meant your food was also food for fungi.

In the old days, roots and perishable food crops were stored in "root cellars," underground structures that controlled temperature and humidity long enough to prevent uncontrolled fungal growth. Our ancestors tried mightily to keep food from rotting by canning, drying, salting and brining (pickles or salt pork anyone?). Compared to starving, the effort was worth it. But in the chemical era, industry had a better idea - make it a standard practice to apply fungicides that could fight off voracious fungi every step of the way, from seed to sprout to harvest to storage. That's how applying fungicides became standard practice. Not only did it increase shelf-life, but it also increased chemical industry profits.

Organic foods are coming of age in the U.S. now, but any large-scale organic food producer must take into account the management or control of fungi in determining their bottom line, just as chemical-using food producers must do. That means the chemical users employ fungus-killing drugs or methods everywhere along the food pathway of growth, harvest, store, process and store again. The alternative for the organic food producers is to develop such an efficient transport system that the *time* from field to shelf in a mega-store is so short that the loss from fungal rot won't destroy all profit. Even now, if you go to the big organic food stores, look at the size of the dumpster out back and see how often that dumpster is emptied of spoiled foods. One of the costs of expensive organic food is *wasted expensive* organic food.

The idea of organic produce sellers to buy "in region," and buy "in season" is one attempt to avoid product loss from fungal growth. Without the ability to prevent loss of product over time, organic food sellers can't compete with the diversity of produce found in the chemical-using mega-markets. When I buy beautiful strawberries from Honduras and gorgeous grapes from Chile in my local mega-store in Maryland at Christmas, I'm buying fungicides.

Fruit of the Fungicide.

The problem of fungi consuming our foods before we do is basically a history of agriculture and civilization. Imagine a culture dependent on wheat for food to carry them year-round: they've got to grow the grain, avoiding crop damage from fungal agents like rusts and smuts. Then they've got to harvest the grain when it's dry, keep it dry and store it somewhere where it'll *stay* dry. Even if the society's agriculture techniques have advanced to efficient harvest and raw material storage, they still have to mill the wheat and turn it into finished foods, like bread, which in turn have to be protected from the ever-relentless fungi attack. Remember that fungi, like taxes, are ubiquitous. They are everywhere, at every time.

Animal feeds are fair game for fungi, too. When the profit from growing one chicken can be as low as ten cents per bird, an increase in feed cost from one cent to two cents per bird isn't a minor consideration. For costs less than one cent per 100 birds, copper sulfate—an EPA Level 1 biocide—can be added to prevent spoilage.

Every time you turn around, there's another fungus waiting to eat your energy source. Don't forget, by the time you can see fungus growing, there will be about 250,000 colonies per square inch and each colony can produce another 250,000 babies (spores) almost overnight. So many agents that could promote fungal attack are faced by crops: insect damage, weather events, physical damage are just a few. If you were sloppy applying topical fungus-killers, or if you weren't vigilant and didn't control fungal growth after heavy dew, "missed a spot" could mean "missed a harvest."

Fungi have been eating plants forever; that's what they do. For the most part, fungal damage to agricultural crops has determined history. For example, as reported in a thorough and thought-provoking book, *Famine on the Wind* (2), the politics of land use in Ireland in the early 1800s essentially guaranteed that potatoes, the staple of the Irish diet, would become vulnerable to fungal attack. Think about the Potato Famine of 1848: what would have happened if potatoes could have been protected from rot? Would we have had masses of hard-working (though starving) Irishmen coming to America? And what did the newly emigrated Irishmen do for the Union Army in the Civil War or on the railroads of the mid-19th century? Without fungal damage to potatoes, would the history of the mid- to late-19th

century U.S. have been different? You bet.

Now take the fungal problem to Central America. Want to grow bananas? The recent history of Central America is colored by the influence of giant banana companies on the economies and politics of the banana-growing countries. If you're the grower, you might choose to run several governments where banana crops are the source of a country's cash; if you "run" the country, the cash will come from you. If you're a giant company - United Fruit in the 20th century, let's say - you have to be able to control fungus damage to banana crops. Since you own the land, with great influence over society and the government, you'll just need to find a way to sell bananas without mushy ends, black spots and odd smells. That means fungicides.

Want to live in France and grow grapes to make lots of wine? You need fungicides or you'll have moldy grapes on dying vines, no wine and no cash. You'll never get the grapes into a crushing vat, much less get the juice into an oak cask to make wine. I acknowledge we need yeast (a fungal agent) to make alcohol from fruit sugars, but only the right kind in the right amount at the right time. We all were taught that the great scientist Louis Pasteur saved the wine industry by keeping harmful microbes, including bacteria and fungi, out of fermenting wine, but don't forget it was the fungicide—copper sulfate in the Bordeaux mixture—that saved the grapes and the vines.

Actually, the copper-based Bordeaux mixture was the first broad-spectrum fungicide that saved crops inexpensively and it's still in use today. Dusting miles of grapevines isn't cheap and if you miss a few spots in your dusting... well, you've got more spots to answer to later.

How about Ceylon and its burgeoning coffee crops in the late 1800s? Because the fungal blights wiped out the coffee plantations there, now we buy *tea* grown in Ceylon. The coffee that used to come from Ceylon now comes from South America.

And let us not forget the blue mold that wiped out tobacco plants and tomatoes in the late 1990s in Maryland, Virginia and North Carolina. My start in environmental ecology began when the use of a combination of copper and dithiocarbamate[2] fungicides along rivers let loose the "Cell from Hell," *Pfiesteria piscicida*[3], which killed fish and made people sick. Who'd ever use such old-fashioned mold-killers on huge cash crops like tobacco and tomatoes? That was because the blue mold was resistant to all other, newer fungicides, so only the old remedies had any benefit.

The cost of disturbing estuarine ecosystems - which is where those fungicides ended up after rain events and run-off - wasn't apparent to governments back then, but the combination of copper and dithiocarbamates altered predator-prey relationships in estuarine ecosystems. The fungicides were labeled as safe to use along estuaries. No one even considered what the long-term effects of the fungicides would be, in part because no one had ever heard of *Pfiesteria*. (3) These fungicides didn't cause mutations but they sure brought along an unexpected effect.

As an aside, now that we know that the toxin made by *Pfiesteria* was dependent on copper, we can again ask the question, "What does fungicide use by population A, do to population B?" Were it not for the blue mold killing tobacco and tomatoes, and copper and dithiocarbamates killing the blue mold, I don't think we would've seen any *Pfiesteria* blooms in the Pocomoke River; no one would've argued with me about the source of the *Pfiesteria*-illness; I wouldn't be interested in biological toxins the way I am now; and this book would never have been written.

Returning to the subject of fungi, even now, State and Federal agencies won't believe that benomyl use created an agricultural disaster. After all, benomyl was always used strictly according to the Federal-determined label, so any and all use was considered safe by the agencies. Besides, it's been years since benomyl was used; it was pulled from the market in the 1990s. But the courts haven't agreed that benomyl was safe, and the total payout from DuPont now exceeds $1.5 billion. More cases are pending.

Think about it: if there's an ecological disaster, with blooms of unusual organisms, whether they're harmful algal blooms in the Pocomoke River in Maryland (a tributary of the Chesapeake Bay) or molds growing in the prehistoric caves of Lascaux, France, and if the problem can be blamed on nutrients or God, then no one - or at least, no one *company* - can be blamed (read *sued*) for causing the problem. (Time Magazine did an interesting story on the fungus that was growing in the cavern in the June 19, 2006 issue, on pages 44-48. There was far more fungal growth after fungicide use than before.)

The *Pfiesteria* problem? Nutrients, of course (*wrong*). The prehistoric cave paintings now being chewed for food by *Fusarium solani*[4] and *Pseudomonas fluorescens*[5]? Clearly an act of God (*also wrong*). The benomyl problem? What problem? The agencies said the product was safe.

Can you hear the manufacturers of benomyl singing in chorus, "Benomyl use was specific to certain crops; it was monitored by its manufacturers, the Environmental Protection Agency (EPA) and the World Health Organization (WHO), all of whom determined it was safe for use, and that it never was a mutagenic agent. Something else caused all those nursery sites—more than 1200 in Florida alone—to have re-cropping problems. It was probably the farmers themselves; they didn't do something right. And we certainly don't agree that using benomyl on bananas had any effect on Ecuadorian shrimp farms." (*Although massive die-offs were documented.*)

It's amazing to me that so many farmers forgot how to farm all at the same time (the agencies implied that idea), and that benomyl labels say nothing about causing memory loss or poor judgment! Agencies just blamed the farmer.

Thus the question arises – once the horseman of the apocalypse has been paroled (received government approval), whose job is it to keep track of him to make sure he truly causes no harm?

The durable legacy of benomyl: mutations, illness and crop damage.

Now let's go to the lakes of Central Florida where nurseries grow ornamental plants for sale all over the USA. We're here to meet Frank Fuzzell. He's disabled and unemployed, the owner of a 20-acre wasteland that was once a productive nursery. Frank lives with 10 years of memories of trying to find some organization whose officials could make a difference in his life. He kept trying without success, as rampant growth of toxigenic fungi closed down his nursery and ruined his health.

Like more than 1200 nurserymen in Florida, Frank used benomyl, which everyone thought was a miracle fungicide. Now all he (and so many others) has to show for a lifetime of labor is an unusable, unsellable barren nursery. He sometimes looks it over on one of the few good days he has between debilitating episodes of fatigue, brain fog and chronic pain.

Frank hasn't given up his fight to let the world know what happened with benomyl. When you go visit him, get ready to see what one *million* documents actually look like. Frank's home is a document repository, an archive of what people knew about benomyl. He has more data than anyone I've ever known.

Frank isn't alone. Listen to Carl Grooms, a strawberry grower from Plant City, Fla. who keeps four cases of what he calls "poison" in the back seat of his first car, a 1969 Chevy Chevelle, parked in a shed. He thinks that chemical - the DuPont Co. fungicide Benlate DF - ruined his strawberry fields in the early 1990s.

"I'll never get rid of it, and I'll never forget," he said Sept. 24, 2006, in the *St. Petersburg Times*, talking about the brand-name version of benomyl.

Reporter William Levesque wrote, "For DuPont (maker of the form of benomyl sold as Benlate®), for the farmers who suffered wilted fields of plants, for scientists trying to fathom an answer, for lawyers litigating in courtrooms from Honolulu to Miami, Benlate is a story without a seeming end, a twisting, confusing, maddening road without equal in the history of American agriculture (4)."

Sadly, Levesque and Grooms don't know how much worse the problem actually is.

An abbreviated time line in benomyl litigation (4)

- **1987** DuPont introduces Benlate, a form of benomyl.
- **1989** DuPont recalls batches of Benlate contaminated with atrazine, an herbicide.
- **1991** DuPont recalls Benlate a second time; pays farmers for crop damages.
- **1992** DuPont says Benlate doesn't produce lasting soil contamination. Soil scientists don't agree.
- **1993** DuPont's first loss in court. Arkansas tomato growers awarded $10.25 million.
- **1994** DuPont wins its first Benlate case. Florida blueberry growers lose.
- **1994** DuPont denies that Benlate is contaminated with sulfonylureas[6], another herbicide.
- **1994** DuPont settles 220 lawsuits (only half of those filed) for $214 million.
- **2003** Florida Supreme Court upholds decision for the plaintiff in a case relating to anencephaly[7]—actually microophthalmia[8]— in the Castillo v. DuPont case (No. SC00-490 7/10/2003; re-hearing denied Sept. 4, 2003).
- **2006** U.S. Supreme Court refuses to consider DuPont appeal in one of numerous Benlate cases pending. This case involved Hawaiian growers who settled with DuPont before learning, they say, that the company withheld evidence that Benlate was contaminated with an herbicide. The court ruling allows farmers to bring racketeering and fraud charges against DuPont.

Eleven years and 560 lawsuits. Just imagine the clogging of the court dockets if the concerns about human health effects of people *living next to* a nursery with crop damage had ever come to trial in just one precedent-setting case.

Frank told me (5) that the biggest agricultural and environmental disaster he knows of came about because farmers needed to find a fungicide that worked *systemically*, that is, throughout the plant, doing its work from the inside. After all, if you give the plant a dose of a broad-spectrum fungus-killer *one time,* even haphazardly, to protect it until the plant died, you'd save time *and* money. Plus, benomyl was so much more effective than previously used fungicides. In the late 1960s, the bottom line improved via advances in chemical fungicides. Instead of *topical* protection, repeated endlessly whenever the wind blew or rain fell, everyone began to use *systemic* fungicides instead. They got inside and spread everywhere in the plant.

The idea was simple: *systemic* fungicides provided protection for cash crops cheaply, using a pesticide that didn't have to be applied repeatedly. One treatment and the plant would be saved from rot, wilt, smuts, rusts and whatever else the fungal world could dish out. Any fungus that tried to grow on a leaf, stalk or fruit would be killed rapidly, before the plant suffered any damage. As *Famine on the Wind* discusses, the search for an effective systemic fungicide was akin to the Holy Grail of agriculture. Preserving crops so they could become food on the table meant preserving society in days of yore. That idea is no different now, when preserving food means preserving large chunks of a national economy.

Benomyl was the first systemic fungicide; it was sold by DuPont and other companies. Benomyl is a substituted azole[9] fungicide that works by disrupting the normal transfer of chromosomes when cells divide. It alters genes and alters the genetics of the few fungi that could survive the onslaught of benomyl. Such substituted azoles are widely used in medicine. The various side chains on the molecule change what the drugs do. Look at oral, topical and intravenous fungus killing medications we use every day: many are azoles. How about the miracle class of proton pump inhibitors we take for heart burn and indigestion? They are now over the counter and they are all substituted benzimidazoles. And the interesting psychiatric medication Abilify? It is an azole, too.

Any one tracking mutations from these other azoles?

No one could have known that such a seemingly minor change in our chemical use would have such profound effects. Now, though, genetically altered organisms are toxin-formers and their by-products flourish in fields, lakes, estuaries, oceans, homes, buildings - and inside us as a result.

Genetic changes followed use of mutagenic fungicides. Resistance to benomyl was the first evidence of the results of chemical selection for a new race of organisms. As it turned out, the genetic changes weren't just in one base pair of DNA here or there: large chunks of DNA were affected and therefore, large changes in the survivors' behavior were introduced.

It didn't take long to find fungi that were resistant to benomyl and thus created problems for farmers. Here's a list (6):

Known Tolerant (Resistant) Strains—1979

Fungus	Year	Crop	Location
Ascochyta[10]	1971	Ornamentals	Holland
Aspergillus[11]	1973	Laboratory	Europe
Botrytis[12]	1971	Ornamentals	Worldwide
Botrytis	1971	Eggplant	Japan
Botrytis	1971	Strawberries	Europe, New Zealand, U.S.
Botrytis	1971	Beans	Pacific N.W.
Botrytis	1971	Grapes	Europe, New Zealand, U.S., Japan
Botrytis	1971	Caneberries	United Kingdom
Botrytis	1971	Lettuce	United Kingdom, U.S.
Botrytis	1971	Cucumber	U.S.
Botrytis	1971	Tomato	United Kingdom, New Zealand
Ceratocystia[13]	1976	Elm	U.S.
Cercospora[14]	1973	Peanuts	U.S., Australia
Cercospora	1973	Sugar Beets	Greece, U.S., Japan

Cercospora	1973	Celery	U.S.
Cercosporella[15]	1973	Cereals	Europe
Cercosporidium[16]	1973	Peanuts	U.S.
Cladosporium[17]	1973	Laboratory	Europe
Collatotrichum[18]	1973	Strawberry	U.S.
Collatotrichum	1973	Coffee	E. Africa
Collatotrichum	1978	Bananas	West Indies
Corynespora[19]	1979	Eggplant	Japan
Diplocarpon[20]	1974	Rose	U.S.
Fusarium[21]	1973	Bulbs	U.S., Holland
Fusarium	1973	Ornamentals	U.S.
Fusarium	1973	Bananas	Central America
Fusicladium[22]	1977	Pecan	U.S.
Monilinia[23]	1976	Stone Fruits	Australia, U.S.
Mycosphaerella[24]	1978	Banana	Central America, Philippines
Mycosphaerella	1979	Citrus	U.S.
Neurospora[25]	1973	Laboratory	Europe
Penicillium[26]	1971	Bulbs	Holland
Penicillium	1971	Ornamentals	Holland
Penicillium	1971	Citrus	U.S., Australia, Japan
Penicillium	1971	Apple	U.S.
Phialophora[27]	1973	Ornamentals	France
Powdery Mildew	1969	Apples	U.S.
Powdery Mildew	1969	Strawberries	Japan
Powdery Mildew	1969	Eggplant	Japan
Powdery Mildew	1969	Curcurbits	Worldwide
Powdery Mildew	1969	Ornamentals	Worldwide
Powdery Mildew	1969	Grapes	U.S.
Powdery Mildew	1969	Turf	U.S.
Rhizoctonia[28]	1973	Laboratory	Europe
Sclerotinia[29]	1974	Turf	U.S., Australia
Septoria[30]	1974	Celery	Australia, U.S.
Septoria	1974	Cereals	Europe
Septoria	1974	Ornamentals	U.S., New Zealand
Ustilago[31]	1973	Laboratory	Europe
Verticillium[32]	1973	Mushrooms	U.S.
Verticillium	1973	Strawberry	United Kingdom
Venturia[33]	1974	Apple, Pears	Worldwide

No one knew to look for resistance to develop in other organisms that might also be affected by benomyl or its degradation daughters, such as carbendazim[34]. The only regulatory inputs came from the

EPA and the U.S. Department of Agriculture (USDA). No systematic search for genetic problems associated with benomyl's use was ever completed.

What would the grower do when faced with a resistant organism? He'd add another fungicide to the mix or rotate fungicides. Benomyl was too good a fungicide to avoid using just because it didn't kill a few species.

What this caused in evolutionary terms was pressure to create new species and for newly made species to flourish in the absence of competition—a powerful force of natural selection, compressed in time from eons to a few years. Even worse, newly formed organisms changed their habitat to increase the likelihood of additional organisms growing where they hadn't been before. Added to the problem were gene-linkages - exchanges between organisms - so that one organism could rapidly acquire a suite of altered genes, much as bacteria share plasmid chunks of DNA that confer antibiotic resistance to many antibiotics. The rapid emergence of methicillin-resistant coagulase-negative staphylococci[35] in parallel with the emerging methicillin-resistant Staph aureus[36] tells us that resistance can spread like wildfire when bacteria multiply every 20 minutes!

Dr. Harry Mills from the University of Georgia has been studying the effects of benomyl on plant growth and on plant and soil microorganisms for years. Together with Dr. David Sasseville and Dr. Robert Kremer, their paper on the effect of Benlate® on leatherleaf fern growth (7) was an introduction to the reasons for crop damage in areas where benomyl had been used. They've written much on the subject that won't be referenced here, but when you ask what these scientists know about benomyl, you'll see a vast database that sends shivers into every environmentalist I've talked with.

The work of this group was incorporated into the work-plan for an investigation regarding the potential role of fungicides in creating an environment that posed health risks for persons living adjacent to lands where benomyl had been used (8). The literature on soil organisms and the effects of applied pesticides is complex, but the work of these three scientists stands out. In work presented to the Florida Department of Environmental Protection in 2000, basically Drs. Mills, Sasseville and Kremer identified microbe populations that were indicative of benomyl having been used previously. Most of the microbial indicators of benomyl use were bacteria, many of which produce plant growth regulators, such as indoleacetic[37] acid (IAA), and toxins. While certain bacteria strains were the primary indicators of benomyl use, one fungus, *Fusarium oxysporum schlecht*[38], also known as FOS, was apparently resistant to the fungicide benomyl. High populations of this fungus are commonly found in the rhizosphere (root zone) of benomyl treated nursery sites and soils. In untreated land areas FOS, was rarely found. Moreover, in treated land wherever FOS was found, a species of bacteria, *Pseudomonas fluorescens*[39], as well as other related pseudomonads[40] were always found.

In untreated land, the populations of these microbes were tremendously lower, or not found at all. One of the characteristics of many of the microbes associated with benomyl application is the production and release of cyanide in the root zone. Some microbes, such as *Pseudomonas fluorescens,* can tolerate cyanide and one of the characteristics of those microbes is that they produce cyanide themselves— they use it as a source of carbon and nitrogen. Most microbes that occur naturally in untreated soil don't tolerate cyanide. As microbes that produce cyanide grow, natural populations diminish, and those that produce or use cyanide become the predominant microbes in the soil. Roots don't like cyanide, so its production in the root zone is one of the problems associated with planting a second- or third-year crop in land treated previously with benomyl.

As an aside, several years ago, some "spray from the hip" drug-eradication officials (not to mention some politicians) thought that a simple solution to the cocaine problem would be to spray *FOS* from helicopters onto the coca plants growing in heavily defended, inaccessible areas of South America. Such use of botanical herbicides might even make sense at first blush: kill the coca plants with toxins and you'll never see anything growing there again—we only have to look at the 1200 nursery sites in Florida to know that. At least until the *Pseudomonas* takes over with its plant growth regulators and a super-race of coca plants, lushly growing in biologically altered soil ecosystems, emerge from the chemical soup. We don't know what would happen if FOS was used as a drug warfare weapon, but imagine the scenario when the coca plants that survive FOS begin to produce. You can almost hear the growers and drug suppliers saying, *"Thank you for the super coke. Fourteen times more potent!"*

Can you imagine the cynical political strategy that might be employed against countries that the U.S. can't attack? If FOS sterilizes most cropland and whichever plants survive FOS makes people sick, then FOS would be a phenomenal biological weapon. Face it: starvation due to crop failure will topple a rogue nuclear power faster than any jawboning about nuclear war ever could. "Bring out the FOS and kiss the foodstuffs goodbye." That strategy could be added to *Famine on the Wind*, as intentional crop destruction with mutated fungi wasn't known when that book was published. Part of the reason that FOS campaigns weren't authorized might have been due to concerns about cancer developing in people exposed to the mycotoxins produced by the FOS!

Is it any surprise then to find out that in the mold-damaged areas of the Lascaux caves, a home too many species of fungi, that we find a pure culture of *Fusarium* and *Pseudomonas fluorescens*? The cave paintings there had survived 15,000 years of natural ventilation until someone decided that he could air condition the caves better than Mother Nature had for 15,000 years. *Wrong.* Change the air, change the humidity, add in hundreds of hot sweaty tourists and look at the new growth of green, yellow, red and black. That new color wasn't new ochre, it was new growth of multiple species of mold. Whenever a stable habitat is altered by any external factor, there will be changes in the ecology. And that means a new fungus will likely be among the new visitors. One can be certain that fungi will be first in line to partake of whatever new food is available.

After several years of being "closed for repairs," the Lascaux caves are re-opened to strictly regulated numbers of people, who can now find a mono-culture of *Fusarium* mold growing luxuriantly with *Pseudomonas fluorescens*. Do you wonder which fungicide was used on the cave paintings to eradicate the multiple species of fungi that then resulted in creating a thriving ecology of only two types of organisms? I sure do.

Based on the Florida experience, we would expect to find benomyl or one of its congeners was used.

As a common theme, we expect that mutated fungi and microbes are more likely than not to be toxin-formers. No surprise then that FOS is a toxin-former—it makes people sick. In the absence of prior treatment with benomyl, the lands there rarely harbored FOS. In my book, *Desperation Medicine* (Gateway Press, 2001), you can read the account of how Fuzzell and others petitioned the Centers for Disease Control and Prevention (CDC) in 1999 to conduct a formal epidemiologic study of 80 people who lived alongside his nursery site, all of whom suffered from an unusual illness identical to that seen in other people exposed to fungal toxins. The CDC quickly referred the petition (hot potato, hot potato!) to its sister organization, the Agency for Toxic Substances and Disease Registry (ATSDR), where officials decided that there wasn't enough information on health effects to investigate.

Wasn't the unknown relationship between fungicide use and human illness the reason why an investigation was requested in the first place? Given that there were at least 1200 sites in Florida alone that experienced re-cropping problems after benomyl was used there, one can only wonder how many people might have been sickened by those exposures. Fungal toxin illness doesn't go away without definitive treatment, but it's often called something else, usually a disease entity that has no biological marker. Imagine how many people affected by mutated fungi are suffering needlessly!

In fact, the syndrome is often misdiagnosed as fibromyalgia or chronic fatigue syndrome, so fungal illness is, in fact, quite commonly found now in the U.S. The CDC says Chronic Fatigue Syndrome (CFS) affects more than one million people in the USA and more than 6 million people here suffer from fibromyalgia, according to the Arthritis Foundation. Those of us in clinical practice know those numbers are absurdly low; illness caused by exposure to biotoxins is not uncommon, but a correct diagnosis is.

Where did all the new toxin-formers come from? Do we know if benomyl had anything to do with their emergence?

Programmed daughter cell death — or programmed mutation?

In a landmark paper from the Genetics Society of America published April 2006 (9), Spradling and co-authors tell us that we're seeing a revolution in understanding the genes of microorganisms, and that new understanding has now has made the *human* organism, not a lab mouse or rat, the one that will be the one most studied in the future.

What we've learned about yeasts, fungi and people is that there's an underlying unity among our cells. Processes that affect cell division in primitive fungi aren't any different genetically from those in our own cells. Spradling looks forward to "diagnosing and treating human disease using information from multiple systems and to a medical science built on the unified history of life on earth," but I'm fearful. That's because I see what we've done to harm other organisms with our chemicals by *altering* the life of fungi, molds and yeasts, as well as an entire host of other creatures, including mites, blue-green algae, earthworms, bees and invertebrates that live in and out of water.

Benomyl leads the list of the chemicals we should have learned to fear. Why worry about a fungicide? The EPA's official stance is that the agency agrees with the manufacturers that benomyl breaks down quickly and therefore, once destroyed, is no longer a source of concern.

Let's go back to high-school biology. Remember all those diagrams of chromosomes doing things in order to let cells divide and still live? Mitosis[41] is what we called one of the cell division processes;

it turns out that mitosis in fungi and in humans is almost identical. Somehow, as the chromosomes that contain our DNA duplicate themselves, they line up in the middle of the cell (we call that stage "anaphase") and then somehow the chromosomes are pulled apart by little threads such that an equal number of chromosomes go to each side of the cell. The cell then neatly splits itself in half and you've got two cells. Do it again and you've got four cells and then eight and so on.

Successful cell division is the source of growth and of continuance of life itself. Cell division seems like a mere formality in the life process of duplication of DNA, with its billions of base pairs, proteins and sugars all wrapped up in a tight bundle. Yet, even after the complexity of duplicating DNA is completed, without successful delivery of the complete complement of intact chromosomes to each new cell, the cell will die. And that delivery literally hangs by a thread - a thread that benomyl can damage or destroy.

Those little threads are called microtubules[42]. They're classic tubular players; they can actually attach to a small protein - the kinetochore[43] protein - in the middle of each of the chromosomes at a structure called a centromere[44]. When it's time to divide, the pull from microtubules is the force that separates each chromosome from its twin at anaphase.

Just imagine what kind of chromosome havoc you could create by disrupting the microtubules' tenuous hold on that little kinetochore protein. If the pull on chromosomes weren't just right, there would be no way for the two new cells to get the right DNA. In fact, the new cells would probably die.

That's how benomyl works (10, 11). Whether the cell is a mold, yeast, earthworm or person, without the right kind of DNA in the offspring cells, the organism dies. We call the benomyl process of disrupting normal cell division, "mis-segregation and dys-segregation."

Five years after benomyl was last used, we're just beginning to see the tip of what benomyl has done. Find a living creature that has chromosomes and uses mitosis to divide and you'll find benomyl mucking up normal cell division. Find disruption of normal cell division and you'll find mutations, Darwin's source material when he wrote on natural selection and survival of the fittest. If benomyl is present in levels of 5 -10 parts per billion (ppb), the precise attachment is disrupted; the daughter cells don't have the right complement of chromosomes and they die. I should say, *usually die*, as the source of the mutations created by using benomyl are new complements of DNA in daughter cells that don't die. Indeed, when they want to create mutant yeast in a laboratory at the National Institutes of Health (NIH), scientists simply add benomyl at 5-10 ppb to the growth media as shown in many academic papers.

Couldn't it be that the mutants made by benomyl are all unrelated, and the chromosomal chaos that benomyl wreaks are just random events?

Nope. We know that the mutations caused by benomyl have a distinct signature, one that's recognized in multiple kinds of organisms, not just fungi and yeasts. Like a molecular fingerprint, that signature is the group of genes that code for beta tubulin-1 (12-21). Such mutation that involves a ring of DNA is a distinctly different kind of mutation than change in a single base pair. Downright *unnatural*. (See Chapter 28, Darwin.)

As early as 1985, benomyl was known to select for resistant fungal species that through DNA-mediated transformation had an altered beta tubulin-1 gene that conferred resistance. If you wanted to study DNA transformation (which is simply another way to say, a new and different form of the parent organism); just add benomyl. Benomyl resistance meant a new kind of beta tubulin gene. Unfortunately, the new beta tubulin gene carried other genes with it, and these "other genes" are ones that not only confer resistance to other fungicides but do so more frequently (22-39).

As it turns out, fungal resistance to benomyl became a diagnostic tool that helped mycologists study fungi. In that group of medically important fungi and actinomycetes[45], many found inside water-damaged buildings (also called Sick Buildings); benomyl resistance is routinely found in *Aspergillus*[11], *Fusarium*[21], *Trichoderma*[46], *Alternaria*[47], *Geotrichum*[48], *Chaetomium*[49], *Acremonium*[50] and *Penicillium*[26], among others.

What did benomyl do to the toxins made by the fungi?

The resistance to benomyl is identified by a unique gene for beta tubulin; the exact DNA differences are now defined in that gene. Moreover, the beta tubulin gene is associated with multiple genetic linkages that markedly alter the behavior of the mutant fungi to include multiple drug resistance. Even worse, as the benomyl pieces of DNA are being pulled every which way, the mutated genes, now linked to each other, have made their way into circular pieces of DNA called plasmids. Plasmids divide at their own rate and are passed on to dividing cells randomly, so the cell division of mutant fungi now results in some genes that can alter behavior of only a few aspects of life to many organisms rapidly. Resistance to multiple fungicides, multiple drugs and the uncanny ability to make new biotoxins can therefore spread quickly.

Even worse, plasmids from one kind of organism can be engulfed and still function in another kind of organism. The mutated DNA (complete with all the different genes, and not just beta tubulin-1), from benomyl can be donated to just about any kind of critter that has cells that divide. The original mutation then becomes a genetic time bomb as it spreads from living creature to living creature. Beta tubulin-1 genes and all the others linked to it by benomyl in other creatures will do what they do in fungi: send out an altered cell message to alter cellular function.

Benomyl acted like the proverbial pebble thrown into a still pool: one ripple leads to one wave that leads a second, which in turn creates a freshet flow that leads to a cascade that then leads to a waterfall and then to the sea.

Simply stated: using benomyl created countless changes in the DNA of countless organisms. Those DNA changes will not disappear simply because the initiating benomyl compound can't be found anymore in our fields, lakes and estuaries.

In other words, benomyl is the worst kind of covert pollutant because it pollutes DNA. Like the handed-down traits of missing teeth or curved fingers, not to mention well-studied illnesses in royal families like porphyria or hemophilia, genetic change that begins with an ancestor can affect countless generations of multiple different living beings to follow.

What we don't know is whether the plasmid DNA containing the mutated genes also "codes for" (causes) toxin production. Based on what I see in my office practice, though, the behaviors of the new mutated fungi include toxin formation. I see it in my patients' symptoms.

Evolution begins with mutations

Let's go to the "Intelligent Design" trials completed in December 20, 2005, in Dover, Pa., to understand more of the source of my concerns. A six-to-three majority of members of the Dover School Board wanted to inject their political and religious spin on what students should learn about evolution. They wrote a resolution: "Students will be made aware of gaps or problems in Darwin's theory and of other theories of evolution, including but not limited to, intelligent design. Note: Origins of Life is not taught."

It seems that every few years, the concepts from the 1925 Scopes Monkey Trial keep popping up, more often than reruns of the great black-and-white movie, "Inherit the Wind," inspired by the play of the same name. The movie depicted the epic courtroom battle over Evolution that followed adoption of a law by the State of Tennessee that made it a crime, "for any public school teacher to teach any theory that denies the story of divine creation of man as taught in the Bible and to teach instead that man has descended from a lower order of animals (40)."

During the trial, Clarence Darrow and William Jennings Bryan slug it out in a verbal fight about Creation versus Evolution. Creationist Bryan wins a small victory for a short time, but Evolution was the final victor, as the Supreme Court of Tennessee reversed the conviction and instructed the state attorney general not to try the case again. The Creationists never have forgotten. Similar state laws were created about teaching Evolution in Arkansas (1928 and 1981), Louisiana (1982); each was struck down. Current legal battles about Evolution and Creationism in schools are still ongoing in Georgia, Kansas and Ohio (40).

The Dover School Board didn't want to risk having Clarence Darrow cross-examine witnesses in their backyard on national TV, with commentary by Mike Wallace and Andy Rooney; they simply wanted their science teachers to present another idea about the origin of life besides Evolution. But since no one knows what happened when the first life forms appeared on Earth, how can anyone deny that a Creator, one with Intelligence, designed life as we know it? Why couldn't they teach Intelligent Design?

Can the Evolutionists say the Creationists are wrong? Can the Creationists say the Evolutionists are wrong? When we find meteorite evidence of amino acids that had to come from Outer Space, how did they get on the meteorite?

Who is to say that the primordial DNA that evolved into all the creatures and plants on Earth didn't come from some asteroid (it takes tough DNA to survive that impact)?

The idea seems to make sense except that when we talk about scientific theories, we'd like to be able to test the hypothesis that *something A* caused *something B*. How can we test the hypothesis that life appeared as a result of Creation, as some religious texts say?

The key to deciding the Dover case came from demonstrating that there are enough evidences, primarily in immunology, of mutations occurring in genetic linkages (41) that provide the raw materials for changes in organisms over time as a result of natural selection, as proposed by Darwin in 1859. The simultaneous development of groups of genetic changes appeared too bizarre to be true to the Creationist expert, Dr. Michael Behe. He described such an occurrence as "a jump in the box of Calvin and Hobbes," as if it were the same fantasy as used by two cartoon characters in their cardboard box.

The determining argument in the Dover case was that of a combined linkage of mutations that created a simultaneous group of genetic changes that, in the end, are no different from what we see in benomyl-treated organisms. The Intelligent Design advocates lost the case and the teaching of Intelligent Design was identified as religion, not science, and therefore barred from the schools. As fate would have it, the School Board members who wanted something other than Darwin's ideas taught were all defeated in the fall election.

The gist of the winning argument in Dover was that Nature can, "glom onto rare, random mutations (42)" and over time, create new life forms from existing ones. Mutations are the raw material of evolution. All that having been said, we all know that the mutation idea doesn't explain how the first life forms came about, so I don't think the religious strength behind this argument will leave the debate any time soon.

But my point here is that mutations aren't always rare and random. Hiroshima and Chernobyl show us that nuclear events act as pulse generators of mutations. Look at the predictable illnesses in the survivors and progeny of the survivors of these two nuclear disasters.

What we did with benomyl is to voluntarily put an efficient mutation generator into ecosystems worldwide so a (relatively) small number of big growers would see increased productivity and increased profits (*Editor's note*: small growers used benomyl too. It was the best fungicide ever made!). Benomyl acted as a pulse-generator of mutation multiplied by its constant use over time all over the world. It worked to make money for the manufacturers of benomyl in its various forms (don't all successful products have the potential to make money fore the manufacturer?), but we're paying the genetic price now. Here on Spaceship Earth we've unleashed a destructive genie from an unknown bottle because we wanted benefits for something unrelated.

Growers like Fuzzell were simply following label directions when they applied benomyl at a rate of *one ton* per acre. We won't stop paying the costs of what mutated life forms do anytime soon, and you'll see the results coming to a produce market near you.

Here the thinking gets worrisome. What if there's been mis-segregation and dys-segregation but the new cells don't die? What do we call the surviving cells with altered genetic code? Mutations. Normally, a mutation from say, radiation, will change one link in the DNA chain—not much for Nature to glom on to. But with benomyl messing up chromosome separation, whole chunks of DNA could suddenly be in the wrong place, and that's not good for *gradual* evolutionary change - the process that created our world as we know it. Gradual evolutionary change allows for other systems to change too – the checks and balances that allow us to live in a degree of harmony with our environment.

What will we do about the fungi and other organisms that laugh at benomyl's attempt to mess up normal cell division? They live despite the fungicide; we call that benomyl resistance. If you've followed my work over the past 10 years, you know that toxin-formers invariably arise out of some chemically based selection pressure that lets them live, despite the fact that they must manufacture biologically "expensive" toxin molecules.

One reason scientists are so interested in benomyl is that it should be a good drug to use in cancer patients, among other illnesses characterized by fast-growing cells. When you can make a cell divide itself to death, and not hurt other cells that aren't dividing much... well, we call that chemotherapy. Indeed, look at current benomyl literature and you'll find a ton of ongoing work using benomyl to stop cancers. Having said that, what can we say about benomyl causing funny distribution of chromosomes? Could it become a cancer *causer*?

Honestly, in hindsight, I can't understand how we allowed anyone to use a fungicide that caused mutations and might be causing cancer. When potential long-term side effects are unknown to regulators, then their regulations and labels are perforce, at best, a guess. A Guess! When it's known that these potential effects are generated by altering DNA – potentially the DNA of *all* living things – then a mutagenic compound is dangerous and should not be applied by the tens of thousands of tons everywhere fungi grow. If our government cannot protect us from such dangers, indeed facilitates them, then the essential contract between the people and our governors has crumbled.

When we look at the chemical structure of some of our most widely prescribed drugs—the proton pump inhibitors that help stop stomach acid production, for example—we find a substituted benzimidazole, almost like benomyl. What are we doing to rapidly dividing stomach cells with our purple pills? How about thiobendazole[51], another substituted benzimidazole[52]? We reach for this drug when we need to cure someone with a protozoan infection. What price do we pay in the future to treat the whole family for pinworms now? And when our ladies take antibiotics for a bacterial infection and then get a yeast infection as part of the deal, we happily prescribe more azole fungicides, knowing that the private itch and discomfort yeast overgrowth causes will go away quickly. Even worse are the people who think that they have chronic *Candida* infections and therefore take more potent azoles fungicides for months at a time.

Someone needs to ask if our unfettered use of azoles in medicine makes any sense. Someone needs to say "Stop, one benomyl was enough!"

Do we know what else the azoles do? Would our prescribing habits change if we found out that azole resistance is increasingly common in *Candida* species, especially those that cause so much misery for immunocompromised patients? Some physicians use antifungal azole medications for extended periods of time to treat patients with fungal sinusitis or illness related to exposure to water-damaged buildings. We don't know what levels that these antifungals reach in blood or urine but thanks to data from Frank Fuzzell's research, we do know that Dutch workers who applied benomyl to bulbs before shipment accumulated urine levels of benomyl exceeding 10 ppb.

Don't you think that some health research group should look at those workers to see how they— and their children —are doing? For a while benomyl was phenomenally successful at clearing fungus problems from the grower's list of concerns. Whether the plants were drenched, sprayed, fed at the roots, it didn't seem to matter, benomyl did the job. So if you're the grower, wouldn't you want to use the same good stuff again next year? Of course, but only as time passed, more growers noticed fungi growing after benomyl use, due to the DNA mutations. We now know that rotating fungicides makes sense, so that no one kind of fungus gets the upper hand. If the *Fusarium* were resistant to benomyl, kill it with chlorothalonil[53] (another fungicide) instead. Then hit the plants with benomyl next year. But that wasn't to be.

What happened instead was that Frank Fuzzell and more than 1200 nursery owners in Florida started seeing plant die-offs and previously successful nurseries no longer could grow enough healthy, good-looking products to sell. The State was called and the lawyers went to work. Who was to blame for the plant deaths?

No one asked what else the mutation king would do to our commensal microbes and us.

Look for resistance to benomyl if you find new kinds of toxin-forming microbes

On a hunch, I asked Skip Goerner, local landowner and President of the Harris Chain of Lakes Commission, for some help with my concerns about benomyl use adjacent to the lakes of Central Florida. Something was definitely unusual here: a blue-green algae called *Cylindrospermopsis*[54] (also known as Cylindro) was growing with another bad blue-green, *Microcystis*[55] in massive levels in Lake Griffin, Lake Harris and most of the rest of the Oklawaha watershed that eventually drains northward into the St Johns River.

Cylindro was a newly recognized organism flourishing Central Florida, first found there in 1994 (did it come from Brazil, China, Australia or maybe Outer Space?). But by 1999, this incredibly well-adapted microbe comprised more than 95% of the *total* algal biomass in many lakes there. Talk about a change in the neighborhood! Imagine a city of millions of people changed in a similar fashion in just five years...

Once home to trophy-sized largemouth bass and the resultant tourism business based on such marvelous fishing, now the previously pristine lakes were pea-green Cylindro mono-cultures, with some *Microcystis* able to compete. Good luck finding a big bass now, much less seeing more than six inches into water that used to be so clear that water ten feet deep looked transparent.

What an unbelievable change. Were nutrients the culprit of the algae blooms? A common theme regarding environmental disasters is that government agencies will quickly blame nutrients. Actually, the nutrient enrichment of the Chain of Lakes had been going on for some time. Nutrient enrichment hadn't happened overnight, but the new toxin-formers had moved in almost that quickly. Think about it: why would nutrient enrichment of nitrogen or phosphate selectively benefit one species of algae to the exclusion of the hundreds of competing species? Or was Cylindro like duckweed that creates its own favorable environment?

Maybe the change in the lakes was simply the new blue-green itself. Just look at what it did: it could fix nitrogen against a background of minimal enrichment, it didn't need phosphate enrichment to multiply and it could survive low-light concentrations. It also had a temperature-sensitive secret weapon, a cyanophage[56], a virus that killed other blue-greens. So, I suggested to Skip that the 1994 appearance of Cylindro might indicate it was a mutant bred from the benomyl basket of DNA changes. If there were anything to the idea that benomyl and genetic change were behind the massive domination of a complex ecosystem by one organism—basically an ongoing bloom—then we should expect to find benomyl resistance.

And we did. Skip's 24 water samples showed abundant Cylindro growth unaffected by either benomyl up to 1 ppm, copper at 1 ppm or a combination of benomyl and copper. Nothing in the algae world had ever been shown to survive such concentrations of biocides. Not only were the new blue-green algae resistant to benomyl, they were resistant to copper as well. Unbelievable! Only in recent years has research into copper resistance in blue-green algae (43, 44, 45, 46, 47) shown how much our agricultural chemical use has selected for toxin-forming strains of freshwater algae.

The new cyanobacteria[57] were competing so well because they were resistant to algae-killers washed into the lakes from the surrounding farmland. All the others, including most of the *Microcystis*, were killed by the benomyl, with copper resistance an unexpected complication.

State apologists/biologists took on the case. No, they said, there wasn't *any human health threat* from the newly changed lake biology. And no, the problem wasn't serious, even though alligators were dying, pelicans were dead and the tiny shrimp were gone from the water column. The algae were just a nuisance, the alligators were dying from thiamine deficiency and the wading birds and pelicans... well, uh, right, "They had Newcastle's disease," or something. Some too-trusting people believed the State's smokescreen explanations. Government is here to look after us – right?

And what about the bass? They were gone. Forget about the glory years of fishing on Lake Griffin.

These two kinds of blue-greens - *Cylindrospermopsis* and *Microcystis*—are major league toxin-formers. The entire Chain of Lakes was under siege from blue-greens, even if all the State agencies involved denied any problems that might threaten tourism. (Don't forget Orlando and Disney World are just south of those Chain of Lakes.)

"Skip," I asked, "is it too far-fetched to think that something has given these toxin-formers a selective advantage? They should be competing with the normal algae rather than becoming 95% of the total algal biomass. And where did these blue-greens come from? Are our mutagenic fungicides at work here?"

I'm not suggesting that benomyl is the cause of copper resistance found in blue-green algae in Central Florida lakes. Copper resistance can arise from other chemical perturbations, too. I'm suggesting that using benomyl on the land, according to label and thus completely legal, was the source of benomyl resistance in the lakes, creating organisms that will change ecosystems forever.

What will happen when the delayed effects of benomyl-induced mutations become widespread? With lots of mutations in multiple kinds of organisms and lots of resistance to the killing effects of agricultural chemicals, what would be the next environmental disaster caused by benomyl? Don't forget that the new cyanobacteria flourishing in benomyl- and copper-enriched lakes are toxin-formers and thus, they're capable of making people incredibly sick, causing permanent damage as well as ruining careers, relationships, marriages and ultimately, lives.

Can you imagine the ad for Disney World just after the Super Bowl if the blue-green algae problem spread south to just a few more lakes?

"I'm going to go to Disney World! Just two days and I might be exposed to some toxins that could make me sick for the rest of my life!"

Mutant fungi in sick buildings

Following another hunch, I wanted to know if there was any basis in fact in thinking that benomyl could be a selection factor in the new toxin-forming strains of fungi growing indoors in water-damaged buildings (WDB). Sick Building Syndrome (SBS) isn't new—even in the Book of Leviticus in the Old Testament of the Bible; readers are warned to stay out of moldy homes. Some people think that SBS began as a result of new techniques in building "tight homes," following the first crisis of oil supply in the early 1970s. I know of no evidence for that theory but it does seem there are organisms in WDB that are different now compared to before. Could benomyl be an underlying factor? Were we blaming human intervention of the 1970s instead of the real 1970's culprit?

Fungal researcher Dr. Chin Yang of P and K Microbiology in Cherry Hill, N.J., confirmed that toxin-forming indoor fungi are commonly found to be resistant to popular indoor fungus-killers like bleach, quaternary-ammonium[58] compounds and copper-based products. Have we killed off so many of the fungi that were sensitive to these fungicides so that only resistant species were left?

One clue came from a local chicken hatchery. Normally, the hatchery works hard to control *Aspergillus fumigatus*[59], a fungus that wreaks havoc on egg production. The hatchery routinely uses a fungicide called enilconazole[60], a substituted benzimidazole[52] similar to benomyl. One worker came for evaluation of his fungal-toxin associated illness after working in the hatchery for more than ten years. He was sick, and so were a number of his co-workers. What he didn't understand was, why wasn't the fungus-killer they used working anymore? And by the way, it seemed that the problem was a lot worse ever since the new paint job. When he put out culture plates with the fungus-killer in them, the *Aspergillus fumigatus*[59] grew quickly. In my lab, the *Aspergillus* was resistant to benomyl, too, at *20 ppm*!

I needed to find a certified lab that could confirm the observed resistances. There are none! Sure, we could find help confirming resistance to miconazole[61], ketoconazole[62] and fluconazole[63], but not to enilconazole[60] or benzimidazole[52].

But there in the stack of documents about benomyl from Frank Fuzzell's library was a statement from a research paint microbiologist from Pittsburgh Paint and Glass, who also was a member of the Paint Research Institute's Mildew Consortium, writing that benomyl had been a useful additive to Lucite Paints *ever since 1970*.

Could we have been using benomyl to prevent growth of *Aureobasidium pullulans*[64] on paint films in an effort to save on maintenance costs by putting a powerful fungal mutagenic agent in buildings at the same time?

What do we know about toxin-forming fungi found indoors in Water-Damaged Buildings (WDB)? Are there transformed fungal species that are benomyl-resistant? Sure: Let the list begin with *Fusarium* (25), *Trichoderma*[46](27), *Aspergillus nidulans*[65](28), and *Aspergillus parasiticus*[66] (31). Add to the list *Penicillium citrinum*[67](36), and one of the most likely organisms to be associated with neurologic effects, *Penicillium chrysogenum*[68](37). The list doesn't end here; include *Acremonium chrysogenum*[69] (50), the aflatoxin-forming *Aspergillus flavus*[70](51) as well as *Penicillium italicum*[71] and *P. digitatum*[72](52). Each of these organisms has been shown to be transformed by the mutant beta tubulin gene.

The story becomes even more incredible. Genetic change is Nature's most powerful force. Let's not underestimate what minor changes can do. Don't forget the change in one horseshoe caused the loss of one horse. Lack of one horse caused the loss of the rider and loss of one rider caused the loss of the battle that caused the loss of the war.

One nail in one horseshoe. One fungicide on one crop in Florida.

The mutant beta tubulin gene isn't just found in indoor fungi. We see it in *Beauveria bassiana*[73], an entomopathogen[74](29), the fruit fly *Drosophila melanogaster*[75](53), various protoctists[76](54), the malaria-causing apicomplexan *Plasmodium falciparum*[77](55), prokaryotes[78] of many types (56), cyanobacteria[57](57) and particularly in *Microcystis*[55](58). Of great concern are the multiple types of yeasts that carry beta tubulin genes, including *Sacchromycetes*[79], which are associated with methotrexate[80] and benomyl resistance (26), as well as other mutations (34). Mutations in medically important *Candida albicans*[81] include methotrexate resistance, as part of multiple drug resistance (59) and azole resistance in AIDS patients (60).

A simple example in *Candida* that parallels that of Cylindro comes from *Candida rugosa*[82], now recognized as an emerging fungal pathogen. Who ever heard of *Candida* before 1997? No one, because an extensive ARTEMIS DISK antifungal surveillance program hadn't recorded its presence. Its frequency of recovery in clinical isolates increased from 0.03% to 0.4% in five years! And in Latin America, *Candida rugosa* is now found as 2.7% of all fungal disease causers. Even worse, when this organism is isolated from the bloodstream, resistance to azoles occurs in nearly 100% of all isolates (49). As azole resistance spreads from *Candida glabrata*[83] to *Candida krusei*[84] to *Candida rugosa*, will we be surprised to find benomyl-induced mutations as the source?

Before we can dismiss the role of benomyl in altering disease behavior in scourges of the 21st Century, from malaria to AIDS to moldy classrooms and water-damaged workplaces that make millions of people ill, not to mention "medically uncertain illnesses," such as Chronic Fatigue Syndrome and fibromyalgia, we've got some work to do.

Someone needs to look for the mutant beta tubulin-1!

Resistance to benomyl has become a fundamental diagnostic test in medical mycology (61), but there's clearly a need for an assay to look for the mutant beta tubulin gene in other organisms as well. When we see health effects from invertebrates that have been transformed by benomyl, how long will it be before we see human health effects?

Birth defects blamed on benomyl

In a hotly disputed legal case (Castillo versus DuPont, 62), the Florida Supreme Court decided that it would permit medical testimony that exposure to benomyl could cause microphthalmia[85], which means a birth defect in the development of eyes in a fetus. A woman seven weeks pregnant walked past a "U-Pick" field November 1 or 2, 1989. The field was being sprayed with benomyl. The wind was blowing and she was completely drenched by the spray. She suffered no adverse effects herself, but her child was born with microphthalmia. British reporter John Ashton contacted her as he was investigating the relationship between benomyl and microphthalmia in Great Britain. She filed a lawsuit against DuPont and the farmer. Medical expert Dr. Charles Howard testified that 20 ppb of benomyl in the bloodstream could cause microphthalmia.

In American courts, experts are held to a standard (variously called Frye or Daubert, depending on the state) in medical testimony that ensures the opinions they express are acceptable to other scientists and are based on reasonable methods. The defendant—DuPont in this case—challenged the expert's credibility, to try to have him excluded from the plaintiff's case. And without Dr. Howard's opinion entered into testimony, there *was* no plaintiff's case and therefore no judgment against DuPont. In the end, the plaintiff's expert opinion was not excluded and a verdict for the plaintiff for $4 million followed. To this day, DuPont feels that this decision was incorrect; they followed every label instruction all regulatory agencies had required.

Two studies would suggest that there's a reasonable need to look at benomyl and the development of brain and other neural tube tissues. Rull and co-workers from the Northern California Cancer Center (63) studied the association of residential exposure to agricultural pesticides within 1000 meters of women of childbearing age from 1987–1991 in California. Rates of neural tube defects were low overall, but after elaborate statistical applications, the only pesticides that were associated with abnormalities in neural tubes were benomyl and methomyl[86]. Another study that reached similar results was performed in rats by Hewitt and his co-workers at The University of Leeds (64). They found a dose-response relationship between benomyl and developmental abnormalities—especially eye abnormalities—occurring at levels of benomyl as low at five ppb.

Still, how do we know if human health effects are related to mutations in the beta tubulin-1 linkage of genes? If we have no assay to confirm the presence of a problem, how can we show there *is* a problem? Conversely, in the absence of proof of causation of health effects from benomyl, we cannot conclude that there's proof of the absence of health effects.

In the health effects discussion, there may be help on the way from Scotland. Dr. John W. Gow, a neurobiologist formerly from the University of Glasgow, now at Glasgow Caledonian, has submitted a patent application for determining if a person has Chronic Fatigue Syndrome based on genetic markers. These genes "provide a rational basis for classifying CFS patients based on biochemical lesions." Dr. Gow postulates that, "CFS isn't a disease caused by inherited gene defects, but it is an acquired condition, involving changes in certain genes associated with infection, immunity, cell membrane function and cell cycle."

That description of cell cycle might involve the microtubule disrupter benomyl. As one might expect from this discussion, the critical gene involved in Dr. Gow's work is beta tubulin-1.

Model for future investigation

We know benomyl can affect and transform many different types of organisms. We know that unusual behavior of just one small group of genes—like those in blue-green algae—can affect multiple different organisms within a freshwater lake ecosystem. The changes in behavior of fungi resident in WDB are more ominous than those in just one lake, as human health effects from moldy buildings are the subject of intense research, not to mention litigation. Couldn't we just assay some of the fungi isolated and look for the benomyl signature of mutated beta tubulin-1 genes? Of course we could. When one sees an unexplained agricultural disaster, look for benomyl lurking in its history.

Let's take a look at the catastrophic die-off of honeybees (*Apis mellifera*[87]) around the world. When a critically important species dies off simultaneously across the globe, it's not likely to be a coincidence. Is there a common mechanism operating here?

A parasitic mite, *Varroa destructor*[88] *Anderson and Trueman*, kills adult over-wintering bees and is thus responsible for a diverse series of crops that never are pollinated by those honeybees (65). No bees, no crops. Talk about an agricultural and environmental disaster! Where did the *Varroa* problem come from? Could benomyl be involved?

We know that fungicides affect mites from work done in citrus (66, 67, 68, 69). Benomyl has been so widely used that the likelihood that there wouldn't be exposure of *Varroa* to benomyl is small. Normally, *Varroa* can be held in control by miticides such as fluvalinate[89](70), pyrethroids[90](71) or essential oils (72). The main controller of *Varroa* populations appears to be the fungus *Hirsutella thompsonii*[91](73). With as little as five minutes of walking exposure to a growing colony of this fungus, there followed predictable physiologic changes in the mites that led to death of more than 90% of them in five days (74). Further research into the mechanism of toxicity of *H. thompsonii* confirmed the presence of a fungal toxin, hirsutellin A[92], that was responsible for the death of the mites exposed to it (75, 76, 77). The injury is caused by ribosome activity inhibition (78), with specific effects on invertebrate cells.

In our model, we're seeing an obvious role for benomyl. Did the worldwide populations of *H. thompsonii* drop out following the use of benomyl, removing the most common control agent of *Varroa*, which then exploded in its growth to the detriment of honeybees? Or did the parasite itself change with exposure to benomyl, developing resistance to multiple types of miticides after its main predator *H. thompsonii* died off? And we must ask, if *H. thompsonii* was affected by benomyl, and not all of those fungi made the toxin that kills *Varroa*, are the surviving fungi *no longer making* hirsutellin A, but other toxins that don't kill *Varroa* instead?

If the *Varroa* are now genetically altered by benomyl, we should be able to find the mutant beta tubulin gene; that should be a straightforward test. Finding out which hives contain *H. thompsonii* and which toxins the fungus makes if it were found would be a more complex task. Almost necessarily, the hive would have to be destroyed to do the fungus and fungus toxin assays.

We must remember the goal here is to enhance population of the bees, not to destroy their homes.

By an unusual twist, if we found that the *Varroa*-killing fungi were altered by benomyl, perhaps the only way we could correct the ongoing die-offs of bees would be to treat the *H. thompsonii* with other mutagenic agents, trying to re-create the ability of the fungus to again kill *Varroa.* Who knows what else *those* mutants would do!

Conclusions

From fungi to earthworms, and from mites to people, genetic change is the source of evolutionary change. Anything that affects the mechanisms of how cells divide that are common to all cells has the capacity to cause genetic changes in those cells. Genetic changes bring about differences in how organisms function, and that change in function of organisms affects all of the other organisms in an ecosystem. An organism change in an ecosystem affects not only that ecosystem but illnesses caused by organisms emerging from the altered ecosystem as well.

"One pebble in an ecological pond changes all the waves in that ecosystem."

Unfortunately, ecosystem changes can't be predicted by a manufacturer or a regulatory agency during the process that leads to the sale of any given agricultural chemical. If officials at a regulatory agency had thought that benomyl would alter the organism that causes malaria, still the dominant illness in the world, would it have been labeled for use on crops growing in equatorial zones? And if anyone had dreamed that benomyl would eventually cause the destruction of honeybees across the world *ten years* after benomyl was no longer being used, would it have been approved? Of course not.

Benomyl does cause such genetic change in multiple organisms. Although its use as a licensed fungicide met a gigantic need and its explosive growth led naturally to its worldwide distribution, nonetheless the consequences of benomyl use are now rightfully regarded as the world's biggest agricultural disaster. As we learn about human health effects ranging from chronic fatigue to birth defects and cancer, we need to keep the mutagenesis from benomyl in mind. Like Pandora who opened her mysterious box, sending scourges of death and pestilence out into the world, our use of benomyl has caused untold numbers of environmental changes. Yet, just as at the bottom of Pandora's Box was a frail creature called Hope, we've developed the ability to detect benomyl-induced genetic mutations. Once we know what our adversary is, we can develop a plan to reverse the genetic damage.

We live in an era in which we've unveiled the genome of countless organisms. Understanding the toxic effects of changes in genomes, the study of the relationship between structure and the activity of the genome and the adverse biological effects of exogenous agents (79), called toxicogenomics, could well begin with learning what benomyl did—and is still doing—to all of us.

Maybe we'll learn from the benomyl debacle. Maybe we can recognize that as stewards of the Earth, our duty is first, do no harm. And after all, "do no harm" remains the first rule for physicians.

Maybe all of us can take that idea to heart.

CHAPTER 12 Frank Fuzzell: My First Mentor and My Friend

Not everyone survives mold; Frank Fuzzell didn't.

I miss him nearly every day.

When he died in 2009, his physical suffering of more than twenty years finally ended. After his death, my sadness was compounded by the knowledge that the new breakthroughs in using VIP and erythropoietin could have returned his quality of life.

How much suffering can anyone suffer before they suffer too much? Maybe the torture endured by POWs in Viet Nam was worse than the torture of daily life that was Frank's life ever since he was injured by toxigenic *Fusarium oxysporum* growing in his wholesale nursery. Maybe it wasn't.

When I wrote about Frank in *Desperation Medicine* in 2001, he was pursuing DuPont via litigation; Frank believed countless people had been poisoned after a DuPont product, benomyl, a mutagenic fungicide, had been applied to plants and to the land (See Chapter 11). Especially ***his*** land. There was a wonderful photo of workers from the Site Investigation Section (SIS) of Florida's Department of Environmental Protection (DEP) on the cover of that book, in full Hazmat suits, investigating Frank's land.

He lived in Leesburg, Florida, too far for him to come to my office in Pocomoke, Maryland for ongoing monitoring. He told me many times that he planned to come for medical care, but he never did. I was a guest in his home and worked with him (and several others) on a special committee convened by the SIS to look into health effects acquired by more than 100 residents whose property abutted Frank's abandoned nursery in the Casteen roads area of Leesburg.

Soil scientists, including Dr. Harry Mills, had identified specific microbes that uniquely identified the change in soil ecology growing in areas all over Florida where benomyl had been used. There were more than 1200 sites where nursery operations had ceased because crop damage occurred after benomyl was used. Frank felt that the microbial problem was coming from a breakdown product of the fungicide, so he focused on the chemical, its structure, the "degradation daughters" (breakdown products) and where those compounds went after they were applied.

Back then, neither he nor I focused on the mutant DNA effects that resulted from forces of natural selection acting on the benomyl application survivors. These pieces of mutated DNA spread through multiple species of fungi, changing the very face of human illness. Such DNA provided new characteristics to previously benign fungi. In the end, we may have to conclude that DNA was the reason we saw "sick building syndrome" appear in the 1970s.

Could anyone imagine what happened to fungi living indoors after they were exposed to benomyl that had been added to indoor paints? New mutants developed that had the beta tubulin-1 gene but they also had a gene that moved an acetyl group (having two carbons, like vinegar) from one part of a toxin molecule where it wasn't detected by ficolins on MASP2 (see Chapter 5, Blood in the Sink) to another location where it hung out in the breeze ready to activate C4 and drive C4a levels sky high. The benomyl in the paint wouldn't last very long but the new toxin formers would; they would breed, too, with the new mutants colonizing an ever expanding circle of WDB. Their offspring now would populate indoor areas where benomyl was never used and benomyl itself would be long gone—and so would DuPont's legal responsibility. But the mutated fungi survived, replacing their benign forefathers with toxin- and inflammagen-formers.

Independent of the source of the human health effects seen in the Casteen Roads cohort, I wanted to be able to diagnose and treat those patients, understanding that DuPont would do all it could to prevent anyone from linking exposure to their product and adverse health effects. Fuzzell and I, along with about 80 residents in the vicinity of the Casteen Roads nursery petitioned the Centers for Disease Control and Prevention (CDC) to look into the unique characteristics of the Casteen Roads patients; they referred the case to their sister organization, the Agency for Toxic Substances Disease Registry, which held a conference call and decided to do nothing more. After years of seeing how the CDC (doesn't) work, and years of developing a well-rounded cynicism regarding government responsiveness to human health needs, I'm not surprised that our inquiry was squashed. But back then, the CDC's utter callousness and absence in fulfilling what I perceived as their duty effectively crushed any illusions I had about agency responsibility to examine environmental issues that might be associated with adverse health effects. Politics - the big kind with lobbyists and the small kind with careers and corner offices - trumps health every time.

My disillusionment wasn't just about their neglecting to investigate the human health effects putatively caused by a compound made by a mega-corporation; their disregard for chronic illness extended to Chronic Fatigue Syndrome, to the *Cylindrospermopsis* and *Microcystis* (blue green algae) which sickened people in Florida's lakes, to *Pfiesteria,* and especially to mold. I have a longer list now. To validate my cynicism and disgust, one simply needs to review CDC actions regarding water-damaged buildings; their approach consistently minimizes the health effects of organisms growing in newly altered ecosystems.

Katrina, anyone?

After 30 years of dealing with families, I look for a person's essential goodness by examining how he acts within his family. Frank passes that test. If Frank could've been treated, we'd see him now as the proud grandfather, beloved by his family. His wife, Lois, still remains one of the kindest women I've ever met outside of my own family. His daughter, Alysha, was one of the first people I knew to be diagnosed with "pediatric fibromyalgia." What nonsense. Her son, Austin, carries Frank's dreaded HLA DR, 0401-3-53. What will happen to his offspring? If his daughter and grandson are exposed to indoor environments with a history of water-damage, they'll surely become ill. If they're exposed to areas that border on abandoned agricultural sites where benomyl was used, will they become ill like Frank?

Frank taught me much. I remember my concern the day he told me that the use of benomyl was the biggest agricultural disaster ever. Back then I don't think he knew just how right he was. The gift of benomyl-better we should call this gift a horror- were the new chunks of DNA that survive with every cell division every twenty minutes now suddenly being available to every microbe that shred DNA. And that was ALL of them. Frank just didn't know that.

One pebble in one pond might make a ripple and then a small wave that ended when the wave was dampened by the shoreline. But one ripple of the effect of mutant segments of DNA, including beta tubulin-1 and acetyl-O-transferase, meant that an entire new world of genetic injury throughout phyla would be shared by DNA-trading microbes from yeasts to slime mold an all the way to us. That is a horror. That is the legacy of benomyl.

Frank knew that.

The hounds of genetic Hell were unleashed after the resistant organisms, those that survived the nuclear-division wars started by benomyl created monsters that we are only now recognizing for the first time.

The Appearance of Good Science (AGS) was Frank's main lesson, and I see it in action every day. It consists of (1) Consensus; (2) Silencing Dissent; (3) Smokescreen; and (4) Burial.

Start with **Consensus**. Convene a panel, stack it with those with whose opinions are either known or can be influenced by self-interest. Maintain control over the panel's "executive summary," so anyone who wants to cite the panel's opinion has a made-for-the-media/courtroom piece in place. Never let out that conflict of interest might be involved or that the executive summary may bear little resemblance to what was actually said. Have co-conspirators refer to the panel's opinion as "definitive," even though the scope of the panel's discussion isn't either thorough or unbiased. Perhaps the Institute of Medicine's 2004 report on damp buildings is a good example. If that one doesn't convince you of the role of AGS in mold, read the opinion of the American College of Occupational and Environmental Medicine published in 2002. After that, the opinion of the American Academy of Asthma, Allergy and Immunology (AAAAI) was published in 2006. The casual reader of both these documents might assume that the authors were pure scientists and that their statements were thorough and truthful. That assumption would be wrong, yet courts all over the U.S. have heard defense attorneys and consultants praise the significance of those two AGS documents (See Chapter 13, Big Tobacco Leads the Way for Big Mold).

The second phase of AGS is **Silencing Dissent**. Should someone attack the absence of validity of IOM, ACOEM or AAAAI, get ready for bared-fangs, "hyena counterattack." Anyone who dares question, "Is the Emperor wearing any clothes?" will find his observations dismissed, his person defamed and his findings discredited. The best example of this approach was the creation of the term "junk science" by Big Tobacco. When a judge accepts such opinion from defense consultants and puts those words into a legal opinion, then silencing dissent about the health effects of WDB has more teeth than anyone would expect.

The third aspect of AGS is the **Smokescreen**, often called Damage Control. Here we see the smooth substitution of "Plan B" for the real problem. No, mold exposure didn't hurt Mr. Jones, it was somatoform disorder, says Dr. Smith. He might have suggested the problem was stress, depression, cigarette smoking, allergy, prior back injury, automobile accident: you can pick any one from a long laundry list. Whenever I see an insurance company's *Independent* Medical Exam (IME) beginning with a selected list of prior medical problems in their first page of discussion about why they're going to deny benefits, I already know what to expect. By the way, has anyone ever found out how such an opinion can be called "Independent?" The "independent" physician contracts with a third party to provide an opinion to the insurance company. The company doing the contracting acts as a buffer to provide the insurance giant deniability; meanwhile, the physician will provide the ammunition the insurance company needs. Oh, yes, very independent.

I'm often asked to speak "peer to peer" —that's the new jargon term for a hired physician's opinion—about the mold illness in an applicant's medical record. I'll ask for proof of permission from the patient to speak to the peer, and then before answering the peer's pre-selected questions, I ask for the peer's qualifications to discuss mold illness. I see many peers who are occupational medicine types or infectious disease types, but I never find anyone who's actually treated a mold illness patient, much less done any research on humans with mold illness. I'll hear their appointments and titles, but it matters little to me if the peer is from Cleveland Clinic or Dartmouth or Boston or Duke (when I talk to peers from my alma mater, I'm really saddened). It doesn't take but one or two questions to get the "peer" to retreat to ACOEM or AAAAI as the basis of his knowledge. And Dr. Peer, what is the academic basis for ACOEM? None?

The phone call doesn't last very long when the empty basis for the peer's opinion is exposed.

When the peer says he doesn't want to argue, he's merely doing his job, and if I'm in a combative mood, I will usually ask him about the officers and officials on trial in Nuremburg after World War II. Those guys were doing their jobs, too. If the peer were truly independent, he'd want to learn more about his "patient's" illness. But no, he isn't interested in any hard facts, just what he needs to do to show his paymaster that he deserves a check.

Asked another way, what can we say about the ethics of a man who takes a lot of money in exchange for promulgation of an opinion that is known to be flawed? Is his act actually fraud? And is the person who solicited the bogus opinion actually guilty of conspiracy? Where does the RICO law come in, with racketeering seemingly the same as what we see being done about mold cases? And if the "independent" guy says his fraudulent opinion under oath, with what he says just one lie after another, wouldn't we simply identify him as a perjurer?

Frank asked me one time how come the physicians (especially the docs employed by Federal agencies) who testified that nothing is wrong from mold exposure aren't all in jail. We laughed, saying that the politicians who are liars (would that be all of them now that Wayne Gilchrest is out of office?) would probably take up all the available jail cells first if someone actually said the legal punishment for lying under oath was going to be enforced in the United States of America. Only Roger Clemens seems to be tried for purported lying these days. Seriously, just pick up any newspaper these days and read what politicians say compared to what is the truth. Then go to the transcripts of Frye hearing transcripts in mold cases and see what defense consultants say under oath. Poor old Roger looks like a choir boy in comparison.

Occasionally, a peer will throw out "Evidence Based Medicine (EBM)" as the basis for his opinion. As defined by *Harrison's Textbook of Medicine*, the physician using EBM will formulate a question for the published medical literature and do a thorough literature search before deciding what's based in evidence or not. Since I use more than 2000 academic papers to support my opinions (see Policyholders of America website), this peer can actually start the process to become a peer by reading the same literature.

Tell me again what "independent" means?

But if that peer is on the witness stand, or is simply writing a report, his opinion must be rebutted. The final decider of which "expert" to believe is the judge or jury but not in disability cases. In those cases, either the insurer or a commissioner will decide. One will not be surprised at massive abuse of truth when the Smokescreen is permitted to stand around.

The last element is **Burial**. Here we see several variations on the theme of getting rid of dissent by acknowledging it. One example will be to say that yes, Dr. Shoemaker's ideas may be valid and in the years to come, with much further study that's already underway, we may have the answers to questions he has raised. But until then, we just have to wait. Thank you all for coming today.

The second is more common when the reputations of scientists are involved. Yes, we agree that inflammation is the key element in mold illness. We've already published this work several years ago. Thank you for coming today.

The most egregious approach is theft. Yes, the use of HLA DR as a marker for susceptibility is valuable. For just $35, "our" group can analyze your HLA and tell you what could be wrong. Credit cards or PayPal, please. This aspect of burial is usually done by shysters and not politicians, however.

Frank wouldn't be surprised to see the AGS in action in the mold arena. His basic human decency would have made him stand up for the rights of those affected by simple exposures. He didn't have much money; he didn't have much support from other growers who took cash settlements from DuPont and went on their way. He never gave up on his insistence that he was right.

Frank said that if you want to be popular, don't hold yourself to truth. If you don't want to be criticized, vilified and otherwise flogged by your enemies, don't stand up to them over the truth. Don't forget: truth loses when it collides with economics. And those lessons are ones he taught me, too.

If we look to popular media, we're raised to believe that truth and justice always win. Superman always gets out of the kryptonite trap, Batman can survive the fall and Tonto will ride away with the Lone Ranger into the setting sun, with grateful residents of the town now able to live their lives in peace since the bad guy's gone (usually killed).

Frank would say no, that's not what usually happens. The effect of ecological modification on microbes and the effect of microbes on human health has *not* been brought to light. There remain too many bad guys keeping the truth hidden. Even worse, for every bad guy who's identified and outed, as long as money can change opinions, there will be three more bad guys just standing in line to take that guy's place.

Still, look at what's happened in the mold illness world in just the few years since *Mold Warriors* was published. Court decisions in our favor and scientific acceptance exist for mold illness victims. Government agencies are now doing research on mold illness and the World Health Organization says that inflammation, caused by a mixture of elements found in WDB, is the illness' source. The illness is an immunological nightmare, not a toxicological problem.

Frank's final lesson— that public policy on chemical disruption of Nature and human health must be dictated by Good Science, not the Appearance of Good Science— isn't yet on the horizon. Dr. Thomas Harblin (see Chapter 23, Battling Mold) writes that public policy must be open and the decision-making process must be transparent.

Maybe I'll see that day before my heart gives out like Frank's did.

CHAPTER 13 Big Tobacco Leads the Way for Big Mold

The Legacy Tobacco Documents Library (**http://legacy.library.ucsf.edu/**) gives us insight into the strategies that defense interests have used in mold litigation. If you have an interest just log onto the website and start typing in names of those who are in Big Mold now that may have deep roots in Big Tobacco. This vast repository of documents shows just how the tobacco industry worked to continue to make huge profits while promoting a product that was known to cause adverse health effects. Part of the approach was to provide a constant stream of misinformation from people that the public would trust, including some highly placed government officials with health responsibilities. Physicians, researchers and authoritative figures basically could be bought. And they still are if we only look at what some mold defense consultants pull down for individual cases. This chapter in American marketing is particularly despicable but the themes used are no different than what we still see every day.

Just look at all the ads on TV for cholesterol products. Surely the drugs are safe and *necessary*? Nope. Surely they are *proven to be needed* by all population groups and not just white males age 40-65? Nope. Surely the experts who perform great research to solve the ills of cardiovascular disease are *impartial and not influenced* by research grant support? Nope. And how about all the estrogen sales back when we didn't know much about clotting, breast cancer and cardiovascular disease? The estrogen push kind of slowed down didn't it. The list goes on: tight control of blood pressure; tight control of diabetes; low fat diets; and more. Marketing drives profits and marketing is most effective when spinning – a bit like a drill.

Ten years ago, Sandy Rose asked me about my advertising budget for *Lose the Weight You Hate*. I talked a lot about insulin, leptin and amylose, the dietary complex carbohydrate that really is the key food that overweight patients must avoid to help in weight loss. I proudly said I would spend no money, the book would sell itself. His comment was that I must certainly be aware that I wouldn't sell very many books.

In mold illness the marketed concepts all stem from the ideas used by Big Tobacco. Mold is safe. It is everywhere. No one has shown any proof of adverse health effects. Any studies that show adverse health effects are junk science. Just look at the authorities who say mold illness is impossible. We have the American College of Occupational and Environmental Medicine, a group that is insinuating itself dollar by dollar into position statements about countless occupational disorders, that in 2002, says mold illness is implausible. Who could argue with them?

And there's the prestigious American Academy of Asthma, Allergy and Immunology saying the same thing in 2006. Who can argue with them?

Actually, all of us can. Just remember Frank Fuzzell's Appearance of Good Science. Consensus, Silencing Dissent (Hyena Attack), Smokescreen and Burial. You will see these elements in their full glory in the pages to follow.

ACOEM and AAAAI used unfounded 'consensus' statements written by essentially the same people using flawed science as a basis of reference. Anyone who argued against those two consensus statements (note not objectively 'correct' but a subjective 'consensus') especially in court, would be attacked with Frye and Daubert challenges. This stratagem trusted that judges would be more impressed by *who was talking* than by the truth of what was said. Sometimes with some judges that worked. When that plot didn't work and judges actually did listen to what was said, we quickly saw the Smokescreen of creating alternative diagnoses used to protect Big Mold. And best of all, Burial comes when the Federal agencies agree to study the issue and what do they use? Parameters that have nothing to do with the illness.

Just look at the $10 million that the National Toxicology Program is spending on rat studies involving complete blood counts, metabolic profiles and allergy testing. Dr. Shoemaker addressed the NTP committee in 2007 in an effort to persuade them to actually study what was wrong. Didn't work. Basically, if the purpose of the study was to identify what adverse effects were caused by exposure to WDB in rats, thereby giving us an indication of what might be wrong with people who were ill after exposure, Shoemaker basically said, "I have some good news for you. We already KNOW what goes wrong in humans."

The message was that if you want to learn about what we find in CIRS-WDB patients by studying rats, he can save you time and $10,000,000 since we know now what we actually do find in **people**. And just look at the NIOSH debacle in New Orleans - See Appendix 7, St. Bernard Parish.

In this chapter I want to show you what we argue about regarding the smokescreen laid down by ACOEM and AAAAI and then show you what we find in the judicial opinion about Big Tobacco handed down in 2006 in US District Court.

Good science has defeated the confabulations of Big Mold. Now the players from Big Tobacco who are still trying to stand with their jackboots on the throats of Americans are getting old-and they are getting exposed. They all still talk about mycotoxins, as if they can't read or think about what the vast preponderance of science says about this issue. Such tactics may have worked in Big Tobacco but not any more in Big Mold.

I don't see too many new recruits coming into the ranks of defense apologists who have anything new to offer (remember the Colorado toxicologist who said that immunology is out of his realm? Of course mold illness has nothing to do with toxicologic monotonic dose responses). The reality is that the current mold apologists (1) have no human studies to back up the idea of implausibility of mold illness; (2) have no research that disproves the vast literature that is now flooding the mold illness field showing immunological abnormalities; (3) have nothing left except a defense attorney's bluster and consultants who are made to look foolish in deposition.

Truth Negates the ACOEM and AAAAI Consensus Statements

This book acknowledges the existence of differing opinions ("Nay-sayers") regarding the subject of human illness acquired following exposure to the interior environment of WDB. Essentially the ideas of the Nay-Sayers can be reduced to just a few concepts, each espoused by two consensus statements published by (i) the American College of Occupational and Environmental Medicine (ACOEM, 10/2002); and (ii) the American Academy of Allergy, Asthma and Immunology (AAAAI, 2/2006).

The differences between the opinions of: (A) treating physicians who deal with real people using objective lab tests showing illness, then predictive treatment protocols achieving real recovery as measured by additional lab tests; and (B) the consensus, without any treated patients, without any tests or any proof. Nay-sayers, are many. These include but are not limited to:

(1) Nay-sayers say it is only mycotoxins that could create illness. [False. It is the multiplicity of components in the WDB that is the actual "dose" of exposure]
(2) Nay-sayers say ingestion is the source of illness and invite the courts to read studies done on sick animals that ate moldy food. [False. Inhalation is the primary source of illness, by far]
(3) Nay-sayers say that because we are all the same fact that some people are not ill means nobody is ill. [False. This denies the proven existence of differential genetic susceptibility to the illness]
(4) Nay-sayers say strange things. The ACOEM consensus paper used a single experimental reference as its basis. High doses of fungal spores were introduced into the nasal passages of rats. The rats became ill. The study author said in so

many words "don't use this work for anything to do with humans." Ignoring this proscription, the ACOEM authors (see Big Tobacco, above) 'concluded' therefore that since a dose of spores injected into their tracheal passages made rats ill, and because humans could not get such a high dose into their tracheal passages/lungs in a WDB, thus it was not feasible for WDB to make people ill. Yes – read that again. A high school science freshman can see the total illogic of this: Rat got sick so humans won't. Yet the US Chamber of Commerce bought this and promoted it.

(5) The truth is that each WDB is a unique and very complex ecological cauldron that is not exactly reproducible in laboratory environments and is neither replicated nor well-modeled by single dose, acute exposure studies. WDB contain a stew of particles and chemicals that are dangerous to some, very dangerous to others and not very dangerous to many others.

(6) Treating physicians identify an individual as a case or not, which may occur within a group that is evaluated. Non-treating physicians cannot.

(7) Patients can be identified with objective laboratory parameters as having a CIRS-WDB that responded to therapies that could help no other illnesses.

(8) Nay-sayers ignore the epidemiologically accepted concepts of causation and causality, particularly with regard to prospective documentation of acquisition of illness and they ignore the immunological basis of the illness associated with activation of exponentially expanding cascades of host responses.

(9) Nay-sayers exclude clinical experience with real patients; they have no research basis on any people to support their comments. They have no research basis to refute the published papers of treating physicians.

Based on the numerous references well documented in the world's literature and opinions of U.S. government and international health agencies, the author of this book feels (i) that the ACOEM and AAAAI papers must be held to the same standard of academic integrity including rigor, thoroughness and transparency as all others; (ii) the two papers are academically without merit; and (iii) neither paper should not be given weight in litigation.

Of significant interest is that these papers have been called into question by two publications (Craner, 2008 and Wall Street Journal, 1/9/07). ACOEM and AAAAI both rely heavily on the fundamentally non-scientific opinions expressed in the Gots/Kelman, 2000 paper (see below). Major errors in these papers include i) the assertion that mycotoxins alone are the agents responsible for human illness; ii) that immune responses are irrelevant; iii) that no human health effects are possible from exposure to WDB; iv) that ingestion, not inhalation, is the primary source of exposure; v) that single dose inhalation studies in rats are comparable to chronic low dose exposures in humans; vi) that there is a linear dose response to exposure; and vii) that it is possible or even scientifically relevant to analyze a single component of exposure from what is found in WDB. Indeed both papers and all Nay-sayers use the "paper tiger" ruse – they erect a false premise and then use it to fabricate and mislead. It is the old sleight of hand trick. If they can get the judge or the workman's adjudicator to pay attention to something irrelevant they can make the pea (truth) vanish.

For those interested in the details, presented below is a point by point refutation of the validity of these papers.

Standard of Academic Integrity

These two documents must be held to a standard of academic integrity with (i) absence of bias; (ii) presence of thoroughness; (iii) presence of academic rigor; and (iv) presence of transparency in decision making, especially including peer review and revelation of conflicts of interest. These standards are found to be lacking on each point in each paper.

A comprehensive summary of the events that led to publication of the ACOEM statement, including decisions made by the executive board of ACOEM that solicited the statement, was written by James Craner MD, and recently published in the International Journal of Occupational and Environmental Health. Dr. Craner's exposé of the flawed process and flawed science used by the ACOEM authors has provided the transparency needed to further discredit not only the incorrect ACOEM opinion but the deeply flawed process that gave birth to it. Per Dr. Craner's article, it appears as though a deliberate attempt was made to deceive not only the members of ACOEM, but also jurists and the public at large. As Dr. Craner and also the Wall Street Journal write (1/9/07), there were concealed conflicts of interest of the three authors of the ACOEM statement – self-described as "evidence-based". There is nothing evidence-based in either ACOEM or AAAAI as that process begins with observation of affected patients.

As of March 2009, ACOEM stated that it intended to revise its 2002 statement, but by the time of this writing in September, 2010, no such revision has been published.

The AAAAI revised their policy of disclosure of conflicts of interests following a series of papers published 9/06 in JACI critical of the absence of conflict of interests statements in the 2/06 AAAAI consensus statement. The AAAAI statement was a subject of intense criticism resulting in publication of 12 letters highly critical of the consensus piece.

Conflicts of interest and academic bias are not the only problems with these papers however. Serious methodological mistakes and leaps of 'logic' to conclusions (not supported by the US GAO and WHO opinions; not supported by any laboratory data obtained from humans or animals; not supported by any study that actually reported thorough baseline studies and results of treated humans; not supported by any current governmental agency report) are found in the studies (animal and others) cited in the paper. The "father of all the Nay-sayers opinions," the Gots/Kelman, 2000 study, will be discussed in its own section. In the paragraphs that follow we list examples of mistakes with others discussed in the ERROR section below.

As discussed in the published paper (Neurotoxicology and Teratology 8/7/06) from the Center for Research on Biotoxin Associated Illnesses, anyone who looks at the ACOEM report must also look at the methods it used before even considering reading the conclusions. The methods of a study set the standard for what conclusions can be drawn. The report relied on one study involving a one-time, short-term exposure of Sprague-Dawley rats to washed fungal spores. No recording of toxin amounts was included and there was no delineation made of the ability of the selected species of "toxigenic" fungi to actually make toxins. Given these lapses, one has to question the utility of the study, considering that it purports to de-link the observed chronic illness to exposure to concentrations of spores, spore fragments and all of the toxins in a WDB, even though it lacks basic baseline parameters required to investigate the link. One may make no conclusions about toxicity when the study involves unknown concentrations of spores and toxins; unknown ability of the fungi in question to actually make toxins; and absence of any validated marker for assessment of toxin effect.

However don't miss the pea in the sleight of hand ruse from ACOEM and AAAAI. The study was not designed to prove anything other than "would some spores in their trachea make rats sick?" Answer "Yes." So the limitations of the study don't matter unless one tries to use it for something else. The ACOEM authors spun this into a statement that WDB cannot make humans ill! No, this leap of science is pure junk. A deliberate fabrication designed solely to be used in litigation. By the way – the rats did get sick. Surprise!

Of course the parameters used in the ACOEM paper bear no relationship to the actual findings in patients exposed to water-damaged buildings and the poor misused rat study author said so in the study – as in "don't use this for people." The level of detection (200x magnification) used to survey for fungal elements is too insensitive (it must be at least 800x), thereby missing a substantial portion of fungal elements that can carry toxins, perhaps by as much as 90%. According to NIOSH/CDC, it is the small elements that carry as much as 99.8% of the total burden of toxin and inflammagens found inside WDB and we know from real studies that these small fragments make it to the lungs more than the larger fragments. Spore counting in the absence of particle counting provides little information. Other than saying yes, a lot of spores were found, such counting cannot say that no particles are found.

> Both the ACOEM and AAAAI reports extrapolate from observation of acute effects, using those data to then **leap totally illogically** (but apparently consensually**) to the conclusion** that there is a threshold for the level of mycotoxins causing injury. There is no basis in logic or fact for this leap. They then compound this error in logic by assuming that there is "decreasing toxicity with longer exposure for a given total dose." Which facts underlay this unreferenced statement? The authors **do not cite any studies** to support this statement because there are none. **In Truth, there is no "decreasing toxicity" with longer exposure**. In fact, as shown by published data, there is recruitment of ***more*** inflammatory pathway components leading to ***further expansion*** of the illness to include injury to production mechanisms of regulatory hypothalamic neuropeptides.

This unfounded leap of logic, concealed in the verbiage of the ACOEM statement, bears no relationship to any known biological pathway. Clinicians who actually treat patients with CIRS-WDB agree that the physiology of the illness involves enhanced response (reactivity) with repetitive exposure. This ACOEM statement to the contrary demonstrates the lack of experience possessed by the authors in diagnosis and treatment of CIRS-WDB; indeed, none have either documented research or even data collection in scientific or public arenas.

The fallacy behind the attempts of the ACOEM and AAAAI papers to deny the reality of CIRS-WDB is well-demonstrated by the report from CRBAI of 850 cases and 132 controls, accepted after peer review for the 9th International Healthy Buildings conference (9/09). The study of these patients, by far the world's largest study of innate inflammatory responses seen in affected patients, expands the older case definition of biotoxin-associated illness to that of a chronic inflammatory response syndrome. This paper, followed by another, yet even more complete study of 815 patients and 130 controls (CRBAI, International Mycology Conference, 8/1/10) increases the total number of patients (adults and children) reported in current literature by CRBAI to 2,030 cases and 450 controls as published in six publications.

The studies cited in the ACOEM and AAAAI paper are not suitable for chronic risk assessment purposes: they are one-time, acute animal exposure studies and do not include adverse effect parameters. The only study (the rat study) the ACOEM and AAAAI opinions cite ***didn't identify any toxins*** or even the specific strain of *Stachybotrys* used in the experiments. The doses of toxic exposures were apparently large with no attempt made to identify the maximum or minimum level of dose tolerated. The rats in the study were sickened by their exposure, with bleeding into the lungs noted. The ACOEM and AAAAI opinions have no toxicity data and no exposure data. There is no logical rationale for applying an extrapolation of animal data to human data in this study. Note that we must recognize that the effects of exposures on human infants and children are not the same as in adults or as in rats. The ACOEM and AAAAI reports never explain the differences between toxins found in indoor water-damaged buildings or whether the toxins were from molds, actinomycetes, mycobacteria or bacteria. Indeed the reader is expected to accept the unproven and unreferenced allusion to "aflatoxin equivalents" as sound science. It is not.

Further, the two statements do not recognize the presence of fine particulates mentioned above that are shed by microbial colonies.

(*Editor's note*: In other words we should accept this consensus statement – "Hey the rats got sick (no-one knows why or how) so it is our opinion that our pals have accepted, that people won't get sick. Yup – the rats did – their lungs were bleeding. We killed them to find out. But you cannot get sick from WDB – so no worries mate.")

The lack of transparency demonstrated by failure to reveal conflicts of interest severely damages the credibility of the ACOEM paper. The AAAAI paper cites the ACOEM opinion. ACOEM cites Gots/Kelman. The questionable methods and the totally unsupported conclusions contained in studies cited by both these consensus statements remove all credibility from both reports.

Validity of Gots/Kelman 2000

The ACOEM statement relies heavily on the scientifically flawed paper written by defense consultants Bruce Kelman and Ronald Gots (see reference to Gots/ Kelman, 2000). The ACOEM and AAAAI opinions make no attempt to re-evaluate the faulty methods and unsupported ideas of the parent paper. Indeed it is important not to forget that the *all Nay-sayers self-serving mold papers followed the erroneous path initiated by Gots/Kelman*. This trail of self-perpetuating inaccuracies resulting in irrelevant conclusions was blazed by the authors. ACOEM cites Gots/Kelman; AAAAI (Bush) cites ACOEM; ACMT cites ACOEM and AAAAI. The validity of all these papers is tainted by the respective authors' working relationship with prominent firms that provide support to the Big Insurance defense interests in mold litigation.

The Gots/Kelman paper reports no use of methods for its conclusions. The paper uses no accepted epidemiologic standards for its approach. Their **assumption** is that the main route of mycotoxin exposure in humans is ingestion. The old paper tiger trick. A quick look at any of the governmental agencies statements regarding mold illness or any of the papers on over 50,000 mold patients definitively demonstrates the importance of inhalation as the mode of acquisition. The statement regarding ingestion in the Gots/Kelman paper **has no cited references for humans**. Looking at previously published studies on one-time inhalation exposure to animals to trichothecenes, significant organ damage in animals is reported. Apparently, observers are supposed to ignore the adverse health effects seen in the test animals from such exposure. Gots/ Kelman readily admits that such one-time *inhalational studies* "do not represent exposure to mycotoxins at chronic, low exposure levels from molds in indoor settings." Yet this admission, one that completely destroys their house of cards, is never discussed by Nay-Sayers.

The AAAAI paper has been extensively criticized by many, including Ammann, Shoemaker and House (NTT 2006) and the GAO. None of its 83 references (the false premise trick again – cite like crazy – use none of them) supports the claim that human health *has not been adversely affected* by inhaled mycotoxins. The ACOEM report and the first CDC statement on mold were both released in the fall of 2002. The Gots/Kelman paper posits as fact that there must be *absorption of a toxic dose over a sufficiently short period of time*. There is no reference for this statement, as there cannot be! **It simply is not true**. Chronic low-dose exposure occurs over longer periods of time, not a "sufficiently short period of time." AAAAI cite studies (Creasia) that show acute toxicity from non-physiologic exposure of animals. Neither of these studies analyze illness-finding in **chronic low dose exposures** in humans. Further, the authors erroneously claim (again, without reference) that the cumulative dose delivered over hours, days or weeks is expected to be ***less toxic*** than single delivery of a bolus of spores. Upon what basis does this illogical assertion rest? Illness acquired following exposure to the interior environments of WDB is clearly based on low level, chronic exposure to humans. The Gots/Kelman paper discussed high-dose acute exposures of animals to unspecified amounts of unspecified toxins found on mold spores. As such, their findings have no relevance to the demonstration of whether or not long term exposure to WDB can cause illness in humans. We note that in litigation, defense "experts" invariably quote these papers (Gots/Kelman, ACOEM and AAAAI) as "proof" that exposure to WDB couldn't make people sick. It is just nonsense.

One should wonder how such "junk science," as labeled in a California ruling (Harold v. California Casualty No. 02AS04291, 2006), based on extrapolating data from one study of acute, high-dose exposure to unknown mycotoxins in rats can lead to the drawing of conclusions about the absence of human illness in association with chronic, low-dose exposure to water-damaged buildings. Yet these misleading and deceptive documents are cited repeatedly as acceptable science by a small cadre of non-treating witnesses who at one time may have qualified as physicians or who sport a Ph.D. Unfortunately, "junk science" leads to erroneous conclusions regarding CIRS-WDB. By citing the first "junk science" paper at the start of a discussion, then adding the ACOEM opinion, itself based on "junk science," and then adding a third "junk science" paper, litigators thereby try to suggest that a "robust literature" supports the idea that mycotoxins alone would be responsible for any symptoms of illness. This idea is not supported by any academic organization and is not supported by any government agency. Specifically the US GAO and WHO opinions directly disagree. The only robust scientific literature is that cited throughout this book and in the Mold Research Committee report published by the Policyholders of America 7/27/10.

The AAAAI report ignores the absence of any support in any literature that looks at human research. The GAO report points this deficiency out by noting that Bush included no data on actual research subjects and included no assessment of the role of immune responses in illness. No one can ever produce evidence that this sham paper represents science at all.

False Premise # 1 - Mycotoxins Alone Responsible for Illness

The ACOEM and AAAAI papers suggest that mycotoxins alone would be responsible for any symptoms of illness. This idea is not supported by any academic organization and is not supported by any government agency. Specifically, the US GAO and WHO opinions directly disagree.

The AAAAI report ignores the absence of any support in any literature that looks at human research. The GAO report points this deficiency out by noting that Bush included no data on actual research subjects.

False Premise # 2 - Innate Immune Responses are Unimportant

The ACOEM and AAAAI opinions would have us believe that immunologic mechanisms, specifically innate immune responses aren't important in pathogenesis of the illness symptoms. This idea is not supported by any academic organization and is not supported by any government agency. Specifically the US GAO and WHO opinions directly disagree. **Bush, in fact, suggests (pg 329) that observers *ignore the innate immune responses* from building exposure. These responses <u>are</u> the illness!** The GAO statement points out this clear attempt at obfuscation by noting that Bush ignored the immunological features that follow exposure to mycotoxins. The WHO report also weighs in repeatedly on the immunological aspects of this illness.

False Premise # 3 - Ingestion of Mold is the Mechanism of Illness

The ACOEM and AAAAI papers tell us that ingestion is the mechanism of illness acquisition. This idea is not supported by any academic organization and is not supported by any government agency. Specifically the US GAO and WHO opinions directly disagree. Common sense says this is just a subterfuge.

This idea comes from the previously discussed paper written by defense consultants Gots/ Kelman in which they *assume* that mold illness is acquired following *<u>ingestion</u>* of mold spores and products. There is no basis for this assumption and indeed no one other than the few collaborating defense consultants suggests that CIRS/WDB comes from ingestion (eating and swallowing mold). All government agencies and researchers in the field agree that inhalation is the primary mechanism of exposure.

The ingestion idea is an artifact from the history of mold contaminated foods, the perception of which is allowed to continue within the artificially created void (shell game) which excludes relevant evidence.

False Premise # 4 - Single Exposure High-dose Inhalation Rat Exposure is Comparable to Chronic Low-dose Human Exposure

The ACOEM and AAAAI statements would have readers believe that a single dose, inhalation exposure study of spores instilled into the trachea of rats (which reported severe inflammatory responses in the test subjects, even after washing the spores with methanol) is equivalent to the long term, sub-acute exposure of patients to the interior environment of a WDB. Consider that the study cited by the ACOEM and AAAAI papers used spores that were washed with methanol to reduce the presence of pathogenic substances including toxins. The data was then extrapolated to support assessment of physiologic findings in humans with chronic low dose exposures. They manufacture the idea of "no effect" by deleting what actually happened to the rats. The "no effect" idea is not supported by any academic organization and is not supported by any government agency. Specifically the US GAO and WHO opinions directly disagree.

An important paper (*Neurotoxicology and Teratology* 2006; 28: 573-588. Sick Building Syndrome and exposure to water damaged buildings: time series study, clinical trial and mechanisms. Shoemaker R, House D.) attacked this study with specific point by point refutation as follows:

1. A no-observable adverse effect level (NOAEL) was not identified in the rodent study. Pulmonary inflammation may occur at dosages below 2.8×10^5 spores/kg body weight. Another study of fungal-induced pulmonary inflammation estimated the NOAEL to be $<3.0 \times 10^4$ spores *S chartarum*/ kg body weight. (Less than 30,000 spores per kilo of body weight)
2. Airborne fungal spores carry only a small fraction of the biologically active mixture components to which people are exposed in WDBs. The concentration of small fungal fragments carrying antigens, mycotoxins

and other biologically active components **exceed spore concentrations by up to 500-fold**. For example, concentrations of air-borne trichothecenes carried primarily on fungal fragments smaller than intact conidia were reported to exceed 1300 pg/m³ in WDBs.

3. Young adult mice were used in the rodent studies. Younger and older populations, as well as other physically compromised populations may be more susceptible to exposure-induced inflammation than healthy, young adult mice. Of course my work is with humans. Thousands of humans.

4. The initial onset of Sick Building Syndrome (SBS) is typically observed in occupants of WDBs following chronic exposure, often many months after exposure begins. Human health risk assessments for SBS should not be based on effect levels from sub-chronic rodent studies without consideration of uncertainty factors. Uncertainty factors include the potential for cumulative effects; toxin accumulation in tissues; and effect threshold shifts to lower levels as protective and repair mechanisms are compromised during chronic exposure. Additional uncertainty factors include interspecies differences in susceptibility and intra-species differences including genetic polymorphisms affecting toxin elimination. Amplification of the pro-inflammatory cytokine response by rising levels of leptin and blockage of the proopiomelanocortin (POMC) response may also be an important factor in the progression of illness during chronic exposure that may not fully develop during acute and sub-chronic exposures.

5. Previous episodes of SBS from exposure to WDB may sensitize humans to subsequent exposures. The hypothesis of sensitization is supported by the observation of relapse within 3 days of re-exposure in the current and previous studies, as opposed to the gradual onset of initial illness reported by the study participants - who are people not rodents.

6. The potential for additive and synergistic induction of a pro-inflammatory cytokine response by mixture components indicates that human health risk assessments for SBS should be based on studies of exposure to mixtures actually observed in WDBs. Studies should determine (a) the NOAEL for development of an inflammatory response during initial acquisition of SBS following chronic exposure to the mixtures; and (b) the concentration and time dependence of acute exposure-induced reacquisition of SBS following cholestyramine (CSM) therapy and subsidence of the pro-inflammatory cytokine response.

In the same rat study the AAAAI and ACOEM statements cited as "definitive," the Rao and co-authors (rat study) specifically state that *no conclusions* about effects of exposure of people in low dose exposure over time *can be drawn*. In addition, the **authors** of the rat study **themselves** say that extrapolation **cannot be made** from the results of the limited, one-time, high-dose exposure in animals to long-term, low-dose exposure in humans. Indeed, they say the results aren't the same as in chronic low-dose exposure (emphasis added).

Despite these obvious and widely acknowledged limitations that would bar any application to human illness and based on one high-dose study in rats looking at unknown mycotoxins only, against the rat study's author's advice, the ACOEM authors conclude that there is no way an adequate dose of toxins from indoor exposure to the complex biological mixtures of toxigenic fungi, bacteria and actinomycetes can cause human illness in chronic exposure. (*Editor's note*: "Say what? The ACOEM endorsed this nonsense?")

This illogical conclusion is not supported by any academic organization and the US GAO and WHO opinions disagree with both ACOEM and AAAAI.

Further, the rat study had more shortcomings. For example, the study did not present any data on what toxin exposure actually does to inflammatory cascades; it didn't report any follow-up on the rats. The methods of the study excluded any reasonable assessment of what material/toxins were delivered to the rat subjects. All the study did was to show that instillation of unknown doses of *Stachybotrys* spores into rat lungs injured the rats severely. The ACOEM report discussed none of the rat study's deficiencies and incorrectly attempted to apply dose-response relationships to elements that could never support such conclusions, especially those from chronic, lower intensity dose exposures.

As would be expected, the rat study cited in the ACOEM report has never been replicated. Indeed, the study has been refuted in multiple studies done by Dr. Thomas Rand's group from Nova Scotia which show significant activation (100,000-fold) of inflammatory cytokines by **as few as 30 spores** of toxigenic fungi per gram in mice following intra-tracheal instillation.

Of note, Dr. Rand didn't wash away the toxins from mold spores before instillation and he did measure pertinent inflammatory markers. His data on hyperacute exposure apply directly to what we see in humans: inflammatory responses are measurable in affected patients beginning within hours of exposure to contaminated structures.

Finally, there is no basis to assume that ongoing chronic exposures in people are equivalent to what was seen in a fixed total exposure in animals in which no other exposure except to mycotoxins was performed. **CIRS-WDB stems from exposure to a complex mixture of inflammagens and toxigenic organisms found in WDB.** The study of Nikulin is also cited. Unknown amounts of toxins were injected intranasally (not intra-tracheal administration, for unknown reasons) into mice one time (only). All mice again had marked lung inflammation. In a twice-weekly inoculation experiment that lasted for three weeks (this is not a chronic, low-dose exposure study design), Nikulin again showed significant lung damage, all the while noting the absence of applicability of the Nikulin study to real world human illness by stating: "intranasal inoculation is unlikely to model the exposure of humans in even very moldy environments."

Curiously, and out of character for the rest of the paper, ACOEM notes, "the issue of mold exposure is important from a health standpoint and can potentially affect anyone in the indoor environment." We don't hear this statement made by defense interests in mold litigation.

False Premise # 5 - Dose Response to Exposure is Linear

The idea that the "the dose makes the poison," permeates toxicology. For a given dose there will be a given and proportionate response. This idea, called a monotonic response, is not applicable to first, activation of pattern receptors following antigen detection; and second, differential gene activation by initiated by such detection. The activation of initial responses leads to profound host innate immune responsiveness that is ***exponential*** and actually *becomes* the illness.

As opposed to the idea from toxicology that the "poison is in the dose," the concept here is that the poison is in the ***initiation*** of the mega-multiplying response to a given signal. As Dr. Lewis Thomas has said repeatedly, "the **response of the host** makes the disease." The host response is the illness and not the antigen. Said another way, Dr. Thomas rightfully tells us that "the reaction of sensing is the clinical disease" (NEJM 1972; 287: 553-555) and, "we are in danger from so many defense mechanisms, that we are in more danger from them than from the invaders." And so it is with CIRS-WDB patients.

We are entering an era of understanding how chronic inflammatory response syndromes impact on the assessment and treatment of chronic fatiguing illness. Dr. Thomas showed us the way to address the importance of host innate immune responses. Results of both animal research and human research studies are consistent with the actual human data revealed (i) in assessment of baseline profiling of cases and then (ii) in responses to therapy. Exceedingly small exposures can set off massive immune responses: this is what treating physicians see on a daily basis in patients with CIRS-WDB. The ACOEM and AAAAI papers reported neither actual human data nor any results of human research. Of course their 'conclusions' are without weight.

No single element can ever be identified as causative in CIRS-WDB illness. Specific causation cannot possibly be assigned to any particular element / elements of the complex mixture found inside the interior environment of WDB, containing as they do, multiple contaminants resulting from the presence of multiple species of fungi, actinomycetes, mycobacteria, Gram positive and negative bacteria and their by-products. Thus, the "dose makes the poison" concept cannot be applied to the diverse conditions of WDB as there is **no single dose; instead there is a multiplicity of components, many synergistic, within the naturally undifferentiated dose.** Finally, genetic variability, age, and pre-existing health conditions of occupants add to the complexity of CIRS-WDB illness.

It is this observation of repetitively (and consistently) observed differential host responses that underlies published repetitive exposure studies.

The final blow to linear dose response arguments:

Consider that an effect or response (X) is related in a linear fashion to dose. (X) will then be equal to the sum of routes of exposure: (A) plus contaminants (B) plus length of time of exposure (C) plus individual genetic susceptibility (D) plus individual prior exposure and change of susceptibility from that exposure (E) plus types of microbial organisms, each potentially acting synergistically with another (F) plus the types of inflammagens causing potentially exponential changes in c-type lectin receptors, especially dectin-1 and dectin-2 receptors. Then in addition, (G) particular compounds, including mannosylated glycoproteins made by fungi can activate mannose receptors (H) that then alter the signal given to antigen recognition cells to respond to such antigens, further altering the processing of antigen in the intracellular components (endoplasmic reticulum and Golgi body) of such cells. X then is equal to the combined effects of A through H, each of which can cause *amplification, not addition,* of the effects of innate immune responses. Moreover, the elements A through H are each themselves variable. The implications of this analysis gets worse for the linear dose-response advocates: there are interactions of A through H, some of which are synergistic and some of which involve differential gene activation as well as epigenetic phenomena. Proponents of a linear dose response cannot account for how their model can withstand the above discussion. *It is impossible to assume that response or effect X will be linearly related to variables, each simultaneously expressed in A through H.*

Assumptions about dose responses being linear, or even worse, attempt to portray dose responses as active and reliable, as they are depicted in the ACOEM and AAAAI papers. This does not have any validity in scientific fact. **Bush, in fact, suggests (pg 329) that observers *ignore the innate immune responses* from building exposure that are actually the illness!**

Error # 6 - Attempts to Assess Specific Causation

No one **can analyze one component of exposure inside WDB**, namely mold spores or mycotoxins, as suggested by non-treating professionals in various groups, and come to any meaningful conclusions from classical monotonic dose-response relationships when the health problems seen repeatedly stem from the exponential, ever-expanding host response to exposure.

Said another way, the complex mixture of exposure and the altered susceptibility of the host resulting from prior exposure to WDB, combined with the complexity of pattern recognition responses of innate immunity results in an extraordinarily amplifying immunologic response by the host to new, even short-term, low level exposures within the WDB. In order to understand the biological cascade of responses to exposure to a WDB, we must assess the potential and actual responses of each **host**. This idea is a fundamental shift in the "Sick Building" paradigm. Prior focus on the individual components of the complex mixture in the building has been superseded by focus on the individual within the building.

Observers must recall that by 2004 we knew that the accumulated mass of fragments of spores was the reservoir in which toxins were found. In 2007, research advances discredited previous erroneous statements about dose-response relationships in CIRS-WDB; these revisions were then incorporated into public policy. Research scientists from the CDC, as mentioned, published a paper in Applied and Environmental Microbiology (March 2007) that confirmed the association of health effects to exposure to WDB. In the acknowledgments to this paper, the authors expressed gratitude to the Inspector General of the Department of Health and Human Services for providing **protection to safeguard the health** of the samplers. One might ask, “Protection from what?”

Moreover, the Ministry of Health of Canada (Canada Gazette 2007-03-31 Part I, volume 141, No. 13) concluded that “results from tests for the presence of fungi in air cannot be used to assess risks to the health of building occupants.” Further, the Minister recommended that individuals, “control humidity and diligently repair any water damage in WDBs to prevent mould growth,” and that all such WDBs should be subject to the directive to, “clean thoroughly any **visible or concealed mould** growing in WDB buildings.” In a major monograph on indoor air quality published in Volume 115 of Environmental Health Perspectives June 2007, editorial writer Bob Weinhold traced the changes in thinking about mold illness and wrote on page A305: “Regardless of the remaining uncertainties, the overall recommendations of many organizations and agencies worldwide are reaching a common conclusion: **Don’t mess with mold. If you can see it or smell it - and especially if health problems are occurring - clean it out, throw it out, or get out.”**

CIRS-WDB involves so many inflammagens and toxins that trying to study the unique ecological niche that is a WDB with just one parameter doesn’t make sense. What occurs inside a building is not the same as what occurs outside the building. A building is not a compost pit or a pile of wet leaves, for example. Moreover, our understanding of just what is in the affected buildings continues to grow with the advance of research. Just two years ago, no one talked about mycolactone-forming mycobacteria and, quite frankly, mentions of c-type lectin receptors were discussed in just a small branch of immunology. Now C-type lectins, especially dectin-1 (and now dectin-*2, how many others?!*) receptors are recognized as critically important in generating an inflammatory response to beta-glucans. Even more important, the response of C-type lectins recognizing glycoproteins, remains critical to understanding how low-dose erythropoietin (epo) reverses the ongoing activation of production of the short-lived anaphylatoxin, C4a. Low dose treatment with erythropoietin lowers C4a and stops its regeneration as described in two papers published by the Center for Research on Biotoxin Associated Illnesses (CRBAI: CDC conference on Chronic Fatigue Syndrome, 1/07, and American Society for Tropical Medicine and Hygiene, ASTMH, 11/07).

Error # 7 - No Human Health Effects Are Possible

The ACOEM and AAAAI papers want to lead us to believe that no human health effects are possible following exposure to the interior environment of WDB. This idea may be comforting to those responsible for that exposure but is not supported by any academic organization and is not supported by any government agency. Specifically the US GAO and WHO opinions directly disagree, as do peer-reviewed published papers covering thousands of sickened patients. The lack of validity of the Gots/Kelman, 2000 study on this point has already been discussed. Again, the ACOEM opinion relies on Gots/Kelman and the AAAAI statement cites the ACOEM report.

Note that the ACOEM and AAAAI papers specifically ignore published human and animal health studies that directly contradict the opinions of ACOEM and AAAAI. Note also that they make no attempt to refute the documented physiologic studies published from around the world showing mechanisms in the CIRS-WDB. Finally, note that the ACOEM and AAAAI reports do not cite any studies reporting any physiologic studies; or any prospective studies done in humans (or even animals) that refute the universally acknowledged documentation of illness reacquisition with re-exposure.

Summary

There are hundreds of academic citations and an overwhelming unity of agency and impartial opinion to support the concepts (1) that WDB host inflammagens and toxigens that are inhaled as the initiating event in the acquisition of illness; (2) that exposure to inhaled, not ingested, inflammagens and toxigens creates predictable human host responses; (3) that the human illness seen following inhalant exposure has epidemiologic consistency among multiple studies published in multiple diverse locations throughout the world; (4) that the inflammatory responses seen following inhalant exposure in affected patients are epidemiologically consistent and are mirrored by experimental data seen in vitro, in humans and in animals; (5) that treatment of affected patients is shown to be effective in peer-reviewed publications, including double-blinded, placebo-controlled clinical trials, and is further supported by experience of treating physicians, **a group whose experience, skill and observations have been omitted to date in the discussion regarding CIRS-WDB**; (6) that re-exposure of previously treated patients re-creates the illness within three days with an accentuation of (i) the ***magnitude*** of inflammatory host responses and (ii) the ***rate*** at which those enhanced inflammatory host responses occurs; and (7) prevention of illness by use of replacement doses of VIP (see treatment protocols) restores control of multiple physiologic abnormalities seen routinely in CIRS-WDB cases.

The reality is that there are no acceptable academic research papers that refute the presence of definable CIRS-WDB. The ACOEM statement is nearly universally referenced by subsequent defense industry-friendly publications without attempts to discuss the unscientific process used by the authors of ACOEM. The only experimental study on mold illness that Nay-sayers can quote regarding long term effects of mold exposure is the single, high dose rat study that is specifically rejected by the rat study authors themselves for use in human illness projections.

Finally, let us not forget the data supporting the comments in this report are generated from actual diagnosis and actual treatment of thousands of real, living patients. In contrast, assertions put forth by Nay-sayers were based upon inaccurate assumptions made by consensus panels comprised of professionals not actually in clinical practice in the field of WDB illness. Our physicians aren't performing isolated experiments in rats, they are treating people. Our physicians aren't guessing what the illness might involve; they are following the established process of science using prospective exposures that give us the right to use the term, "caused," as validated by the peer-reviewers. Over 50,000 patients from 14 countries have been reported in peer-reviewed scientific articles. That is compelling evidence for the existence of CIRS-WDB. Any "academic statement" which rejects this truth, while simultaneously not providing any human physiologic data, must axiomatically be considered "irrelevant."

On August 17, 2006 Gladys Kessler, United States District Court Judge issued the FINAL OPINION in the Big Tobacco case. Some of her findings are relevant to WDB and Big Insurance. These are paragraphs from that opinion.

785. Beginning in 1986, Brennan Dawson, spokesperson for the Tobacco Institute, reiterated in numerous television appearances the Tobacco Institute's public position that the links between smoking and disease had not yet been established. The Tobacco Institute's position that it had not been proven that smoking caused disease was not shared by a single public health organization during the entire time Dawson served as spokesperson for the organization.

788. In a January 11, 1989 appearance on the television show "Good Morning America," Dawson stated that "all the links that have been established between smoking and certain diseases are based on statistics. What that means is that the causative [sic] relationship has not yet been established."

This was twenty-five years after the Surgeon General announced a causal relationship between smoking and lung cancer.

789. Similarly, in an appearance on CNN's "Crossfire" on April 18, 1989, Dawson claimed:
     Statistically there are associations. In terms of biological causation that hasn't been found which is why I came to the conclusion that smokers have to make up their own minds.
800. Sandefur admitted that he was unaware of any studies showing that whole smoke does not cause disease, and he was unable to name one scientist or medical doctor totally unconnected with the tobacco industry who said that it had not been established that cigarette smoking causes cancer.

(*Editors note*: This is what the epidemiologists call the "counter-factual" approach. It applies to big mold as well. The Nay-sayers can point to no human research study that supports their claims.)

808. Although Philip Morris expressed "difference" in opinions between it and the public health authorities, Senior Vice President and General Counsel Denise Keane could not, while testifying in this litigation, cite any peer-reviewed article, study, or consensus report from 1977-1997 that disputes the scientific conclusion that smoking causes lung cancer.

810. Finally, on October 13, 1999 when Philip Morris launched a corporate website, it changed its public position on smoking and health issues. The website stated: "There is an overwhelming medical and scientific consensus that cigarette smoking causes lung cancer, heart disease, emphysema, and other serious disease in smokers." Steve Parrish, Senior Vice President of Corporate Affairs for Altria Group, acknowledged that the overwhelming scientific consensus referenced in the October 1999

statement had existed for decades. Parrish further conceded that Philip Morris' refusal to acknowledge prior to October 1999 that smoking caused disease had damaged the company's credibility because there was no support for Philip Morris's view outside of the tobacco industry.

824. From at least 1953 until at least 2000, each and every one of these Defendants repeatedly, consistently, vigorously and falsely denied the existence of any adverse health effects from smoking. Moreover, they mounted a coordinated, well-financed, sophisticated public relations campaign to attack and distort the scientific evidence demonstrating the relationship between smoking and disease, claiming that the link between the two was still an "open question."

Finally, in doing so, they ignored the massive documentation in their internal corporate files from their own scientists, executives and public relations people that, as Philip Morris's Vice President of Research and Development, Helmut Wakeham, admitted, there was "little basis for disputing the findings [of the 1964 Surgeon General's Report] at this time.

CHAPTER 14 Liana Jones, My Wife: Winston Churchill Said "Never Give In"

By Tom Jones

Liana, my bride of 27 years, is a mold survivor. She's survived the inflammation that her workplace exposure caused. She has survived the infectious process that was caused by the workplace. And she's survived the misdiagnosis of cancer by some of the best practitioners on the East Coast.

If I'd listened to the advice of "experts," and their suggested treatments of "cancer," when a mold infection was her problem, I think Liana would have been long gone.

Her story isn't just the usual "I was exposed to mold and now I'm desperately ill." The treatment for her illness nearly killed her. She went blind, and then got the cancer diagnosis; she received chemotherapy and radiation therapy for an illness she didn't have. Some of the best physicians on the East Coast were all wrong about her.

Thanks to Dr. Michael Gray in Benson, Arizona, Dr. Dennis Hooper of Dallas, Texas, and Dr. Shoemaker of Pocomoke, Maryland, Liana is here today. Dr. Shoemaker asked me to tell you about Liana, and explain my role in this case. Dr. Shoe says that my actions saved my wife's life.

Liana's story has so many twists and turns that Mystery Diagnosis will feature her case in one of its episodes in the spring of 2011.

And now I can truly tell you that with her life safe, the use of VIP is now restoring quality of life. Seeing my vibrant bride as active today as she was years ago was never my real idea of what was going to happen. I simply wanted her alive, but now our lives together are like new; our new miracle is getting up every day with energy, love and optimism, smiling as the sun rises in the East. I'm not ashamed to tell you I get emotional about what blessings we have received from the darkest hours of the past years. Liana's courage in the face of all adversity, her faith and the strength of our marriage are personal, yet Dr. Shoe says we are an example of the best of Winston Churchill's approach to finishing problems. He, and we, never gave up until the job was done.

I think what I did is what anyone would do when faced with the final question: will I fight for the life of the love of my life or not? I fought.

Yet the fight Liana and I faced together was one no one should have to see. At the most critical times, we had to choose to either believe in our own review of obscure medical literature or what we were told by respected medical authorities. I think there's something about molds and fungi that cause respected physicians to turn a deaf ear, to stop being objective and to stop being professional. As a research scientist, that kind of attitude is completely antithetical to every ounce of my training. Such blind, irrational bias is at best disgusting; it is repugnant to all tenets of the objective evaluation of data that scientists must have.

As I'll tell you, if we'd listened to the experts, Liana would be dead from an untreated fungal infection or from the surgery, chemotherapy and radiation treatments she didn't need. I can still remember Dr. Shoemaker telling me that he had gone as far as he could take us but that I should trust my data. As a scientist, I trust data, too, but when the data I collected— from photomicrographs to cultures to arcane assays— disagree with highly paid consultants, and my wife's life hangs in the balance, how can I ignore what I see objectively? How does what I can prove compare to the (mere) opinion of an expert?

Everything I did made good sense to me even though a voice deep inside asked/accused me every night, "but what if you're wrong?" I followed my gut instincts and searched for more facts. I never gave up, but I cannot describe the private agony I've felt. I think that Liana and Dr. Shoemaker know, and I believe that my actions were the result of either an indomitable sense of purpose similar Winston Churchill's or a desperate gambler's wild acts of love, lost in the twists of the game, "You Bet Her Life."

It never ceases to amaze me how life throws curves at us all. But the Good Book, which I read often, says that you're never tried more than you can stand and there's always an escape. I have to admit there were times during our ordeal that I couldn't see a way out. We seemed cast away in a deep gorge, unable to glimpse the top no matter how hard we strained to see.

Our story is like many others in the war with mold. My love's unexplained symptoms - fatigue, weakness, aches, headaches, sensitivity to temperature, skin sensitivity, shortness of breath, unusual muscle cramps, numbness, tingling, mood swings, excessive thirst, increased urination, night sweats, loss of memory, difficulty focusing on tasks, concentration problems, difficulty finding the right words or names - appeared gradually over three years. Her return to health required a sequential series of therapies much like climbing a sheer cliff. One misstep and the valley floor 4,000 feet below waits for you to make an impression.

Liana first had symptoms after she returned to work for a property management firm she'd worked for seventeen years before. In the interim the company had moved to an old row house just down the street from their old location. The building was about 100 years old. It had had been converted into offices not long ago. It had that old building smell to it, but it had been upgraded with air-conditioning. If I knew then what I know now, the smell, the old building, and the second story porch's jack-leg carpentry were factors that are markers for risk of water intrusion. And that means microbial growth. I should have heard warning bells.

Before I tell you "the rest of the story," I should tell you a little about myself. My name is Tom and I am an employed research scientist, for

38 years now. Uncle Sam's Army trained me first as a combat medic, then as a clinical laboratory technician, and finally as a research technician. I had a working knowledge of good scientific method and I knew how to perform a differential scientific analysis. I graduated from St. Mary's University's pre-med course with a bachelor degree in biology. In the private sector, I was responsible for research and development of a number of FDA-approved clinical diagnostic kits. For example, I was a lead member of a research and development team that developed, qualified and manufactured the Western Blot confirmatory assay for HIV (AIDS). I've acquired extensive knowledge in immunology, bacteriology, and virology as they relate to diagnostics. I am not an MD.

At one point, I worked on a Defense Advanced Research Program Administration (DARPA) project to detect haptens. A **hapten** is a small molecule that can elicit an immune response only when attached to a large carrier, such as a protein. The target that was chosen was the mycotoxin aflatoxin B1, as it's thought to be the most potent natural carcinogen known. Though I'd studied some mycology in the Army lab school, had seen several examples of systemic mycosis there, and had taken the mycology portions of the microbiology courses in college, this project was my first real foray into the subject matter.

For three years after Liana became deathly ill, I threw myself into studying mycology. I'd seen firsthand the effect of yellow rain in Vietnam, namely the use of the mycotoxin T-2 toxin as a biological warfare agent. I'd seen the effect it had on a few soldiers. I am aware that there are some politicians who said mycotoxins weren't used as a biological weapon.

Now the stage is set. My wife became sick and I had enough scientific knowledge to make me wary of unsupported opinions from others, especially physicians who use assumptions as their basis for decision making.

Now I can tell you the rest of the story.

Liana had worked as a children's pastor for fifteen years and a private school administrator for five of those years when the church decided to close the school. Insurance costs were too high. Liana and I decided that the sixty-hour work weeks she was putting in between the school and full-time children's pastoring was taking a toll and it was time to let a younger individual take the lead.

She'd not had any health issues while working for the church. Sure, her cholesterol was a tad high and her blood pressure was on the high side, but both problems were easily controlled with medication. Our five kids were all grown and out on their own. We'd purchased the farm of our dreams. We both like to ride; quarter horses and cow cutting have been a passion of mine for years. We were all set to put our retirement plan into place, using a horse breeding operation to supplement our income. Fifteen years out from retirement and all was right with the world.

Liana started working for the property management company in 2001. In 2002, she became the accounts manager and moved into a second-floor office. Occasionally, she'd come home and tell me about the animals in the low-ceiling garret above the office. She could hear their little feet as the animals walked across the ceiling, and on several occasions, I was troubled when she told me she could smell a dead animal. But because of the roof's steep pitch, no one could get up into that area from the outside to clean it out. In fact, later on, the only way to get to that area was to take the roof off.

Then there was the gutter problem. A bad gutter had caused a water leak around the windows on the building's north side. I have no idea how long the problem had been there and Liana didn't remember either. The owner's crew got the wet-vac, cleaned the water up and fixed the gutter. Then the symptoms started. She developed a skin rash mistakenly thought to be shingles. She had severe cramps in her legs and feet at night. Her fingers would lock up while she was typing.

"It must be my age—we're getting older," she'd say. The chronic fatigue was there, too. On Monday, she'd head into work feeling pretty well. But, you know I almost could mark my watch because by the close of business Wednesday, she'd be so tired that she'd have to take a two-hour nap after work. She'd be in bed by nine. Saturdays were a wash because she had to spend half the day in bed to recover from the workweek. Sundays were better but cleaning and especially vacuuming, her joys, were out of the question. Couldn't do it and have enough energy to work on Monday or Tuesday. Then on Monday, the cycle started all over again.

In Spring 2003, a mole on Liana's left leg began to hurt. She went to the dermatologist and the mole was found to be Stage 1 melanoma. It was removed, and they found no sign of the cancer having spread to the lymph glands. But— there's always a "but"—she got a post-operative bacterial infection at the wound site, which resulted in a wound that was three inches wide by seven inches long by three inches deep on the calf of her left leg. What a nightmare. It took nearly ten months of bandage changes and wound debridement to allow the wound to heal secondarily. Liana continued to work through it all, but that's all she did during the healing process. The symptoms were still there, too. By February 2004, the wound finally closed over.

Then she began to have trouble seeing. The left eye was worse than the right. She did a lot of computer work—that must be it, too much computer work. Liana had her eyes checked and got a stronger prescription for her glasses; problem solved. Or was it?

Then there were the short lapses in memory at work and the brain fog. One day at the end of August, Liana had stopped off at Wal-Mart to do some shopping. When she got home, she said she'd driven into the parking lot and suddenly couldn't remember why she was there. It took her a few minutes to figure out where she was and what she was supposed to buy. This was not like Liana; her memory was like a steel trap. She'd been able to recall property addresses, the owner's names, the tenant's names from when she worked for the management company years before.

From August through November 2004, Liana had an ever-increasing sinus problem. This wasn't your grandmother's runny nose. Over the years, Liana had sinus problems that were usually controlled with antihistamines, but not any more. September brought uncontrolled hiccups and sneezing. She'd lose her voice and have trouble breathing. She took more antihistamines, thinking that her problems were allergies. The meds helped, but she'd still lose her voice, but only at work. At home, she'd feel a little better.

I didn't know how to understand her new symptoms at home and at work: the night sweats and the severe muscle cramps. I've never heard of such cramps. When she'd get agonizing muscle spasms in her calves, she'd leap out of bed desperately trying to extend her ankle, stretching the calf muscle, but the muscle would *twist and knot* and then actually tear. She'd sit on the cold bathroom floor in a cold sweat with her back against the wall, pulling on her big toe to stretch until the muscle storm would pass. She'd plead with her out-of-control muscles, "Stop," but the spasms just did what they wanted to. Some mornings she'd have blood appear under the skin along the outside of her foot from the muscle injury. (*Author's note*: such muscle spasms are not unusual in those with high C4a like Liana had, but interestingly, none of her physicians asked about this horrible symptom until she came to Pocomoke.) The vision problems just didn't make sense. October brought more trouble with her eyes, another stronger prescription and a new pair of glasses.

In late November, she developed sinusitis with severe upper and lower respiratory congestion and flu-like symptoms. Our family doctor gave her meds. By Christmas, her chest symptoms were gone but the sinus symptoms persisted. Her eyes, though, didn't improve; in fact, her sight was worse.

For three months she suffered and nothing helped. I'm not a physician and I had nothing to offer. In February 2005, Liana was hit with two debilitating episodes of severe headaches that lasted between 12 and 24 hours. All the symptoms were consistent with a migraine. Then in April 2005, Liana went blind in her left eye. Was it a stroke? Optic neuritis? MS?

I took her to every physician I could find but full work ups brought no results. All tests were normal. She wasn't making up her blindness. Was I asking too much for some expert physician to give me an answer?

I was really upset. What could cause her to go blind so suddenly? Liana knew how bad I felt for her, yet she never lost her sense of humor. To try to make me laugh, she'd look at me and say "you're there," then close just her right eye and say, "you're gone" as fast as she could. "Don't do that!" I'd tell her, then laugh. The laughs in the face of such loss were sort of empty and we knew it.

That same month, Liana found a dead bat one morning at work in the kitchen sink and a few more of the little critters were flying around the office. Her co-worker caught the bats and removed them, but where were they coming from?

Early in June, a worker for a heating and air conditioning firm came to the office to drop off an invoice for a job he'd just finished. As it turned out, this young man was very sensitive to mold. He told the whole office staff, ***"There's mold in this building."*** When Liana got home that night, she told me about the guy's certainty that the building was moldy. WHAM! Those studies from my DARPA days came back to me. Could all her medical problems be due to mold?

I told her that the company had to check the building for mold and surprisingly, they did. It took a couple of weeks to get the testing done. They took one air sample from Liana's office and one outside air sample. The mold report, dated Aug. 23, 2005, confirmed the presence of toxigenic fungi Asp/Pen 520 CFU/m^3 with a total count of 960 CFU/m^3 in her office. The report plainly said: *A mold spore count in indoor air of any one genus of mold greater than 50% of the total count is indicative of vegetative mold amplification.*

The *Penicillium/Aspergillus* concentration in the office was 54% of the total spore count found in the office versus those same species being only 7% of the total outdoor air spore count. We told our family doctor about what we found out and he sent Liana to an allergist. In the history recorded the doctor said:

"Since April, she's had daily sinus pressure headaches when at work. She denies nasal reactivity on exposure to noxious odors, but barometric pressure changes trigger sinus headaches. Liana dates the onset of this to an environmental issue that is going on in the building in which she works. She has worked at her current workplace for the past four years. She works in a 90-year-old building that has a musty odor without visible mold growth. It has also recently been noted that bats, squirrels, and birds are in the attic of the building. The building is slated to have remediation done beginning in October. A mold study was done by Pro-Lab, which was remarkable for significant levels of Penicillium and Aspergillus molds.

"Skin Test Results: Liana underwent prick/puncture and intradermal testing for inhalant allergens. She had positive reactions to dust mites; tree pollens; and mold groups with positive reactions to ASPERGILLUS FUMIGATUS, Fusarium, and PENICILLIUM MOLDS when done individually. She had negative skin tests to grass, ragweed, and minor weed pollens; cat; dog; feathers; and cockroach."

The allergist had also ordered a CT scan of Liana's head and had put her on Nasonex and Astelin nasal sprays and a tapering dose of prednisone starting at 40 mg, tapering to 10 mg over the course of a month. (The prednisone and the length of time she was on it is an issue later on. Getting her off the prednisone wasn't easy.)

Our heads were spinning. I knew mold was dangerous to a person's health. It's in the building were Liana works and she had a pretty strong allergic reaction to those same molds. Was this illness all just an allergy? I haven't heard of allergy causing blindness. A few days later, our family GP called and wanted to see us. The CT scan is troubling. When he read the results to us, we asked "What does this mean?" I pleaded with him to not forget about mold.

"CT OF THE SINUSES: A CT scan of the sinuses was performed utilizing coronal imaging. The examination demonstrates

> significant opacification of the sphenoid sinuses bilaterally. In addition, there is thinning of the adjacent bony structures in the skull base region adjacent to the sphenoid sinus. This finding could indicate associated destructive changes. The possibility of a soft tissue mass arising in this region should also be considered."

He told us bluntly that he wanted Liana to see an ENT doctor as soon as possible. Another office, another recitation of the now-familiar history, only now a mass was there. We couldn't hear anything more that day. Mass means cancer. Cancer means death. Liana's blindness meant death.

The "Big C" had crept into our lives. Liana didn't appear to be worried about living or dying, that's just the way she is. But I knew that brain cancer basically meant death. We made sure that the ENT doctor knew about the work environment. An MRI was scheduled for the 29th, just before biopsy surgery scheduled to done on the 30th, to know what we were dealing with. The STAT MRI report read:

> "Findings: There is a complex mass, which does involve the nasopharynx; however appears to have an epicenter in the clivus, and expanding the sella. The optic chiasm is displaced, and may be partially involved. The pituitary gland cannot be identified, because of the complex mass that has mild effect on the pre-pontine cistern. **It measures approximately 5.3 x 5.6 x 6.0 cm.** The mass is **heterogeneous** in signal intensity on T1 as well as the axial FLAIR and the T2 weighted images. Post Gadolinium enhancement it demonstrates **homogeneous** enhancement..."

This was not good at all. The days are a blur now. The ENT's report to our family doc read in part:

> "CT scan of the sinuses was personally reviewed revealing a skull base tumor involving the nasopharynx and complete filling of the sphenoid sinus. An MRI scan was obtained, and this was reviewed again with Ms. Jones and her husband.

> This revealed a skull based tumor filling the sphenoid sinus with extension into the left nasopharynx and extension to the left optic nerve. There was a second lesion in the left parietal lobe. The situation was reviewed with radiology and with neurosurgery. It is my impression that Ms. Jones suffers from a skull base tumor with extension to the sphenoid sinus, nasopharynx, and questionable secondary lesion in the left hemisphere. The differential diagnosis would consist of metastatic malignant melanoma, primary neurologic tumor, pituitary tumor, or extensive fungal disease. We plan to set up for a biopsy in the Operating Room. Additionally, a skull base university evaluation is being *initiated."*

No one really paid much attention to the idea that extensive fungal disease could be the problem, but there was the concern in black and white. Please look for mold, I asked. Just be thorough, OK?

The morning of the surgery arrived. Liana is a pillar of strength but I'm nervous. After Liana's experience with the postoperative infection with her leg, she pleaded with the ENT doc to give her an antibiotic after the surgery. He told her it wasn't necessary but he'd grant her request. The surgery only took a few hours but it felt like an eternity. The doctor came out to tell me that everything went well. Liana had very little bleeding, less than a thimble of blood, and the sample was sent to the pathology lab to be tested. By two that afternoon, she was ready to go home and wait for the report. The next day passed. Somehow as we tried not to talk about the surgery. Big C takes some getting used to.

Before dawn Sunday, Liana excitedly shook me awake. "I can see the cat."

"Have you lost your mind?" I asked.

"No, you don't understand. I can see the cat out of my ***left*** eye!"

An answer to my silent and continuous prayers was here. She could see out of her left eye! For six months, she'd been totally blind in that eye. Now she could see just 48 hours after the biopsy. No one had given us hope: after six months of blindness, blind means blind. Did the biopsy relieve pressure on the optic nerve or did the bacteria killing antibiotic have an effect?

Back to our ENT. He reviewed the pathology result. This comment came from the University of Maryland Medical Services (UMMS) review of the pathology done by Frederick Memorial Hospital (FMH).

> (Pathology report 1 from UMMS of biopsy done at FMH)
>
> "Comment: The neoplasm is comprised of relatively uniform small round cells with, in areas, a fibrillary background. Perivascular pseudo rosettes and a richly vascular and hyalinized stroma are also present. Outside immunostains demonstrate immunoreactivity for synaptophysin, chromogranin, and keratin. These results are consistent with olfactory neuroblastoma. We also agree with the point that Dr. (--) alluded to, that it is difficult to exclude a non-secreting pituitary adenoma."

There it was: Neoplasm. New cellular material. In other words? Cancer.

He proceeded to explain that an olfactory neuroblastoma was serious and that this was a condition he couldn't fix. He recommended that we go to UMMS to an ENT surgeon. We also reviewed FMH's fungal culture result by. The culture was done on SAB agar (only). It was negative and stayed negative even after four weeks. Clearly, therefore, I am told, this couldn't be extensive fungal disease. I again asked them to be thorough.

LIANA

During this time, Tom was constantly in his office downstairs on the computer. We were both frightened by the cancer diagnosis. He was sure that the mass, "Fred," was caused by the mold exposure in my office. Yes, I gave the mass a name, Tom kept saying, a fungal mass hasn't been ruled out. I was so sick; I wanted an end to this madness. Despite all the questions going through my head—was I going to die, should I write letters to my children and grandchildren, what's the right thing to do?— I knew that Tom knew what he was talking about. But what if he was wrong? The thought of dying of cancer, when no one could explain why or how, was overwhelming.

I knew what having cancer meant. What if this mass was actually a mold or fungal infection like Tom thought? Why won't anyone listen to him? "Fred" had really caused a major problem. What do we do and who can we ask? Who in the world can treat a mold or fungus infection? I just wanted to get away from myself, but I knew I couldn't. I had to face this with all the strength I had. I was so confused. I remembered what God told me at the beginning of this madness: "It won't be pretty, but you're going to get through it." God's message is what I held onto.

TOM

As I said, "I'm a cause-and-effect scientist." I didn't understand these results. Here we have data that shows the presence of fungi in the workplace. The symptoms that Liana had were worse at the workplace than at home. The allergy testing showed the immune response to exposure to the fungi in the workplace. I don't believe this. 1+1 doesn't equal 5. After returning home, I literally ran to the computer.

So what's an olfactory neuroblastoma?

> "Olfactory neuroblastoma, also called esthesioneuroblastoma, is a cancerous tumor believed to originate in the olfactory cells. The olfactory cells, located in the upper rear of the inside of the nose, are responsible for the sense of smell. Olfactory neuroblastoma often responds well to radiation therapy, but the tumor has a high tendency to recur after excision. The cancer was first characterized in 1924, and there have been less than 1,000 recorded cases since then.

> Olfactory neuroblastoma can cause loss of smell, taste, and vision, as well as facial disfigurement in advanced cases."

Wait a minute. This cancer is predominantly found in children, I read. It's also misdiagnosed in more than 80% of cases. Liana and I'd been given all of the medical records, reports and CT and MRI films to take to UMMS. Our life now was entirely consumed by reading the records, defining the terms and comments, and trying to make sense of something that seemed nonsensical. Hope was all we had and I'd cling to any glimmer. Just look at the pathology reports: *"No mitotic cells seen."* I'd already read that part of the definition of an esthesioneuroblastoma is that there invariably are more than two mitotic events in the microscope field of view at high power. No dividing cells versus at least two? Here again, I felt that fragile Butterfly of Hope flitting about in my mind.

The pathology results don't fit the definition; that much was clear. What could this be? If it has a fungal cause, why didn't the culture show the presence of fungi?

I read the published work of Dr. Donald Dennis from Atlanta, and his findings on fungal sinusitis led me to more information from the Mayo Clinic's published work. I found work by Dr. Denning from the UK, who wrote about the pathogenesis of *Aspergillus* and the treatment of infections. Fascinating. The bottom line was that undiagnosed and untreated fungal infections *are 100% fatal*. Sixty to 70% of cultures attempted from infected sinuses using SAB agar are negative, or so said the Denning papers and Mayo clinic studies. There's that butterfly again. And yet the voice deep down says what part of the word negative *don't I understand*. Maybe I was the irrational scientist.

Then I went back to the medical records. They'd run flow cytometry on the biopsy, which could tell me the cellular makeup of the mass.

> "Flow cytometry results: The total WBC count for the specimen is 12,300.

14% are lymphocytes, 1% are monocytes, and 85% are neutrophils. 5% of lymphocytes co-express CD4 with typical antigen density and CD8 at low antigen density. This can be found with immature thymocytes or in lymphocyte activation. Granulocytic cells are mature PMNs."

The biopsy simply showed pus! This was an infection, not cancer! Fifteen years in the clinical lab had given me enough experience to know what that report showed. What's the body's response to cancer? Not an outpouring of neutrophils. What's the response to an infection? Neutrophils.

More study shows that without sufficient, clear, definitive testing *no one* can distinguish the response to cancer from other causes, such as a fungal infection. Back to the work of Dr. Dennis, who'd shown that the primary cellular response to fungal sinusitis is activated lymphocytes. And there they were in the biopsy along with the pus. But not that many - that's a worry.

I'm not a doctor. Now the butterflies were in my stomach. Did I dare to challenge the diagnosis of the "experts?" And if I do, will anyone listen to me? I need to study more.

I went to the pathology reports. What were they basing their diagnosis on? First FMH: what did they see and what immunohistochemistry tests did they run?

FMH itself said (Pathology report 2):

SKULL, BASE TUMOR, FAVOR OLFACTORY NEUROBLASTOMA

"Comment: The histological features are consistent with olfactory neuroblastoma. Clinical history of malignant melanoma of leg in 2003 is noted. The differential diagnosis includes pituitary adenoma and small cell carcinoma.

Immunoperoxidase studies show tumor cells negative for S100 and melin A and positive for cytokeratin, synaptophysin, and chromogranin. This profile excludes malignant melanoma, neuroendocrine markers do not exclude pituitary tumor or small cell carcinoma, the morphologic features are more in keeping with olfactory neuroblastoma. At the patient's request, the case will be reviewed at the University of Maryland (*Editor's note*: that is Pathology 1)."

From the report from UMMS I saw

(Pathology 3):

"SKULL BASE TUMOR, BIOPSY #2:

RESPIRATORY MUCOSA AND SUBMUCOSA WITH FOCAL INVOLVEMENT BY ESTHESIONEUROBLASTOMA.

Comment: The tumor is comprised of uniform small cells with finely stippled chromatic and small nucleoli. Perivascular pseudorosettes and cytoplasmic fibrillarity are noted.

Immunohistochemical stains for synaptophysin, neuron-specific enolase, CD56, and Cam 5.2 are positive in the tumor cells; CD99, cytokeratin 903, S-100, and HMB-45 are negative. Immunostains for pituitary hormones (prolactin, growth hormone, LH, FSH, TSH, and ACTH) are negative aside from minimal non-specific staining."

Our oncologist in Frederick had had the first biopsy reviewed at Johns Hopkins. Their report read (Pathology 4):

"SKULL BASE: PITUITARY ADENOMA.

Note: Immunohistochemical studies performed at JHH show that the tumor cells are immunoreactive for synaptophysin, ACTH (strong staining in scattered tumor cells), and human growth hormone (weak staining in scattered tumor cells); they are not immunoreactive for AE1:AE3, EMA, chromogranin, or S-100."

With the discrepancy between the labs, I asked Liana's GP to have one more lab take a look at the first biopsy. He agreed to have the block sent to AFIP (Armed Forces Institute of Pathology; Pathology 5). Their report was:

> "DIAGNOSIS: Sphenoid sinus, sellar, and suprasellar mass, biopsy:
>
> Pituitary adenoma.
>
> COMMENTS: The anatomic (radiographic) location of this tumor is *absolutely inconsistent* (emphasis added) with the diagnosis of olfactory neuroblastoma. With the number of different people apparently involved in the care of this patient, it is surprising that this inconsistency has not been perceived. From the histopathologic standpoint, the wide spread keratin reactivity *would not* (emphasis added) be expected in an olfactory neuroblastoma.
>
> Scattered tumor cells are strongly reactive for GH. Stains for other pituitary hormones (ACTH, PRL, FSF, LH, and TSH) are negative.
>
> Regarding the question about fungal infection, together with the staff of the Infectious Diseases Pathology Division, we find no evidence of such infection. Our GMS stain is unrevealing. In view of this negative result, the submitted request for stains related to factors that might be involved with such an infection becomes moot."

Four laboratories, five pathology reports, two diagnoses, and no consistency in the tests run with varied results. Think about it: cancer is cancer, right? Nope.

Every test done to confirm olfactory neuroblastoma didn't confirm it. That Butterfly of Hope I told you about? Now it's Madame Butterfly. This cancer diagnosis can't survive critical review. Liana can't survive without a diagnosis that withstands critical review.

I decide to keep reading and searching. I'll never give up.

What are we really looking at here? None of the four pathology labs see fungi in the tissue, yet the environmental exposure was to fungi growing where they shouldn't be growing. I really wished I could get my hands on some of those tissue samples. Our lab could find the fungi!

Investigating the positive results from UMMS, who did the most immuno-histochemistry to date, I looked at the results in the light of an infection rather than cancer. What I found was very interesting:

> "Review of Pathology report from UMMS- literature search findings:
>
> Paper by Dr. Donald Dennis shows T-cells/NK cells as the major cellular response to fungal sinusitis.
>
> <u>Synaptophysin</u> – Serologically homologous to leukophysin and granulophysin found in activated T-cells, NK cells, macrophages and platelets.
>
> <u>Neuron specific enolase (NSE)</u>-74% of normal cells of the skull base are NSE positive. As per product insert sheets from multiple producers of antibodies directed against NSE, T-cells can be NSE positive depending on their origin.
>
> <u>Cam 5.2</u> –Immunocytochemistry – Cam 5.2 are antibodies that react with CK8 and CK18 cell surface proteins. These cytokeratin markers are normally found in the simple epithelial cells. The simple epithelium lines the sphenoid sinus.
>
> <u>CD 56</u> – cell attachment factor. NK cells, T-cells, mast cells and epithelia cells can be/are CD56+."

So what have I learned from all this? *There are other diagnoses*. I'm not just making up ideas because I'm in denial. It could be a pituitary adenoma? What about the mold? Two additional facts about mold detection:

- the GMS standard stain for mold is inconsistent and difficult to interpret
- all of the positive histology tests can be attributed to the *innate immune* response.

I have to find tests that are more specific for mold. What stain kits are available for fungal detection? What stains are predominantly used in medical mycology? Are there practicing medical mycologists?

We trek back to Baltimore Oct. 7th for our appointment with the UMMS ENT surgeon. Liana goes through a whole gamut of physical, MRI and PET scans over the next two weeks. The only abnormality found was in the sphenoid sinus. We're told that after the doctors had a chance to present the case to a "Tumor Board," along with a consultation with the oncology group and a neurosurgeon, they would have a plan.

On the 13th,, we went back to see the UMMS oncologist so he could examine Liana and review the MRI and PET results. Another doctor, another opinion. I tried to discuss the diagnosis in the light of the mold exposure. Guess what? No results. I asked about the cell morphology and Dr. Dennis' work and got no response. These guys just don't want to talk about the science. It's like they're following some kind of script by the numbers with a preconceived opinion. Frustrating!

October 17th found us back at the UMMS ENT office with a package of MRI films and all the final results from FMH. The doctor comes into the room, claps his hands together and says "OK, let's get started. Baby needs a new pair of shoes." I'm not kidding you; those were his exact words.

He proceeds to tell us about the proposed surgical procedure. The approach is cranial-facial, and he explains that getting to the tumor isn't easy because the face and brain are in the way. He goes on to explain the potential downsides of the procedure: a cerebral spinal fluid leak, loss of smell, blindness. Whoa! This guy is talking about cutting my wife's head open like she's a mackerel on ice, lifting up her brain and scooping this mass out of her head like ice cream out of a box. I'd already brought to their attention the mold exposure. Was that possibility discounted? You bet.

My years of work in infectious diseases told me that if they were wrong, this surgery would kill her. No way; I couldn't allow this procedure without more data. I told them right there that face-opening and brain-opening surgery was not our first option. The oncologist arrived, saying the other option was chemotherapy followed by surgery then radiation. His experience was that the chemo would be very successful. Next, we were sent to the neurosurgeon.

The neurosurgeon was attentive. He set the films onto the viewer in his office and said, "Tell me what's going on." I told him my scientific background and all of Liana's symptoms and about the mold exposure. I told him about my concerns and about my studies to date concerning fungal sinusitis. After about a 30-minute discussion with him, the neurosurgeon said, "I think your attendings should listen to Mr. Jones. This mass is inoperable as it is now."

Apparently, I'd shaken things up things enough for the ENT doctor to put in his notes:

> "Mr. Jones is a virologist and is concerned regarding potential concomitant fungal infection with this malignancy. I did agree that it is certainly reasonable to get titers on her for common fungi."

Maybe they'd really consider the mold exposure, or would they? The phrase "with this malignancy" showed they were convinced that this was cancer. It still made no sense. I needed to study more. Sounds like a theme for my life.

The decision was made: they'd give Liana chemotherapy first, and then re-visit the issue. Liana and I felt we'd be better served to have the chemo treatments done at the Frederick Cancer Center. In the back of my head, I was thinking, "If this isn't cancer, what am I doing allowing these guys to do this to my wife?" But if I'm wrong about the fungal infection and don't allow them to treat her, she's going to die. I remember thinking: I wish I had more to go on.

I needed help. One of my old research co-workers from USAMRIID, the U.S. Army research facility in nearby Fort Detrick, now headed up the Frederick Cancer Center (FCC). He and I had worked together on a couple of research projects before he'd opened his medical practice in Frederick. He had taken care of Liana's melanoma and I knew that he wouldn't steer us wrong. I could at least discuss Liana's condition scientifically. Or could I?

I asked the simple question, "What could cause this?" The answer didn't surprise me: "We don't know." I began questioning everything. Could this be viral? Why didn't we see a clonal expansion of a cell type if this was cancer? Could the cells we were looking at and calling cancerous be natural killer cells (NK Cells), Langerhans cells? The answers were short and basically vague. What am I doing, allowing these guys to do this to my wife? But if I'm wrong about the fungal infection and don't allow them to treat her, she's going to die. Damn, I wish I had more to go on.

Was I challenging the almighty medical standard? The FCC doc wrote:

> "Of note in her history, and as a confounding factor, the place in which she works was recently found to be infested by mold and her husband, Mr. Tom Jones, who is a laboratory technologist, was convinced that this was all fungal infection."

One week after the first treatment with cisplatin and etopside, an MRI showed a slight decrease in the size of the mass. We returned for the second cycle of chemo, which brought a 20% reduction in the tumor mass. Liana was incredibly ill from the treatment but continued to work.

I continued to bring up the issue of mold exposure. Here's what was in the Frederick oncologist's notes:

> "Her husband Tom, who is a laboratory technician, felt that she had been exposed to fungal infections and raised the question as to whether or not this could be a midline fungal infection of undetermined etiology. For this reason, a CMS stain was performed at Frederick Memorial Hospital on the block, and no fungi were identified. In addition, flow cytometry was undertaken and did not show clonal expansion or an aberrant phenotype."

OK, there's no fungi, but no clonal expansion either, and therefore, no cancer?

The next step in this odyssey was a surgical consult with the UMMS doctors. On Dec 12th, we met to discuss the procedure with the ENT surgeon and the neurosurgeon. The mass was simply too large for a cranial-facial procedure. They proposed an endoscopic resection of the mass through the nose. I once again brought up fungal exposure (no one had answered my questions adequately) and was told to "forget the mold." On Jan. 3, 2006, they used an endoscope to open up the sphenoid sinus through the nose after first removing bone at the base of the nasal passage. There were complications that kept Liana in the hospital for a week.

After returning home from the hospital, Liana went back to work two half days each week over my strenuous disagreement. She promised to take it easy. After the first day, Liana had difficulty swallowing and nasal congestion, but now, each time she blew her nose, pieces of tissue (I could see them!) came out through the surgical opening.

As time would tell, the endoscopy didn't resolve the problems. After less than a week, she couldn't swallow water and had a fever. I rushed her back to UMMS, even though at first she refused to return to a hospital. After a week, she was discharged with a diagnosis of a viral infection. Once again, she was told to take prednisone, which relieved some of the symptoms, but not all.

In the meantime, Liana keeps expelling these pieces of tissue, some of them larger than a quarter. The tissue is multi-colored, from black to a peanut butter yellow. At work, I asked the histology tech and the department supervisor if they would cut some slides and do some stains of the tissue. I amassed some 10% buffered formalin (formaldehyde) in 50 ml conical centrifuge tubes and went home to start my collection.

The next step in Liana's treatment was to be radiation—five days a week for 32 days. But before we can make it to the first appointment, Liana started to complain of soreness in her left leg and had difficulty walking. What next? We called our GP and he ordered an ultrasound of her left leg. It turned out that she had blood clots at the site of the scar from the melanoma surgery!

They admitted her and immediately started her on an IV heparin drip. No sooner does the first milliliter of solution go into her arm than her blood pressure goes down far enough that she starts to pass out. Further tests in the ICU unit showed a severe reaction to heparin. Well, that explains why each time the UMMS nurses gave a shot of heparin, her blood pressure would drop. They replace the heparin with Coumadin and four days later, she walks out of the hospital having beaten this hurdle. How many more were to come?

Finally, the radiation treatments can begin. Liana would work everyday from 9 a.m. to 2:30 p.m. I'd pick her up from work and make the run to Baltimore for our 4:30 p.m. appointment. Here's the final therapy report:

> "TREATMENT SUMMARY: Ms. Jones is a 51-year-old Caucasian woman with a history of esthesioneuroblastoma with loss of eyesight in the right eye which improved after biopsy. She underwent neoadjuvant chemotherapy with minimal response, followed by debulking surgery. She was then treated with postoperative radiation therapy to the base of skull.
>
> The radiation treatment is summarized below:

Tx Area: Base of skull
Dose: 57.6Gy
Energy: 6MV photons
Fx: 32
Elapsed Days: 46
Daily Dose: 1.8Gy
Dates: 2/6/06 to 3/23/06

TECHNIQUE OF RADIOTHERAPY: Ms. Jones underwent AcQsim simulation followed by intensity modulated radiation therapy treatment planning. RESPONSE TO RADIOTHERAPY: Overall, Ms. Jones tolerated the treatments quite well. She complained of congestion, some chills, and green nasal discharge, which had occurred for approximately two weeks at the start of therapy. She also complained of fatigue at the start of therapy. At 30Gy, she complained of nausea and vomiting which has occurred since starting her Coumadin. We suggested that she continue to take Compazine around the clock. Her nausea subsequently resolved. At 36Gy she complained of dry mouth especially at nighttime as well as loss of taste and smell. She had a mild sore throat but denies any dysphagia. At 45Gy she complained of increase in tearing. All of her other symptoms were stable. At the end of therapy she stated that the left hearing loss, which she has had since chemotherapy, was getting worse. On examination she had some mild erythema of the skin. She, otherwise, had no other major complaints.

PLAN: We have asked her to follow-up with us in our Clinic in one month's time."

What the report doesn't say is how devastating the treatment had been. Liana had lost all of her hair along with her ability to taste and smell. All of the symptoms that she had were still there to varying degrees. She'd lost more than 30 pounds, which she said was the only positive outcome of the whole process. On top of all of it, the MRI done in May showed little change in the mass:

> "FINDINGS: On today's examination there is a persistent mass with the epicenter apparently in the clivus expanding the sella. It measures approximately **2.5 x 3.8 x 3.1 cm**. It demonstrates heterogeneous enhancement."

While we were going through the radiation therapy, the pathology slides were stained. During March, I grabbed the slide box with the various stained tissue samples and sat for hours at the microscope. What I saw was very disturbing. In just about every field, I saw what looked like fungal elements. I'm no mycologist. I'm an immunologist/virologist. I needed someone with more experience to look at this.

It turned out that a number of my co-workers had been teachers at the Army's advanced medical lab school and two of them had taught mycology. When I told them our story and my suspicions, they were willing to take a look. Sure enough, they saw the same thing I saw. There was no way to tell what type of fungi it was, just that ***there was fungus in the tissue***. When I told Liana about my findings, she asked what should we do? I'd taken photomicrographs of the various slides and e-mailed them to the UMMS doctors. The bottom line was, they wouldn't even talk to me about it. One commented I should "get off the mold kick." We had to find a doctor who understood mold and what it can do. My studies continue as night after night, I search for answers.

Time flies by; it is now July 2006. Liana had been to see our oncologist in Frederick for a follow-up and he orders another MRI:

> "FINDING: On today's examination, persistent mass lesion within the clivus is again identified. The lesion appears stable in size and configuration from the previous study. The lesion measures **4.5 x 3.4 x 4.4 cm** *(Editor's note:* much bigger*)* in greatest extent. The lesion on today's examination appears much more cystic in nature, particularly near the optic chiasm and anteriorly, subjacent to the frontal lobe. There is continued mild mass effect against the optic chiasm."

Wait a minute, this mass is getting bigger. When we started, the mass was measured at ***5.3 x 5.6 x 6.0 cm.*** After all the treatments, it measured ***2.5 x 3.8 x 3.1 cm***. Now the thing is **4.5 x 3.4 x 4.4 cm.** The mass is growing again. We have to get someone to help us investigate the mold.

I had to do something. In my studies, I'd come across Dr. Dennis' treatment regiment. After the primary anti-fungal nasal spray treatment, he'd prescribed an all-natural, over-the-counter, citrus seed extract nasal spray with anti-fungal properties to prevent a re-infection. Liana and I talked about it and decided to try it. It might help kill the fungus. She'd take two sprays per nostril three times a day. This is Desperation Medicine.

Over the past few years, I'd heard that there was a doctor specializing in mold cases, but I couldn't remember his name. I asked my soon to be in-laws, who live near Ocean City, Md., if they know about a doctor like that on the Eastern Shore and was told that Dr. Ritchie Shoemaker had a practice in Pocomoke City. Around the middle of July, with Liana not getting any better yet still working, I called Dr. Shoemaker's office to make an appointment.

August 10, 2006, Liana had a first appointment with Dr. Ritchie Shoemaker. For more than two hours, Dr. Shoe went over all of Liana symptoms and heard the whole story. His evaluation, including the VCS and symptoms roster, strongly suggested a biotoxin reactivity now known as a "chronic inflammatory response syndrome (CIRS)."

Blood tests were done to confirm the CIRS and Liana began the biotoxin medications. Three weeks after she started the meds, her symptoms greatly diminished, but didn't go completely away. Dr. Shoe told Liana that her workplace could be making her sick and that she shouldn't go back to work. She offered to work from home but the request was denied.

I'd taken along the photomicrographs that I had. Dr. Shoemaker suggested that I send a tissue sample to a mycologist at Duke University to try and get confirmation of what I was seeing, so I sent them. That physician wrote back to Dr. Shoe and said he'd found septate hyphae and included pictures of what he found.

Dr. Shoemaker writes in his report:

> "In 2006, we didn't have the advantage of a government-approved case definition for "mold illness" like we have now. Our group published in 2003 a series of 156 patients compared to 111 controls (Appendix 11) that established a case definition of what people with illness had and what people without illness didn't have. Liana met all aspects of the required two-tiered case approach, but the fact that she had a CIRS didn't prove where the illness came from. In order to prove definitively that the building made her ill, she'd have to be re-exposed to it. In science, these repetitive exposure trials are called prospective studies and a prospective study is required to show causation. Such a study requires strict adherence to another published protocol, one in which Liana would have labs done at her baseline, followed by labs at her end of treatment with cholestyramine and whatever else was a first-line therapy. She'd then stop all medications for her mold illness and stay away from the known WDB, with labs drawn after the three-day "waiting period." She'd then go back into the building, again without use of protective medications, for a full 8-hour shift. She would have labs done after 24, 48 and 72 hours. Then she'd re-start on CSM and we'd treat her again until she felt better."

Liana followed the protocol exactly, even though she became quite ill after re-entering the building. The results of the blood tests were exactly what Dr. Shoemaker has published. The causation argument was shining forth in full luster just like the lights on the National Christmas tree. The building made her sick alright, and her symptoms changed exactly as hundreds of others Dr. Shoemaker has recorded doing this protocol. With this indisputable evidence in hand, she could prove to anyone (well, those with biases who make their living twisting the truth in litigation wouldn't necessarily agree) that the building made her ill. After this step was taken, we could logically say that the time period of three days was too short for the mass to make any difference: the mass didn't cause her biotoxin illness, the building did.

Liana had made a follow up appointment with her Frederick oncologist for Sept. 8th. As usual, I came along. Only this time I was armed with the photomicrographs and a detailed annotated summary of each slide and what it showed. I also carried the review of the pathology as well as some of the blood test results that Dr. Shoe had done. To say the doctor wasn't pleased with me would be an understatement. We sat and chatted and the only advice he could give was a referral to another oncologist at Johns Hopkins. From his notes he recorded the following:

> "Today Liana returns to the office for follow-up, accompanied by her husband Tom. Since I've last seen her, they have seen Dr. Ritchie Shoemaker at 410-957-1550, who's done a variety of testing, including HLA typing, and she is said to have an HLA-A 17-2 and 314—5 (*Editor's note:* the doc butchered the HLA like crazy*)*. A variety of other studies were undertaken. He did obtain a cortisol level on 8/11 which was 22.8, but time of collection was not specified. Tom comes with a large body of literature and numerous photomicrographs. He does have some immunofluorescent studies from Duke University which are said to be done on Liana's nasal discharge which does show branching septa and hyphae. He also has numerous H and E stains which do show, in fact, that she has some hyphal elements which can be compatible with either a secondary fungal overgrowth and/or primary.
>
> Tom is absolutely convinced that she has an original allergic sinusitis with secondary fungal infection and that she has no tumor at all."

I wasn't ready to understand what Dr. Shoemaker was really saying at our next follow-up. Sure, we had the proof that the inflammatory conditions she had were due to exposure her workplace. But then he told me that CSM and all the protocols he had for inflammatory effects wouldn't help Liana if her mass was truly a fungal infection.

He looked at the photos in detail and talked with his mycology colleagues. One mycologist at Duke reviewed all the materials I sent but he wouldn't commit to saying he found enough to say treat with anti-fungals. I don't mean any criticism of Dr. Shoe, but when I asked for anti-fungals, he just wouldn't budge. He told me I didn't have enough data.

During November, I searched the internet for someone to review the photomicrographs and slides as well as the photos from the Duke mycologist. Prior to Thanksgiving 2006, I sent the collection of histology photos to Prof. M. J. Dumanov from the "Mycological Institute for the Study of Fungal Mold in Human Habitations," a certified medical mycologist, for his review and evaluation. He returned the photos and the slides with a report that stated,

> "There are clearly discernable septate hyphae, stipe, vesicle, metula, phialide, and conidia present in all these photos. These structures and morphology are taxonomically associated with the genus *Aspergillus*. Although speciation can not be confirmed from these photos, in human tissue these forms are most often with *Aspergillus fumigatus*."

Finally! Here was a scientist of excellent credentials and reputation telling me what I felt was true all along. There's the data Dr. Shoe said he needed. I had confirmation of what I'd seen in the tissue from Liana! She has a vegetative fungal infection, I'm sure of it. There's got to be some way to get her on anti-fungal medication. I called Liana's GP and told him about what I'd found. I took copies of the photomicrographs, including the photos from the mycologist from Duke and the reports. He suggested that we see a doctor at George Washington University Medical Center who supposedly specialized in environmental exposures to infectious diseases. I looked up the doctor on the Internet and found that she specialized in bacterial infections, particularly Staph. On Dec. 19th, Liana and I went to DC for an appointment. We gave her the allergy report, office mold report, all pathology reports, digital copies of the pictures sent to Dr. Dumanov, and a summary of what Liana had been through so far. I was told to collect a piece of tissue if Liana expelled one and

to take it to GW's lab for bacterial and fungal culture. O
26, 2006, I submitted two samples Liana had expelled fr
for lab testing at George Washington University Medic
heard from the doctor Jan. 3, 2007.

She reported that Liana had a Staph infection and placed her on dicloxacillin, 500 mg per tablet four times per day. After a seven-day fungal culture, she also reported the fungal culture as negative. From my studies, I'd found that, depending on the species and strain of fungi, it could take up to eight weeks for the fungi to grow. Also, the media needed to grow many of them aren't the normal culture media used by most clinical labs. After all my conversations with the various doctors that had been involved in Liana's case, only two had bothered to listen and help: Liana's GP and Dr. Shoemaker. Liana and I talked about all of this and decided that GW was a waste of time. We cancelled her next appointments.

At this point, our heads were spinning. During our last visit with Dr. Shoe, we'd talked about the prednisone that Liana was taking. He told us that taking prednisone with a fungal infection was like pouring gasoline on a fire. Liana had been taking 20 mg per day for nearly six months. The medication alleviated some of the symptoms and seemed to give her a little more energy. But what else was it doing to her?

In essence, due to the lack of MSH and the lack of VIP, the only inflammation-regulating pathway Liana had left was ACTH/cortisol. The more she took prednisone, the more impairment there'd be to ACTH production. Once ACTH secretion ceases, Liana had nothing left for the next inflammation insult, which in her case, could be a trip to the antique shop in Emmittsburg, just up Route 15 from us.

Even worse, if you want a fungus to grow, just feed it prednisone. Oh, my God!

So prednisone treats the symptoms but does nothing to treat the cause! *The medication has allowed the fungus to grow unimpeded.* It's basically shut down or severely compromised her immune ystem. Added to the chemo and radiation, this was a serious problem. Liana has to get off the prednisone. Dr. Shoe had given us a protocol to follow to taper off the prednisone. On Dec. 22nd we started the tapering protocol. At the same time, based on the results of an MR spectroscopy, Dr. Shoe prescribed low-dose Procrit, and so Liana took both meds from Dec 19th 2006 through Jan 17th 2007. Results: her symptoms greatly reduced.

The reason for the Procrit protocol was simple: Liana's C4a was far too high, creating capillary hypoperfusion. She was burning her sugar (glucose) reserves for fuel, but without enough oxygen (due to decreased flow in the nurturing capillaries) she couldn't get the full benefit from all the possible energy she might otherwise get if her starving cells were allowed to have enough oxygen to kick in their mitochondria. The Procrit knocked out the sky-high C4a, and the lower C4a allowed blood flow into capillaries, the way it was supposed to. Suddenly, Liana had extra energy. This protocol is so simple and so elegant.

For all of my life, I've been involved with animal husbandry. I can't remember not having animals. For the last twenty-five years, we've had horses. Understanding veterinary medicine is an important part of having healthy livestock. I've been extremely lucky to have known and worked with some of the best veterinarians. If you want to learn how to diagnose, follow a vet for a year. Their patients can't tell them what's wrong or where it hurts; instead, the vet has to watch and diagnose based on the observations and tests that he has at his disposal. Animals surely suffer from many of the pathogens that humans do. The experiences in the barn have been invaluable to me.

Vets also rely on technicians like farriers who, in many cases, have the same talents as a vet. Henry Heymering has been my farrier for more than 12 years and has been in that business for more than thirty. His knowledge of the equine is outstanding. It just so happens that he has Lyme disease and multiple chemical sensitivities. When I told Henry about Liana and all that she was going through, we began to compare notes. In late December 2006, Henry introduced me to Dr. Larry

Plumlee, who's associated with the Chemical Sens
Association. Dr. Plumlee had developed a chemica
while working for the EPA. I called him and told him L
symptoms of chemical sensitivities and biotoxin ill
identical; both are a chronic inflammatory respor
asked about natural remedies for fungal infections and if he knew of any doctors that could treat this type of fungal infection. He mentioned a number of doctors that he knew. Among them was Dr. Michael Gray of Progressive Health Care center in Benson, Arizona.

With Dr. Gray, I discussed virgin coconut oil and the antifungal properties of the caprylic and caproic acids. We talked about the therapeutic properties of colloidal silver and its anti-bacterial and anti-fungal properties. At the end of our conversation, and after internet study on each, Liana and I decided that using these substances would be safe and could only be beneficial. In March 2007, I went to the health food store and got a can of virgin coconut oil that was the consistency of lard. Liana agreed to take two tablespoons a day with a meal. I also purchased a colloidal silver nasal spray to take the place of the citrus seed extract. Once the Procrit treatment was done, we started with the anti-fungal treatment. (Desperation Medicine, again).

So, what would be the effect of Liana taking these new treatments? I called Dr. Rea in Dallas, Dr. Dennis in Atlanta, and got onto the internet to once again look into the mycotoxin production. What happens when there's a die off of fungi? When fungi are killed, they release mycotoxins. If these new treatments work, there would be a wholesale release of toxins and Liana's going to feel worse. The treatment Dr. Shoe had put Liana on had worked—her C4a level in January 2007 had been a little over 1000 (down from over 20,000, with normal being less than 2830) and her VCS had improved. Now if we were going to kill Fred, the mass, she was going to get sicker. We had to expect that the test results for MSH, C4a, etc. would all go bad again. It was imperative for her to stay on the CSM four times a day, inconvenience be damned.

The stained slides confirmed that there was fungus in the expelled

360 | tiss

sue. But what about the surgical samples, the samples taken aseptically? What other diagnostic tools could we use to confirm that this is fungal? The answer was polymerase chain reaction, PCR. I went back to the internet to find a laboratory that did that type of testing on human samples and found Real Time Labs in Dallas, Texas. I called our GP and told the doctor that I wanted to have this lab do some testing for us. "Tom, it sounds like you know more about this than I do, go for it" was his comment. I called the lab director, pathologist Dr. Dennis Hooper.

I told him our story and asked if he'd be willing to do the testing and discussed the various other testing they could provide. Dr. Hooper agreed to request samples from UMMS and FMH and we agreed that I'd send the formalin fixed samples that I'd collected from Liana and that he'd submit a request to the UMMS and FMH labs to get the surgical specimens. Throughout February, we gathered the samples. He'd perform PCR for the fungi, one strain of *Fusarium*, one strain of *Stachybotrys*, and a strain of *Penicillium*; all were negative. Given the fact that there are twenty-three strains of pathogenic *Aspergillus*, though disappointed, I wasn't surprised by those results. The cultures that were performed on fresh samples I collected were negative but the mycotoxin level was 0.83 parts per billion (ppb) for tricothecenes. The urine mycotoxin results were another story. The aflatoxins and ochratoxin are mycotoxins made by *Penicillium* species and *Aspergillus* species. Tricothecenes are produced by *Stachybotrys* species and *Fusarium* species. The initial first morning urine tested was positive for aflatoxin and tricothecenes at *five times the detection limit of the test.* The final diagnosis from all of the testing was that Liana had a mycotoxicosis. The report was sent to Liana's GP.

Next, I spoke to Liana's GP about the results and convinced him that we should be giving her antifungal therapy. He agreed to write a prescription for Diflucan, beginning June 3, 2007. Liana collected a first morning urine sample for mycotoxin testing. That same week I got an e-mail from Prof. Dumanov concerning samples that I'd sent to him. From my studies, I'd found that Diflucan wasn't effective against *Aspergillus*. June 11th we talked with the doctor about itraconazole.

Not much help here.

Our general practitioner and I talked about what my research had revealed, and he agreed to give Liana 600 mg per day of itraconazole. He wasn't as receptive to following Dr. Dennis' protocols using the Amphotericin B nasal spray. At this point, I was happy just to get her treated systemically. Liana's chronic fatigue had worsened. And the other symptoms—weakness, some tremors, balance, and memory issues—were worse, too. Forty-eight hours after starting itraconazole, Liana collected a first morning urine and we sent it to Dr. Hooper. The results showed aflatoxin and tricothecenes to be very elevated. She was still taking the virgin coconut oil and CSM. It seemed we were on the right track. But Liana was still expelling multi-colored tissue from her sinus.

By now, it's July 2007 and Liana had another MRI done. The mass now measures 3 x 3.4 x 2.9 cm. When compared to the dimensions of the mass in October 2006 at 3.7 x 2.8 x 4.1 cm, it's obvious that the mass was getting smaller, but not that much smaller. On July 10, 2007, the results of another first morning urine that had been sent to Dr. Hooper were positive for aflatoxin and tricothecenes (12 ppb aflatoxin; 0.73 ppb tricothecenes).

I went back to my research and found myself convinced that the best results would be achieved by following Dr. Dennis' protocol using the anti-fungal nasal sprays. All of the doctors that I talked to couldn't help. Liana's color was a sickly shade of gray. She was the sickest she'd been. We talked about our options. Though Liana didn't relish going to Arizona, we decided to call Dr. Michael Gray and at least get some idea about how we could kill Fred. On July 31st I called Dr. Gray and told him the whole story. I told him that Dr. Shoemaker had diagnosed Liana with a biotoxin illness. I faxed him Dr. Shoe's reports, the results from Dr. Hooper and a list of the medicines that Liana was taking. He said he'd call back after he had a chance to read through the reports. Later that evening, he called me back and said he could help. He was aware of the work of Drs. Shoemaker and Dennis. He agreed to prescribe the anti-fungal nasal sprays after he saw Liana. He also added activated charcoal and bentonite clay to Dr. Shoe's CSM protocol (it's also the U.S. Army's protocol for T-2 toxin exposure). He prescribed meds for his oxidative stress reduction protocol. The next day, I called his office and arranged an appointment with him at his office at Progressive Health Care in Benson, Arizona on the 28th and 29Th of August.

Prior to going to see Dr. Gray, another urine was analyzed. On August 21st, the results were aflatoxin 5 ppb, ochratoxin 1.2 ppb, and tricothecenes 21.37 ppb. —the itraconazole was working, or at least I hoped it was.

August 28th and 29th, Liana and I saw Dr. Gray in Arizona. Neurological tests revealed that Liana had 12 neurological deficits caused by the mycotoxins. Dr. Gray uses Dr. Shoe's blood tests in his differential diagnosis. The labs showed the same abnormalities seen in Dr. Shoe's blood workup. Dr. Gray gave Liana scripts for anti-fungal nasal sprays and other medications to improve her lung function. The pulmonary tests had shown her to be challenged. After returning home on Sept. 1, 2007, Liana started all the prescribed medications.

Throughout September, Liana began to show some improvement. She was starting to get back to how she felt in December of '06 after Dr. Shoe's initial treatment. In October, she showed significant improvement. Follow up appointments were set with Dr. Gray on November 27th and 28th and with Dr. Shoemaker on Dec. 4th.

The last week of October and the first week of November, Liana expelled two additional multi-colored pieces of tissue the size of a quarter. I placed them in 10% formalin as I'd done with the other samples. I'd been looking for a specific fungal stain and had found a product called "Fungalase F." It was a patented product invented by Dr. Roger Laine at LSU. I called him and discussed the test in detail. He kindly sent me a kit to try on the most recently collected samples. The results of the test were another affirmation of the fungal infection. There were clearly fungal structures in all of the tissues.

I'd called Dr. Hooper and asked him to see if he would stain some of the slides that he had gotten from FMH and UMMS with GMS and

PAS fungal stains. I also asked him to send the stained slides to Prc Dumanov. I also asked him about the Fungalase F test, and asked whether he'd be willing to do the stain on the surgical samples. He agreed. Both the surgical samples were positive for fungal elements. Prof. Dumanov also reported back that the results of his review of the slides from Dr. Hooper also had clearly discernable vegetative fungal elements.

After our second appointment with Dr.Gray, the neurological deficits are down to seven from the twelve observed in August. Prof. Dumanov had sent me the slides from Dr. Hooper; I showed them to Dr. Gray. UMMS slides as well as the standard pathology slides of samples collected during September and late October, and the first few days of November were left with Dr. Gray so that he and Dr. Maluff could review them. Dr. Gary Maluff was a long-time associate with thirty years of experience in mycology. He and Dr. Gray had worked together on a large number of fungal cases together. Their review of the slides was the same as Prof. Dumanov's.

Dr. Hooper had developed and validated a polyvalent PCR assay that detects most strains of *Penicillium* and *Aspergillus*. He tested Liana's samples and they were positive. We knew from previous PCR assays which strains were not the causative agent. So which strain of *Aspergillus* was it? We knew there had to be *Aspergillus* because of the presence of the aflatoxin in her urine samples. The DNA from the samples was sent to a reference lab to discern the strain.

Now it's July 2008. We had lots of irons in the fire. All of these new diagnostic tools were slowly bringing into focus what we were dealing with. And all this time, Liana and I adapted the slogan, "KILL FRED, MUST KILL FRED." Liana was doing better, if she could just get over the chronic fatigue and the cognitive deficits. The last three MRIs had shown the mass to be stable, with no change. Dr. Gray had found that some *Aspergillus* strains didn't respond well to itraconazole. The decision was made to add voriconazole in pill form. Liana had a hard time with the medication because it upset her stomach. After a month and a half, she stopped taking the pills. Dr. Gray had come up with an alternative. We'd use voriconazole as a nasal spray. He'd also found that if Liana added "Restasis," also known as cyclosporine, the effects of the anti-fungal would be amplified by 100 to 1000.

In August 2008, Liana began using voriconazole nasal spray. She'd take one drop of Restasis in each nostril, then 15 minutes later administer the nasal sprays and she'd repeat the process every four hours. At around the same time, Dr. Hooper informed us that he'd found the strain of *Aspergillus*: it was *Aspergillus terreus*. The results were confirmed by two labs. It turns out that terreus is a strain that doesn't respond to itraconazole.

September brought another MRI. The three prior had shown little or no change. The report from this most recent showed a dramatic change. Though there was still some of "Fred" left, it was greatly diminished. At our next appointment with Dr. Gray, Liana's neurological deficits were still down at six, but the degree of abnormality had continued to improve. The mycotoxins were still positive, but the quantity had also decreased. Dr. Gray decided that an endoscopic examination by a surgical team at the University of Pittsburgh Medical Center was needed to examine the sinus and remove anything remaining of the mass. He also requested that they lavage the sinus to wash out any contaminating spores that might be present. They were to send samples to Dr. Hooper for PCR and histology work.

On November 10, 2008, Liana had the surgery. The report came back, "No Tumor Seen." The final pathology came back as NO CANCER. Dr. Hooper's PCR detected *Aspergillus terreus.* As before, the fungal stains done by UPMC were negative for fungi. The mass is just about gone! Liana continued with the voriconazole nasal sprays and the other medications. She seemed to be doing a whole lot better.

Through the holidays, Liana did fine. For the first time in three years, we actually had a good Christmas. Had we finally killed Fred? Dr. Gray explained that we had to continue all medications until Liana's PCR results were negative twice after sinus washes. She'd need to take the sequestering agents probably, for the rest of her life, to remove the mycotoxins. Prior to going to see him again in February, Liana would have another MRI. He also arranged to have Dr. Cravens, an

ENT in Tucson, perform the first sinus wash and for the collected wash sample to be assayed via PCR by Dr. Hooper.

The MRI results were outstanding: there was no sign of the mass. The PCR results were negative, and even the mycotoxin levels from the first morning urine sample was very low. The tricothecene mycotoxins were only slightly above the detection limit of the assay. At the appointment with Dr. Gray, the neurological deficits were at two. But the MSH and other blood tests were still out of whack. WAHOOOOOO!!!! Fred is dead!

In June, Liana had another appointment with Dr. Cravens and Dr. Gray. The procedure went well. Once again, the PCR and the mycotoxin assays were negative. Some of the deficits were a little worse: they'd gone back to six. But the ones that had been normalized were only slightly out of the normal range. Liana could finally stop the anti-fungal nasal sprays. We could finally say we'd won a major battle. She was alive and doing better. But (there's always a "but"), the symptoms of the chronic inflammatory response syndrome still persist to varying degrees.

In February 2010, Dr. Gray prepared a poster on Liana's case and presented it in Rome, Italy, at the *Aspergillus* Conference. We'll be going back to Dr. Gray for future testing and evaluation to make absolutely sure that the infection is cleared. One result of everything Liana has gone through has been that a small portion of the bone of her sinus has been lost and her brain has herniated into the sinus. We're going to have to watch this one.

This is the best it has been in five years. We're really looking forward to it getting even better.

Liana and I have had the blessing of God to have found Dr. Shoemaker, Dr. Gray, Dr. Hooper and Joseph Dumanov. We want to thank each of them for helping us get through these years of medical torture.

Yet, in this back and forth saga, we aren't done. We knocked out the initial inflammatory responses and the fungus ball, but we haven't ıe final step of restoring regulation of inflammation. Have e through this journey in the desert only to see that there is Promised Land that we can't enter?

Meanwhile, the science and understanding of the effects of exposure to biotoxins is growing. Dr. Shoe has found a treatment using "Vasoactive Intestinal Polypeptide" (VIP) that lowers the C4a and TGF beta-1 levels (See Chapter 16, VIP). It corrects the symptoms and normalizes some of the abnormal biomarkers. We planned to begin VIP treatments with Dr. Shoe as soon as Liana qualified to join the clinical study.

What VIP does is correct abnormal pulmonary artery pressure response to exercise, but it won't do much for people who have ongoing exposure to moldy places, as shown by ERMI scores over 2. What Dr. Shoe wanted to know was if the initial benefits seen from use VIP would survive in a strictly structured clinical trial. Sure enough, Liana had the expected pulmonary artery pressure problem from her low VIP. She started the drug and with her first dose, I'm not kidding, I could see results in 15 minutes *(Author's note:* these changes are recorded on all participants because the benefits are routinely seen in 15 minutes.) and her use of VIP was an almost instantaneous breath of life. I couldn't believe what I was seeing. But there were her lab results getting better. The exercise improvement was really obvious and the ongoing benefit of VIP defeating that nemesis of normal physiologic functioning, the pulmonary artery pressure rise in exercise, was easily shown.

With VIP as the last step, her fatigue was gone and her brain was working just fine. We are allowed to walk in the Promised Land!

We haven't proven what the right dose of VIP is going to be forever, and maybe she'll only need two doses a day for awhile, but the results of her intervention trial are both outstanding and typical.

As Liana sang to me in June 2010, paraphrasing James Brown:

> "I feel good

Like I knew that I would now

I feel good each day

With just two little sprays

I feel good,

I feel good,

So good, but WOW!"

What the building stole from Liana, regulatory control of inflammation has now returned.

I can't say what Liana will be like in three years, but with the improvements in her now, the future isn't my concern right now. I want to just look at my bride, tell her I love her and show her I love her.

Dr. Shoe says that Liana's case is one no one has recorded. Her illness was both inflammation and infection and despite clearing those monumental hurdles, even when the twin anvils sitting on her shoulders were fixed, she wasn't right. The final anvil sitting on her shoulder was the absence of neuropeptide regulation of inflammation. And now that hurdle is gone too.

How many will learn from Liana? I honestly don't know but if only one person learns how to save their life like Liana saved hers, then writing this chapter is worth far more than just jotting down notes.

What happened to us needs never be repeated. But it will.

What I have learned and what I know might help some who read this story and maybe that will be swift justice delivered to those persons who say water-damaged buildings aren't a problem.

They are.

CHAPTER 15 Laura's Story: When Mother/Physician Becomes Patient

By Laura Mark M.D.

Shannon wasn't thriving as she could have in a world without water-damaged buildings and without long-delayed diagnoses; nonetheless, she was surviving. With painstaking precision and planning, she created a life for herself. Her main focus was to plan and build a safe (mold, moisture, and chemical free) place for her and Jonathan to live. She was still disabled and she couldn't hold a job because the kind of accommodations she'd need were impossible to find in a bank or an insurance agency.

Me? I'm not doing so well. Now that I know what inflammatory responses can do to my children, I see the telltale signs everywhere, including in my patients and in myself. I know now that understanding Shannon's illness has changed my life. It may even save it.

How I wonder which buildings hurt me as I grew up. During the first seven years of my life, the family lived in many places in the U.S. and abroad with humidity and no central air conditioning. A series of sore throats and ear infections led to a tonsillectomy at age five. When I was six, I was forced to switch writing with my dominant left to writing with my right hand. Maybe the attempt to "rewire" my brain at that age contributed to my subsequent stuttering. I recall crying a lot, although I wasn't generally unhappy. I just seemed to have a low threshold for the "faucets being turned on"; I still do. I share that observation because my recent experiences have made me more aware of the need to revise my notions of cause and effect.

Let's look at the ongoing debate of nature versus nurture in forming temperament and personality. Later, as a Board Certified psychiatrist impacted by the ravages of my brain's inflammation, tears would still gush out with minimal provocation. Observing this phenomenon, I was struck by how similar this was to the "disinhibition" and "personality changes" that accompany a variety of my patients' brain injuries. Of course, the whopping toxic encephalopathy I developed by July 2007 was a brain injury. But what about more subtle, chronic, and milder forms of this same process? Might my genetic susceptibility to a whole host of toxic triggers (particularly musty, moldy, moist places) have contributed to the years of stuttering, crying, shyness, intermittent brain fog, irrational anxieties and sleep disturbance? Were these things due to an underlying inflammatory response syndrome? Was I "a crybaby" or were those things a result of where I lived and went to school, what foods I ate, or what air I breathed?

Imagine what was happening to my patients' children, so often misdiagnosed and mis-labeled, pigeonholed. If a teenager had an episode of suicidal gesture at age 15, that history was the first thing recited in a subsequent psychiatric evaluation. Yet if her gesture was due to frustration, sadness, chronic pain and cognitive issues from exposure to a moldy bedroom, would we still label her as an adult with depression just because she had ideation of suicide years before?

I'm learning. I see so many themes now, starting from the need to educate physicians, the need to get the word out the need to protect my patients from moldy buildings, and to preserve my own health. The learning has so many dimensions.

Looking back at my years as a teenager, I was likely already experiencing the effects of pituitary/hypothalamic dysfunction. I was a late developer: my menstrual cycle began when I was sixteen and throughout my twenties and thirties, I noticed that "stress" would quickly shut it down completely for months. In hindsight, when I was "stressed" consistently coincided with living in "moist" buildings and wasn't there when I wasn't exposed to wet indoor spaces. At age 35, four kids and one medical degree later, I had a tubal ligation. Within two weeks, I dropped into what felt like a depression. It wasn't until later that I'd learned I'd entered menopause. I was so close to the edge of hormonal axis shutdown that this minor surgical procedure snuffed out the last bits of my ovarian function.

Talk about an immersion course in the teachings of the widespread effects of estrogen and progesterone! (*Editor's note*: MSH controls gonadotrophin function, which, in turn, controls estrogen and progesterone.) Hormones are incredibly potent, complex creatures. They're biochemical messengers. When any of your hormones (or biochemical messengers) are knocked out of balance, *you don't feel good*. Worse, you feel helpless because you have no control over what's happening.

At first, my doctors and I thought my symptoms were caused by depression because I'd had bouts of "depression" in college, medical school, after giving birth and related to stress, so I was prescribed an antidepressant. But my symptoms following the tubal ligation persisted— sleep disturbance, drenching hot flashes and night sweats, cognitive impairment (boy, was I worried about dementia), fatigue, overwhelming anxiety and irritability, jitteriness, and a perpetual PMS personality.

I returned to my internist and gynecology doctors and visited an endocrinologist as well. Imagine my confusion: I was post-menopausal, even though a tubal ligation doesn't cause menopause. I'd also developed osteoporosis. The good news: by adjusting my estrogen and progesterone levels, within 3 days, I regained my sense of wellbeing. The restoration of my brain power and energy was wonderful. Yippee! By simply replacing those two depleted hormones, a whole host of cells in my brain, nerves and blood vessels were once again receiving the correct instructions from those two hormone messengers. Balance was restored. I felt good.

Had both my doctors and I known back then what I know now, a lot of needless suffering, wasted time, and expense could've been avoid-ed. During my medical evaluations, a key line of questioning wasn't included: asking a simple environmental exposure history would've elicited descriptions of buildings in which I lived and worked during periods of my life that I experienced "psychiatric symptoms" or "endocrine dysfunction." During most of my adult life, I lived and/or worked in buildings that clearly had previous or current water intrusion, faulty ventilation, and moisture control problems. I didn't suffer from allergies.

For the first 35 years of my life, even with my health issues, I considered myself to be generally healthy and definitely normal (right, kids?). I never could've anticipated that the cumulative deterioration of my health would soon accelerate to a crisis point in the not-too-distant future. It may have also been a "point of no return" ... only time will tell.

The first major assault on my health took place in Virginia. It was 1995, and at age forty, I joined the staff of a state-run psychiatric hospital. I loved working there - I felt personal and professional satisfaction from getting to know the folks who wound up in the care of this state's mental health services. I'm upbeat and always see the good in bad situations. My health, however, was *not* loving it. Over the next five years, I gradually acquired quite a collection of specialty physicians and diagnoses.

I developed frequent stabbing headaches, itchy rashes, generalized joint swelling and pain, unbearable fatigue, irritable bowel syndrome, carpal tunnel syndrome in both hands, red irritated eyes, abdominal pain, frequent sinus and respiratory infections, unusual pains, mood swings and irritability, and trouble concentrating. Each problem resulted in doctor and testing appointments and time away from work and family and endless bills. I became increasingly suspicious that there was something more systemically wrong with me, more than just some "itis" showing up. (*Editor's note*: in medical diagnosis jargon, if one sees "itis," think inflammation.)

Over time, I noticed that I wasn't alone in these experiences; many of my coworkers had similar complaints. At home, what began as a series of off-hand comments from my kids about my forgetfulness soon became an outright concern. I noticed more frequent "brain glitches," like saying or writing one word when I meant another, having trouble doing simple calculations in my head, asking the same question multiple times during the day and losing track of what I was doing if even minimally distracted. Naturally I tried to explain away these short-term memory lapses.

And that worked for a while. Sure, I was simply multitasking and not really paying attention, right? Privately, I wondered if I was developing dementia. But like an ostrich sticking its head in the sand, I marched onward; I just had too much to do. If I could just keep doggedly plodding on, I'd be fine.

Eventually, it became apparent to a number of us at the hospital that there must be something in the building that caused so many of us to be sick. Organizations have a way of developing powerful survival instincts, and our expressions of concern went unheeded. Later, an indoor air quality assessment of the building was done. My memory is fuzzy on the details but I recall that there were problems involving the HVAC system that included a lack of adequate fresh airflow. But, of course, "there was no health risk." The report contained suggestions on the types of cleaning products we should use.

Corrective action was recommended, and while some steps may have been taken, we never received additional feedback from the administration. My health continued to deteriorate. Finally, I knew I had to get out. When I spent extended periods of time away from that building, I began to feel better, but with re-exposure, I was worse. Fortunately, I was offered an outpatient position in the state system, within commuting distance. Little did I know then that I'd jumped out of the frying pan into the fire.

For the first few years at my next job, I experienced no significant health issues. Gradually, though, medical staffing needs changed and I began to work in the Mother of all Mold Castles —a 200-year-old satellite clinic that had been a schoolhouse, set in a lovely small rural town. It had been previously remodeled into a series of offices. The daily drive passing through beautiful countryside was peaceful and relaxing, and varied with the seasons. Unfortunately, within that structure laid the ingredients that would eventually disrupt my life's very foundations– my health, my occupation, my relationships, and my financial security.

But great pain and suffering often create the opportunity for growth and transformation, once you get beyond the aspect of merely surviving. I'm continuing to work on those aspects of change.

And so, following in the wake of Tropical Storm Ernesto in September 2006, I joined my daughter in becoming Dr. Shoemaker's patient. The fierceness of that storm resulted in a major roof leak in rear section of the building where I worked, creating the perfect environment for rapid growth of mold and other microorganisms. Those little buggers had a feeding frenzy on the charts, papers, carpets, walls and furniture. I started thinking that the only thing that wasn't mold food was me. The products of all of that biological activity - microbial volatile organic compounds (mVOCs) - quickly injured staff working in the area, and the rest of us were affected as well. The inadequately functioning HVAC system distributed those harmful molecules and whatever else that was biological in the air (bioaerosols) throughout the building.

You've heard that the solution to pollution is dilution, right? But fresh air wasn't being moved through the building to dilute and remove the toxins and inflammation-producers. Crucial daily dehumidification hadn't been incorporated into the HVAC system. And who knows if there was a HEPA filter to remove the problem particles?

Soon, mold starting growing on moisture-damaged records. Solution? Move them to the front of the building, guaranteeing efficient cross-contamination. We were instructed to share office spaces, double-stacking more moldy materials for us to share in ever-more-crowded conditions. Remember, when impacted by a host inflammation response from the toxic triggers, only 20-25% (the genetically susceptible folks) will develop an illness due to the excesses of that inflammatory response. And, the unique past exposure history and genetic expression of the subsequent effects would determine how an individual would feel as a result. Of huge concern, though, is the potential for the cumulative and devastating effects on susceptible individuals *over time*.

Many staff complained about the smell and developed headaches, respiratory, and "allergy" problems. "Corrective actions" had been taken but several were quite seriously sickened and due to ongoing health problems, several decided to leave. I wasn't yet informed about the serious health effects of poor indoor air quality. Of course, I was familiar with Sick Building Syndrome (SBS) from an academic perspective. For years I'd chosen to live a relatively "organic" lifestyle, and, as is typical of human nature, I'd pushed memories of my SBS from the previous job into the recesses of my mind. After all, I had patients to care for and a family to support. In the hectic nature of my own schedule, I didn't pay too much attention to those events. It was that same September that Shannon came home to visit and asked me to read *Desperation Medicine*. In October 2006, Shannon first saw Dr. Shoemaker. By December 2006, I was experiencing my own precipitous health decline. By the time Shannon had her second appointment with Dr. Shoemaker, we decided that I should have my lab tests drawn.

In January 2007, my personal bombshell struck. Life was suddenly surreal. The family is adrift, on some boat floating in uncharted waters. I was working hard to learn a new "language," a new way of understanding my family's health and future. Drastic measures, including remediation, purging of contaminated belongings, and moving a safe space didn't seem odd. Those measures were no different from moving away form the Dust Bowl of the Oklahoma in the Depression - we had to survive. With appalling speed, activities I previously engaged in daily - jogging, bike-riding, gardening - became unbearably difficult. My body seemed to turn to lead, my brain seemed to be filling with molasses, and my life as I knew it was coming to a crashing halt.

I had a resurgence of symptoms, ones I knew from my days at the state hospital. Several times a month, I thought I was coming down with a "virus" of some sort. My muscles ached, I felt feverish, my skin tingled and was hypersensitive, my intestines cramped and hurt, my throat was sore, my eyes and sinuses burned, my head hurt and

I couldn't focus. Increasingly, I noticed that within 20-30 minutes of entering the building, I felt jittery, nauseous, apprehensive, light-headed, and had vertigo. It felt like a veil of fog was being draped over my brain. I was concerned I had a problem with my blood pressure or blood sugar, but quick checks by my nurse revealed no abnormalities.

By the time I received my own lab results back from Dr. Shoemaker in January 2007, I'd developed some familiarity with the various inflammatory markers. I didn't need an explanation to recognize the extent to which my own health was impaired.

The next six months blurred by with a flurry of both excitement and apprehension. The excitement is the most fun to write about so I'll share that first! It began with a dawning realization as I viewed retrospectively my own past history of mood episodes. I realized that the very same cluster of "psychiatric symptoms" was identical to the current cluster of "biotoxin-associated" inflammatory symptoms, minus the assigned attribute of depression or anxiety. Once I could see on paper, in black and white, the lab abnormalities and physiologic derangement of my various organ systems, I considered, which came first – the chicken or the egg? Did the toxins trigger inflammation, which resulted in acute/chronic biological reactions, which I and other doctors interpreted to be psychiatric, i.e. depression and anxiety, because that's what we were taught? Or did depression or anxiety somehow "occur," and those conditions "caused" the very same physical symptoms?

I wondered how many of my current patients could benefit, as I did, from a "diagnostic reassessment." My caseload then was comprised of more than 500 patients. For a good number of them, I'd been their psychiatrist and effectively their primary care doctor for several years. Most had a significant variety of both of psychiatric and medical diagnoses. Dominating the list were neurologic and cognitive, cardio- and cerebro-vascular, gastrointestinal, and autoimmune diseases. With great zeal, I developed a plan to educate my staff and patients about the underlying physiologic processes resulting in these multiple symptoms, as well as their potential causes and treatments. I discussed this strategy with administrative supervisors and my psychiatric and nursing colleagues. During repeated educational sessions, I distributed informational handouts. The correlation between my family's pattern of symptoms and our significantly abnormal lab results provided important cause and effect data. The symptoms were obviously there, and not just in theory. Just seeing the dysregulation of the various systems exactly as Dr. Shoemaker's model predicted they would be, gave me comfort as well as the confidence to share the information with others. With this revised approach to evaluation and potential treatment of these medical problems, I addressed the bottom line of my life and job: improving my patients' quality of life.

The next step was simple: determine which of my patients "fit the bill." Which folks could benefit from a reconsideration of the cause and perpetuating factors for their health issues? In our specialty-dominated medical world, some people forget that all fields of medicine aren't distinct but are instead highly intertwined and interactive. I developed an initial list of those patients who carried certain diagnoses. I started by considering medical diagnoses and then went back to reassess their corresponding psychiatric diagnoses. Specifically, once I knew their underlying inflammatory process, I went back and again questioned the specifics of the documented "psychiatric symptoms" as well as their history of exposure. The more I learned, the bigger the list grew. I gathered as much family history as possible. It was no surprise that symptom clusters dominated many family histories.

From that group of 500 plus patients, I identified a group of at least 100 who were prime candidates for more diagnostic testing. My goal was to at least be able to reassure them that this illness really wasn't all in their head. I wanted to be able to show them with abnormal lab results that there really *was* something medically wrong with them. Just as important would be providing better symptom management by addressing the underlying problem rather than just treating the symptoms *caused* by the problem. I invited other people involved with the care and treatment of my patients to participate in teaching sessions. Attendees included case managers, spouses, children, therapist, pastors, and friends. This information had to be shared with their physicians at follow-up appointments. My patients gave me permission, both verbally and with signed consents, to contact these doctors and provide relevant information about their conditions, testing, and possible treatment.

For the most part, these community physicians were appreciative of my efforts and expressed an interest in learning more. I'll never forget, however, one specialist's response: "Do you really expect me to change my whole way of practicing medicine at this stage in my career?"

Next I gathered evidence of their diseases. With the greatly appreciated assistance of my nursing and clinical staff, laboratory testing to identify genetic susceptibility began, followed by laboratory testing for evidence of disease. With jaw-dropping amazement, I received my patients' lab results. Initially, I consulted with Dr. Shoemaker to confirm my interpretations, but I quickly discovered I could predict fairly accurately which labs would come back abnormal. Seeing for myself was an epiphany. I can never go back now that I have seen the hidden face of chronic illness.

What gave me the greatest pleasure was the expressions of relief and gratitude of patient after patient when I confirmed for them what they already knew – that they were really sick! As it turned out, because of my own declining health, I was only able to refer for lab testing and get results back on the first 20 out of the initial 100 identified patients.

What was remarkable to me at the time is now merely predictable: <u>every</u> patient had, as predicted by Dr. Shoemaker's work, an identified problem genotype completely contrary to the general public (and frequently it was the "dreaded" one) and each one had marked lab abnormalities. Wow! Now both my personal and my professional lives had been turned topsy-turvy. I couldn't wait to share the news with my peers. I spoke with a number of them, beginning with my patients' primary care and specialty physicians. I contacted the state's

Department of Mental Health administration. Fortunately, there was an upcoming medical directors meeting: Dr. Shoemaker would share his material with the group in June 2007. His presentation was well-received and stimulated an interesting discussion. Unfortunately, again for health reasons, my efforts to educate fellow practitioners about these important concepts and their clinical applications were placed on hold for a while.

Now, for the down side of this little adventure. I'd rather not remember most of it, but I'll present a few highlights to illustrate just how quickly ignorance, fear, and insecurity can result in manipulation, distortion, and deceit. Particularly when the truth about an individual's or organization's motives are called into question. And especially when greed and power are inevitable players in the battle, and perceived survival is at stake.

For those of us willing to stick out our necks - and there are many - the desire to take action for the "greater good" is a powerful driving force. Clearly, the courageous individuals, families, and organizations described in this book are stark examples of what can be accomplished with perseverance and unity.

When I'd explain the dramatic breakthroughs in my daughter's health to my colleagues, at first my explanations were met with genuine curiosity and compassion. Interest continued as I explained its relevance to what appeared to be a substantial number of our patients. Our clients were enthusiastic and grateful. Over the next few months, many staff members approached me with their own health concerns. They were eager to ask personal questions, confidentially, and described their own symptom pattern. Several staff could, like me, identify rapid physiologic changes soon after entering the building; similarly, their symptoms subsided with extended periods away from it. Increasingly, I shared my concerns with the administration. Progressively, the initial positive reception transformed into skepticism, disinterest, and finally to denial. I was shocked at how quickly administrators moved to "circle the wagons" and exclude a perceived threat from access to "the loop."

Is anyone surprised that the administration told me that the agency had corrected "the problem" after Tropical Storm Ernesto brought water from Florida to the inside of our exam rooms in Virginia? Naturally, by that time, I wanted to review the damage report as well as the corrective action taken. With a newly educated and informed perspective, that's just what I did. There were no surprises: both the initial assessment and subsequent remediation steps were grossly inadequate.

I told them they were out of step with any and all modern thinking about buildings and human health. I followed up my criticism with further educational attempts. The administration wasn't happy with me and when nothing was done, my unhappiness accelerated. I voiced my concerns repeatedly that not only was I continuing to get sicker, but that co-workers were as well. I was keeping a list. Little old me as a complainer? Easy to brush aside. Two people as a conspiracy? Almost as easy to discredit. Nope, this was an *entire cohort* of sickened people; good luck making a case that all twenty of us were making up stories.

Maybe the last straw was my simple statement about our duties as caregivers to some of our most disadvantaged patients. I stated we were likely subjecting many of our most vulnerable patients to the effects of that problematic indoor air quality. In my naiveté I presumed that this declaration would result in further testing and potential remediation efforts. I believed it was the duty of the agency to protect both its employees and patients (consumers).

Boy, was I wrong! It was enough to turn your stomach. What actually occurred was that I was informed by an administrator that I was to speak no further with anyone at work about mold. *Gag.* Conversations were whispered, due to heightened fear and anxiety about both insinuated and stated threats to outspoken staff members' job security. *Gag*. Slanderous remarks made by one of my administrators in a public interagency forum trickled back to me: "I don't really believe that physician is sick at all. I think her illness is a fake." *Gag.* Those statements set the stage for what was to happen next.

During the month of July 2007, Shannon and I both had appointments with Dr. Shoemaker. Shannon went first, and clearly she was doing much better. But when Dr. Shoemaker asked how I was doing, she forcefully interjected, "Mom is doing terribly. She sleeps all the time, can't remember a thing, hurts all over, and doesn't want anyone to know how bad she really feels."

Dr. Shoemaker looked over at me and inquired, "Is that true?" When I started mumbling that maybe it was, he looked over and said simply, "You know, if you walk into that building without protection one more time your brain might never recover. Just like in the Broadway show, *Phantom of the Opera*, there is a Point of No Return."

That was the first hammer blow. I wasn't fully convinced of the urgency to put my own health first (isn't it amazing how we caretakers just don't get it!). I told Dr. Shoemaker I'd consider further the idea that I may need to go on disability for awhile. Shannon and I left and drove home. By the end of that three-hour car ride, I was convinced and my Jericho wall of denial came crashing down.

The first six months of what was to become a 15-month hiatus were a nightmare. Because of my brain's impaired functioning, even the simplest tasks were too much. I was overwhelmed. I needed help organizing and then implementing what needed to be done. I'm eternally grateful for the daily support and help I received from various family members and friends because I couldn't keep the ideas in my head long enough to follow through in the afternoon with carefully made plans made that morning. Just as frustrating was the speed with which my energy evaporated. I slept through much of the first six months. When I could stay awake, I was pacing, making lists, making more lists, looking for the first lists I made and then forgetting about the second lists I made.

The assaults to my integrity and professional survival came faster. Workplace administrators accused me of "practicing outside the realm of psychiatry," telling me that the testing I was doing was not what we were taught in medical school or residency. "Our malpractice insurance carrier will not support your doing this kind of work," they said. The same attack came from more than one administration source, as if stating the same untruth more than once would give the idea credibility.

These attacks really offended every sensibility I had about medicine. Ours was once a dignified profession that put patient care ahead of dogma; it put integrity ahead of deniability. Aren't we trained as physicians, mandated to undertake continuing medical education for the purpose of learning new developments in our fields and incorporating them into our clinical practices? Aren't we charged with the responsibility to perform a thorough differential diagnosis and refer for additional testing if indicated to identify possible causes for the presenting symptoms? Don't we have a duty to offer our patients the best chance at improving their health and quality of life? Didn't we all take the Hippocratic Oath to "never do harm?"

Isn't it harmful to ignore, or not consider, a potentially treatable condition? Isn't it harmful to be providing often ineffective treatments that themselves can be harmful when a more cause-specific approach to treatment and prevention is known and available? Aren't we charged by the state to treat and protect our patients' rights? Don't our patients have the right to visit their physicians, attend groups and day programs in a safe indoor environment? And, by the way, don't fellow employees have the right to work in a safe building?

As you might expect, administrators responded to these questions with, "the problem has been taken care of." Sure enough, after many, many months, and many, many thousands of dollars, the problem was taken care of. *I* had become the problem ... and the problem was eliminated. *Gag.* Through a series of legal maneuvers, I was ultimately closed out of my job, and denied the ability to prove with a prospective exposure study that the building had made me ill, thereby blocking the possibility for me to consider a possible Workers' Compensation claim. *Gag.* I was ordered to stay out of the building because—get this - "what if you get sick again? Then we would have liability," they told me. No kidding. At least they understood that if no re-exposure occurred then I couldn't prove causation. (*Editor's note*: this attempt at a cover-up always fails when the medical evidence was as strong as it was in this case.)

My friends and colleagues on staff were prohibited from having contact with me. *Gag.* Attempts to deny me using leave time were prevented only with an attorney's assistance. *Gag.* Ultimately, because an agency typically has more resources than any individual, the agency won. Unfortunately, the rights of both the employees and patients that the agency is charged with protecting are trampled upon. The legal system in Virginia, as elsewhere throughout this country, is encumbered by antiquated laws, amoral and sometimes unethical lawyers and judges who, by definition are part of the power structure, which results in bad things happening to good people. When will that ever end?

The next big battle was dealing with my private disability insurance company. I purchased their product while in my psychiatric residency, never expecting to have to use it. I'd purchased a top-of-the-line product, presuming I'd receive top-of-the-line service if and when it was ever needed. *Wrong, again.*

I quickly learned that the company I'd purchased the product from had been bought out by an industry giant. I also learned quickly that the primary goal of that giant insurance company was to make it as difficult as possible to actually obtain the benefits that were paid for and promised in the event of a disability. Again, immoral and unethical tactics were used in an attempt to deny payment of benefits. And, again, only with legal representation and many thousands of dollars in expenditures was I awarded my benefit.

By making the process as difficult as possible, the true goals of our legal systems and insurances businesses becomes crystal clear – lining their own pockets with as much money as possible by denying us the benefits to which we're entitled, having paid for them in good faith. We keep our word – we pay. The insurers try to avoid paying back.

I can't imagine how much more difficult it must be for people without the supports I had to survive this kind of battle. That's why it's critically important to take action now, in the form of education, treatment, and prevention; of improving the legal and judicial systems so that they truly protect the rights of this nation's citizens; of holding insurance companies accountable for their corrupt behaviors and disastrous consequences; and of assisting our governmental agencies in carrying out their assigned tasks of protecting their citizens.

Time passed; the journey continues. Gradually, my health improved. My physicians monitored my laboratory studies regularly, and by the fall of 2008, I was approached by a former colleague about the possibility of returning to work. I had gone inside no buildings except three thoroughly remediated homes (each with low ERMI scores), my doctors' offices, and the hospital or satellite lab. I considered what type of work I could do that would include safe working conditions. I needed to consider that although I "felt better," the condition was tenuous and could deteriorate almost immediately if I were exposed again to a hidden. My inflammatory markers had returned to normal. My executive cognitive functions (brain functions) had not fully recovered despite all I'd done, and fatigue would suddenly creep in after I'd been awake for 6-7 hours, requiring an immediate two-hour nap, which would rejuvenate me so that I could remain awake for another five hours or so.

I still had nowhere near my previous levels of energy and brain acuity, but at least I was functioning. And that meant I wanted to get to work, and if possible, move in the direction of educating consumers and providers about these common and devastating conditions and their potential causes.

In September 2008, I was approached by the medical director of the state hospital where I'd previously worked. There was a newly-constructed geriatric treatment facility, and he was actively recruiting for a geriatric psychiatrist. I described my residual short-term memory and attention difficulties, problems with multi-tasking, persistent need for a daily nap, and the need for flexibility in any potential work schedule. After consideration by the executive, medical, and human resources directors, I was hired as a "physician with a disability." I wanted to work. I began the job with a modified workload and schedule, gradually increasing the number of patients I could manage. I was extremely fortunate to have the support of the administration and the staff on my unit as I fumbled my way back into a work routine.

Although I thought I'd be able to function fairly well, I was surprised by the ongoing difficulties I experienced learning new skills and routines. Prior to this job, I had absolutely no experience with computers. In large part, that was due to the annoyance and discomfort that headache, vertigo, and nausea caused when I looked at a computer screen. Fluorescent lights have the same effect. I'd long been aware of this difficulty and therefore structured my life, including the workplace, in a way that didn't require using computers. I'd consulted with neurologists, ophthalmologists, and computer experts to figure out a way around this problem, but there were no good solutions offered. Apparently, this is a fairly common problem among inflammation-affected people.

For most of the first six months on the job, I ended up in tears due to the lingering effects of brain irritability and to the frustrating consequences of not being able to multi-task. I educated my staff about my limitations. I found that I could function best if I had minimal interruptions. Obviously, this is usually not the norm for a physician, but accommodations were generously provided. My previous outpatient job was so incredibly different!

I felt my new life wasn't fair. It wasn't fair that Shannon's potential to be a physician won't ever be realized and mine might not as well. I didn't afford myself the luxury of asking "why?" Yet the questions are still unanswered. Why did this happen to me? Why did my life have to be one of such suffering? When does the torment stop?

Given the drama at my previous job, I was determined to "have my ducks in a row" at this one. I arranged for baseline labs prior to starting, and have been regularly monitoring my inflammatory response to exposure to the building. I meticulously continued for the first nine months "living in my bubble" so that I could keep confounding variables out of the equation when interpreting response. I documented that, unfortunately, I'm again reacting to triggers in the building. My inflammatory markers progressively increased from their normal baseline to current high levels.

It remained unclear to me exactly what I was reacting to, but likely candidates include spores and toxins transferred from the old buildings to the new building via charts, cardboard storage boxes, fabrics, etc. Simply handling charts extensively when I first started resulted in tremulousness, headaches and brain fog fairly quickly; was it huge amounts and varieties of chemical exposures, i.e. cleaning, disinfectant, and deodorizing products; lotions, perfumes and hand sanitizing products; waxing and buffing products; and off-gassing from carpets, furniture, printers and copiers? These chemicals were overwhelmingly pervasive in every nook and cranny of the building. Given the elevated levels of humidity for extended periods of time, and from roof leaks that occurred on several occasions during strong storms with water intrusion into living and working spaces, was it new fungal growth and other problem microorganisms?

In response to that trend, I discussed with the hospital administration further interventions, including: establishing a work station at home, where I spent 60% of my time; moving from an office with no access to outside fresh air to an office with an exterior door I could leave open; and (most recently) purchasing a basic van without carpeting or fabric, from which I could work on the campus without having to go into the building.

I'm implementing the third step now. In my hospital offices, I have personal HEPA filtration units. Due to HVAC difficulties maintaining humidity control in the new structure, my first office frequently had humidity levels of 60-73% during the summer. I brought in my own dehumidifying unit and was surprised to empty a full 3-gallon collection bucket daily (including needing to empty daily on the weekends). As I expected, the first two steps weren't successful in preventing a resurgence of my inflammatory markers. The Biotoxin Pathway had been fully triggered back in September 2006. Once activated, the systems of innate immunity (See Chapter Four, *Blood in the Sink*) bring forth a life of their own. It will take a long period of continued treatment and trigger avoidance to get my hyper-reactivity turned off. I can only hope research continues to find answers and maybe, cures.

As you may have surmised by now, I'm not performing the quality and quantity of work I did previously. I've adjusted to this scaled-down capacity. Still, I continue in my efforts to share the things I've learned about inflammatory disorders with my patients and their families, my coworkers, and now, in this chapter, with you.

I joke with my staff, family, and friends about being a glorified secretary. I brainstorm with my medical director about how I can be more useful to the hospital. I try to help my colleagues in making the secretarial components of their responsibilities less cumbersome. I make choices in my life that are best for my family while attempting to minimize risk to my own health. I've got kids in school and unbelievable medical bills to pay: I'm no different from anyone else whose life has been ruined by water-damaged buildings. I simply cannot afford not to work. I struggle with the same issues millions of other global citizens struggle with – providing for my family and protecting their health. Like many parents, I put the welfare of my children first. I have to keep reminding myself that I must take care of myself as well in order to do that.

Like a seesaw, as my inflammatory markers have gone way up over this past year, my health has gone way down. The "itises" have reared their ugly heads again. The improvement I noticed in my memory and concentration has slipped a bit – likely a milder form of the encephalitis I experienced previously. Mental tasks have become more difficult again. Compensatory strategies I've developed have helped. Cholecystitis resulted in my gall bladder being removed; it may have over time also contributed to the chronic pancreatitis I've developed. Episcleritis (red eyes) has recurred; arthritis is more pronounced; tendonitis and bursitis are intermittent and tend to aggravate me most at night. Neuritis keeps my legs twitching and fasciculating, again most noticeable when I try to sleep. Plantar fasciitis makes the bottoms of my feet hurt so that standing or walking for even short periods of time cause moderate discomfort. Irritable bowel is also no fun. (*Author's note*: when so many inflammatory conditions occur, you almost know that TGF beta-1 has gone sky high. Indeed, as Dr. Mark's illness has exacerbated over time, her 2007 normal TGF beta-1 now exceeds *23,000*; over 10-times normal. If you hear "-itis," think TGF beta-1. You'll rarely be wrong.).

My vision of the future includes helping to educate (and that may mean continuing to be a glorified secretary, shucks!) at many levels in the health pyramid, from consumers to health-care providers, from teachers to legislators, from builders to remediators - wherever my assistance can be used to stem the tide of injury, disability, and death from this scourge. So, once again I arranged for Dr. Shoemaker to give a presentation of his work, this time at the state psychiatric hospital. Staff are interested; they're suffering, both from exposure at work and likely from other sources. I teach, give support, and refer for follow-up where possible. I'm excited about the recent development of an organization whose goal is education and bringing about legislative change. Global Indoor Health Network (GIHN) is a newly formed group that will produce material for the agencies to consider, including a new consensus statement on human health effects caused by exposure to the interior environment of water-damaged buildings. My life will continue to be subject to the forces of this condition – some I can control and many I cannot. I expect to continue making adjustments to both my personal and occupational lives as I continue to learn from the experts who dedicate their lives to the search for truth and answers.

I've wondered frequently in my "re-education" process just how often my family, friends and patients may have been misdiagnosed and therefore inadequately (or incorrectly or incompletely) treated. Obviously, in my previous outpatient position, attempts were initiated to clarify, and then correct this. The guilt I still carry with me in regards to those folks who were offered a glimmer of hope, only to have it cruelly removed, weighs heavy on my spirit at times. I can only hope that at least, in the early stages of my learning curve, I was able to provide reassurance to those 100 or so patients that they truly were medically sick.

What about my current inpatient position? It breaks my heart to consider that so many patients who were admitted to the state hospital over the years were likely unintentionally harmed. The State of Virginia doesn't run a Cuckoo's Nest, yet when I think of all the wrongheaded things that have been done in psychiatry over the years when the problem is actually inflammation, I could cry, with true sobs of sadness that so many people were harmed needlessly.

The illness could have occurred by triggering an underlying predisposition or aggravating an already activated inflammatory pathway. Either way, there's a high likelihood of the "psychiatric symptoms and behaviors" actually emerging or intensifying. How much of patients' anxiety, apprehension, irritability, mood swings, explosiveness, cognitive dysfunction, neurologic (ie. seizure and movement disorder) symptoms may have resulted from pre or post-admission exposure? In the worst-case scenarios, how many of our long-term folks inevitably became long-term care patients, with their life confined to an institution's four walls as a consequence of exposure to the water-intruded, chemically saturated indoor air quality of the very institution charged with facilitating their recovery in a "safe" setting? Adding insult to injury, through our well-intentioned efforts, we physicians likely added further to their toxic load by administering potentially neurotoxic varieties and doses of medications to already inflamed and fragile brains.

In my current position at the geriatric treatment center, I work with folks who have moderate-to-severe cognitive impairment, moderate-to-severe movement disorders, and multiple medical problems. Clearly, for these individuals, there will be no improvement in their overall outlook. They are provided with total care, treated with respect and dignity, and made as comfortable as possible. I ask myself daily how, in the context of this work, I could continue to make a difference in this world, to reduce overall pain and suffering. Once I settled into the routine (and boy, did that take a while, given my difficulty with new learning), I began to look carefully at my patients' psychiatric diagnoses. Like a detective, I explored further for factors that may have contributed to onset of symptoms and progressive deterioration. For the patients themselves, this effort unfortunately has no benefit. But by considering the role and long-term complications of inflammation, family members may get benefit. How? Let me give a few examples.

One gentleman diagnosed with early onset Alzheimer's dementia and Parkinson's disease actually had an acute onset and rapid progression of cognitive dysfunction after sustaining an injury whereby hydraulic motor oil was accidentally injected into his hand. That resulted in a massive inflammatory response that required multiple surgeries and amputating one finger. When meeting with family members and eliciting this history, they were reassured to learn about the effects of inflammation on brain and nerves. Specifically, the adult children had been worrying about their own chances of developing Alzheimer's and Parkinson's diseases and passing those genes on their own kids. I did caution them, however, that it may be worth checking their HLA at some point to establish whether or not they carried a highly susceptible genotype that would warrant ongoing caution regarding exposures.

A patient in her early sixties was diagnosed with early-onset Alzheimer's disease in her early fifties. When I learned that her parent was also a patient in our geriatric facility, I contacted her sibling to get further details of possible exposures, as well as a family medical history. It turned out that the patient, her parent and her sibling had worked for years in a tobacco packing factory. The building had no heat or air conditioning, was always moist and dank, and probably played a large role in the development of their inflammatory-based illness. I wasn't surprised to learn that the sibling had developed three serious (and potentially lethal) autoimmune diseases. Again, it was too late to make a difference in the lives of my patient, the parent and the sibling. But certainly for the patient's and sibling's children (and grandchildren), identifying susceptibility, establishing a current laboratory baseline, and striving to minimize future exposures could be critical in preventing another tragedy.

There's no shortage of examples; every day I meet someone with a similar story to tell. When I first began on this journey, because of the problem's enormity I concluded that maybe the best that could be offered to these folks was recognition of their condition, education about causes, treatment, prevention, and assisting them in obtaining benefits to which they would be entitled.

It's only been recently, through the efforts of the group of very talented individuals (GIHN) that my optimism is returning. I'm hopeful that their hard work and dedication to "getting the word out" will bring about the necessary changes to our health care, political, judicial, and governmental agencies and catapult the massive impact of poor indoor air quality from microbial and chemical contaminants into the spotlight of reform.

CHAPTER 16 VIP: Just What the Doctor Ordered

Before I started writing *Mold Warriors* in 2004, I still had the fanciful idea that there had to be an unknown silver bullet for mold illness. Or maybe a Holy Grail that would be a balm for all the inflammatory problems I couldn't correct. I don't think I'm alone in hoping for simple answers to complex problems, and mold illness can be incredibly complex.

So many people got better with CSM alone, but others needed something else, a "step two." What were the differences between those who recovered easily compared to those who didn't? Still others—a smaller number— needed steps three, and even fewer needed steps four and five. Sure, we had to start with CSM or Welchol, but couldn't all the rest of the steps be short-circuited, just add the missing potion? One illness, one drug, next case? Nope; trying to skip steps just frustrated all involved, especially me, because that leapfrog approach just didn't work.

My practice has changed over the years, and I rarely see people who just need a dash of CSM these days. The people I usually see now have been treated already with steps one, two and three, or so it seems. What is that magic potion, anyway? Are we bereft of silver bullets?

In 2010, mold illness (and many other chronic fatiguing illnesses as well) is revealed to be so complex but surprisingly *each of these disease entities shares a common final pathway*, one in which lab abnormalities are consistently similar from one illness to the next. Yes, I want to know what these diseases are respectively, but more importantly, my job is to *fix a given patient's problems* here and now, so the source of the problem, while important, is *less* important than fixing what they have now.

Some days I almost long for "the good old days," when all I needed to say to people who were treated successfully in less than one month with CSM was, "OK, now stay away from mold." That was all the advice they needed; the treatment was simple. Those days are gone. Yet the patients who need steps four, five and up through nine aren't being treated as commonly, so it's no surprise that, in the patients I usually see - a cohort of persistently ill people for whom we have no Holy Grail of treatment - we need to take them through steps that I didn't even have to offer last year.

All scientists know to always challenge yesterday's hypotheses today and today's hypotheses tomorrow. What do we know now? I've given up yesterday's hypothesis - that there's a silver bullet. Instead, what we must do is use good science and disciplined medicine (that means no Ass^2) to slog through the inflammatory mechanisms, diverse as they are, to bring us the final Holy Grail: a cure.

I know that while CSM helped an initial layer of patients and that was all they needed, other patients had a second layer of illness — the coagulase negative staph layer, chock full of antibiotic resistant, biofilm-forming organisms that silently hurt the host, never letting the unfortunate patient know that an illness-perpetuator was sitting on their mucus membranes, deep in the back of the nose. But if those bacteria, living in a community of bacteria and protected by a thick layer of impervious "slime" weren't eradicated, there was no hope for improvement.

An example: imagine having a boat with an anchor rope coated by biofilm after the rope had been in the water a few days. The anchor-lifting mechanism is two rubber hubs spinning, and slipping, against the slimy rope. There's no traction; the rope doesn't get wound up, the anchor stays in place, so the boat stays in the same place, too.

If the biofilm-former isn't eradicated, the patient stays ill.

A different problem for low MSH and chronic fatiguing illness patients was that they never knew about the slime formers that ruined their innate immune response because the bacteria didn't cause sinus infections, a runny nose or anything they'd notice. And if the physician didn't do an API-STAPH culture, now done routinely by Cambridge Biomedical, no one in the microbiology lab would be able to report the very slow-growing biofilm makers, either. The routine cultures simply showed the bacteria that grew quickly, something the biofilm formers could never do.

As time passed, high MMP9 was unveiled as the next hurdle. Then all the people with antigliadin antibodies who were proven to have an additional layer that required an additional intervention. No gluten for them. And then there were hormone dysregulation people—the ADH/osmolality crowd and ACTH/cortisol sufferers, all mixed in with the low androgen folks. All those hormone abnormalities had to be fixed before we could clear the illness.

Don't forget that some physicians love to use steroid hormones (prednisone and androgens) almost as much as terrorists like to plant road-side bombs. If the hormones are *truly* needed, fine; if they're guessed at (and my goodness, do I see incessant and unwarranted guesses here, especially from the "fibromyalgia" physicians), just forget any decent result over time. Like the explosions of road side bombs, the medical results of such hormone misuse are catastrophic. The "worst hormone patients" were ones who'd seen somebody who thought that adrenal fatigue or low testosterone or low thyroid was the problem, even when results from commercial labs showed nothing confirming the need for hormone therapy.

Look out (and run away as fast as you can) when some doc who sees you only occasionally says you have fibromyalgia and offers you some kind of adrenal steroid, some thyroid meds and some kind of testosterone, usually garnished with a cachet of supplements (10-20 being added all at one time aren't unusual) that they sell by the front door of their office. Of course, the supplements you buy there, while *perhaps a bit more expensive*, are invariably of so much *finer quality* that the extra value is worth the added price, so they'll tell you. The bigger problem is that the docs who profit by such sales are usually giving ALL of the hormones at the same time. How does one know what ten interventions do what to whom when and in what company of pills already on board do the added ones help or hurt?

Maybe I have become too sensitized to seeing outright theft being sanctioned as good medical care. How I wish that last scenario weren't true so often.

Later, I felt as if treating mold illness was like attacking a Medusa crossed with a Hydra. Cut off one snake-haired head - one layer of illness - and there were a couple more snaky heads immediately growing out to turn the patient's life into stone. Even more important, it became pretty clear, pretty fast that the ***order*** *of correcting these problems* mattered a great deal. By 2006, it was clear that C3a had to be fixed and *only then* could we bring C4a levels out of the stratosphere - steps seven and eight. After all that, what's left to fix? Aren't there a finite number of things that can be screwed up by inflammation? Hmmm.

By 2008, TGF beta-1 was unveiled as a key player that hurt mold patients (and others, especially CFSers). Then, after our group was able to show two interventions that lowered TGF beta-1, that previously unknown driver of a "new kind" of immunity (Th-17) surrendered its previously ill-defined stranglehold on chronic illness. Soon however, we found out that if the patient was still exposed to a WDB, then whatever we did to TGF beta-1 failed. **Stay away from exposure is Rule # 1!**

Great, we can fix TGF beta-1, but only if there isn't ongoing mold exposure. Now what was left? You'd think that after twelve years of unwinding the strings of illness from this amorphous ball of twine called chronic fatiguing illness (that all of us were looking at) we'd at least see some of the ball's central core.

Enter the Missing Piece

Maybe we were. Since 2005, I'd seen a constant stream of blood test results that consistently found low levels of vasoactive intestinal polypeptide (VIP) in my patients. Sure, VIP deficiency appeared almost in lock step with low MSH. And we showed that helping VIP with tadalafil (Cialis®) also helped MSH in some people. And then there was the role of vasopressin: abnormal vasopressin, combined with low VIP and low MSH, was the most commonly seen combo in the worst patients (See Chapter 9, the Man With No Hypothalamus). I called it the "hypothalamic trio" since vasopressin, a hormone we usually think of as being made by the posterior pituitary, actually has important regulatory effects in the suprachiasmatic nucleus of the hypothalamus. (*Editor's note*: don't be confused by the fact that there are four names for vasopressin, including antidiuretic hormone, ADH and AVP. We can add arginine vasopressin as garnish to the synthetic form of AVP called DDAVP. They all make the kidney hold on to free water and regulate the interaction of hypothalamic regulatory hormones MSH and VIP.)

If VIP was a big deal in controlling inflammation, then could it be that VIP's *absence* was even more important? As I saw it, VIP deficiency was really no different than insulin deficiency for a Type I diabetic. Give the Type I diabetic some insulin and he lives like a normal person. What happened when people deficient in VIP took VIP?

Perhaps, no one will be surprised to find that restoring VIP helped people get better.

And look at what VIP did to regulation of blood pressure responses to exercise in the pulmonary artery! They fell back to normal making it easier to pump blood when we needed it to climb up stairs or carry a bag of groceries, not to mention putting up fence posts. Was this compound ***too good to be true***?

As I learned just how important VIP was regarding regulating inflammation, I also found that VIP could be legally prescribed as a compounded medication for people. Here, using IRB guidance in bringing VIP to a clinical trial was invaluable.

But there was another concern: the worst of the VIP-deficient patients were the multiple chemical sensitivity (MCS) people. Would VIP replacement fix *most* people with chemical sensitivity, since VIP controlled the inputs of olfactory nerves in the master gland part of the brain, the hypothalamus? Let's face it, while no one knows the cause of chemical sensitivity, every one of the chemically sensitive people I saw was dead low in VIP. Would fixing VIP fix the MCS?

As it turned out; yes. But the benefits of the VIP didn't accrue to those who still had exposure to WDB or who still had untreated inflammatory problems. No magic potion there: the steps to correct the illness *had to be taken in order*. No exceptions.

Actually, the problem with the kinds of patients I see, who have such complex illnesses, is that they can *rarely wait* for "one step at a time." Here's a real problem: patients who frequently have limited mood control and concentration abilities, as well as a limited ability to assimilate new knowledge have to be disciplined in their approach to a medical intervention. *Good luck.*

For those whose patience allows them to adopt a scientific approach, we can give them step by step documentation of lab results to show what was wrong physiologically; what happened with intervention; and what will be done next. Such transparency only comes from doing one intervention at a time. Follow this protocol through nine, ten or eleven steps and out comes a healed person!

But you've got to have *absolutely no exposure to buildings with a fungal index level that is too high*. The index I have talked about, the Environmental Relative Moldiness Index (ERMI)) can't be higher than 2 if your MSH is less than 35 and your C4a is less than 20,000. The cut-off ERMI falls to (-1) if the MSH is less than 35 and the C4a is over 20,000. ERMI of (-1) isn't impossible to find, but those buildings aren't common.

There are people I see who say, "I'm not better anymore. CSM didn't fix me like those people in the book." My response is usually, "Did you have the ERMI test done?" *No.* "Did you remove yourself from exposure?" *No.*

Yikes. Am I being too aggressive here to say *why not*?

This next item was a big one for me. The idea of reducing MASP2 activity underlies *Surviving Mold*. Stop the reactivity caused by MASP2 auto-activation and the illness can go away. The idea is key for all those with C4a that went nuts after a mere 5-10 minute exposure to a WDB, like mine used to. Would people become *less reactive* following exposure to WDB? This idea is crucial to everything I wanted to achieve in this 13-year odyssey - save the survivors from relapse! Restore neuropeptide regulation of inflammation. When a series of my initial VIP patients didn't get sick with re-exposure, like they had before, can you imagine my excitement?

Less reactive means no more "sicker, quicker," after even trivial exposure. Instead, for the same exposure that hurt people badly when their MASP2 was on auto-pilot and they were exposed during a quick sit-down lunch at a WDB restaurant (who knew about the flood in the women's bathroom three months ago?), we'd see a "tiny bit sick over a longer time" and a trivial rise in C4a after the same exposure. If the inflammatory forces driving the illness were blunted, then the therapy needed to permit people to recover from the exposure to the WDB would be much less intense and much shorter in duration.

Years ago, my brother Wells, a Stanford-trained pediatrician and now a consultant to a number of national health care reform think tanks, cautioned me never to be the first or last to use new ideas in medicine. I wasn't the first to use VIP, but I was the first to use it to treat chronic fatiguing illness.

How did I know I was going to cause "No Harm," as Hippocrates would have us swear? Would treated people die immediately, grow three heads, hemorrhage, have twenty diarrheal stools an hour or become manic depressive or demented? Nope to all of those.

In fact, so-called adverse events just didn't happen at any rate over 2%. Would VIP cause antibody formation against the normal human form of the compound? Nope. How about exercise tolerance? Would people be able to do more, show less fatigue and suffer less delayed recovery from normal activity? Yes, yes, and yes.

It sounded *too good to be true*. Was VIP replacement my panacea for those at the top of the "pyramid" of persistent illness?

Think about it: if I just knew the answer to the last question, I could stop the abnormal innate immune activation, restore regulatory control to a damaged system and provide a mechanism to correct all the damaged secondary phenomena found in chronic systemic inflammatory response syndromes (that would work for so many, not just for those ill due to mold).

I knew that Paul Cheney, MD was seeing people with long-standing Chronic Fatigue Syndrome get better with stem cell replacements obtained out of the country (talk about being the first to try something!). I was seeing the same results from simply correcting the inflammatory pyramid and then adding the final VIP touch. I knew that VIP was safe, was readily available through standard medical processes and didn't cost a fortune. Surely this was the Golden Age of treatment!

The ideal mold survival drug would be:

1. Available
2. Affordable
3. Effective
4. Safe

And it would:

1. Stop excessive reactivity
2. Reduce inflammatory changes, especially C4a
3. Restore neuropeptide control to inflammation
4. Provide correction of secondary hormone problems involving vitamin D and androgens.

What I didn't know to ask for in my dreams was that the same drug that would also:

1. Lower TGF Beta-1
2. Raise VEGF
3. Raise MSH

VIP was all of those. It was *too good to be true*!

After reading just about everything that Drs. Doina Ganea and Mario Delgado had written on VIP (and others too), I knew that VIP could restore neuropeptide control of inflammatory responses and thereby should bottle up the inflammation genie that was wreaking havoc on my patients.

Then I read that VIP deficiency affected just about every cell (Dr. Mark Poli wouldn't let me say *every* cell) in the body via its role in regulating a vital cellular communication system tied to cyclic AMP (cAMP). This compound, called the second messenger, takes the signal coming from outside the cell to the guardian of the cell, the cell membrane, and then transfers the message to structures deeper in the cell, including the nucleus. If there were a receptor for VIP on the cell (and there are several kinds of VIP receptors), then cell function would be affected by lack of VIP.

Basically, messages from the "outside world" arrive at the outer membrane of a cell constantly. Some messages simply fail to be received; others are greeted as important and are ushered directly to the cell's command center, the nucleus. The vast majority of chemicals are subjected to further processing using the "second messenger" system. The idea is that cell-to-cell signaling must be reinterpreted with a built-in delay mechanism. Events inside the cell contribute to signal-processing with cAMP having a key role. Since VIP regulated cAMP, then VIP had a key role in communications.

Having intracellular regulation and therefore, intracellular function depending on extracellular regulation (VIP) is a phenomenal concept. Here's some Big Brother (VIP) basically deciding which outside messages will be successfully processed inside, therefore controlling what the cell does, by controlling the *inside message system*. We aren't looking at outside control from some monolith; we're looking at cells freed to do what they're supposed to do. VIP jacks up cAMP such that those messages that involve cAMP are preferentially efficiently transmitted to the nucleus.

Imagine if our enemies were able to disrupt every cell in our body by disrupting VIP physiology. Horrors! Frankly, this VIP story is a bit scary because biotoxins do that every day to everyone who's genetically susceptible. Our military and civil defense forces need to know this fact (and so do you).

But it isn't our enemies who tear apart the regulation of systemic inflammation, not to mention cell communication, for so many innocent people. It's our homes, our schools and our workplaces. Take it one step further, and you'll agree it's the people responsible for protecting us from indoor microbial growth in those buildings who are responsible. It's the people whose duty it is to clean up and repair things properly in those buildings after the water intrusion event when the alarm goes off, indicating there's microbial growth. Let us not forget the ultimate responsibility rests on those vested interests whose cash runs and insures those moldy buildings. When they don't order the maintenance or the correction and/or they do order the cover-up, well, couldn't you say that they're domestic terrorists?

Not everyone becomes ill from a moldy building, but when enough people lose VIP regulation from unopposed inflammation, we might as well have been attacked by some foreign terrorist using a handful of *Aspergillus penicilloides* with some *Actinomycetes* thrown in as an accelerant. At least if a foreign terrorist did the dirty deed, we'd have someone to blame besides those here in the US who decried mold illness as insignificant. But so far, it isn't a foreigner who's hurting so many people every day in so many buildings. Pogo creator Walt Kelly had his character tell us: "We have met the enemy and he is us."

The Holy Grail

VIP as a Big Brother concerned me. Command and control is needed for every military, business and every family operation. And that's the way it should be for a single entity, too, as a person is the organized sum of a trillion (or so cells) acting as a coordinated unit. The individual cell is out of luck. If it wanted to be free, like one-celled bacteria, fungi and dinoflagellates, it should've picked different parents!

Imagine if we had a set regulatory behavioral system for each person in a population controlled by a Big Brother type of external agent. We wouldn't see a diversity of response to external factors; we would see responses as if all of us were acting as clones instead. George Orwell and Aldous Huxley worried about loss of individual freedom, as have countless Sci-Fi writers. But an individual must act as a sum of cloned cells because we *are*! Every one of our cells starts with the exact same genetic information. We must have our Big Brothers, regulatory neuropeptides like VIP that coordinate all our cells, or we die.

Just look at the free-swimming bacteria, an idyllic one-celled creature, swimming in a planktonic pool of life. No controlling Big Brothers, no limit to what its DNA can do, and no limit to how quickly the helpless little fellow can be killed in the hard life that is reality for such small, "primitive" creatures. Imagine having to breed every twenty minutes just to keep your DNA flowing to offspring. Maybe Big Brother doesn't sound so bad. If one or two bacteria were too tired to reproduce, they'd die.

Or consider what would happen to the planktonic bacteria if they could rest without having to breed every twenty minutes. What if their life style let them rest a bit? Imagine if individual cells traded individual freedom to live in a community - one that provided safety and security as the struggle to find food, drink and places to reproduce and raise young was assured. Sound like a good plan?

Bacteria live just this way in biofilm. Each individual cell sacrifices freedom for security. Each is a clone of its neighbor but doesn't necessarily act like a clone because DNA messaging, the response to extracellular messaging (controlled by Big Brother neuropeptides), has another layer of regulation built in and that is differential DNA replication. Basically, some genes are activated and some aren't.

If we throw in differential gene activity controlled by Big Brother, which means some DNA is activated - open for business - and some isn't. Such differences in gene activation provide the biofilm bacteria with an extra margin of diversity that adds safety and enhanced productivity while reducing the demand for constant reproduction to ensure survival. Biofilm communities are still a collection of single cells and aren't a multi-cellular organism.

This fundamental biological difference between a single cell pulling one oar versus a multi-celled creature pulling multiple oars simultaneously is associated with a change from no Big Brother for the single cell critter to the presence of a Big Brother in the multi-cell creature. And differential gene activation is one way Big Brother does the job.

If we want our vessel to perform better in the seas of life, we could add more rowers, more kinds of oars and more ways to row. All of these activities demand more command and control—more Big Brother.

A simple question follows: did the fundamental leap from a single-celled creature to a multi-celled creature only occur *after* Big Brother arrived? And was there applause for that occurrence? What is the sound of one cell clapping?

MSH is just such a Big Brother. It's long been known to be part of the primordial innate immune system. MSH is manufactured deep in the hypothalamus (among other areas) in some of the evolutionarily oldest pathways of control of hormone activity, inflammation activity and nerve functioning pathways. If MSH production is impaired - and it usually is in mold illness patients - patients don't have an important Big Brother. Without command and control, individuals lose coordinated hormone responses, inflammation chemicals and nerve response. And yet, despite such foul-ups, we don't die. We need MSH for a good life, but its absence isn't fatal.

Restoring MSH production should lead to better control of hormonal function (it does); better control of inflammation (it does) and nerve function (it does that, too). Simple examples of the importance of MSH deficiency are the hypothalamic symptoms of mood swings, appetite swings, sweats (especially night sweats) and loss of control of regulation of body temperature (always hot; always cold; fluctuating from hot to cold). Given the importance of MSH deficiency, having access to replacement MSH should be followed by a new vibrant quality of life, thriving with Big Brother at the helm.

So when I asked the FDA for permission to use MSH in people, I thought of the benefits of giving insulin to those with Type I diabetes and giving thyroid hormone to those without it. My Holy Grail became MSH, because with MSH restoration, restoration of functional life would follow. The solution to nearly all chronic fatiguing illnesses was just around the corner. I forgot only one thing: the FDA hadn't approved MSH for human use. *Oops.*

The FDA was perfectly willing to let me use MSH once I submitted the animal data they needed, complete with toxicology studies, cardiovascular studies and more. I could have my Grail and use my Big Brother for the benefit of so many millions who needed it just as soon as I satisfied the FDA that my idea passed their regulatory and safety requirements. That would cost about $20 million to start with, so my naïve ideas about fixing illness rapidly fell flat.

No wonder the big drug companies had to spend so many millions of dollars to bring drugs to the market. Surely, there'd be an answer to the MSH problem.

MSH deficiency: wandering in the wilderness

Back in 1999, fixing MSH deficiency became my new purpose. If I couldn't overwhelm the inflammation resulting from MSH deficiency with replacement MSH, the next best thing was to stop the process that destroyed the MSH production process in the first place. What are the destroying factors? Inflammatory cytokines that would bind to leptin receptors in the brain would block MSH production, so if we could fix the inflammation, then we could stop the assault on MSH production.

But correcting inflammation didn't always result in restored MSH production, because MSH is vulnerable to attack from exotoxins, too. Made by unusual, biofilm forming, multiply antibiotic resistant, coagulase-negative Staph species (MARCoNS), exotoxins are proteins that cleave MSH, effectively destroying it. The MSH/exotoxin battle was epic warfare on a cellular basis. If MSH is present, the biofilm is destroyed and the Staph species vanish. Without MSH, the Staphs quietly thrive, protected by their igloo-like biofilms from other mucus membrane defenses. So the strategy was straightforward: Compounds released by MARCoNS could destroy MSH, so get rid of the MARCoNS and MSH is preserved.

What about stopping the adverse effects on MSH by knocking out the sources of the inflammation, especially exposure to the interior environment of WDBs? As it turns out, this step must be the first. Get out of the building and use Welchol or CSM, without fail.

As it turned out, way back then, finding the source of MSH destruction unveiled basic mechanisms of the innate immune inflammatory responses that kept patients chronically fatigued. Their cognitive symptoms (you know, the ones others assumed to be some sort of mental illness) were no different. Fix the inflammation and the cognitive aspects cleared. MSH controlled inflammation until inflammatory responses overwhelmed the MSH production mechanism. Like wolves caged to protect the cage-keeper from becoming wolf-supper, once freed from control, the inflammatory process turned on its *controller,* providing more freedom from control. Inflammatory anarchy, if you will. No Big Brother. And look what happens.

By 2002, we knew that MSH deficiencies didn't guarantee a terrible clinical prognosis, just a tremendous risk for developing illness if and when another exposure occurred. Having no MSH was an open invitation for unchecked inflammation in lung, GI tract, skin and anywhere blood circulated.

By 2004, we knew that a group of patients didn't show the expected improvement in MSH with any kind of treatment. These people had the now infamous HLA of 4-3-53 or 11-3-52B (or less commonly, 13-3-52A and 12-3-52B). There was just something awful about those HLA haplotypes; I called them "the dreaded," as I dreaded seeing them. Exactly why these haplotypes are dreaded still isn't clear, though TGF beta-1 is woven into the answer (see Chapter 4, Data Shows us the Way).

And yet, even allowing for ongoing exposure and inflammation, colonization with MSH-splitting biofilm-formers and genetics, there was another missing factor: there had to be something else keeping MSH low. The answer didn't come until 2005: VIP.

2005: Enter VIP

The name, vasoactive intestinal polypeptide, doesn't give one much more than a few clues that VIP was a big deal in the body's inflammation regulation business. Who would think such a non-descript protein would become so "magical" in restoring life to those with ongoing chronic fatiguing illnesses and lack of regulatory control of innate immune responses?

That VIP deficiency turned out to be really common in those with mold illness shouldn't now be much of a surprise. People who weren't sick - those without chronic fatigue and without a multisystem, multisymptom illness - had normal VIP levels. As the work of researchers like Drs. Ganea and Delgado made clear, global beneficial effects of VIP replacement on inflammation and cell physiology followed its use in the test tube and in animals. There was a developing literature of studies done in animals that, when taken as a whole, supported the entire idea that VIP was a master controlling peptide, similar in overall magnitude to the importance of MSH.

Question: Was VIP the missing link to refractory MSH?

Answer: You bet!

By 2006, we knew that VIP and MSH mutually regulated each other and that improvement in levels of one of the two peptides improved the other. When the additional role of antidiuretic hormone (ADH, also called vasopressin) became clear as the third member of the hypothalamic regulatory triangle, we knew that with improvement of any one of the three helping with deficiency of the other two, we had a "system" approach to regulation. Looking at either MSH or VIP as an integral part of a systemic whole was exactly correct.

As we can now predict, the ongoing model of lack of regulation of systemic responses failed if we only looked at one regulatory element. When we look at all three (VIP, MSH, ADH), the model worked precisely. I started calling this hypothalamic triumvirate the "three-legged stool," as cutting one of the three legs resulted in the fall of the chair. Similarly, if the lack of inflammatory regulation was due to lack of repair of any one of the three key hormone regulators, the risk of reacquisition of illness with re-exposure increased rapidly.

The scientific literature provided guideposts for me in this research, but the only solid human data that I could rely on came from my patient practice. It didn't take data sets from all 2000 patients to see the simple pattern of MSH, VIP and ADH repeat itself countless times. The systems approach, also called a landscape approach, gave intelligible answers to complex questions. I insist to this day that no clinical intervention be a multiple: do one thing at a time. Yet I also insist that each result of any single intervention be analyzed systemically.

VIP as part of systems analysis

If inflammatory responses were going wild in mold illness (they were) and lack of regulation of inflammation let those inflammatory responses go wild (they did), then systems analysis predicted that: (1) correction of the source of inflammation; followed by (2) correction of the excess of "downstream" inflammatory compounds; followed by (3) maximal normalization of MSH and ADH/osmolality, would lead to a final, single targeted therapy using VIP. "It will work," said the all-encompassing model. If I had the right to use VIP (I did) and I knew the correct VIP dose (I did), then restoring regulation should work rapidly.

It did.

Too good to be true

We began before Thanksgiving, 2008, giving VIP by nasal spray. VIP is broken down quickly, so the risk of over-dosage is basically non-existent. If we gave the drug four times a day, as the Europeans had done successfully in sarcoid patients, we should see improvement in inflammation, improvement in exercise tolerance and improvement in joint symptoms almost overnight.

We were wrong. The beneficial effects, including reduction of cognitive problems, appeared in *15 minutes.*

Side effects? Less than 3% of our first 150 patients had any.

Benefits? At least 95% had dramatic improvement.

Reducing inflammation? Yes, C4a and TGF beta-1 fell.

Adding control? Yes, ADH/osmolality improved, as did MSH.

How about vitamin D, the new buzzword compound in chronic fatigue? For the relatively small number with problems, benefits accrued that vitamin D replacement (and its specific compounds, especially vitamin D3) hadn't corrected.

How about androgen problems? Excessive activity of the enzyme aromatase, one that converts testosterone to estradiol, is blocked; androgens rise, estrogens return to normal, but we don't see too much androgen or too little estrogen.

How about VEGF? This incredibly important compound normalized after VIP and with it we saw improved oxygen delivery and correction of capillary hypoperfusion.

Over time, patients decreased the dose of VIP to two sprays a day and then to one a day. A few no longer use VIP, except as an urgent drug for acute exposure or one used in preparation for exposure or exceptional exertion.

"I had a mold hit yesterday so I went back on my VIP for a while. I know how to use it now. For me, it's like Tylenol and a headache. I'll never be cured of all headaches, but if I get one, I take a Tylenol and go on with my life."

Can you imagine VIP as "Tylenol"? Just ask Sisyphus

Basically, VIP brought me to consider a four-letter word for chronic fatiguing illnesses, especially mold illness. The word is one I haven't used in my lifetime for this illness until now: Cure.

Cure has been my medical mountain. I labored (as do many others) like Sisyphus, pushing our interventions up the mountain just like the tortured mythological Greek figure did, only to see the results of our labors dashed as the boulder bounds back down the hill. Like Sisyphus, however, as treating physicians, we don't have the luxury of *not* pushing the boulder back up the mountain.

And this time, with VIP as my trusty companion, the boulder stayed up! Imagine my joy, my utter ecstasy, at seeing VIP do what it can do! I felt like a proud parent whose child was marching at the head of the parade.

Oh, I needed to tell my friends and my colleagues. Well, on second thought, maybe not yet. Imagine the possible responses from "colleagues" across the country.

Colleague 1: "What is VIP?" *Again, I felt the boulder on my back.*

Colleague 2: "Great, I'll just use VIP first, before all those other steps are finished. My patient has been so sick for so long, we can't wait.

Besides, I'm really uncomfortable doing all the things *you say* I have to do." *Another boulder on my back.*

Colleague 3: "I'm going to dose the drug differently, because I think I should, and I don't care about mold exposure. Besides, look at all the money I can make."

Boulders 3-10 now loading. And the mountain just got higher.

Colleague 4: "I used VIP twice a day, and not only didn't I get a cure, I felt even worse."

Here's the rest of the mountain of boulders on my back. We know that mold exposure, hidden or obvious, *must* be cleared before using the drug. In fact, a rise in TGF beta-1 and/or C4a after a test dose of VIP is warning to look for hidden exposure. And don't skip the diagnostic test dose, either.

"Oh, I use VIP for headache; and I use it for menstrual cramps; and I use it for my androgen deficiency; and I use it for...Data base? No, I don't need no stinking data base."

I guess it is human nature to abuse whatever good new thing comes along. And I guess it is human nature to be creative and independent about exciting new therapies, so I really shouldn't be complaining about VIP misuse by fellow physicians. Still, I worked so hard for nearly five years to bring this golden goose to life that I don't want to see someone cut off its head just because *he thought* it might be a good idea.

What would you do?

When Skookum Jim, Dawson Charlie, Kate and George Cormack filed their claim on Rabbit Creek in August 1896 (soon renamed Bonanza Creek), the four-letter word spread like Oklahoma wild fire: Gold. Soon there would be an endless string of men climbing Chilkoot Pass in the snow, waiting for the ice to melt to let them into the Yukon Gold fields.

How was this overwhelming wave of human demand going to be different for my four-letter word: Cure?

So our group decided to wait. VIP deficiency studies had to be published. It had to be studied and it had to be controlled. We had to do a double-blinded study and we had to publish those data. Hiding the Tagish find of the yellow riches wasn't easy; enforcing another four-letter word (wait) wasn't easy, either.

During the Gold Rush, people stole the gold. They killed others in the hope of having the gold. More gold became a yellow fever that gripped a nation's poor, the desperate and the shrewd. The gold fields of the Yukon could have provided wealth for many, but so many came to prey on those with the easy money that parts of the Yukon became lawless and unsafe.

I had to think that if VIP were let loose on those with chronic fatiguing illness that those who didn't follow "usage laws" would ruin the wonderful benefit the drug can bring. So we are taking time to get the science right. Our study with collaborating physicians will be finished soon. We'll have our good science and our good studies, all approved and in order. There's no room for error, as when used exactly as our protocols require, we show the restoration of life to those with VIP deficiency and metabolic inflammatory disasters. They change from being the walking dead to what they were, though with a loss of time and a host of scars to show for their journeys.

We'll continue to look at the top of the treatment pyramid, for biology is an unforgiving discipline. Just when I think we've reached the top of the therapy mountain, we'll find some other unexplored rocky height above us that was hidden by perpetual clouds, but there nonetheless all the time.

So, we'll just try to figure out how to get up the next cliff, even if the way isn't clear at first.

And in the end, I'll trust science in its purest form to show me the way.

CHAPTER 17 The Novice Pilot: CFS and Other Medical Mistakes

By Erik Johnson

Hearing the distinctive whoosh of a hang glider, I looked up to watch. Just overhead, the glider banked into 180-degree turn, reversed course, and headed back along the coastal ridge, hugging the cliff to stay in the narrow band of upward lift generated by the sea breeze.

Momentarily distracted, I turned back to look at the fog-drenched metal link hurricane fence I'd been perusing.

What strange attraction did this fence have for hang gliders, I wondered?

Since 1980, I'd been honing my novice pilot skills on the beach, 150 feet below the face of this precipice, accumulating the necessary "sign offs" and eagerly looking forward to soaring the winds deflected up by the cliffs along the coast at San Francisco's Fort Funston flying site. With my 20 calm air practice flights to the beach completed, I'd finally achieved a flying rating, an observer signoff attesting to my soaring and launching qualifications. I now had my "Funston Rating" and was cleared to spend endless hours soaring in the smooth, reliable coastal winds, high over pounding surf.

Yet there was still one thing that challenged my confidence and held me back. Many experienced pilots with thousands of launches under their hang straps had flubbed a launch at this very spot, stalling into a right-hand turn, crashing directly into this fence. How could I expect to evade a similar fate, with only my few hours of soaring time?

Other novice pilots also new to the flying site shared my apprehensions, yet decided that the best thing was to simply "go for it" and hope to work out the kinks along the way. They nearly always made it. Occasionally, one or two would perform that amazingly quick right-hand tip-stall, but perhaps it paid to be a beginner, for they spun in so quickly that they didn't go more than a few feet, lessening the impact of the crash and minimizing the damage.

Ever mindful that discretion is the better part of valor, I chose to be patient, find out what was going wrong if I could, perhaps to avoid smashing up my brand new Wills Wing 209 Raven, and possibly spare my own neck as well.

I stood and watched as glider after glider launched, but nothing stood out as a precursor to who might lose control. Some fairly inexperienced pilots never had the slightest difficulty, while sometimes even the most accomplished veteran would suddenly pitch up to the ragged edge of a stall. "Pull in! Pull in!" people called from the ground as the pilot fought to regain airspeed. In sheer desperation, some let go of the vertical "down tubes" which is the normal launch position, flinging themselves out prone to grab the base-tube of the control bar and "bury the bar to the knees" as they pulled their weight forward to lower the nose. Whew, another one saved!

It was totally baffling that an experienced pilot with the latest high-performance glider would have so much trouble, when the wind hadn't changed, yet the novice just a few moments before with a vastly slower training glider hadn't had any trouble whatsoever. I wracked my brain trying to makes sense of it all, acted out the pilot's arm motions that I'd just witnessed, and then it just hit me. In shock, almost disbelief, doubtful of what I'd just figured out, my brain rebelled at the notion that such a simple thing could have brought so many good pilots crashing back into the cliff. In growing excitement, I sat and waited to see if another scary moment would happen, and if this problem was a part of it.

It didn't take long before I had a confirmation. Virtually every pilot who struggled for airspeed was making the same mistake.

Soon I was happily soaring amongst my fellow flyers, feeling secure with the confidence that comes from knowing that the silly oversight that had resulted in so much disaster wasn't a mistake I was going to make. I shared the information with my novice pilot buddies, and we could all see how easily anyone who didn't know about it could get themselves into trouble.

Then one day, a professional pilot showed up at Fort Funston, armed with all the latest gear. He'd won trophies at several contests and was skilled at aerobatic maneuvers. We greatly admired his super-modern, streamlined racing harness, faired tubing to reduce drag, double-surface highest performing glider, and gathered eagerly to see this guy in action. After carefully positioning for launch and balancing his wings, he plunged forward into the upward moving air, and then everything went wrong.

The nose of the glider pitched straight up, climbing into a high-angle stall. In horror everyone screamed, "Pull in, pull in!" He called back down to us in a voice tinged with fear, "I am, I am!"

But he was *not*. His weight didn't go forward, his body stayed behind the control bar, too far aft to get proper airspeed. Poised on the edge of whipstall disaster, he lunged for the base-tube and managed to pull himself forward. Fortunately, he'd stalled facing directly out to the ocean and hadn't drifted back above the ridge, for his glider dropped into a full-luff dive, disappearing below the face of the cliff before recovering airspeed and righting itself to fly away.

Stunned at what I'd just seen, I reeled in disbelief that this highly experienced pilot had nearly succumbed to "the mistake" like a rank amateur.

Later, after a good flight and a dazzling display of aerobatics, he circled in for a perfect landing, and people gathered around to hear what he had to say.

A mere novice, I held back as the veteran pilots pressed close around him for answers. "It must have been the wind," I heard him say. "Or maybe the glider; it's brand new and I only have a few hours on it." A few more theories were ventured before the entire incident was consigned to some kind of mysterious fluke.

What a bunch of malarkey," I thought. *"The wind is as smooth as glass, and he's got the fastest glider here. He doesn't know about the mistake he's making."* Emboldened by urgency, and thinking I might do a bit of good, I stepped forward, maybe half a step, before I held up short at the audacity of what I was doing. *"He won't listen to me, I'm just a novice pilot. What am I thinking?"* I thought to myself. *But then again, he could kill himself and how would I feel if I knew why it happened, and said nothing? This is ridiculous. He needs to know and I'm going to tell him. Once the explanation is on the table, it speaks for itself. He may dislike the fact that a novice is telling him something he didn't know, but even so, once someone hears about this, they can't help but take it into account. It makes so much sense, if you see it even once."*

I stepped forward and introduced myself. After a moment discussing his fancy glider, I said. "You know, sometimes during launch, if you leave your hands up too high on the control bar, there's no way to pull your body forward, because at shoulder level, you can never get your hands any further back than your shoulders. This is about as far back as they go (tug, tug) but if you slide them down to waist level, then you can hook your thumbs and *really* pull yourself forward... like this!" demonstrating with my arms straight behind me as I talked.

I didn't exactly expect that my input would be warmly received, and he didn't disappoint me one bit.

With eyes widening in shock, his face took on a haughty glare, he looked me up and down, taking in my low-performance training harness, low-speed glider, and just one or two flying site permits stuck to my helmet, which is a dead giveaway for little experience. He stuttered, "Just WHO the HELL do you THINK you ARE?"

In a huff, he turned and strode away, refusing to even look in my direction again.

Well, now let's see the results of my little experiment, I thought. After all, I demonstrated what I'd learned, made my point. He's got to see how the mistake works, so now it just remains to find out how his next launch goes.

It was very much a replay of the first. Not quite as dramatic, for now he was prepared for the pitch up, but he still kept his hands too high and didn't get his body forward.

People on the ground again called for him to pull in. After a few long seconds, seeing that he wasn't picking up more speed, he abandoned the standing position and grabbed for the base-tube, from which he could better pull himself forward. I heard people beside me murmur that it must surely be some design flaw with the glider, for the wind wasn't the problem.

I stepped forward with my rag-bag 209 Raven student glider and plowed forward into a launch. I made a point of pulling in the bar so much that the glider didn't gain an inch after launch, just flying straight forward, then, without shifting my low-hand position, I put my glider into a dive for a few feet, then used my excess speed for a "Chandelle," a climbing turn that showed off just how much extra kinetic energy I had to spare. Had the matter been one of performance, then a training glider clearly outshined the racing glider.

From up high, as if a glider could reflect the pilot's attitude, his glider almost seemed to swiftly turn tail as he sped away in the distance.

How odd that he must have understood the meaning of my arm motions, yet he didn't bother to incorporate it into his launch technique, not even with his own life at stake.

If you show someone who desperately needs it a simple "trick of the trade" could it really be possible that they'd turn it down because they had no respect for who delivered it?

When I was teaching hang gliding, I found I had to demonstrate it to get my point across. When I told people about it, they didn't actually seem to get it in a way that allowed them to put the hand position into practice. They had to *see* it, and mimic the motions before they could make use of the difference. Otherwise, they rarely slid their hands down far enough for it to work.

Which is just how it is with the mold. I can explain it endlessly without success, and then lead that same person into a mold zone and they go, "Oh, you mean *that*?" as if all that explaining had been for naught.

I think it'd be worthwhile to ask the reader to put down the book for a minute and try the hand motion for themselves, to get a graphic representation into their heads: "Imagine you're holding a metal bar in front of you, and you're trying to get as much weight in front of that bar as you can. Raise your imaginary bar to shoulder height and try to pull yourself forward and put every ounce of weight that you can on the *other* side of that bar. Now lower it to waist height and try again. Your entire upper body is now on the other side of that bar.

It's as simple as that. It's famously said in aviation, that "what you don't know will kill you," and that might just as well be said about mold, too.

Crazy

Wondering what mysterious thing was making some experienced pilots boff their launch wasn't the only thing bothering me at Fort Funston.

Hang gliding is pretty exciting stuff. When I got up in the morning to check if the winds were favorable, the signs of a good day for soaring were enough to make me rush packing up my glider, harness and helmet, and head out for the coast with all due haste. So it was strange to feel my excitement waver and diminish as I got close to arriving at the flying site.

Was it some subconscious fear? That didn't make a lot of sense, as I felt pretty confident in my skills now, and didn't feel uncomfortable while launching or flying.

Yet as I sat in the parking lot, it seemed like all my energy and enthusiasm had just drained out of me. I was overcome by the blahs. Sometimes I felt so wrung out that I didn't even want to fly, or do much of anything else, for that matter. But this had happened often enough that I'd learned that if I sat out in the wind for perhaps an hour or so, all the enthusiasm I had felt while loading my rig would suddenly fire up again, and I would jump for my gear and start setting up.

I didn't have a strange loss of interest in flying at other sites. Why this one? I began rating my "Want to go flying" level, to see if I could find some point along my drive at which it suddenly began to drop a few points.

Certainly I had my suspicions about what it might be, for I'd experienced these episodes before, but whatever was along this particular route was so strong, and I hadn't really felt anything quite like this while driving. It was putting quite a damper on my favorite flying site.

It was about this same time that a co-worker started having some serious marital problems and just didn't know what to do, or who to talk to about it. He said his marriage was a good one, and they weren't going through particularly difficult times, but his wife had begun acting strangely. He'd wake up and find her gone, out to sleep right on the ground in the back yard. She had no explanation other than not feeling right inside their house. But in the last few days, this behavior had escalated - when he couldn't find her and she was nowhere in their home or on the yard, he searched the neighborhood, only to find her curled up in various places, further and further from the house.

Some helpful people said it was probably drugs, an affair, or she was nuts. He vehemently defended her, saying they had an honest relationship and talked freely about what was happening to her. It was none of that, he insisted; she just didn't feel good in the house anymore.

"Is it me? Do you want to leave?" he asked her.

"No, it's just the house. Something isn't right," she replied.

Totally perplexed, and obviously in pure desperation, he opened up to more people, hoping to find some answers. I told him about my own strange circumstance, and volunteered to come to his house. Relieved that anyone took him seriously, he eagerly took me up on my offer. On the drive, he explained that it was a quaint little "mother-in-law" apartment that had been converted from a garage. It had once served as a carriage house next to the barn.

"But it's very comfortable, and just right for my small family," he said.

My friend's wife was far from the raving drug addicted lunatic that some had predicted she must be. In fact, I found her to be perfectly lucid and reasonable. She told me that while most times, the apartment felt acceptable there were moments that swept upon her like a nightmare, when she felt that her very life depended on getting out of the house, immediately.

"Why don't you wake your husband?" I inquired, and she said how odd it was that at these very same times, he was seemingly so deep in repose that he almost seemed in a trancelike coma. Even if he almost seemed to be waking, he would just lay right back down and instantly fall asleep. She told me that the backyard hadn't given her relief, and she'd found it necessary to get farther away before feeling better. Her intention had only been to get out for a few hours, maybe for a long walk, and then return, but when she lay down in a neighbors' doorway or on a park bench, she'd lose all sense of time until the daylight woke her. She stressed that she meant no harm by any of this.

The calm demeanor with which she spoke, and the blazing reasonableness of everything she described was enough to convince me that here was no crazy person, but there was more. My heart pounded, but not with excitement, for whatever it was that I encountered along my drive to go flying, it seemed to be here, too.

While my friend was relieved when I told him that his wife was perfectly rational and had excellent reasons for the way she was behaving, he still seemed reluctant to accept my insistence that "something" was causing this. My own growing unease at the effect I encountered was enough to make me change my plans. Why drive for an hour to get to a flying site, only to consistently feel so beat up that I didn't feel like flying? Although Mount Tamalpais was a far less predictable flying site, and had much greater logistical problems, at least I didn't run into that strange problem, and my joy in hang gliding was normal there.

For the first time, I began going to doctors, trying to describe the weird effect I got from various locations, which resulted in debilitating bouts of fatigue and depression. It seemed to shake up the doctors a bit when I said this would hit while on my way to go hang gliding. The whole idea that something like this would happen while engaged in such an activity caught them a bit off guard, and it wasn't so easy to put this down to some normal kind of depression. They had no answers.

Sick Thermal Syndrome

Despite the fact that I could only carry three gliders, my car had unique advantages. My "Hang Mobile" was a Subaru that I kept polished to a high shine. All the upscale pilots had Toyota trucks that could carry a huge load of gliders and a bunch of people with all their equipment. A desirable thing indeed, when so many want to fly, and so few want to drive the vehicle back down. Being of the more impoverished class of pilot, the only way I could keep up my bragging rights, which was to say that the ladies prefer to ride inside my shiny little car instead of hunched on a pile of parachutes and harnesses in the dusty bed of a pickup truck, and it was often easier for me to obtain a volunteer driver.

Although I didn't have much ground clearance, in low gear, the four-wheel drive Subaru could get me to most launch sites with little difficulty. I arrived at our campsite a bit after dark, weary from the long drive, but looking forward to the campfire parties that we enjoyed on all of our flying trips. The beer flowed as freely as our tales of hair-raising adventure, which typically grew more hair-raising as the night went on.

Setting up camp didn't go as smoothly as it should have. I carelessly put my lantern on my car, where it promptly slid off the hood and shattered. Then my tent wouldn't set up easily - strange, it had never been a problem before. During my numerous stumblings back and forth between tent and car, a painful awareness was gradually dawning: something wasn't right.

And the something that wasn't right was all over my car. Feeling woozy, weird, and totally drained, I found that my car was the source, almost like the center of a dust devil of swirling particulates.

Something "bad" that I'd driven through had stuck to my car like some kind of bizarre sticky fog.

I lay down for just a brief moment to shake off this wretched feeling before heading off to join everyone at the campfire, and that's the last thing I remember before sunlight beat its way into my eyes. The morning found me, as it usually does, long before I found it.

I woke up totally drenched in sweat, but I seemed little the worse for wear.

Under the light of a new day, it seemed like just a bad dream, and I couldn't feel anything amiss as I got breakfast-makings out of the cooler in my car. I may have been a bit less talkative than usual while riding up to the launch area, but, no worries, it looked like a good day for thermals, and hopefully we were all going to "Sky Out."

Setup and launch went smoothly, and soon, I was out thrashing around in small ragged thermals, seeking to find the hottest "lift" to carry me up to where champion pilots were already circling high above. Scanning the terrain below, I looked for signs of areas that might have a thermal advantage: rocky cliff faces; bare patches; areas that had a good angle to the sun, shielded by mountains to keep wayward breezes from "triggering" until a large volume of hot air overcame its ground-bound inertia, pulsing upwards in a thermal column of fast-moving rising air.

Checking to see how the other guys were doing, I looked northward to a barren stretch of area deforested by a wildfire. We'd discussed this area on the way up. With it's vegetation denuded and devoid of greenery, the bare ground on the side of a steep mountain had the perfect angle to the sun for capturing heat and channeling thermals up to its peak.

There were no roads in this area - the fire had been solely contained by the strategic placements of fire retardant from fire bombers.

A gaggle of gliders had already found their way to this area and were ascending at great speed, so I turned my nose northward and got ready to join the fun. Drawing close, I could see the blackened limbs of trees pointing upward like sharpened spikes, definitely not a good place to go down.

But I had plenty of altitude, and the place was "cooking." I didn't even have to search, because the moment I arrived, a "Boomer" caught me right between the wings.

The heat from this thermal fanned my face as I cranked off a hard left turn to search for the "core." Fifteen hundred; 2,000; 2,500 feet per minute up, this baby was a monster, and I was ready for the ride to Angel City.

Suddenly, without warning, it felt as if the Devil himself had sent up blackened Hell from those charred trees far below. Engulfed in a wave of badness, I gasped for air, almost uncomprehending that here in a clear blue sky, whatever wretched substance that was tormenting me on the ground, could reach up and touch me, even way up here.

In another turn or two, I realized just how much trouble I was in. Nauseated and head spinning, I turned southward, trying to get out of the badness and back out to calmer air, where I could rest and think. Although I soon reached smoother air, I didn't feel much better. Admitting to myself that my day was done and there was no way I could go on, I turned toward the landing area, seven long miles away. In smooth air now, I waited to feel better, but it just didn't happen.

As I got closer to the landing zone above the green trees, the air was so calm that I rested my head on the control bar, until I realized that I was on the verge of passing out, and would probably wake up in a spiral dive, or perhaps even make impact without waking up, if the blood-flow to my brain had been so deprived. I concentrated on breathing, on staying awake, and wondering what had happened.

Air sickness? How could that be, when I'd spent so much time bashing around in thermals, and never felt like this? The way I'd been feeling in various places seemed like the same feeling I got from mold, but here? Those dead trees below, the ash, maybe combined with the fire retardant - *something* was very bad for me.

I'd flown into that bad air, and I'd felt that badness decrease the instant I left the thermal, but the aftermath on me was really a butt kicker. It took all my remaining strength to plan out and execute a decent landing approach, and when I touched down, I didn't even try to stand up.

Buckling onto my knees, I just stayed there, gasping for air, until a friend who saw my distress ran out and helped me back to a log, on which I could prop myself. He brought me some water and asked "Are you OK?" I lied, "Sure, just got airsick," I said, knowing I could never explain what had just happened, for I didn't understand it myself.

Small-town Doctor

Things weren't the same after smacking into what I think of as a "Sick Thermal." Perhaps I'd been building up to that moment, as shown by my reaction to the car, but that had been a particularly hard hit. My life was knocked significantly out of kilter thereafter.

Buildings now had areas in corners of certain rooms that bothered me. More highways had "something" rolling right across the roadway that left me drenched and drained even after just momentarily passing through. Sections of some towns had a lingering effect that convinced me that I was better off for never going there, if at all possible.

When our Gliding Caravan stopped at a bar for pizza and drinks after a hard day of sky sailing, a painful experience stopped me at the door. Does it have to be this? I'd make my excuses, for I dared not enter. When all else is ruled out and the common element remains complicit, your choices are winnowed down, just as surely as wheat from the chaff.

Flying was still good. Being out in Nature was restorative, but civilization was definitely becoming more of a problem for me. Not all civilization, just the interior water damaged parts, but that means an awful lot of it. So what to do?

I'd shifted into teaching hang gliding full time, and an opportunity presented itself. The teaching site was many miles north of San Francisco, while the Sky-school business was to the south, which presented a horrendous commute.

I bought a small RV and arranged to remain with the gliders and equipment on site, right on the coast in a camping area at the base of the training dunes. It was an amazing lifestyle and suited my needs, seemingly a perfect solution to my problems.

But the condensation along the coast is very high, and my setup had some serious flaws. Headaches and fatigue slowly crept back into my life that no amount of bleach seemed to solve.

Mold in the RV just seemed unstoppable.

I did what I could, even removing the cabinets and sanding blackened areas out of the plywood floor, but still, things just kept getting worse. Throwing away the old cabinets and rebuilding the interior helped reverse the trend, but only for awhile.

By the end of summer in 1984, headaches were blinding, the tiredness never completely left me, and my former endurance was ebbing away, making teaching hang gliding, which involves a lot of running, very difficult and painful instead of the joy it once was. Still, I looked healthy to all outward appearances. One day, I turned to look up at the top of the sand dune just in time to watch someone training themselves fall back to the ground into a horrible crash. We all began to rush toward the victim, and I had only gone a hundred yards before I collapsed and could go no further. Out of breath, out of juice, out of options, brought to a total stop. Completely crapped out.

"What's happening to me? This cannot be! I'm almost unable to move." Who to go to, with a problem like this? I'd seen several doctors already, and just like my friend's "crazy" wife, nobody recognized or knew what I was talking about. Scarcely a week later, I readied for a short training run, holding the flying wires of a student for a quick hop down the "bunny slope," Ducking beneath the crossbar, I bent over, turned my head for a final check and to signal the student to begin his run, and the world went sideways.

What a peculiar thing to watch, like a television being knocked on its side. My horizon simply upended itself and everything went askew.

In wonderment at the amazing world-spinning spectacle, I'd not the slightest sensation or awareness that it might actually be me that was turning sideways, until the side of my head met the soft sand. Splonk! "Wow. *That* never happened before!" I didn't even know it was possible to drop like that, without being aware that you were falling; it was a totally alien concept. Learn something new every day!

OK, I said to myself, now it's definitely doctor time.

My normal routine was to teach gliding during the summer and shovel snow at Lake Tahoe during the winter. My family lives there, and it was time to go home. I high-tailed it back up to the mountains, my head pounding, internal pressure seeming to grow worse with the altitude. By the time I arrived, I was grogged out and losing touch, and having a hard time making decisions.

My mom took a look at me and bundled me off to the nearest doctor, who happened to be right across the street. This physician seemed like a pleasant enough fellow, never once appeared to doubt my description, and gave me a lot more confidence than other doctors I'd seen.. At first glance, it seemed like I might have a brain tumor. He ordered a number of tests, including a brain scan, and over the next few months, tried to figure out what was going on.

But each new round of tests failed to show what the problem might be.

Like other doctors, he didn't attach much importance to the mold, except as an effect of something deeper, which sounded reasonable enough, since I didn't exactly see lots of other people felled by mold the way I was, although a few *did* seem to be pretty uncomfortable around it. Finally, in November of '84, he called me into his office to explain that he had run out of tests and ideas, and was totally stymied by my case.

"I've never seen anyone quite like you. We're at an impasse, and I really don't know where to go next," he admitted. Advising that perhaps one of the big clinics might be in a position to help, he recommended that I try the large medical centers down in Reno.

I'd hoped that my situation wouldn't be so difficult, but I suppose that's just my luck. I should've known better than to go to a small-town doctor with my problems, when the San Francisco ones hadn't been able to deal with it. So I left Dr. Paul Cheney's tiny Tahoe practice to try elsewhere.

The Truckee Crud

I decided to concentrate on the black mold growing on the plywood in my RV and sanded it until all was gone. This seemed to help, but not as much as I would've liked, so I continued going to doctors. They each told me that my complaints about mold didn't make sense, since mold was incapable of causing my symptoms.

But they weren't interested in finding out how someone could be so reactive, either. They either told me that no such reaction existed, or that anything beyond the point of a reasonable allergy must be some kind of mental problem.

The winter of '84 was a heavy one, and despite headaches and fatigue, I was able to earn a good living shoveling snow. When the "Sierra cement" reached a point that houses were starting to collapse, the money became even better. It was about the middle of winter when word of a strange "flu" began to circulate. Word came from Truckee about a horrible flu that never ended. Rumor even had it that some people weren't recovering, and had been stuck with flu for months. It seemed that if someone got this flu and recovered within about four weeks, they were pretty much out of the woods, but if it went on for any longer than six weeks, those unfortunate ones just weren't shaking this thing off. I had no reason to think that if I were to get this illness that I'd be among those who didn't recover. We certainly had never heard of any illness like that, but it seemed reasonable to think that only those who were in exceptionally poor health would remain ill, and I was in my late twenties, in pretty good shape other than for this weird headache and fatigue situation. Then something happened that put the first twinge of fear into me. A member of my snow shoveling team caught it, and left work to go home. Every time someone called to see how he was doing, it was as if he was just getting progressively worse instead of better.

Then he stopped responding to calls. When someone went to check on him, they found him dead. The cause of death was officially determined to be heart failure.

I asked if the weird flu could be part of what caused his death, but was told that "flu doesn't kill so it doesn't matter whether he had it or not. He must have had a weak heart." That's when I began thinking that until this strange flu had run its course, it might be best to work on my "antisocial" skills and stay away from people.

I was able to return to my hang gliding job for another summer, but my stamina had taken a definite hit from whatever had happened. I was winded easily and the headaches crept back to full force. Black streaks resurfaced where I'd completely sanded away the mold. Could that be the problem? But how could all the doctors be wrong? With my health wavering, I decided to push harder on exercising and getting out in the fresh air.

More instances of this strange lack of recovery were being related in our community at North Lake Tahoe. People were scared of this particular flu, enough to give it a name. Since the first cases seemed to center around Truckee, we called it "The Truckee Crud."

It seemed to me that helping my brother collect firewood would be the perfect way to get myself out of the way of people with this weird flu, and so one morning we drove out to get some. My brother ran the chainsaw while I loaded the logs in the truck.

Everything was going great until I picked up a log that was covered in fuzzy mold. A cloud of musty mold smell draped itself over me as I dropped the log and stepped back, but it was too late, I'd already taken a deep breath. Suddenly the world looked foggy to me, like a place of wispy grey misty shadows. I sank to my knees, as my legs no longer seemed to work right. My brother stopped the chainsaw and stepped over to where I knelt.

"Are you OK?" he asked.

"I don't know," I replied. I figured in a few minutes, I could give him a better answer, but I wasn't better later. Things seemed to be getting worse. It felt like the glands in my throat were swelling, and a sore throat was starting. I felt feverish, but didn't appear to have a temperature. My brother helped me back to the truck and drove me home.

Something was definitely wrong. As the hours passed, everything continued to get worse. Within two days, I had the worst sore throat I'd ever had in my life, blinding white-flash brain-crushing headaches, felt sick to the very core, had a constant hacking cough, and there wasn't one part of me that didn't hurt, right down to the roots of my hair. I could barely stand up. We'd never seen mold do this to anyone, so it seemed pretty clear that the Truckee Crud had got me, and I began to mark time, hoping that in four weeks, I'd be among those who got better.

When five weeks had passed, and I hadn't improved, I really started to worry. By the end of six weeks, it was plain to see that I was losing even more ground, as now I couldn't stand up at all. Even crawling took so much out of me that I had to plan out my time frame for going hands-and-knees toward the bathroom, starting out a long time in advance before it became a matter of urgency, because I had to take plenty of breaks along the way. Every time I lay down flat, I would experience wild vertigo, with the full-on "spins," but if I tried to sit up, I'd pass out. At night, the dim light from a single clock would fill my eyes with blinding glare, and all I could do is wrap a pillow around my head in some kind of futile attempt to soften the crushing pain. Twenty hours a day were spent in a strange coma-like state that others would mistake for sleep, but it was like no sleep I'd ever felt before. This was more like lapsing into unconsciousness after getting hit on the head with a croquet mallet, which had actually happened to me, so I know what that's like.

There was a kind of weird exhaustive component to the illness, but it felt more like some kind of toxic poisoning than any kind of fatigue, because no amount of rest ever reduced it. Despite the short-term memory loss, the confusion, killer sore throat, brain fog, sense of unreality, aches and pains, and strange sensation of being poisoned, the absolute worst part was trying to reach for a fork to feed myself, and finding that my arm did not respond.

With a growing sense of panic, I tried harder and harder to "force" my arm to move, and finally, sweating from the sheer mental effort of will, I got my arm to reach out. My fingers were pointed toward the fork, but I missed the fork by about six inches. By dragging my hand along the tabletop until encountering the fork, I tried to grasp it, only to find that my fingers were similarly confused and unresponsive. I compromised and gripped the fork like a child, all fingers wrapped around the handle, and finally, I was able to avoid the embarrassment of asking to be fed. Between the pain of trying to force food past a tormentingly painful sore throat, and sweating with the sheer exertion of mentally willing every movement, I could only eat a little before I ran out of the ability to do any more, and I remember thinking, "If this goes on for very long, I might starve to death with food set right in front of me."

I'd already seen Dr Cheney when my problems were considerably less severe, so there didn't seem to be any point in going back to a doctor who was clearly unfamiliar with weird illnesses. My mother drove me to doctor after doctor, while I lay flat as possible, with the passenger seat as low as it would go to keep from me passing out.

None of these doctors had any idea what was going on, and most didn't even believe me, despite the visible infection in my throat, visibly swollen glands, dark circles under the eyes, and a chronic cough. It seemed that as far as they were concerned, I must've been a heroin addict who smoked too many cigarettes and drank wood grain alcohol, just to have a reason to get a doctor's valuable attention.

It was during this phase when a peculiar aspect of this illness came to my attention.

My brother was willing to take care of me, and I was staying in a bedroom/laundry room, when I noticed that I felt slightly better when I was out in the living room than while I was trying to do what passes for sleep. Sometimes, after enough hours in the living room,

I could even stand for a minute or two.

Determined not to be a totally useless lump, I tried to lean against the sink to steady myself, and do my own dishes in short one-minute bursts of activity.

But no sooner had I steadied myself in place, leaned forward to begin washing, when the sensation of being on the verge of total collapse would wash over me, forcing a hasty retreat to sit down somewhere. Over and over, I tried, with the same result. I got up and walked to the living room window and steadied myself on the windowsill, in the closest physical approximation of being at the kitchen sink, and there I stood, conspicuously failing to collapse, for minutes at a time.

Retesting by going back to the sink, I got the same sense of blood leaving my brain and impending collapse. I tried a few other things. It seemed that when I lay totally flat in the living room, the spins weren't as bad as when I was in the bedroom/laundry room. With great effort, for crawling is no easy task, I staggered-leaned-crawled my way back and forth between the living room and bedroom, testing to see if the difference was consistent, and similar to the way I felt between the living room and kitchen sink.

What I felt scared the heck out of me, for it was totally perceptible and predictable. I honed in on the drainpipe of the washing machine, because that area was where the difference in my disablement was the strongest. Moving back to the kitchen, I hovered around various places until I felt the worst of it was against the back wall behind the sink.

Hard to believe, but these tiny areas no more than a few feet in diameter were having the most amazing effect on my ability to stand up. Mindful of my previous experiences—the black patches in my RV, the inception of my illness being a blast of mold from a musty old log—it seemed like these water damaged areas were having some kind of "mold effect" on me.

What could I do? It seemed to be coming at me from all directions.

I made my way outside to sit in the fresh air to ponder my options. My RV was no good, despite my efforts to clean it up. I had to think of something else.

You Knew

Since it seemed that finding a Cheney-like doctor was a total crapshoot, there was little point in undertaking the journey on carsick-inducing roads over the mountains to Reno, which only made my condition worse. I might as well just look locally.

I heard that there was a "nice" doctor and made an appointment. Sure enough, this guy was kind and never once contradicted me about my illness. At the same time, even though I spoke of physical symptoms, he seemed to reinterpret them as something induced mentally.

It was hard to pin him down on this, as he spoke convincingly of how seriously he took my disease, but instead of suggesting testing or treatment, the discussion always came back to the importance of keeping a positive mental attitude and taking care of one's self. Not that these are bad things, of course, but it wasn't quite what I had in mind. His reassurance that I should be getting better any minute wasn't exactly working out, either, as I continued on in a kind of limbo. I wasn't steadily declining anymore, as long as I managed to stay out of these odd "bad" places, but I wasn't really improving either. And I still wasn't getting the answers I wanted about what could possibly be in these places that might be capable of doing such a thing.

Then the newspaper headline hit. I read the article in the North Lake Tahoe "Bonanza" about a mysterious fatiguing illness. I eagerly read how two doctors in town had identified about twenty patients with a strange condition that seemed connected to that weird flu that had been going through Lake Tahoe. The list of symptoms matched mine perfectly. Rushing to bring this to my nice doctor, I was shocked when he barely glanced at the paper, and distastefully cast it aside.

"Yes," he said, "I've heard about that." Surprised at his reaction, I countered, "But this describes all my symptoms perfectly. This is what I've been talking about." To my dismay, he replied, "We know all about that, but those two doctors are incompetent quacks. They need to be tarred and feathered, and run out of town on a rail."

In growing rage, I roared, "You knew about this? And you weren't going to tell me?" There was nothing more to be said. I walked out and immediately went down the street to those two "quack doctors" and made an appointment. It wasn't far to go, only one street over, and about a block away. Dr. Cheney remembered me instantly and welcomed me back.

CEBV Syndrome

I learned that since I'd been away, Drs. Cheney and Peterson had found out about a disease that appeared to match our symptom descriptions. This newly described condition in 1985 was provisionally being called "Chronic Epstein Barr Virus Syndrome," or CEBV syndrome, and they now tried to diagnose our small group of "mystery illness" patients with it. We were hopeful that this would resolve the mysterious malady, but we ran into trouble immediately because our illness wasn't acting quite like a chronic EBV infection. Several illness clusters had emerged in people from our local schools, and the manner of their onset, although somewhat similar to the description in the medical journal, wasn't exactly the same. As Dr. Cheney said, the paper didn't describe such groupings. Fearing the possibility of an epidemic, Dr. Peterson had called the local health department for assistance, and then the CDC, but neither took the illness seriously.

As the numbers of "Mystery illness" patients grew, so did the alarm in our community.

Some patients remained bedridden, while many others like me were functioning at a greatly reduced level, in constant pain and with a lot of neurological dysfunction. We had a strange inability to do even the slightest exercise without "crashing" afterwards. The time I was spending in my camper had helped considerably, and I'd become somewhat stabilized at a fraction of normalcy, but I wasn't able even to go for a walk. I became one of the patients who "haunted" their offices, returning again and again to the only doctors in town that would even listen, trying to find answers.

Lake Tahoe is a destination resort, and rumors of this illness began flying around the world, adding economic ruin to the threat of contagion. People began to be truly afraid of this mysterious phenomenon. Drs. Cheney and Peterson became the leading symbols of this threat, and their patients were the tangible manifestation, we became the targets of our neighbor's fear and loathing. We were alternately blamed for being a physical threat to community health, or spreaders of a false epidemic that existed only the minds of hysterical malingerers who'd found two doctors that we were able to influence by our chronic complaining. Strangely, even though the two concepts are fundamentally opposed, someone would accuse us of both. At any rate, there was general agreement that our presence was destroying the tourist trade, and business owners complained that profits had fallen considerably as a result of our presence. Newspapers across the country began to report stories of what was happening in our little region in the mountains. Our friends and neighbors went wild and accused us of "having an insane agenda to bring down the Tahoe economy." We were asked to leave, and not nicely.

We were told that regardless of whether our illness was real or imaginary, we were ruining the area. "If you cared about anybody but yourselves, you'd go somewhere else for the sake of us all." But this was our home. What had we done to deserve this? Couldn't they see that nobody could predict who was becoming ill, and it might just as well have been them instead of us?

And then finally, a break; the CDC agreed to send an investigative team. This was exciting news. Just the idea that the CDC was sending someone was enough to make it seem like answers were already on the way. Yet at the same time, the idea that the mystery illness was important enough for the CDC to come out had also raised the fear of contagion to a fever pitch, and the tension in town was palpable as we waited to hear what the verdict would be.

Then the bubble popped. CDC epidemiologists Jon Kaplan and Gary Holmes had conducted a cursory and swift investigation of only a few patients, but this didn't bother us, as we took it as a sign that the evidence was so compelling that no more was needed.

We couldn't have been more wrong.

They simply left town, saying that they weren't seeing evidence of an unusual fatiguing illness, and had no intention of doing any follow-up studies. (*Editor's note*: NIOSH and CDC are linked.)

This simple statement compelled the entire community attitude to change from taking the illness seriously to one of pure scorn and derision.

Drs. Cheney and Peterson bore the full brunt of the "righteous wrath" of the townspeople, who felt they'd been fully vindicated in saying this was nothing more than a bunch of hysterical hypochondriacs who's managed to match themselves up with suggestible doctors. The name that people had been using switched from "Tahoe Mystery Illness" to "Incline Village disease," as a direct slur against those two "quack doctors in Incline Village who'd made the fatal mistake of listening to their crazy patients."

In a weird way, this helped defuse the anger and hatred that had been building up against our group. Since we'd become objects of derision and scorn, it was a relief from being subjected to outright expressions of hatred and intimations of possible violence if we didn't leave town.

Two local teachers who'd become sick in a common teachers lounge were especially disappointed at the CDC's abrupt departure. They'd implicated the school in making them ill, and had pointed at several suspects for toxic exposure, especially the ancient copy machines that spewed clouds of fumes. They happened to be in Dr Peterson's office while Kaplan and Holmes were conducting their investigation, and they jumped at the opportunity to describe the situation to the CDC epidemiologists. Strangely, the epidemiologists hadn't asked to talk to patients; they wished to get their information from perusing medical charts. Nine teachers who'd been using this teachers' room became ill with the mystery disease, with the sole exception of one lone teacher who'd decided the air was intolerable there, and who had taken refuge in his camper whenever possible. Their description of the burning sensation they got when they were in that place fell on deaf ears, as nobody recognized that this was anything of importance. It seemed that whatever it was must be some kind of secondary reaction to the mystery illness itself.

The teachers persisted, speculating that perhaps some viral or microbial agent was lingering in the schools air filter system, but as the teachers described their experience, the CDC epidemiologists made it clear that they thought this was sheer lunacy. Surprised at the level of disinterest, one of the teachers later said that you'd think that they'd have asked at least some questions.

Prototype for a Syndrome: How Did THAT Happen?

Using the camper as a refuge had given me the most effective relief so far, and I felt sufficiently better to take a camping trip out to the Nevada desert. I went to a place called Sand Mountain, a 500-foot tall sand dune some people were using as a training hill for Hang Gliding.
All I encountered out there were roaring motorcycles and Dune Buggies, throwing sprays of sand as they raced around the giant dune.

There was something else though. I tried to take a short walk, and found that I wasn't running out of breath. I could stand for long periods of time, and I felt far better than I had in a long time. Now that was strange. Back in Incline Village, I was wrung out after walking a few hundred feet, but out here, I started out slow and kept on climbing until eventually I was at the top. This was the most exercise I'd been capable of in more than a year. Amazing. Returning to the camper, I crashed for several hours, but the crash was so much less than what I expected. I told Drs. Cheney and Peterson how the camper was helping me a lot, but they thought I was talking about having a place to rest. It didn't enter their minds that my increased energy was connected to being away from some kind of substance, and that this was an effect worth pursuing.

My brother bought a new mountain bike around this time, and I made the mistake of trying to ride it around the block. I think I only made it about halfway before I felt like I was might drop. I walked it home, leaning heavily on the bike to support myself. What a comedown after feeling so much better out in the desert.

What was happening to me? What was this stuff that was in town, but not out there?

Could it really be mold?

Although our community and the CDC seemed to consider our illness a closed case, Drs. Cheney and Peterson continued their tests, trying to find out what our "mystery illness" was, and sent blood out to various labs in an attempt to determine the role of EBV in this peculiar phenomenon.

Other doctors across the country and some support groups were agreeing with the CDC's concept, but the tests seemed to point at a deeper phenomenon in which EBV had either been reactivated from a latent state, or that somehow adults had been rendered strangely vulnerable to a new, extremely rare infection. Word went around that tests were showing that some people had no immune response to EBV.

This didn't look so much like the CDC's concept that EBV plus some other factor kept it reactivated as a condition in which the normal immune response to EBV had been disabled, regardless of whether someone had EBV or not. But as long as people kept showing up with EBV, there was still the possibility that the CDC was right, so the tests couldn't definitively rule EBV out as a possible cause.

"CEBV Syndrome" was a pretty good term, and we preferred it to "Mystery Illness" or Truckee Crud, and especially "Raggedy Ann Syndrome." So it was a bit of shock when I got my results back from the Nichols lab in April 1986 and Dr. Cheney told me that I was entirely EBV negative. He drew a bold circle around my results and wrote "neg" below, underlining it twice, seeming to think it was important.

For the next two years, I was placed in the rather peculiar position of having to explain to people that although what we had was what others were calling CEBV syndrome, in my case, my illness wasn't caused by EBV at all. I'd never had the virus, so having CEBV syndrome was completely impossible.

The "subsequent development" of patients without EBV lent weight to Drs. Peterson and Cheney's arguments, and the battle to get the CDC to recognize that CEBV Syndrome was an incorrect diagnosis began to take shape.

For the next two years, I was placed in the rather peculiar position of having to explain to people that although what we had was what others were calling CEBV syndrome, that in our strange flu-like illness, or at least, in my case of it, wasn't from EBV at all. I had never had the virus, so in my case this was completely impossible.

The contradiction of patients without EBV lent weight to Dr Peterson and Dr Cheney's arguments, and the battle to get the CDC to recognize that CEBV Syndrome was wrong began to take shape.

An article in the June 7, 1986 Los Angeles Times by Robert Steinbrook publically ignited the controversy: "160 Victims at Lake Tahoe. Chronic Flu-like Illness a Medical Mystery Story." The article described how a local marathon runner became ill and was forced to quit work. This runner was among 160 residents of Lake Tahoe's North Shore who had described the same type of illness to Dr Cheney and Dr Peterson, who believed that the chronic mononucleosis-like "CEBV syndrome" was the most likely culprit.

Yet here is where the first cracks appeared, for while the CDC epidemiologists claimed that they could neither "prove nor disprove" whether EBV played a role in the Tahoe illness, Dr Cheney disputed this conclusion, He said the investigation was too narrow and has been overtaken by subsequent developments.

When Dr Cheney asked me to volunteer to be in the Holmes CFS definition study group the next year, he explained to him that my participation was necessary, because I was the only member of the original Tahoe group who had been EBV negative all along.

This was the critical reason why Dr Cheney selected me to represent the new syndrome as a prototype. My "exception" had disproved the rule.

I've Helped To Start a Syndrome. Now What?

I set about gathering stories from other Tahoe patients, to see if any of them shared my strange "allergy", and wound up finding that this complaint was nearly universal. With growing interest, I soon found out that there were few people who didn't have some "mold story" to tell. One member of the Tahoe "mystery malady" cohort who worked in the casino where everyone became ill was aware that it was a sick building, and told me that other employees were afraid to mention it, for anyone who did so was fired. The schools where the clusters occurred had been knocking me flat for years. THAT was certainly a worthy clue. Wasn't it worth a look? Maybe do something simple and accurate, like an ERMI? But that test didn't exist then. The concept of chemicals in the buildings was given some attention but was discarded if the same substance were not in each and every place. The notion that ubiquitous mold could have accounted for any more than a few sniffles and sneezes was completely scoffed at.

I accompanied other patients into areas that I had painfully learned to avoid, and they would complain bitterly that they felt ill too, but most were doubtful when I proposed what it might be. I was told, "At first, I thought it was mold too, but I asked my doctor and he told me this was impossible...so I guess that can't be it."

Discarded, disbelieved! Only because the doctor didn't believe it. Odd, wouldn't you say?

One by one, story after story, the clues would point at a strange commonality of mold exposure, but since common household fungi were not known to do this, any suggestion that mold could have more than an ancillary role in this phenomenon just seemed unthinkable to people.

So unfamiliar and foreign, it was as if the very idea of mold exceeding its customary effect just couldn't take shape.

The disbelief took an odd form, like saying, "I'm not saying you're wrong, but I don't think you're right," a self-contradictory sort of logic.

That never made sense to me, for if someone says they don't think you are right, how can that be anything but another way of saying, "You're wrong," just expressed another way?

My point was that if this environmental component was consistently present where so many people complained and fell ill, it should be researched regardless of whether it was a cause, an effect, or just something particularly troublesome for patients. The resistance to such a simple notion took me by surprise.

It was as if the "whatever" could not be proven present where each and every patient succumbed to sickness, then it must be only a random result of the illness, and unworthy of investigation... a waste of time and effort.

I never wanted to be put in the position of arguing with a doctor, especially not ones who were trying to help us, but I did so now. I inquired of Dr Cheney "Why would a virus care about mold?" When he replied that a virus would not have any special aversion, I countered, "Then it must be bacterial, for whatever's got hold of me seems to care a GREAT DEAL about mold."

I insisted, "There is a specificity to this reaction, and it bears looking into."

Dr Cheney just grew more emphatic that a reactivity of this type was a general result of the illness, and that my focus on one substance was misdirected and unwarranted.

Naturally I did not want to alienate any doctors working so hard on our behalf, and I just let it go. I just told him, "Then while you are finding out what the cause of the weird flu is, I'll just stay away from the mold," and I counted upon Dr Cheney and Dr Peterson to remember my words later.

Surely when they heard about more similar sounding clues from other patients, reconsideration would be in order. These were smart doctors, and I was certain that when that time came, they would recollect my emphasis on mold, and would want to discuss it with me.

Other patients were amazed that I would leave Dr Cheney and Dr Peterson and go elsewhere seeking advice from other doctors, when these two physicians were at the leading edge of unraveling this illness, indeed, among the very few who even believed that this wasn't all in our heads, but it wasn't like I was discounting their concepts and had no intention of going back. I was just putting my bucks where I got the most bangs.

When and if they uncovered something important, I wasn't far away, and I could easily return.

In the fresh air up in the mountains above Incline, my "brain fog" would clear, and I had plenty of time to think about the bizarreness of what had happened. I had stumbled over an "effect" that was directly located to a "mystery malady" that had gained national attention, and nobody was listening.

A quote attributed to Winston Churchill says that many men stumble over the truth, but most pick themselves up and hurry on as if nothing had happened. The truth of the matter was that if I didn't pay careful attention to this one, I wasn't going to hurry anywhere. This stuff was laying me out flat.

Paging Dr. Semmelweis

The famous story of Dr. Ignaz Semmelweis sprang to mind. Nearly everyone knows the story of the Hungarian physician who observed the clues about the high mortality in the doctors obstetric clinic as compared the vastly lower one in the identically run clinic staffed by the midwife trainees. He wound up introducing systematic hand washing to the medical profession to control the spread of something that still hadn't been discovered. Something he provisionally referred to as "Cadaverous material' from the dissection rooms used by the doctor-trainees.

Although bacteria were suspected, it wasn't "proven," but sure knowledge of the cause wasn't necessary to take advantage of the "effect." Regardless of what it was... hand washing could effectively prevent spreading it.

This situation seemed remarkably similar. Something bad was in certain places. This substance could be carried to other places and have a lingering adverse result. I wanted to know what it was, even if it wasn't necessary in order to practice avoidance of it.

I had uncovered a singular clue (avoidance); agreed to participate in research by 2 local doctors; and felt I was a personal prototype of this first "cluster" - so hoped that Chronic Fatigue Syndrome ("CFS") researchers would talk to me. Now all I needed was to find a doctor like Semmelweis to help me figure it out, since it most certainly appeared that others could undoubtedly benefit by knowing of this. I was stunned that Dr Cheney and Dr Peterson could fail to be interested, but they seemed so professional and firmly convinced that a virus was the culprit that it didn't seem right for an uneducated patient like me to dare argue with them, so I had to try my luck elsewhere.

After the way local doctors had battled us over "Tahoe mystery malady," I didn't expect to find a researcher-type doctor straight away, but surely if I kept putting out the clues as if I were "Paging Dr. Semmelweis," eventually a physician or researcher with scientific curiosity would want to hear it. I began contacting doctors, telling my story, using my status as a prototype for CFS to get my foot in the door.

"You're just one patient;" "It doesn't matter;" "That is purely impossible;" "Even if this were true, you can't prove it;" I was told, over and over. What? I had thought proving it was THEIR job? Aren't doctors looking for clues, anything that helps? Why are they acting this way?

I even randomly picked doctors names out of the phone book, "Are you familiar with Chronic Fatigue Syndrome? Would you care to hear the story of an Incline Village survivor?" None did. How do doctors ever discover anything new, if this is their attitude? They can't dismiss clues that there might BE some clues.

This cannot be right. There must be something fundamentally flawed in the medical mindset. Trying to look at this objectively, what right do doctors have to refuse to listen? If a doctor claims to have any expertise in CFS, how could he be disinterested in hearing from a CFS prototype who was at the very inception of the syndrome. If a historian wants to make sure they have their facts straight on how the syndrome began, what right do they have to turn away from my story? One thing I knew for sure. If any patient felt that they were subject to this mold effect and wanted information, they certainly would never get it from a doctor who refuses to accept information and input about its existence. To my astonishment, I could see that the obstacle to getting doctors to listen is that they had defined "research" strictly as "Peer reviewed literature", and nothing more.

In a kind of anti-science irony, nothing that has not already been written by someone with "researcher" on their name badge can ever get researched, for without credentials, it remains in limbo, removed from consideration. Without research, clues stay anecdotal, discarded for the very reason that they have not yet been researched.

Not even the force of publicity around a "new dynamic" of unexplained illness and a brand new syndrome was enough to break through this ironclad philosophy, and I was completely stymied. I started collecting "rejection slips" from all the top-name CFS doctors and researchers. Mention of my participation in the "original CFS cohort" did open a few doors, yet I soon saw a pattern to the progression of events. The moment I mentioned that I had clues from "ground zero" and how much I had benefited by mold avoidance, it was like watching the lights go out, and a door slam shut. I learned to see by the look in their eyes, the exact moment when mild curiosity suddenly turned to concerted disinterest and conspicuous indifference. It was at the exact moment when doctors perceived that I wasn't looking for their medicines, but was handing them a clue which demanded action on their part. I gave them a chance to help their patients by telling them how I had helped myself, and they wanted no part of it.

Many doctors like to think of themselves as patterned after the renowned diagnostician, Sir William Osler, who famously advised, "Listen to the patient, he will tell you his diagnosis," yet no doctors were listening to me, not even when I attempted to tell them how important this was to my recovery and how it had affected others.

I had started out passively, just quietly suggesting that I might have an interesting clue, but when this was easily brushed aside with mild dismissals, I tried an increasingly aggressive approach. The doctor response was absolutely commensurate with how hard I attempted to break through the blockade. The harder I tried to tell them, the more they refused to listen, and the angrier they became. It would become a shouting match, and nothing was accomplished.

Even recounting the Semmelweis story and reminding them of Osler's words failed to make a difference. These doctors were bound and determined to never hear this clue.

They wanted to know absolutely NOTHING about conditions at the inception of the Tahoe mystery malady that didn't already confirm their preconceptions, and I suffered from the disability of not having "Dr." next to my name, and no "Researcher" on a name badge.

If information didn't come from a credentialed source, there appeared to be no way to break this impasse. In an effort at "shock value" I even said to a group of doctors, "But you are looking for clues, and I am a prototype for CFS who is out climbing mountains. Only a brain-dead moron could fail to ask at least a few questions, just to see if such a thing could possibly be true." Nothing worked, and I had tried everything. If I hadn't seen this with my own eyes, I'm not sure I would have ever believed this hidden medical conceit that would even deny information from an observer and participant of the VERY "CFS" phenomenon that they purported to study.

Now I know that this sounds terribly arrogant, but the test of scientific curiosity is really quite objective, laid out just the same as when Dr John Snow, "the father of modern epidemiology" noticed the high incidence of illness in a certain location during the 1854 London "Broad Street cholera epidemic" and made a map which showed a public water source was at the very center of the cluster.

Dr. Snow didn't know the bacterial cause of cholera, but he didn't need to, in order to effectively intercede. The experiment of removing the pump handle from the well and shutting it down had the effect of stopping the epidemic. In an uncannily similar way, I had seen locations where groups of people fell ill, learned to stay away from these places and it was paying off, but doctors failed to see how this might possibly apply and wouldn't take my word for it when this was explained to them. I used every story I had, every means of persuasion to try and get a doctor to listen, and they all failed to respond in a Semmelweis-like manner. In effect, they had all flunked "The Semmelweis test" of willingness to give a discordant bit of information its proper due.

There is no "multiple score" to this test. It is a case of "pass/fail." One either responds with the appropriate scientific interest, or they do not. There is no "But Dr. Semmelweis, I know all about hand washing, and do it several times a day... whether they need it or not."

I won't bother for now with the very long list of these famous names that turned their backs on these clues. Maybe I shall list them in another book, but they ranged from local general practitioners, allergists, to the most prominent CFS researchers and doctors in the field, from all corners of the Earth. It would make an interesting psychological study to determine how doctors could practice such a universally consistent, concerted indifference to potential problem-solving information, especially so soon after Dr. Barry Marshall and Dr. Robin Warren had humiliated the entire medical profession by demonstrating this exact same level of antiscientific disinterest to clues regarding their concept of H. pylori causation of ulcers. I told all these tales, gave examples of other patients who had noticed this effect of bad-locations, made it clear that doctors have no right to ignore the clues, even explained that when mold-illness becomes known, those who refused to examine the situation will fall into the same category as the doctors who mocked Semmelweis, denigrated Snow, and blocked Marshall and Warren.

Of course, I only took it to this antagonistic level as a "last shot," to possibly shame them to change their resistant stance after it was apparent that they had no intention of EVER looking into the circumstances underlying the inception of CFS. I didn't expect it to work, but at least when "Toxic Mold" becomes known, they might have a lingering memory of their malfeasance, to remind them that perhaps the next time a clue presents itself, better to reconsider their obstinate behaviors.

For years, my story hit a brick wall of medical indifference... then, a lucky break. A friend of mine, another Tahoe patient, gave me a recently (2001) published book that rocked my world, *Desperation Medicine*, by Dr. Ritchie Shoemaker. It was with a shock that I read that all the clues I had tried so desperately to get all CFS researchers to listen to, were explained so clearly. Dr. Shoemaker even had a chapter that outlined how he believed that our Tahoe phenomenon was caused by biotoxins. "This is IT, now they'll HAVE to listen to me." I thought, but... I had made that mistake before, and even this additional backup to my claims was not enough to induce doctors to listen. How could it be possible that a doctor all the way across the country could so clearly see this effect, see how it applied at Tahoe, while doctors who were right on the scene could be so completely oblivious to every clue that resoundingly presented itself?

I managed to join an internet group where Dr. Shoemaker sometimes posted, and told my story... and that was it. That was all it took. ONE time of seeing the same story I had fruitlessly offered to so many others, was enough to make Dr. Shoemaker instantly contact me back, wanting to hear the details.

Far from demanding that I make expensive appointments (pay the lady at the front desk...) while I fruitlessly begged doctors to listen to mold clues, Dr. Shoemaker took his own valuable time to get MY story and even paid for research that would verify the mold connection to CFS.

Now, I am not a patient of Dr. Shoemaker's and cannot speak to his bedside manner with his clients, but I have heard rumors that he is an arrogant man. If that is so, then I judge it must be the same type of arrogance that allowed Dr. Semmelweis to stand up to his dismissive and antagonistic peers, the "arrogance" that gave Dr. Snow the conviction in his beliefs to create general annoyance by locking down a public water supply, or perhaps the kind of "arrogance" that caused Dr. Robin Warren and Dr. Barry Marshall to defy the entire entrenched medical establishment, winning the Nobel Prize for their efforts, because when all was said and done, they stood up for the facts.

In the years since Dr. Shoemaker told my story of "Mold at Ground Zero for CFS" in *Mold Warriors,* I know that not one CFS doctor, CFS advocate or researcher has seen fit to contact me. This is no oversight, I know... for I've tested their manner of response in CFS seminars and meetings in the same way practitioners of medical hand washing surely must have presented this concept at every suitable opportunity. Even with *Mold Warriors* in my hand, the many, many doctors and researchers that I contacted, turned and fled when I tried to engage their interest in the biotoxin connection to CFS. Not even after toxic mold became household words and a national scourge, did so much as a single doctor respond anything like Dr. Shoemaker did. I wish I could say their indifference was by accident or oversight but it is not. This was matter of disinterest by design. That was plain enough to see by the precise timing of the moment so many doctors choose to walk away when I related my story. I learned predictably to see that when the framework of my story demanded a response from them, the loss of interest was near-immediate.

I had hoped for better from the medical profession and my disappointment cannot be more profound.

The times, they are a changin' though, and now I see that many of the very same doctors who refused my story, took no interest in my, or other patients' descriptions of this effect, never followed up on reports of this type, and didn't bother to revisit the inception of the Chronic Fatigue Syndrome to see if any clues might have been overlooked, are now eagerly jumping on board with the mold factor, vaguely implying that this was something that was "well known" to doctors, and never in much doubt (*Editor's note*: see Burial, in Chapter 12). Now, at last, are they prepared to proffer their professional services?

Where was their interest, when it should have been dictated by science, by clues, and by the interests of patients' health?

This is one more cautionary tale in a long line of repetitive demonstrations that the fundamental failure of the medical mindset is not related to any particular epoch, but is just as forceful a hindrance today as it was in the time of Semmelweis. If we are to make progress, edify ourselves, and learn from past mistakes, these problems must be brought to light. The medical profession must be admonished to pay more than lip-service to their own oft-repeated famous quotation regarding the importance of listening to the patient. They must return to the old ways and put the words of Osler into actual practice.

With so much discussion and confusion about CFS, it had seemed to me that there must be at least a few doctors who would want to hear from someone who was present at the inception of the defined syndrome.

In the final analysis, after presenting my story to so many, there was but a single doctor in the entire world who responded in the way that most doctors claim they would if ever they were presented with such a situation.

One single physician whose words were more than rhetoric, whose deeds speak for themselves by acting as Sir William Osler surely would have done in a similar circumstance. One is not a large number.

This scares the heck out of me. What happened to the "foot-soldiers of medicine" we counted on to act as guardians of the national health? Is CFS such a minor matter that the details and circumstances of its "first sighting" are of no interest to anyone?

If the medical profession is incapable of perceiving, or worse, refuses to even admit yet another dismal failure to respond in a meaningful manner to relevant information, we have been given one more warning that the next time an unfamiliar paradigm emerges, history will surely be repeated.

This must change. It is up to us to make it happen.

If the medical profession is incapable of perceiving, or worse, refuses to even admit yet another dismal failure to respond in a meaningful manner to relevant information, and not just about CFS and mold, we have been given one more warning that the next time a unfamiliar paradigm emerges, history will surely be repeated.

Recently, I received another reminder of how ones pre-existing conceptual framework and personal prejudices against an unrespected social standing can cause even the most obvious clues to disappear from plain sight.

In August 2010 I attended the Grand Opening of the Whittemore

Peterson Institute for Neuro Immune disorders in Reno. I made a point of showing up on my bicycle to show that CFS does not have control of me. Annette Whittemore saw me circling the parking lot, admiring the new addition to the University of Nevada's medical complex, and personally invited me in. Noting that I had to chain my bike to a tree, she said the institute needs a place for bicycle riders to lock up, since many students use them to get around the campus.

I circulated among the CFSers, few of whom seemed capable of riding a bicycle, yet only the Whittemore Peterson Institute science director, Dr. Judy Mikovits, expressed interest that I could ride all day long.

My response to the usual, "How are you doing?" was, "Totally awesome, thanks to mold avoidance," but it failed to lure any follow up questions from the people present. I told people straight out that I had found something that seemed to assist in my recovery, and this only caused them to disappear across the room even faster.

It seems that ***if*** the causes and cures for CFS are not totally to be found in a bacteria or virus, and a pill is then in a bottle waiting to affect cure by eradicating a living foreign invader, ***then*** the universal desire appears to be that *it would be better to not find it at all*.

As stated in the beginning of my chapter, an *uncredentialed person has no means* to convey anything to a mind that is unprepared to accept it, not even by demonstrations of efficacy. What good do even the most impressive of medical edifices serve, if the minds within express disdain for potential clues? I have never challenged or rejected an infective cause for "The Yuppie Flu," so why are they so quick to dismiss the concept that mold, and the inflammatory response that its exposure brings, is capable of learning a new trick? If they would just look at their labs, would they then listen to me?

I wonder how many times some useful knowledge has been inadvertently consigned to oblivion by the very people searching so diligently to find it?

This morning, a month after the official inauguration of the Whittemore Peterson Institute and Center for Molecular Medicine, I rode by to see how things are progressing.

A brand new, high tech looking set of bike racks have been installed.

The main reason I was riding a bike may have been utterly overlooked and lost, but the fact itself did leave some small impression.

That much at least, came out of my visit.

Amused, I thought to myself that like the CFS syndrome itself, I had played a small part in how and why this came to happen.

- *By Erik Johnson*

"That men do not learn very much from the lessons of history is the most important of all the lessons that history has to teach." Aldous Huxley

Author's note: More than 25 years have passed since the Incline Village outbreak of CFS. Erik Johnson still maintains that he is alive today because of the avoidance skills he learned, beginning with the plumes that affected him while hang-gliding and followed by exposures on land that made him ill rapidly. He has worked hard to stay alert, functioning and self-supporting. I have referred a number of people to him with severe reactivity to WDB. They have learned the importance of observation and response to environmental cues.

What these patients learn is that they can regain control of their own health by avoiding exposures and by taking rapid actions if they are exposed. Most of my patients use a pharmaceutical approach to treatment of consequences of exposure, similar to what we would expect from the Greek figure, Panacea. Erik is by far the most successful person I know who treats his exposures, now evermore severe and ever-more common, with non-pharmaceutical interventions, such as what we would associate with the Greek figure

Hygeia. In a classic demonstration of the benefits of his approach, Erik climbs Mount Whitney once or twice a year, taking photos along the way. His ongoing theme is that *what he has learned has potential benefit for so many others*. In some ways, Erik is perplexed that more people don't listen to his theme: Show me a 25-year survivor of CFS who climbs Mount Whitney, continues to work full time and writes as if he never had been ill.

As *Surviving Mold* goes to press, Erik has suffered some serious setbacks by the emerging unpredictable sources of biotoxin exposure, with a greater adverse impact on his wellbeing than he had experienced in years. Fortunately, and once again, each of these setbacks responded to aggressive mold avoidance. He is more vigorous than ever now, as ***Survival for Erik*** begins with careful observation.

We cannot go back in time to prove that Erik's successful avoidance strategy was associated with the expected correction of abnormal laboratory findings that we can see routinely in others sickened by exposures to moldy environments. If circumstances were different, we might be able to show what happens physiologically to Erik when he is exposed and sickened. Indeed, his future health might depend on it.

Still, Erik's observations may be more telling than any of us realize. It was his clear demonstration of benefit by reaching for high altitude that directly led to use of low dose erythropoietin, discovering its ability to lower C4a and restore capillary perfusion to normal, all the while reducing symptoms and markedly improving cognitive problems often termed, "brain fog."

I have yet to be able to show that Erik has been wrong about any of his ideas. I have tried. Regarding his thoughts of mold exposure being the cause of the CFS in Incline Village, Drs. Cheney and Peterson will likely not agree. There certainly was amplification of indoor mold growth in the areas where people became ill in Incline Village but no ERMI or similar testing was done. We routinely see people with CFS becoming terribly sickened after exposure to moldy environments such that these questions emerge – "even if there is some viral source of CFS, and perhaps the XMRV virus has a role in that initiation of illness, does mold exposure make their illness more complex? Or has 'mold' (an innate immune system run amok) now *become the illness* and identifying the initiator really is of ***limited*** importance?"

We cannot look at anyone with a chronic fatiguing illness, initiated variously by ciguatera, Lyme or Borna virus, for example, without noting the universal appearance of the laboratory features seen in mold illness. One may reasonably ask if the lab findings aren't simply the end result of an unremitting inflammatory attack on regulatory neuropeptides such that these illnesses all present with a "final common pathway." And if so, each person affected in this way can NEVER afford to be exposed to the interior environment of a WDB.

For those, like Erik, who remain "in the dark" about the exact inflammatory pictures that hang in their bloodstreams, *Surviving Mold* offers light in that darkness. Extreme Avoidance is helpful, but when even the best practitioner of that art form linked to Hygeia (Erik) becomes disabled, we need to turn to Panacea for solutions.

CHAPTER 18 Spouses, Choice and HLA DR

I Want a Girl Just Like the Girl That Married Dear Old Dad

I started looking at HLA DR in 2001 in the hope that knowing more about these immune response genes would explain why some people got sick after biotoxin exposure but others didn't. I didn't know then that HLA held secrets far beyond simple susceptibility to biotoxin illness.

In 2010, I'm not yet making a big deal about HLA's curious association with other elements of human life beyond antigen recognition, but some statisticians are intrigued by all the curious linkages I've recorded. Does anyone really know why some people are attracted to each other right away? Or why so many marriages fail? Wouldn't you agree that any data on such topics would be worth a few minutes under the bright light of speculation? Mere speculation mind you – but that is always the first tiny seed that may grow into musings which might foster a theory that could lead to experiments which may grow into knowledge.

Or are you wondering, why bother thinking about HLA DR, anyway?

For four years, since the outbreaks of *Pfiesteria* illness in 1997-1998, one question kept gnawing at me: How come only three of ten people swimming and frolicking at the same place for the same amount of time at Williams Point the day before the big fish kill were sickened? I found no obvious answers, despite looking since then. I looked at so many variables, including diabetes, cigarette smoking, alcohol use, illegal drug use, weight, hair color (I'm not kidding), cancer, age, gender, race, you name it, but nothing made sense.

And then there was HLA, a buzzword acronym for Human Leukocyte Antigens. Bingo!

The answer had been there all along: all of those who became ill had one grouping of HLA (called a haplotype). None of the well but similarly exposed people had that haplotype. I could hear the ghost of Francis Bacon, father of inductive reasoning saying, "I told you so," as I failed to find the answer using traditional deductive reasoning. Taking a look at what I knew, figuratively placing those observations on a problem solving table and then making an inductive leap into genetics, led directly to the answer. Others might call such an exercise a lucky guess, and I suppose they'd be right, too.

It was pretty clear very early that HLA DR was more than just a casual mixture of several genes (DRB1, DQ, DRB3, DRB4 and DRB5); there were also linkages between the various genes found on chromosome 6. As I started my registry, seeing who had which HLA, every time I wrote down a DRB1 of 7, for example, there was a DQ of 2 or 3 and then there was DRB4 of 53. It was uncanny. For every DRB1 of 4, there was a DQ of 3, 7 or 8, and always these groupings had the same DRB4 of 53. For every DRB1 of 11, there was a DQ of 3 or 7 and a DRB 3 of 52B.

The early key to recognizing these patterns came from the curious finding of *so many* people who had two copies of the same gene type. I mean, the odds of finding a homozygous —or double copy— of the same gene were really low. Since we found 54 separate haplotypes, and we're looking at Mom and Dad having their own two copies of genes, one each from their Moms and Dads to begin with, that would mean that Mom and Dad had two chances each to have one of their parent's given genes. That means you get two cracks at having any one out of 54 genes being present. The odds of that particular gene matching the one from the other parent are 1 in 27 times, making the total chance of a double being 1 in 729 or about $1/10^{th}$ of 1%.

That's not what I saw. My patients are doubles (homozygous) about 15% of the time or about 115 times more than the average population. What gives? Something wild was going on here. Still, every time I looked at these numbers, they were too different to be real. I needed a statistician to figure out the odds.

Statistician extraordinaire Dennis House was happy to help.

And the children in the study group also often had a homozygous presentation of immune response genes (also doubles). Given that we have 54 haplotypes, the likelihood of finding one match for Mum's 4-3-53 would be one in 54 or 1.85%. But the odds of finding a homozygous *pair of any predetermined gene* (here 4-3-53) is one in 54 *times* one in 54. That number is 0.0003429 or slightly more than 3 in ten thousand, he said.

I asked some others to do the calculations as Dennis had. Some kept on changing what it was they were calculating, so the numbers were different. Oh my; try to imagine arguing with statisticians. The first concept to keep in mind is that you will be wrong.

Answering science questions invariable involves defining the questions carefully first. For me it didn't matter too much if the numbers were one in 100 or 1 in 10,000; there were a whole lot more kids who were homozygous than any statistician would predict.

We could follow these genes across generations so that we could make a family tree of who had what gene make-up, just like they told us about in genetics class when the Royal Family of the House of Tudor was being looked at for porphyria. Let me be clear: the expression of the kind of genes people have —the appearance— is called *phenotype*; I was looking at the actual gene make-up, which is called *genotype*. Each member of the grouping of genes is called an allele and each distinct gene grouping is called a haplotype. Confused yet? Indeed, one of the HLA reference books kept talking about the need for the many diverse people working on HLA to use the same language. So true! I kept trying to figure out what everyone meant by the terms they were using, but they kept changing the nomenclature.

There are some links we can make in science that are "true but unrelated." I'm not interested in guessing about the hidden meaning of immune response genes, I'm just looking at what occurs. Why? Because what matters to me is what is going on in my job trying to *fix the illness*. My job is not so much to be an expert in HLA gene structure so much as it is to know that there is a structure and whether or not that structure plays a role in antigen presentation.

There wasn't any roadmap for interpreting the HLA values I started to compile. Some of the major organizations that studied immune response genes such as **www.anthonynolan.org,** devoted a sizable amount of space to the different nomenclatures used by diverse organizational entities looking at tissue transplantation, the main use at that time for HLA DR typing. Talk about a communication problem. Three languages, each with codes of numbers and letters, case sensitive; order specific and nothing similar between the three. Even the World War II Navaho Code Talkers wouldn't feel comfortable talking in three languages at once.

Back then learning about HLA was an exciting task for me, and as I turned each page in books that can be appropriately described as "HLA for Dummies," they brought brand new information to me. Back then, medicine was transitioning from using HLA simply for matching people for tissue transplantation. Tissue typing (matching) was required to reduce the risk of transplant rejection; HLA scientists didn't know back then that HLA DR would be associated with so many different illnesses including biotoxin-associated ones.

There were many references for HLA serological assessment. Those old-fashioned tests had been around for a long time, at least four years (anything older than one year is getting wrinkled and a bit grey around the edges in immunology). But those discussions weren't about the new DNA testing, a technique so exact that serology was out of date *even before* all the interested participants could agree on how they should name the different kinds of genes they could identify in HLA.

Still, after the "obvious" answer to "why did she get sick from Pfiesteria but he didn't?" actually became obvious, a simple thought of differential genetic susceptibility based on antigen recognition held my attention. These were *immune response genes*. How did they accomplish their recognition duty? What is it about an immune response gene that does that? The genes code for amino acids (building blocks for proteins), which are assembled into a unique shape for each haplotype, making a glove that can only fit over one kind of hand. The shape of the hand is actually the shape of an antigen. Antigens are described by the NIH as any substance that causes your immune system to produce antibodies against it. An antigen may be a foreign substance from the environment such as chemicals, bacteria, viruses, or pollen. An antigen may also be formed within the body, as with bacterial toxins or tissue cells. Only once held by the uniquely shaped HLA glove, can the antigen be processed by specialized white blood cells that are devoted to antigen recognition and processing. These cells, called dendritic cells, are so good at what they do that they're called "professional antigen recognition cells."

Reading over and over about HLA helped to visualize just what it was. Here were a series of amino acids, joined together to make a series of proteins that created an enclosing cleft, like a valley in between surrounding mountains. Into the valley could ride the 600 (or however many proteins that the body wanted to identify) as foreign antigens. Once the protein fits into the valley, it could be recognized as foreign. Immune responses involved HLA.

If there were different amino acids coded for by the HLA genes, then the shape of the final protein (the valley) would be different. And to make the point, if the valley (glove) had a different shape, then the protein didn't fit into it. The action of the protein activating the valley would be lost. Each name (4-3-53, for example) for an immune response gene would mean different amino acids in a different order, creating a gigantic number of possible shapes of the hand or the valley. Bring the statistician in again, please.

Now throw in a mutation or two in the sequence of the amino acids, thereby changing the shape of the valley and the problem of matching valleys and proteins really gets complicated. The people who figure out these answers are a different breed!

Immune response genes are so important, I think they'll be a key to further research in many areas. Just think about the basics: how can we help our immune system recognize and kill all cancer cells, not just the 99% that they already do? How about metastatic cancer cells? And why not look at cells that begin to die when they shouldn't or don't when they should (defective apoptosis)? How about making organs available from stem cells? Do you think Type I diabetics might want a new collection of pancreatic islet cells? Look for HLA research to lead the way.

We already know that certain kinds of HLA are disproportionately represented in various illnesses. I've talked briefly about transplantation: there we look for HLA Class I genes as a whole, including DR, DQ, A, B and C. HLA B27 is perhaps the one physicians have heard about most, as that one shows up often in cases of ankylosing spondylitis and spondyloarthropathy. The HLA B 5701 is a common test used in evaluating HIV. Celiac disease joined the HLA testing fraternity some years ago, when the DQ loci of 2 and 8 showed up like huge blips on the association screen. For some reason, narcolepsy is associated with DR 15 and DQ 0602; DR 15 also shows up with DR2 in the immunosuppressive response association seen in myelodysplastic syndrome. The specific HLA haplotypes 15-6-51 show up in multiple sclerosis and Post-Lyme. Please look at the "Rosetta Stone," an appendix in *Mold Warriors* and here (Appendix 2), to read about your own HLA. Don't pay an unscrupulous internet physician to read your HLA for you! The reading's always been free to anyone here, and may it ever be so.

Look out for the "dreaded" genotypes of 4-3-53 and 11(12)-3-52B. We know a lot about those genes and how they show up like crazy in some of the worst biotoxin-associated illnesses.

We know some people got sick from Pfiesteria and some didn't. The split was mirrored by their differences in gene structure. Did the appearance of all cases with illness compared to all cases without illness give us "susceptible genotypes?" When I say, "susceptibility" the epidemiologist thinks in terms of "relative risk." What that term means is that when we look at people who are ill and compare their HLA types to the HLA types found in people who *aren't* ill (incidence in cases divided by incidence in controls), we get a number. This relative risk number has no units: it isn't per square inch or a percent or anything—the number is pure. Any time that number equals or is greater than 2.0, the statisticians all agree, that value is significant. Some organizations, like the Centers for Disease Control and Prevention (CDC), have used relative risk as low as 1.5 to make their points about genomics in 'Chronic Fatigue Syndrome.' Our published work uses the more restrictive standard of 2.0 for relative risk.

Still, all I had was a statistical association between parents with similar genes. The relative risk was astronomically high that a guy with a given gene - let's call him Levi - would find a girl with the same gene. We'll call her Lee. Could there be something - an unknown shared genetic element - that could theoretically increase the chance that Lee would fall in love with Levi? Was it love at first sight? Did Lee make some kind of genetic pheromone that attracted Levi from across the room? Vice versa?

I started to look at the genes we had in our data set from multiple generations. I had to ask, although Lee and Levi didn't have a shared HLA DR, might there be something else involved? Maybe it was just coincidence that my rare HLA DR was shared by my *father-in-law*, but then my bride had a gene that I didn't have but that *my mother did* have. The idea was to look at so-called discordant gene couples, people that *didn't* share a particular HLA DR gene. What did their parents have? We actually had more than 200 families in this group. Fully 40% of the discordant couples had a bride with an HLA gene that was possessed by the groom's mother and vice versa, with the groom sharing a gene of the bride's father. Fascinating.

Just for fun, playing around with the word "discordant," I wondered if the 40% of discordant couples *actually* had marital discord. I didn't have a tool to use to survey these folks, just a question: "are you divorced?" a question on the demographics page of each chart, so simply pulling their answers from their charts gave us that response. For the 60% of discordant couples without shared parental genes in the group (remember this means groom's Mom and bride shared, and bride's Dad shared with groom), divorce had happened in more than 75% of the cases, and in the discordant with shared parental/ spouse genes, divorce had happened less than 10%.

In countries like Iceland, genes have been tracked across generations and exist in a national data base, so we should be able to look at gene frequencies there, understanding that there might not be the same diversity of origin of people in Iceland that exist now in San Francisco. Then we could look at marriage survivability and see if the discordant non-parental people had more life discord than the discordant-parental group. Iceland has the data; we just have to obtain permission to access it.

And then we could look for measures of societal unhappiness such as spousal abuse, child abuse, violent crime and all the rest, looking for HLA associations. It might be that discordant marriages without parental sharing of genes underlie much unhappiness and trouble in our society.

Or we could just find this to be nothing more than an interesting parlor game.

However, don't forget that it was our statistical analysis that found people with autoantibodies were linked to a wingspan (measuring outstretched left hand to outstretched right) greater than height. For some time this stood out as the only red flag that something was going on with body shape, immune responses and illness. This was at least five years before the link between DQ-2 and D-8 and celiac disease was published by others. Those DQ genes were invariably linked to either DRB1-7 or -17, and those in turn were linked to either 53 or 52A, respectively. These linkages of 7-2-53 and 17-2-52A were mold-susceptible haplotypes, although the celiac authors didn't talk about that. So does the DQ 2 group show up with autoantibodies more in the long-wingspan group? Sure, and independent of celiac disease! And do the autoantibodies show up before they're sick from exposure to mold, microbes and inflammagens in a WDB? Nope. How about after? Yes, indeed, in fact incredibly common.

So for years I hummed a song about wingspan and misbehaving autoimmunity caused by environmental exposure and fueled solely by genetic susceptibility. Even now, the NIH and NIEHS want to run studies on environmental illness and genetics, hoping to find something juicy, yet I presented these data in 2003 to the American Society for Tropical Medicine and Hygiene, then presented different data along the same lines in 2004 to the American Society for Microbiology. "Very nice," "Interesting," "Thanks for sharing this," they said politely.

All I needed was a mechanism that explained this seemingly curious biological reality. The missing link to the puzzle was TGF beta-1. This compound has been found to have a lot to do with the absence of normal collagen cross-linking in Marfan's patients. It went through the roof in mold illness patients and seemingly went up on its own in the Ehlers-Danlos III people, those long-limbed hyper-flexible people you see on volleyball courts, basketball courts, model runways and contortionist shows. Just about all of those folks are 11-3-52B genotypes (less than 1% of the population). If they get exposed to moldy buildings, look out for their marriage if the spouse is a discordant non-parental share, as high TGF beta-1 sucks the energy and the functional life from its victims!

We can speculate for a while why TGF beta-1 is so much higher in 11-3-52B. Could the answer be linked to absence of production of a defining antibody to an antigen found inside WDB? Or did adequate antibody formation regulate cellular immunity to lower the production of TGF beta-1?

Do I think that couples should swap HLA tests of themselves and their parents? Don't forget, I'm the guy who doesn't like Big Brother (see Chapter 16, VIP). But I'll keep recording data. Maybe the analogous "son of TGF beta-1" will show up to link HLA to societal ills.

And maybe it won't. But the parlor game is fun. Besides, when you're sitting for hours in Terminal A in Seattle waiting for the US Air flight to Philadelphia, you can watch the people going into the trendy restaurant/bar and wonder what their genes are as well as what the likelihood is that they'll still be together 30 years later.

Or you can always go to the barbershop chorus songbook and belt out that wise old ditty:

"I want a girl, just like the girl, that married dear old Dad.

She was a pearl, and the only girl, that Daddy ever had.

A good old-fashioned girl with a heart so true

One who loved nobody else but you

I want a girl, just like the girl, that married dear old Dad."

Once again, I show my age, but the old-timers know what I'm talking about.

CHAPTER 19 After Katrina: Mold, Politics and Government in St. Bernard's Parish, New Orleans, Louisiana

Maybe we could agree that in 2010, just about every public health concern ends up being mixed up with politics and government. Some agency will have an opinion, a committee, a list of regulations and a slate of underpaid scientists who work in a system where any research they do or appearances they make must be approved by a manager who's usually a career government employee. The manager has a boss, who also has superiors, who (at some level of the chain of control), provide the link to the political opinions of and from the executive branch.

If the top rung of control wants a topic to be funded or discussed, the chain of directives goes from the top to the bottom quickly. If the bottom rung, occupied by the people who actually do the work, wants to pursue a topic, the idea must pass a political litmus test first before the idea can be pursued.

Am I saying that the U.S. government agencies provide the public with a highly managed voice about human health issues? Am I saying there's some spin, maybe? You bet. But if the Feds don't have much useful to say, they won't say anything.

Face it: the Feds just don't want to talk about moldy buildings.

What the Feds said about mold disasters, like the damage that followed the hurricanes and floods of Katrina, was pure poppycock. Listen in 2005 to CDC physician, Dr. Stephen Redd, who said, "There is no evidence of toxic mold causing illnesses, except when eaten or touched. They won't produce these toxins at all times but under certain circumstances, like (when) the nutrient supply is getting short or some environmental issue, they may start producing toxins and those can be dangerous if they're eaten or if they're touched. There's up to now not been evidence that airborne mold toxins have produced disease."

The Feds' opinion was quickly supported by two people who make a lot of money testifying against mold illness plaintiffs, Bruce Kelman PhD, writing in the Los Angeles Times and Emil Bardana, MD, who wrote for WebMD. Although memory impairment and immunological complications have been reported by many Americans who've been exposed to mold in prior flooded buildings, a WebMD Health Alert regarding toxic mold syndrome issued to the public in September 2009 quoted allergist Bardana saying, "We know that mold can make people sick if they end up in the foods they eat. But there's little evidence that inhaled environmental mold exposure can cause the serious illnesses that are attributed to it."

Note the repeated use of the idea of illness coming solely from *eating* mycotoxins. This manipulative emphasis on fungal toxins alone might be overlooked when people are talking about "Toxic Mold," or "Black Mold." And then they tell us that molds only make their mycotoxins when they're being stressed, either by competition for living space or by their own reproduction. We know that mycotoxin production occurs when there's adequate fuel available for "secondary metabolite production," an ecological finding routinely seen in WDB but not in compost pits and piles of wets leaves in fall. So the Nay-sayers' strategy back then was to promote false premises about human illness and mold growth alike.

The basis for the ingestion comment comes from a 2002 ACOEM report citing a 1947 paper from Russia about animals eating moldy hay and grain. The basis for litigation was that inhalation couldn't possibly cause illness, so therefore if there *is* illness, it comes from a silly little allergy or eating many kilograms of rotting moldy food every day. Don't forget, this ingestion idea is part of a carefully designed legal ploy cooked up by the defense some time before 2002, and probably as early as 2000. If a judge lets the defense say that ingestion is *the source of* illness, the bad guys can walk away with a win.

Actually, the "ingestion as source of illness" idea is worse than poppycock, it's a plain lie. Ditto for the "only the mycotoxins can make you ill" fable. This has been disproven by eminent researchers from various countries and universities too often to support any further debate.

Yet those confabulated opinions were so close in content, you'd almost think the Feds were working with some of the bad guys. Come to think of it, the Feds did hire the company that Dr. Kelman started, Veritox, to testify against mold plaintiffs in a case from Fort Sill (Mitchell, 2006), and FEMA hired Veritox to testify against the victims of formaldehyde exposure in all those dangerous New Orleans trailers. And the ACOEM paper appeared on its website almost to the day the CDC website talked trash about mold illness in October 2002. That kind of coincidence would make folks around here think those two outfits were in bed together behind the scenes. Wonder what deleted files their computer hard drives still have on them?

Dr. Kelman isn't a physician; so of course, he's never treated a patient. His partner, a former senior manager at the same CDC (See Redd above) is not a physician either.

And don't forget that each of these physician "authorities" Drs. Redd and Bardana had, between them, never treated a patient for a complex syndrome acquired following exposure to the interior environment of water-damaged buildings. That might be partly because they say these people do not exist, or it might be that these non-existent patients don't beat a path to the Redd/Bardana door.

Consider how it is that the U.S. government will authorize more than $200 million in research money (for five years) beginning in 2010 to investigate harmful waterborne algal blooms (HR 3650, establishing National Harmful Algal Bloom and Hypoxia Program, passed the House and goes to the Senate for final sign-off March 10, 2010). This resolution looks at marine biotoxins as the lethal threats to our society that they are. There is no argument as well in 2010 that the interior environments of WDB pose health threats even worse than marine biotoxins, but despite that, there will not be even $10 million in research funding directed toward finding answers to the massively documented problems of indoor biotoxins and inflammagens that affect 25% of our buildings and as much as 25% of our populace. I agree that funding research on biotoxin-associated human health is a terrific idea; please see our group's definitive paper on the chronic inflammatory response syndrome that characterizes and defines chronic ciguatera, one of the world's least commonly diagnosed but quite commonly found chronic biotoxin illnesses. But just so I make this point clear: the government agrees that **biotoxins are harmful in the estuaries, lakes and the sea** (that's true, no doubt), and worthy of millions of dollars of research work, but **biotoxins found in our closets, attics, basements and board rooms aren't** worthy of research interest? How can that be? What's really going on here?

It's not a coincidence that no one has insurance protecting them from damage acquired when swimming or eating fish, but at one time every property owner had insurance protecting users from sickness caused by damaged homes, apartments, offices and factories. Insurance companies, however, don't like to pay out. Thus arises the big business insurance industry boondoggle, complete with paid witnesses. Unfortunately, big business is often bad business for the rest of us.

Defense consultants in mold illness litigation like Dr. Bardana, and perhaps Dr. Redd (who in fairness may have changed his tune since 2005) try to say that biotoxins are a figment of my imagination. But if you look at the world of biotoxins globally, there's a gigantic public health problem, especially in water-damaged buildings. Despite that, the moldy buildings problem is largely ignored by the Feds in the U.S.

As you might expect, senior Federal managers serve important control functions, not just to support the original political connections that *got them their jobs* years ago, but also to maintain the current political structure that helps them *keep their jobs*. As government agencies age, like sedimentary rocks, there's an ever-reproducing multi-layer structure of politically petrified managers who sit on committees (yet they are called standing committees) and keep their managers (who also sit on committees) in line, who in turn, then keep their scientists in line. Many of these scientists actually work and care and are concerned. In science I trust; in managers I have no trust. Nor should you.

If agency employees who sit on standing committees didn't tell the truth, would they be members of a lying committee?

If there's a public health concern that affects the public impression of the President, for example, someone will be directed to do something about that concern, understanding that public concerns can be easily manipulated into *public fear* of unhealthy water, air, spinach, mice, ticks, Tylenol, Toyotas, jalapenos, and peanuts, with the results filtered upstream before their public release. The U.S. public health system can't alarm the public, you know. Well, maybe they do; see my list of "unhealthys" above, where true public concern was magnified by media attention. Those people who are fearful might vote for the other side if they were worried about how the party in power handled the crisis. The elected officials might lose their jobs, but not the managers. Of course, the elected officials will often go to work for private industry and make some real money (on top of their lifetime pension and insurance benefits).

Still when we had 1000 cases of Salmonella from a few egg producers, the nation was in an uproar, complete with experts on the Morning Shows telling us to make our cooked eggs bouncy and not runny. Put the same experts on the Morning Shows to say that 10 million American kids are being harmed intellectually every day by the moldy schools the public paid for and I think you'll see some national outcry. Instead of spending billions on special programs for math and science enrichment, just fix the leaking roof and shoddy HVAC systems to start. Imagine President Obama congratulating some fictitious CDC manager charged with suppressing the truth about mold in years past by saying, "Heckuva job, Brownie!" splashed across MSNBC breaking news headlines when the truth about mold illness is opened up to sunshine.

It used to be that science, like art, was independent of the influence of politics and all the things (and people) that influence politicians. The idea of the starving artist comes from lack of patronage and of financial support. Scientists used to have to scramble to find patrons for their work when funding was tied to a private entity. Sure, there was influence exerted on the scientists by their patrons (think DuPont and Big Pharma), but government regulatory influence wasn't the original prime mover in research then like it is today.

Perhaps the change in funding for science in the national interest can be traced to the development of the National Institutes of Health (NIH). Government had an interest in health from the early 1800's, but it was only after the Civil War that the government interest went beyond Marine Hospitals to looking at infectious diseases such as cholera and yellow fever. The real boom in the government providing money for career scientists began in 1946. A newly built complex of office buildings in Bethesda, Maryland, included a Research Grants Office that extended the role of government-funded research beyond rickettsial diseases, diseases of sanitation, the illnesses of sailors, the National Cancer Institute and more. By 1950, the NIH had its current name and more divisions were rapidly developed (allergy, neurologic diseases and many others). Government control of funding and therefore *control of the direction* of research in these fields has grown exponentionally since.

And that may not be an altogether bad thing, unless you happen to be an outsider. Forget funding the unknown scientists not associated with a research institution. I often wonder if it were true that the fraternity of researchers with life-time grants from NIH coined the phrase "Not Invented Here," to show what some wags think of NIH. Perhaps "coin" is the active word in this concept.

As a college student and then as a medical student, I always wondered why Duke University kept building so many research buildings among the hills where the soul-satisfying pine trees, Euonymus and dogwoods once grew. The answer was money. What a surprise. By having research facilities, the university could recruit excellent researchers, who brought in grant money, of which about 50% went to the university. The presence of the excellent researchers enhanced the university's reputation, which then helped the university attract more excellent researchers who not only needed lab space but had the NIH grant money to pay for the space. The more labs they built, the more money the University made. Just imagine having something like lab space, and selling its use, and still having it to sell, to be used again and again. Once a critical mass of users was recruited and retained, the buildings helped fuel the boom of Duke and countless other institutions.

The presence of the critical mass of researchers creates an academic sanctuary fomenting open discussion and new research, right? Isn't that the noble idea? Except for NIH. Not invented here.

Such noblesse is inoperative when the problem is human illness acquired following exposure to the interior environment of water-damaged buildings. Government is more involved with us and our money than ever before, so much so that it now seems any science performed without funding from a governmental agency is melting away from public view faster than ice along the Northwest Passage. Lest anyone forget: government funding essentially goes to those who pass political correctness tests administered by grant managers, personnel managers, and managers of managers. The elected representatives are at the top of the management pinnacle and they all need election funding. Therein lays the weakness of the approach.

I wonder if you're thinking as I am, that something isn't right with our government's response to mold; maybe we could agree that the best use of a politician isn't what we have them doing now. Maybe we could agree that every layer of government managers should have an expiration date (actually, it does: it's called the pension date). The ultimate bosses have to be protected from public criticism, so the process of problem solving must provide "deniability." And maybe we could agree that in government, a certain percentage of employees really do little for their pay.

When I see the same government people involved with mold issues now as I did in 2002, and those people are still in the same government positions, saying the same things now as they did then, I've got to wonder what they're doing for their wage and benefit package. In that group are more than a few who knew little about the subject then and still know little today. How could they know, though? They don't treat patients; they have no responsibility to the spouses and family members of those sickened people they might see testifying in front of some committee (that's the only place the Feds see real patients and then it is across the room). They have no passion for finding the answers to primary and specialty care of people sickened by exposure to the interior environment of water-damaged buildings. Think about it, what in God's name gives *them the right* to speak at the same table as those who actually do the medical work? Nothing, except they're the Feds.

Still, these are our regulatory authorities and heaven help the poor miserable scientist who suggests to the five-foot tall, 220-pound Federal official that he might have a weight problem. In the darkness of no "Sunshine of Truth" in Federal decision making, his grant opportunity just disappeared. Or if the aggrieved and insulted official (maybe the scientist didn't call him fat, just arrogant) is a super-manager, and the underling is actually trying to do the right thing, his next assignment might be to provide a personal, first-hand assessment of the effect of wind chill on bare fingers and toes on top of Mount Washington in February.

Maybe someone thinks I exaggerate?

Don't forget that when the administrator of GIHN (then called Action Committee on Health Effects on Mold, Microbes and Indoor Contaminants) discussed the Mold Experts report (released 7/27/10 by Policyholders of America) with an official of EPA in August 2010, the suggestion was made by EPA to explore holding a conference to discuss the report in fall/winter 2011.

It's true my views on government are pretty cynical, but I don't think I'm alone. Of the three common lies we're warned about, isn't one, "I'm from the government and I'm here to help you?" Surely someone else thinks that "gubmint" makes no sense these days. I heard on NBS news March 16, 2010 that a Wall Street Journal poll claimed that **77%** of Americans want to toss out the incumbents from this Congress. I'd say the gubmint has some image problems to repair. I wonder what would happen to government responsiveness to the needs of the populace if the agencies lost 77% of their life-long staffers as well.

By 2006, as far as mold illness goes, the U.S. government had done precious little. The two agency–sponsored reports (CDC/Institute of Medicine, 2004, IOM) and EPA/UConn (2004) were in one case ill-informed and in the other stale when released. Interestingly I see the IOM misquoted often by the defense in litigation, but never the EPA/UConn. That report was the first to include a roster of symptoms that physicians actually see in mold illness patients, but it was stuck far in the back as Appendix D. And, no surprise, the EPA/UConn report also had nothing on the pathophysiology of the illness or on treatment, since none of the authors had *ever* treated *any* mold illness patients, forget successfully or not; and none had documented before and after lab parameters. Asking non-treating physicians to describe what treating physicians actually see is akin to asking the little boy standing outside the fence around the baseball park what's happening on the diamond. He jumps up and down, trying to see over the fence because he really wants to see the game, but in the end, all he can do is guess at what happened. For a physician to have any credibility in this field, you've got to treat the illness and prove that you're competent.

The arguments about health effects caused by exposure to the interior environment of water-damaged buildings were brought to the U.S. Senate Health Education Labor and Pension Committee (HELP) in January 2006, largely through the tireless efforts of Sharon Kramer. She'd provided Senator Ted Kennedy's office with an overwhelming amount of data to show that the current U.S. government approach to mold illness was not only shortsighted and biased, it was plain wrong. Senator Kennedy of HELP and Senator Jeffords of the Senate Public Works Committee called for a legislative staff briefing, with invitations provided to all Senate members. The meeting was held in the Dirksen Building in January 2006. Thank goodness that it wasn't held in the Rayburn Building; (see Chapter 21, Tourists' Guide to Mold DC).

Panelists were Vincent Marinkovich, MD; Chin Yang, PhD; David Sherris, MD; and Ritchie Shoemaker, MD, with Mrs. Kramer organizing and moderating the briefing. The EPA, CDC and HHS were supposed to send speakers as well so that an informed dialogue could take place for the benefit of the Senate legislative staffers, and therefore the U.S. citizens. The agencies cancelled their appearance at the last minute. I can only imagine how some of the staffers attending must have felt as they were bombarded with words like Type III hypersensitivity, interleukin 13, eosinophils and innate immune responses. That's why there was a question-and-answer session, but it was getting close to 4:30 and the meeting broke up without much further discussion.

Understanding that (a) most elected officials aren't comfortable with potential threats to vested financial interests (in the case of water-damaged buildings, those interests involve building ownership and the property and liability insurance industries); and (b) discussion of human health effects due to exposure to water-damaged buildings exposes such threats to those interests, it was curious that such a conference could be held at all. No videos or minutes of the meeting were permitted to be taken so the Senate staffers could feel comfortable to ask questions. I expected that there would be some sort of maneuver surrounding this scientific and political event, so it was no surprise that government agencies, including the EPA, pulled their representatives at the last minute, though no explanation was given.

However, I'm told that super-managers were in attendance. A few Senators showed up; one staffer from Senator Jeffords' (an Independent from Vermont) office came in late and asked me for materials about the pathophysiology of mold illness. I gave her a color copy of the Biotoxin Pathway, an effort that distilled into one diagram information derived from thousands of hours of research. She asked if there was anything more. Yes, there is, much more.

The upshot of my talk on the reality of human illness from exposure to the interior environment of water-damaged buildings (available as a free download on **www.biotoxin.info**) was that several Senate staffers, especially Senator Kennedy's, wanted information about illness that could be identified in areas of New Orleans, which had been hard hit with catastrophic damages after flooding from Hurricanes Rita and Katrina just four months before. Specifically, they wanted to know if human illness caused by exposure to water-damaged buildings actually existed. And if so, was it being covered up?

That area of inquiry subsequently led to a request from Senator Kennedy's office in October 2006 to the General Accountability Office for a review of the Federal effort. Again, Sharon Kramer's incredible effort was instrumental in the GAO request that led in turn to the 2008 US GAO report that completely destroyed the defense or government Nay-sayers' credibility in mold illness issues. Thanks to Sharon and Senator Kennedy's staff, the longstanding idiotic arguments about mycotoxins alone being the problem from WDB have now been put to rest, with the exception of some really primitive defense attorneys who don't know that the old ACOEM-quoting defense and the old AAAAI-quoting defense are a prescription for a loss in court.

As for New Orleans and Katrina, I said "Yes, the illness actually exists wherever water-damaged buildings exist and illness will be found in rough proportion to the genetic susceptibility of those exposed. The answers are easy enough to obtain. All we need to do is to go there, interview some patients like we always do in our screening clinics, do visual contrast testing, and then draw blood and run the labs. Treat the sick people—that won't be a small number—and show that they get better on medication and don't get sick off medication *as long as they stay away from their source of exposure, often sadly their homes*. Since they won't get sick from the 'ubiquitous fungi' of the world, then let them then go back to their homes, basically into harm's way without medication, provided they give informed consent. They'll be right back into their horrible illness within three days. Guaranteed. Or, keep them on medication and use a whole lot of common sense regarding what to do with exposure to the incredibly toxic (microbial) waste dumps that are now their homes. But whoever goes to do this screening and testing would have to have a Louisiana medical license or some kind of waiver to do the treatment."

I really didn't expect much to happen from the meeting. I'd basically said that the government was intentionally making people ill by not acknowledging:

1) the obvious microbial problems in WDB that can cause human illness;

2) that the illness was recognizable beyond any reasonable doubt; and

3) that the illness, once recognized, was treatable.

I ended by saying that the current "management" by FEMA, CDC - and whoever else was standing in the way of effective disaster relief - was cursing a sizable percentage of the next generation of New Orleans area residents to grow up moldy. (Maybe that's why the CDC managers like me so much.) Can you imagine, I thought, Tuskegee II, when instead of *not treating syphilis* in black men (Tuskegee I), now the Feds were *not authorizing treatment* of mold illness in a residual population that was largely poor, black, Cajun, Latino and/or under-employed? It was nice to see some attention from Senator Kennedy to a huge disaster, since it was clear that the nonsense we heard coming from the CDC was useless. The CDC had clout; I didn't. However, I had human health effects data on thousands of patients: the CDC didn't.

Isn't the CDC supposed to protect the poor, black, Cajun, Latino and under-employed as part of what they do for all America?

But in *The Morbidity and Mortality Weekly Report* dated January 2006, in a piece titled, "Health Concerns Associated with Mold in Water-Damaged Homes after Hurricanes Katrina and Rita - New Orleans Area, Louisiana, October 2005," the CDC just said that people working in the areas were advised to use respirators to avoid developing cough and respiratory problems. That's it. The name of the piece was "Health Concerns." Who's the CDC trying to kid? And how about hypocrisy, because what does a respirator do to protect health if, as Dr. Redd and the CDC had already said, ingesting moldy foodstuffs was the only possible route to illness? Did they mean that wearing a respirator would make it hard to eat 30 pounds of moldy drywall a day.

The CDC looked at 112 homes and found a massive increase in molds, endotoxins and beta glucans, acknowledging that these compounds cause human health effects. The CDC team looked at 159 patients between October 18 and 23rd, 2005 in St. Bernard's Parish. No symptoms were reported as being recorded by the team of CDC investigators, just data on the use of respirators.

[*Editor's note:* Dr. Shoemaker saw 212 patients in St. Bernard Parish in 48 'office' hours in February 2006 and each was provided with thorough symptom recording and VCS testing See Appendix 7, St. Bernard's Parish.]

Given the CDC's concerns and the massive outpouring of money for Katrina victims funneled into charitable organizations, don't you think the CDC staff should've been directing others to distribute respirators to those who didn't have them? The CDC was unequivocal: using respirators was required. Yet the CDC didn't make respirators available except to themselves.

And the Red Cross wasn't a big help to those cleaning up the soggy mess of what was once a place to live. They gave the survivors a broom, a beach pail and told them to sweep up their homes and place the toxin-laden debris beside the few FEMA trailers that were being hooked up to existing sewer systems. Back then, FEMA knew that formaldehyde or "something" was wrong with the trailers, because recipients were told to keep the trailer doors open when they weren't sleeping or using the bathroom. So the "lucky people" — those with a trailer hooked up to the sewer system - were ventilating their formaldehyde polluted sleeping quarters with massive amounts of bioaerosols wafting out of the pile of moldy drywall, insulation and building materials placed just outside.

By the way, do you think some of the debris became airborne from first sweeping to make the pile and then filling the pail? The Red Cross advice was beyond the pale. Personally, I don't think we can say the Red Cross *meant to do harm* here. They were just ignorant and for that we can thank the insurance industry, its lobbyists, its paid "experts," lawyers "just doing their job" and the CDC and NIOSH. Just for sake of illustration regarding the health hazards created by this kind of advice, imagine a situation - a school project, let's say - where some 4th graders are helping clean up old houses:

"Children, first scrape all that lead-based paint off the windows, you can brush it with your hands if you have to, and drop it into a pile on the floor. Johnny will come along and he'll help you sweep the lead dust and chips into this little bucket. If it makes a puff of dust when you dump it into the bucket, it's OK because the only way to get sick from the lead is if you have an allergy to it or if you eat it. So don't lick your fingers. I know that is true because that's what it says here in this statement from a witness in a lead poisoning case. Yes, he was paid by the defense. But never mind, children. Don't take too long; the dust might make you cough. Now let's get started. We're going to do 30 hours of volunteer work this week and then 25 more hours next week. Yes, the health department says what we're doing is just fine. The newspaper will be by later to do a story and to take pictures of all you have done. You are all making a difference!"

Can you see why the defense attorneys thought they were in mold heaven with the Katrina advice being a nationally accepted policy? Here were the Federal authorities buying the "ingestion is illness" mantra and only a little bit of cough might happen if you were in a moldy building.

Look at the politically disenfranchised who were affected. Who were they to argue with NIOSH and CDC? What voice did the people reaching out for help from the Mississippi River bridges for three days actually have? How about the people lying next to the broken lavatories in the Superdome? If anyone acted after listening to their cries for help, the people would have been off the bridge or out of Sewage Central. I think you and I will agree that advice like this from our trusted health agencies ends up being harmful.

Yet that's exactly what people everywhere were being told to do by the Feds before February 2006. And it's also why some people think *still* think that a little bit of mold might make you cough, but that's it.

Sharon Kramer called a few days after the US Senate conference. "Kennedy's office has been talking with FEMA and Homeland Security. You're cleared to go to New Orleans to do a clinic."

I had to pay my own way, of course, and find the patients that I said must be there. Sure, I'll just waltz down Bourbon Street, at a time close to Mardi Gras, and do VCS tests on whoever isn't staggering.

Matthew Hudson had a different idea. Soon after the disaster struck New Orleans, there were three Carnival cruise ships that offered to house the newly homeless at cruise ship prices. FEMA filled three boats right away. There were plenty of homeless. Hudson's ship, the Scotia Prince, was on its way to Cancun, so he offered it to the Navy who he had done business with before. Housing, recreation, three smorgasbord meals daily, unlimited beverages (no booze), showers and safety from robbers and molds: not a bad deal for $50 per person per night. About 1/3 of cruise ship prices. Right away the Scotia Prince was docked on the Mississippi at Violet, Louisiana, housing nearly 1000 people. They were just around the bend downstream from the infamous Ninth Ward.

I'd be permitted to stay on the Scotia Prince and do my testing there, provided Homeland Security and FEMA agreed. They did. The US Public Health Service had three physicians on the boat; they arranged to set up a clinic for me at their local facility so I could teach them about mold illness. St. Bernard Parish agreed with the plan as well. Led by Charlie Reppel, associate of Henry (Junior) Rodriguez, President of the Parish, local officials agreed we would make a report to the Parish to publish on their website.

Everything was ready to go. Still, would you be a little bit suspicious if you were in my shoes? Basically, if I found illness in the homeless ship residents, I was mounting an informed attack on entrenched U.S., and state (local?) government opinion. And that means a whole series of opinion managers might want me to disappear into the Mississippi mud.

Jumping ahead for a minute, if our team could show that *there was* an illness that was easily identifiable among the homeless people on the ship, an illness that was readily distinguishable from a control group of M/S Scotia Prince officers and crew from foreign lands who'd never had any exposure to the buildings in the Parish, then we could plan a screening and treatment program for the thousands of sickened people still at risk in the Parish nearby. I don't think the Parish was expecting to find employees from the Courthouse and the firemen to be so terribly sickened. The firemen were the most poignant group. These muscular young fathers, fiancées, and men in the bloom of their lives were largely French fried. Yet despite their increasing disability, these young studs continued to put themselves in harm's way every day. And the women were at risk for spontaneous first trimester loss of pregnancies.

The flight to New Orleans was uneventful and there was very little traffic on the way from the airport to the ship. I'd never seen such an incredible sight in my life. I was scheduled to meet Dr. Richard Lipsey, a world-renowned toxicologist also invited by Matthew Hudson, coming in from Jacksonville, Florida. He'd do environmental sampling in homes where people had been before they found safe haven on the Scotia Prince. I'd do their work-ups, so that we could correlate exposure with illness, or not. Except there were only two sites where there was medical care available (*five months* after the hurricanes hit), so we couldn't get any labs. There were only three Public Health physicians within 50 miles, and they were on the M/S Scotia Prince, too. Compare the international response to the needs of earthquake victims in Haiti, to the U.S. response to health care needs in New Orleans. Staggering.

The land was a wasteland. I always look for local birds and local vegetation when I travel. *Gone*. This area is pure wetland/delta habitat, so it should be teeming with wildlife. *Nope*. I didn't see any egrets, well, maybe one or two in transit high overhead. Just a couple of pelicans. No birds in the few remaining trees. No sounds of amphibians at night, just stillness and quiet. Mold makes no sound as it digests everything in its path. No children; no rock and roll from the high school kids' cars booming in the spring afternoon. I didn't see many cars with four wheels upright on the road either, though there were plenty of cars and boats and parts of houses turned up one side on another. No homes were spared in the wake of the flood.

Everyone who first saw the devastation that Katrina and the failed levees had caused was awestruck at how bad it was. No film of tsunamis in Indonesia or earthquakes in Iran, Pakistan or Chile or volcanoes in Krakatoa or Pompeii can match what I saw.

The salt water line was nine feet up on some homes, others fifteen feet. Fifteen feet of storm surge and this, in an area that had endured prior hurricanes for years. Trees were laid down to sleep; only their blanket now was an amorphous green and pale yellow and black. A few daffodils tried to stunt their way through the salty soil to bloom, but the azaleas were gone. Even the swamp maples, incredibly hardy and resilient wetland species, were sparse in their spring bloom. Was this town a modern Carthage, as after the Romans won the second Punic Wars they salted the fields of the losing side as a final insult to future survival? Not much grows well in salty soils.

There were few people. No one was serving the boat that was wedged going the wrong way in the hamburger drive-through. A few trailers were selling Po' Boys they brought in from somewhere, but forget buying any hot food. I didn't hear Al Hirt, Harry Coniff Jr. or the Preservation Hall Jazz Band, just the soft sound now and then of a gentle breeze, often lifting some plastic bag aloft as a bizarre kite that seemed to be saying, "we aren't dead yet." I was relieved that there wasn't a pile of human corpses rotting in the sun, just a few bloated dogs. I decided not to eat one of the Po' Boys. There were a few brave souls who came back to try to start over in a now-poisoned landscape. What could they be thinking?

As I write these paragraphs, I try to imagine the sci-fi movies where the nuclear disaster wipes out everyone and everything except for a few. I was seeing one right there in Violet, Louisiana. And if there were mold illnesses here, and if the illnesses were misdiagnosed and blown off as "Katrina cough," the human health scene was going to get worse. My treatment protocols, even back then, could largely prevent the human suffering that was otherwise *guaranteed to occur.*

As it turns out, the Scotia Prince was the only safe haven left for the homeless. I'm told that the Carnival ships had taken on so much personal moldy property that those boats were now contaminated. So those who went to the Carnival liners went from the frying pan into the fire.

Eddie Elovsson, the Swedish captain of the Scotia Prince, greeted me warmly when I arrived at noon. "We're ready for you. How many people can you see in the time you'll be on board? We have a thousand who need you right now; there are people lined up already. Please get your identification photo and I'll show you to your berth so you can get started."

We started at one p.m. on Friday. Fortunately, this was a typical Hudson plan and all the advance preparations had been made for an efficient clinic. People had already filled out our intake forms and were waiting in an hour-by-hour queue. FEMA stopped by to make sure all was well, as did Homeland Security. We couldn't draw blood, which was their main concern it appeared, and we couldn't prescribe medication, as there were no pharmacies. By three p.m., when the staff of the *New Orleans Times-Picayune* and the camera crew from ABC News, Channel 6 TV, WDSU, showed up, the results from just the first 60 patients were astounding. Compared to the crewmembers, who were never onshore, the homeless people living on the ship were an incredibly sick group.

We saw a variety of patients; there were some obvious sub-groups.

There were firemen, led by Fire Chief Stone and Union President Eddie Appel; Parish employees from the Courthouse; a few health care workers who volunteered to come to the flood zone; homeless adults and homeless children (see report in Appendix 7). As it turned out, 54.5% of all the 189 displaced people living temporarily on the Scotia Prince met peer reviewed, published criteria for illness. Only one of the 23 crewmen was ill, with his problems stemming not from mold, but from untreated ciguatera acquired after eating barracuda in the Caribbean. The findings in the people *who didn't meet criteria* for illness were indistinguishable from those found in an established roster of 239 control patients.

So the Parish people's illness was no different from what I saw from 27 foreign countries and every state in the US. True, I was missing patients from a few provinces of Canada (no one from the Northwest Territories and Prince Edward Island), but this illness knew no racial, cultural or economic class bounds. It didn't matter if you owned Gretna, Louisiana real estate or just worked on the grounds of local estates or worked in some kitchen making Po' Boys (heaven forbid, hotdogs) or if you were too ill to work at all. Mold illness destroys everyone's innate immunity and thus, human health.

Screening for mold illness is fast, accurate and reproducibly reliable. As we've seen repeatedly, when our symptoms and VCS testing is backed up by blood testing, we find that the screens are incredibly accurate. My data completely destroyed the false façade constructed by government agencies and whatever mold defense company was in cahoots with them. So if they can't stop me with data, not to mention truth, what will they try to do? They always try something.

The ABC report at 6 PM brought a swift government response. We were given *five minutes* on the evening news, an unheard of amount of time. The newscaster reported the 24% figure for the percent of people exposed who were likely to have been sickened by the water-damaged buildings. By 7 p.m., FEMA was on the phone to Captain Eddie and Charlie Reppel. All I could hear was one side of the phone call. Eddie is talking, his Swedish voice becoming more animated and his fair-skinned face becoming redder by the minute:

"No, we can't do that. You've already agreed. We don't go back on our word on this boat. You'll have to talk that over with Homeland Security. No, this isn't Federal property, this boat is owned privately. You're simply renting space. I can't stop you from pulling the physicians off the boat, but won't you even wait until they have had their supper?"

Eddie turned to me and said, "FEMA wants you to go home tomorrow. If you won't go on your own before the next clinic starts, they'll arrest you for trespassing on Federal property. Homeland Security is protecting your presence here. You can go, but if you wish to stay on board, you're welcome on this ship. FEMA can control the three physicians and they're being removed from the boat. I honestly don't think they can control you. I'm curious; what do you do that makes FEMA so angry? Do you somehow poison doctors' minds when you give them information that corrects this illness?"

Eddie is quite the unflappable guy. Let not any turmoil disrupt his dinner. "I'm sorry for the interruption. Now we're preparing for the waiter to bring us dessert. May I recommend the cheesecake and blueberries?"

For the next few minutes, I had all sorts of movie scenes flash behind my eyes. The themes were basically the same even if I didn't know the names of the movies or the actors. Here were the forces of good, finding illness where the Feds said none existed. How many times have we seen dramatic stories of courage in the face of oppression, even if what we were doing wasn't the same as say, smuggling the needed documents past the roadblocks, sentries and U-Boats to safe hands to help the Allies win World War II?

Yikes, how imaginations do run wild. This is America. No one tries to keep the truth from the people. Especially about a common, life-sapping, but treatable illness.

Or do they?

Maybe the blueberries had some sort of Swedish after-dinner cognac in them. I'm still thinking about the FEMA boys wanting me gone. The reality of the situation was simple: we needed to protect the data I had collected. There was nothing to stop FEMA's black SUVs (did they really have helicopters and storm troopers?) from coming aboard in the morning and seizing all my data (and me). We needed to make copies and together with the maps of prior residences of patients that pinpointed where the flood waters and the flood of illness met, keep them secret and rush the copies to the Post Office in the very early morning.

Really, who could I trust? I've traveled to the Heart of Darkness to destroy the idea that mold is just a nice warm fuzzy growth that rots a strawberry but couldn't hurt anyone. The CDC had already said so, just ask Dr. Redd. What can one use to successfully argue with the CDC? *Data.* So you understand my need to protect my data.

In data, I trust. In science, I trust. In FEMA, forget trust. Heckuva job, all right.

Eddie agreed. He'd seen the forces of evil all over the globe for too many years. This threat from FEMA wasn't the same as being captured and shot in Libya, but who wanted to take a chance? The poor copy machine in the purser's office on the boat wasn't made for high-speed work, but when dawn came, the copies were in a plastic bag inside a box covered with the ship's debris going to the landfill, with just a short, unscheduled stop at the Post Office. Saturday's and Sunday's clinic data were duplicated and saved from suppression as well.

Meanwhile, the second arm of our approach to identifying a serious health threat in St. Bernard's Parish was the work of Richard Lipsey, PhD. FEMA said he could stay. They clearly didn't care if someone knew about the mold, but the effects of the mold? Yes, that was a concern. Dr. Lipsey's report is further proof of the massive government cover-up of the intensity and severity of illness from Katrina and across the U.S.

REPORT ON THE INSPECTION AND SAMPLING – ST. BERNARD PARISH

Dr. Richard L. Lipsey, Toxicologist

February 24, 2006

I inspected the ship, the Scotia Prince, docked in St Bernard Parish at the request of Parish officials, and talked with many of the residents on board the ship Feb 11, 12. I saw the preliminary data from Dr. Ritchie Shoemaker's study regarding the health of ship-board workers. The Scotia Prince workers had few if any symptoms and all of the St. Bernard Parish residents who were living on the Scotia Prince said they felt better on the ship than when they were walking in the neighborhoods in St. Bernard. They were breathing filtered, safe air on board the Scotia Prince and getting showers and three meals a day.

The results of the sampling I did of six homes are attached. It shows that the homes I inspected had the **highest levels** of pathogenic molds I have ever seen in **my 35 years** of investigating sick buildings. Every sample showed pathogenic molds in the millions per square inch of wall space or millions per gram of insulation. These levels of these pathogenic molds are sufficient to make anyone sick, whether they have compromised immune systems or not.

I was shocked to find out that the Habitat for Humanity volunteers that I toured with, who were trained by FEMA, had been told that mold cannot hurt you and you do not need any protective equipment. This was totally wrong and I gave them the phone number of the lab that could overnight the proper safety equipment to them c/o the Scotia Prince, since none of them had mailing addresses. Highly pathogenic endotoxins from gram-negative bacteria were found earlier by NIOSH scientists and personnel from the Louisiana Dept. of Health; the levels were 20 times above normal, on average.

The levels were not only high inside the flooded homes but also in the ambient air in the neighborhood. Gram-negative bacteria all produce endotoxins, similar to mycotoxins used in germ warfare at very high levels to kill people.

I inspected almost every area and neighborhood in St. Bernard Parish on Friday and Saturday and took samples. I also took pictures. Most of the residents doing remediation of their homes were not wearing protective equipment and were dragging contaminated debris to the curbside, especially if they had a FEMA trailer. I inspected many homes. The stench of rotting materials was in every home and there was significant water damage with high levels of pathogenic molds and bacteria in every home, since the homes had been under water for days. The homes have been growing mold for months. There were snakes living in some of the homes and marsh grass in most of the homes. Many had marsh grass on top of the roof, indicating how deep the water was in those areas of St. Bernard Parish. None of the homes were safe to occupy or even be inside for any length of time without personal protective equipment, including a HEPA respirator, rubber gloves, goggles and a Tyvek suit.

The levels of mold spores, not counting fragments of spores, were all in the millions of spores per square inch on walls and furniture and millions of *Stachybotrys* spores per gram in insulation. There was *Stachybotrys* mold in every sample. *Stachybotrys* is the most toxic of all toxic molds and can produce mycotoxins. It's sometimes called the black mold. It's 10 times more toxic than the most pathogenic of the *Penicillium* or *Aspergillus* molds that also can produce mycotoxins. *Stachybotrys* produces trichothecene toxins; highly purified forms were developed by the U.S. Army as weapons and never used and have since probably been destroyed.

None of the Scotia Prince residents should return to these neighborhoods without proper protective equipment,

> including HEPA respirators and Tyvek suits with rubber gloves, much less return to their destroyed homes and try to live in them. Most of the homes I saw in St. Bernard Parish should be bulldozed and burned because they cannot be salvaged.

I rarely get to talk with Dr. Lipsey any more as he's considering downsizing his practice and I understand that he *still* hasn't traveled to every country in the world. Should he continue to choose trips to the Falkland Islands over the sickened homes of Faulkner's South, I will miss him. He tells the truth which is why he is Matthew Hudson's toxicologist, and there are precious few like him.

Word of FEMA's abuse towards me reached Congress. Mary Mulvey Jacobson wrote of my experiences to Congressman Barney Frank (D, Mass). Frank's been noted to have a liberal spin to his politics; his words from Feb. 23, 2006 are cogent:

"Thanks for sharing with me the information from Dr. Shoemaker. The more I hear about this administration's (Bush) approach to Katrina, the worse it is revealed to be. The work that Dr. Shoemaker is doing does not fall directly within the jurisdiction of the committee on which I serve, which deals with housing and flood insurances, but rather has to do with health matters, which is why he was appropriately in touch with the Senate HELP Committee. *But this transcends jurisdictional lines and I will be raising with FEMA the question of why he was treated in this manner - or rather I will register my objection to it, since we know why he was treated this way: namely that they don't want to be embarrassed by people who (un)cover their failings (*italics added for emphasis*).*"

Margaret Maizel went to work on the graphics for the report to St. Bernard's Parish. She completed the data collation within two days, and the Parish website carried the report from my office and Dr. Lipsey on Feb. 15, 2006.

The upshot of all of the work done by so many people was that <u>both reports were removed from the St. Bernard Parish website the next</u>

day. Matthew and the people in the Parish received their "thank you's" as well - FEMA ordered the Scotia Prince to vacate in 14 days on March 1st, despite impassioned pleas from Henry Rodriguez and Charles Reppel, thereby returning the homeless boarders to an unknown fate. I changed my mantra about FEMA from "in FEMA, distrust" to "in FEMA, disgust."

At first angry, and then frustrated that the Feds and FEMA could treat human life and dignity as a political commodity to be twisted any way they liked, I had to wonder about whether our Public Health authorities were honest regarding such an obvious human health crisis. I would accept truth, compassion and honesty as such evidence. I thought that both true reporting of human health effects and proper documentation of microbial contamination were basic parameters that needed to be presented. NIOSH had been in some of the same places I had been; they'd seen (before me) some of the same people I saw. Maybe they'd show the decency required in academic reporting. *Nope.* Surely NIOSH would report on the adverse health effects and some of the terrible contamination that Dr. Lipsey had confirmed was present. *Nope.* Surely NIOSH would confirm that the "S" in their name meant safety. *Nope.*

After NIOSH visited the site October 5, no reports were written until March 2007, when the CDC published a remarkable paper in the microbiology journal *Acquired and Environmental Medicine* (AEM 2007; 73(5): 1630-1634), entitled, "Characterization of airborne molds, endotoxins and glucans in homes in New Orleans after Hurricanes Katrina and Rita." It was the first indication that the CDC managers would agree to publish what the scientists had known for years. Written by lead author Carol Rao, PhD[3], with co-authors Ginger Chew, beta glucan expert, and Clive Brown, director of the CDC Indoor Air group and several others, this paper signaled a complete about-face in the CDC's opinion about human health and exposure to biologically active fragments of materials found inside

[3] a researcher whose earlier work in which she instilled washed fungal spores into rat's trachea and showed injury, had been bastardized beyond belief by the 2002 ACOEM consensus statement, which was written by two principals in Veritox and a physician who testifies for defense in mold litigation

water-damaged buildings. Suddenly, mycotoxins, beta glucans and endotoxins that the CDC once said couldn't hurt anyone were found to be associated with human health effects! The references for establishing health problems were all published much earlier and were readily available before Katrina hit.

What had happened so that old data that showed the CDC and Dr. Redd (and Dr. Bardana and Dr. Kelman and their fellow cadre members) were dead wrong in 2006, were now pillars of insight in 2007?

One might wonder which super manager told the CDC to change its approach. This paper *expressed gratitude* from the authors to agencies that had helped them with adequate protection from illness during their investigation *so they didn't get sick.* If the researchers were given such protection, why not also give protection to the homeless who were dumped off the Scotia Prince on the wharf at Violet? What about these poor wharf rats breathing spores into their trachea every day?

Finally, NIOSH published their findings in June 2007 (HETA #2005-0369-3034; Hurricane Katrina Response) with Chandran Achutan, PhD listed as the lead author listed as responsible for all content.

The NIOSH Report link is:

http://www.cdc.gov/niosh/hhe/reports/pdfs/2005-0369-3034.pdf

I contacted Dr. Achutan by phone to ask him about some bizarre comments in the executive summary. Unfortunately, I didn't ask him who actually wrote them.

"Except for a limited number of noise exposure samples above the NIOSH recommended exposure limit and carbon monoxide levels above the NIOSH ceiling limit, environmental sampling for a variety of substances including asbestos, metals and dust did not reveal anything above recognized occupational exposure limits. A summary of the findings was shared with workers and employers. Safety hazards such as broken glass posed a risk to workers. Worksites in the flood-ravaged areas had varying need for the readily accessible, pertinent, understandable information regarding workplace hazards and exposures. This was apparent throughout the response, and distribution of information proved challenging."

I didn't see mold or bacteria or glucans listed anywhere. Wasn't this a health hazard report? What did they do about testing for mold?

Buried on page four was a short paragraph on what the Occupational Health group (Team 1) had done about the potential health hazards from exposure to water-damaged buildings. I asked Dr. Achutan if he felt the activities listed on page four comprised a health evaluation. It says on page four: "they provided information and recommendations to various groups. They participated in discussions on how to safely clean homes; provided technical assistance on mold cleanup; provided recommendations for cleaning HVAC. They also held a press conference with the Mayor of New Orleans."

That's all it says. "Did they actually look for health effects?" I asked him. "Couldn't they have actually done a health survey—maybe take a history from a cross-section of 200 people? I mean Team 1 was there longer than I was and there were eight of them. Couldn't they interview 200 people? And what about the Courthouse, since I had health effects data on 31 Parish Courthouse employees?"

No, that wasn't his part of the project. Dr. Achutan graciously said he would have someone call me, which they did one week later. On the speakerphone was Kenneth Meade (his credentials are listed on the report as MS, PE). He wouldn't identify who else was listening, but I could hear papers rustling.

"No, we didn't do a health survey. We didn't do any testing for molds, since the buildings were massively contaminated with mold. We performed a 60-minute walk-through in the Courthouse."

That was all. NIOSH submitted Appendix F for the report (page 22). Sure enough, the inspectors told people about personal protection and remediation.

I asked why the executive summary didn't talk about the health threats from *exposure to mold*? Well, they couldn't, since no sampling was done. Why weren't samples taken? That one was simple: *There was so much mold* that no samples were needed. Wouldn't it been reasonable for a health effects report to mention the massive amount of mold and put that in the executive summary rather than burying it on page 22? Or wouldn't it have been academically honest to say in the Executive Summary that there was so much mold that it wasn't measured? I know this sounds extreme, but since you were there with physicians, couldn't you let them do something or say something about health effects, such as, "Because of the massive amount of mold noted by the CDC earlier this year, we conducted health surveys and found 66% of the Courthouse workers were seriously affected and should be treated right away." No, they weren't physicians. OK, then why were non-physicians doing health effects analyses?

Deniability, of course.

Nothing accomplished. And this is what we're supposed to believe is the best protection from the American public health system? What a disgrace.

Then there was the article from the *New Orleans Times Picayune* December 12, 2009, written by Chris Kirkham that reported the Courthouse was now closed because of the mold that was *never remediated* from the 2005 disaster. How much moldy paper - every legal document from St. Bernard's Parish - was in the long-term repository? Nothing had been done since the NIOSH "health assessment." And what health assessment was that? The one not even done by the engineers who held a news conference. How many of the people I saw in February 2006 were still able to work? How many were now disabled? How many had chronic health conditions? If we don't look, we cannot see. Deniability!

All of the misery from the acute inflammatory response syndrome caused by acute exposure to water-damaged buildings can be diagnosed rapidly, with treatment for the early cases almost trivially easy to correct with medications, beginning with cholestyramine. The longer one waits, the more complex the host responses become and therefore the greater the problem bringing health back to chronically ill people.

Maybe Sharon Kramer said it best when she contacted Dr. Redd on Feb. 27. 2006:

"Hi, Dr. Redd. When are you going to do something about mycoses and mycotoxicoses? This is not right. If people could get treatment, these illnesses would be no big deal."

Sharon

OK, so the Feds won't do anything about sickened people. Ahh, but what about bigwigs? Governors know when to get out of a moldy mansion; they don't listen to the CDC.

Doug Haney sent me an unreferenced article in March 2006:

> While the average citizens of New Orleans are being encouraged to return to the city with little warning of potential health hazards from breathing mold and the toxins they produce, Governor Kathleen Blanco has been residing outside of the Governor's mansion. The mansion has been undergoing an $800,000 renovation, $500,000 of which is for mold removal.
>
> After a $5.6 million renovation of the South Carolina Governors Mansion three years earlier, First Lady Jenny Sanford said *Stachybotrys* was causing health problems with her family. *Stachybotrys* is a known toxin-producing mold. According to Mike Sponhour, spokesman for the South Carolina State Budget and Control Board, which oversees maintenance on the building, "We understand the concern the First Lady has for the health of her family and children. We take that very seriously. We're committed to doing everything we can to fix the problem and make sure it doesn't happen again.

> Like numerous other government officials' families from across the U.S., Governor Sanford's family is apparently highly susceptible to mold-induced illnesses.
>
> The article continued, "Another government official who may suffer from the susceptibility to illness following exposure is North Carolina Governor Mike Easley. In August of this year, he and his family were forced to move from the governor's mansion because of mold. This is the second time in four years that the 114-year-old mansion has been invaded by mold. According to Secretary of Administration, Gwynn Swinson, the governor and his family needed to clear out for health reasons.
>
> And then there was the Governor's Mansion in Texas that was cleaned up in 1999 after high levels of *Aspergillus* and *Penicillium* were found. Laura Bush wrote me in 2000 that no one in the Mansion was sickened, but I still wonder.

How would our world have been different if President Bush were moldy and was treated? What would've happened to Governor Sanford's marriage? What about the 24% of 450,000 people in New Orleans who were at highest risk but were never evaluated properly?

All I know is, when good-hearted people sit by and do nothing, the bad guys will make sure that bad things happen. Let us not forget the lost Battle for the People of New Orleans. Let us not knuckle under to the bad guys any longer.

CHAPTER 20

Teaching in a Water-Damaged School Fighting for Our Lives

By Lee Thomassen

It was Tuesday, May 18, 2010, 10:55 a.m. when I woke up on the floor of my classroom. The bell had just rung. The last thing I remembered was grading papers before the start of the last period. One second, I was leaning over my grade book and papers; the next, I was sprawled on the floor.

There were dirt marks on my pants, and I frantically began brushing them off. Students were entering my classroom as I collected my grade book and scattered papers. Did I pass out? I'd never passed out before. I didn't know what happened.

Could it be that fumes from a moldy custodial room undergoing remediation were wafting into my classroom? How did I know the room was moldy? Simple. The mold was right there, visible for all to see. Just smell it. Just look at the environmental report I had that proved that what I could see and smell wasn't soot or old chalk or something I made up.

I looked up; the gap in the dropped ceiling was still there, waiting for another fresh one to become soaked in the next rain. The custodial room also had a missing ceiling panel.

My gifted and talented students were ready to present their oral history projects, an intensive six-month research assignment. Period 4/5 was my top class, but you wouldn't have known it today. A dozen students were holding their foreheads in pain. One student put his head down on the desk. Students were rubbing their eyes and shaking their heads like something was wrong. I remember little about class that day.

"What's that horrible smell in the hallway?" asked one student. "It smells like my wet basement," said another.

Nick was the first presenter in the next class. He was confident and knowledgeable. I marked on the blue rubric paper that he earned 24 out of a possible 25 points for his excellent presentation.

Suddenly, Nick stopped talking. He swayed for a brief second and his eyes rolled upwards, like he was having a seizure. He toppled to his left, whacking his head on a desk as he fell. He slumped into a seated position between the wall and the desk. Before he could fall forward, I grabbed him. I held his head upright and stared at his blank white eyes that neither closed nor blinked. Was he conscious? Then his eyes returned to normal and he looked at me.

"What happened?" he asked.

THE EARLY BATTLES

Dumbarton Middle School was built in 1957 in Towson, Maryland. It was (and still is) a premier school serving neighborhoods of middle class homeowners. The nearby houses are brick, just like the school. It's a well educated neighborhood with a strong sense of community. Almost all of the students I've taught for the last twenty years have parents with college degrees.

In the 1980s, many of the teachers in the first floor art and science wing of the "T" shaped school became seriously ill. Teachers in room 120 and 125 contracted Legionnaires' disease. Cancer was a frequent visitor, too, though no one ever proved the wing caused so many unexplained fatal illnesses. Those who worked there called that part of the building "the death wing." Some faculty members wanted to move out from that wing, but had no such luck.

Indoor air quality was a problem. In 1989, after so many faculty members complained about foul smells, stuffiness and so many symptoms that occurred while at work but not at home, the principal announced that an indoor air quality study would be conducted. They were happy that something was being done, but deep down, they expected that "nothing wrong" in a study was just about guaranteed. As if an air sample would silence our health concerns.

Don't forget the late 1950s were the Cold War years and schools were often made into public fallout shelters. Sure enough, there were still the yellow fallout signs in the school, relics of the time when we worried about nuclear injury from airborne attackers. Only now the dirt-floored fallout shelter bore the insignia of a different kind of airborne attacker. Beneath the school there was a rising water table; moisture in this basement was creating microbial growth in the shelter and throughout the first floor. Teachers frequently reported papers and books were growing mold. Corrective measures included simply throwing away moldy paper-based products and some furniture, too. Nobody looked at water intrusion as an insidious enemy. Nobody agreed that the teachers were employees to be protected. Nobody questioned the wisdom of teaching children in an environment that created the kind of inflammation that stole away cognition and respiratory function.

In 1989, a newly hired art teacher was surprised to find workmen in her classroom removing the sink units from her room and an adjoining art room. They told her that the unit was full of dangerous black mold. It was *not safe* to scrub the units – they'd need to be removed and destroyed. Don't forget this was in 1989. The Department of Facilities covered 90% of the dirt floor in the basement with a black plastic tarp to contain the moisture, as if 90% coverage equaled 100% protection. We all know now that this half-assed attempt at microbial containment measure was ineffective: it provided no protection from the mold reservoirs, yet allowed the administration to say they were helping keep us safe.

My own history as an employee in the Baltimore County Public School System (BCPS) began in 1989 at Woodlawn Middle School. At first glance, the building appeared safe, but so many staff members had sinus issues, sore throats and bronchitis. I have allergies, including those to insecticides, but in the building, my reactions were so much worse. You don't use insecticides for the roaches, do you? I asked. Nope, the administration said.

It was a lie.

In 1991, I was in the photocopy room when a stranger without either identifying clothing or protective gear began spraying around the room. I identified myself as a person with allergic sensitivities to insecticides. The man assured me that this was "non-allergic insecticide" and continued to spray. I tried to leave the room almost immediately, but the man was faster and he sprayed insecticide on both of my shoes.

I was furious. I was sick for a week because of that incident. The State of Maryland doesn't allow its educational employees to organize and form unions with the right to strike, so we have an "association." I joined the Teachers Association of Baltimore County (TABCO) when I was hired in 1989. TABCO negotiated a "Master Agreement" with Baltimore County Public Schools (BCPS), which stipulates that employees can take grievances to the administration via a committee of faculty members called the Faculty Council. My grievance was simple: BCPS was using insecticides inside the school building, sickening the faculty.

The Faculty Council was notified that a high-ranking member of the Facilities Department would attend their meeting. The gentleman admitted that the building was being sprayed weekly to control the roach population. He assured the Faculty Council, however, that the faculty couldn't be getting ill from the insecticide because it was non-allergic. He presented a notebook two inches thick to reassure all of us about the safety and effectiveness of this non-allergic product. He held it out for someone to take and skim through. "I'll take it," I called out. The product guide contained three pages showing the molecular composition of the chemical insecticide. Then there was a statement in large bold capital letters: "WARNING—THIS PRODUCT

CAN BE HIGHLY ALLERGENIC. IN CASE OF EXPOSURE, CONSULT A PHYSICIAN IMMEDIATELY!"

At first, this story seemed to have a happy ending. The council commissioned a health survey that showed that every teacher working close to the home economics rooms was sick. The BCPS Department of Environmental Resources arranged to spray only on Friday evenings after 5:30 p.m. to give the chemicals a chance to dissipate before teachers re-entered their rooms on Monday.

You'd think that the school system would show some compassion for the poor home economics teachers who were sickened and who filed a Workers' Compensation claim. They knew the Facilities Department had verified that there was also an indoor air quality problem. Yet the school system aggressively fought their Workers' Compensation cases, despite presence of obvious illness. One teacher become so sick that she had to leave teaching, as did her colleague. Her health was ruined.

The air intake vents in the school were all closed and couldn't open. The vents were finally opened, letting fresh airflow into the school again. Even as the air circulation improved, the indoor school climate worsened. Something new was making people ill. Teachers who claimed they had health problems because of the insecticide were threatened. The retaliation spread to teachers who'd put in for transfers - fully one-third of the faculty. Most of the sick teachers wanted out, but so did teachers who were unhappy with the discipline situation in the school. In a particularly nasty faculty meeting, one of the bigwigs in the upper echelons of the school system lambasted the faculty for their disloyalty to the school because they'd requested transfers, myself included. He stated that teachers were using Woodlawn as a steppingstone to get to Dumbarton (which was considered the Cadillac of schools in the system). That afternoon, Charlie Springer in the BCPS Social Studies Office called me and asked if I was interested in filling a position at Dumbarton. I said yes.

Teaching at Dumbarton Middle School in the 1990s was one of the greatest joys in my educational career. The students were polite and said “sir” and “ma’am.” The administration was supportive, not nasty and threatening. Schools can be places of happiness, with mutual respect among teachers, staff, administration and students, which leads to achievement.

I suffered two life-altering spinal injuries during that decade. The first was in 1995, when I herniated a disk in my neck while breaking up a fight between two girls at Owings Mills Summer School. I lost much of the use of my right hand and arm. Over-the-head traction restored some function to the injured arm and hand, and I learned to write on the board with my left hand. I created a device that let me grade papers by tying a rubber band around my thumb and two fingers so I could hold a pencil. Then in 1997, a student punched me in the lower spine, snapping my head back so hard that I re-injured my herniated disk. I suffered nerve damage around the area of the lumbar vertebrae, too. Six months of physical therapy made a major difference. Today, I walk much easier and I’ve regained enough use of my right hand that many of my students and fellow teachers don’t know I have a disability.

You’d think that the school system would support me in 1997 with regard to my Worker’s Comp case. Nope. I learned a painful lesson then that serves me well today: when you’re injured, you’re disposable. For four years, BCPS disputed my claim. By 2001, I was being sued for non-payment of medical bills.

But then BCPS personnel were caught lying. After four years without resolution, the Workers’ Compensation Commission contacted BCPS and asked why my case was still on the books. Were all bills paid so that they could close the case? Yes, assured Baltimore County, all bills had been paid. Of course, the commission also asks the same thing of the victim and his attorney. I asked my attorney to notify the Commission that BCPS had filed fraudulent paperwork and that my bills hadn’t been paid. It took less than a month to settle after that. I was awarded 10 percent permanent partial disability: five percent for the new condition (punched in the spine) and five percent for the previous condition (the herniated disk).

My experience with insecticide at Woodlawn and my spinal injuries at Owings Mills Summer School and Dumbarton reminded me of when I read *The Call of the Wild.* One by one, the sled dogs died. No one cared, no one played fair; the dead dog was simply replaced. BCPS is the 26th largest school system in the country, with over a hundred thousand students. Big school systems are like mega-corporations with two rules. Rule # 1: there are no problems. Rule # 2: if you claim that there *is* a problem, then *you become the problem*. Replacing teachers is easier in the American educational system than replacing sled dogs in the Yukon. We stand in the employment line and plead for the job of running in the harness.

Meanwhile, Dumbarton had developed a big problem of its own in 1998, and it was growing underneath the inadequate vapor tarp in the basement. Teachers started having trouble with their voices and their throats. It seemed like everyone had a sinus infection. The closets had strange smells. One teacher said her asthma, inactive for years, was now out of control. A science teacher complained about fumes coming up from the room below her floor.

THE 1999-2000 SCHOOL YEAR

Things didn't seem right in late August 1999. This is high-risk time for schools, as the summer of no kids and therefore, no air conditioning, usually creates microbial growth somewhere. Some of the closets smelled worse than the year before. A child was hospitalized with asthma, though she never wheezed before except during the spring tree season. When glaziers caulked the windows outside, there were complaints from teachers that the smell was burning their throats. I wrote a letter to BCPS on September 1, 1999 saying that I thought that we had an occupational exposure. I asked if any chemicals, especially "non-allergic" insecticides, had been used in the school. The Department of Environmental Services (DES) came to the school two weeks later, dismissing the complaint. I wrote another letter September 20, 1999, requesting air quality testing. I recommended that BCPS interview the first floor teachers. None of the letters were answered. More teachers complained about sore, irritated throats. After the Facilities Department started painting the lobby, a math teacher with paint sensitivities was repeatedly ill, finally requiring care from the emergency room. Almost every teacher above the vapor tarp on the first floor was sick. DES sent its investigator to the school.

By Sept 30, both I and the math teacher could breathe again after we'd both been placed on prednisone by our physicians. Big mistake! Later when I read *Mold Warriors*, I cringed when I found out that needless use of steroids can suppress the last component of the anti-inflammatory hormone response systems.

On the morning of October 12, the sounds of hammering reverberated through the school. No one knew what was going on. Construction was starting in the fallout shelter, but the Environmental Services investigator hadn't even briefed the Assistant Principal in charge of the building. BCPS was encapsulating the basement.

The concerns about air quality and health effects festered. On November 16, our administrators announced a required meeting to discuss the indoor air quality situation. Attending were the Assistant Superintendent for the Central Area, the Director of Secondary Education, the director of the Wellness Office and the head of Environmental Services. I was surprised that no one from Risk Management was there.

What we got was a lesson in the tactics of betrayal of trust, complete with tactics one through six (below). At the end of the meeting, I was seriously looking for little wooden boys recently come to life with newly growing, really long noses.

The Area Superintendent started by saying that the faculty did the right thing by communicating its concerns (there's the throw-away line, I thought). He said that teachers could communicate their concerns today, whether they were the concerns of one person or many. Although he noted that only three people had communicated concerns, the real number was closer to twenty. (**Tactic one**: trivialize and minimize the number of complaints.)

The head of Environmental Services noted that there were "only a couple" of complaints. He mentioned "sporadic" problems (see tactic one above), but this was "not surprising" because the ventilation system wasn't functioning the way that it was supposed to (no kidding). It was never designed to accommodate such a large number of students and staff, he said. The school needed more fresh air from the outside, and he encouraged us to open the windows to supplement the air intake. (**Tactic two**: Don't admit that the school has a problem, in this case, excess CO2 in the classrooms. **Tactic three**: blame the staff for either causing the problem or not fixing the problem). Just as I was thinking the same thing, a teacher whispered to me, "And what do we do for make-up air during the winter?" He promised that the system would be upgraded during the following summer. The art and science wings were home to problems, since both used chemicals (oh, those bad teachers using chemicals. See Tactic three, ignoring the use of hoods to vent away fumes). He admitted that it was a mistake to introduce paint into a setting where teachers were complaining about the air quality; however, there were "no serious problems." (There's Tactic one again; let's ask the math teacher how she liked her hospital vacation). He noted that some improvements had been made: the heating coils had been power-washed, new filters were put on the heating units, and the basement had been encapsulated. (**Tactic four**: praise trivial changes as magnificent and useful). There was no actual problem in the basement (another lie!), but because there were complaints from faculty members, the administration actively decided to fix a potential problem. (**Tactic five**: state that there are no problems, but fix the "nonexistent" problems because you're saint-like and truly looking out for the staff.) When asked to explain what was going on in the basement that would require encapsulation, he replied, "There is no mold. There is no humidity." (Right. There was no Watergate break-in. Nixon didn't record conversations. The Congressman didn't put $100,000 in his refrigerator. Any other lies to discuss today?)

Teachers asked questions. Environmental Services said that because the Federal government doesn't require radon testing, it wasn't done. When a teacher said that five female teachers had either died of cancer or currently have cancer and suggested that testing be done, the reply was that it takes years to develop cancer from radon. At that point, art teacher Christine Goldman, who was in remission from cancer, got up and left the room. When a second teacher asked if radon testing could be done, he repeated, "It's not mandated." When a third teacher asked, she was answered with silence. [**Tactic 6**: ignore questions that you don't like.]

I concentrated on the encapsulation project. There was evidence that teachers and students were responding to mold spores coming up out of the fall-out shelter. I discussed the history of mold and moisture problems in the 1980s. I noted that a vapor tarp trapping moisture above a dirt surface was creating a perfect biosphere for mold growth. Students and teachers had been hospitalized with breathing problems. "Oh? We don't know anything about that," said the head of the Wellness Office. The school nurse had told her that nothing abnormal was happening medically among the school's population. I responded, "I wasn't asthmatic before the occupational exposure began this year." At that point, the Director of Secondary Education declared the meeting over.

The wooden noses fell off, leaving nearly a wheelbarrow load of kindling. And then the noses re-grew.

The teachers were angry. One left the room saying, "They must really think we're stupid." Another said, "Of course, there's mold. That's how nature causes things to decompose!" One teacher thanked me for being so open. He and another teacher on the second floor had horrible sore throats for three months, but they'd never known that anyone else had the same condition.

That evening, the investigator phoned the father of the girl who'd been hospitalized twice. He said the air ventilation system was operating at 50% capacity. He stated that the filtration system of the ventilation system was completely gone and that the system could not filter out contaminants from outside the school (*Editor's note*:

or inside the school). He stated that no mold had been found except under one of the sinks in the Art Department.

Beginning that night, a team of facilities workers began tearing up the mildewed carpet in room 109, which had been there since 1957. It was glued to the underlay, and the underlay was stuck on the floor. The facilities people used ice scrapers to pry up the underlay, while others ripped the carpet into one-foot wide strips. A cloud of heavy particulates filled the air, spreading through the hallway. The maintenance staff had already swept the halls for the night, so this cloud settled overnight, until the students came in the next day to make it airborne again.

The next day was like a scene from a Stephen King-type horror movie, "Bioaerosols from Hell." Students were hacking and coughing. Half of my first period class had bloodshot eyes. The water fountain had line twenty-students deep at 8:30 a.m. The school nurse was overrun. The teacher across the hall from me had it worst of all. I recorded in my journal, *"At 12:15, during the changing of classes, [she] went out on hall duty. She looked and sounded terrible. Her eyes were red and her breathing labored. [I] got a [facial] respirator out of [the] closet and recommended that she use it. She looked at the respirator and started to cry. She turned, welling-up with emotion, and went into her classroom. She came back into the hallway a few minutes later and she took a cup of hot tea and two orange lifesavers from [me]. She returned to her class and wrote on the board for her students that she was upset and could not talk. She left school at 1:05 clearly in asthmatic distress and another teacher took her period 9 class."*

That day, I also met with the three art teachers and told them that there was black mold under one of their sinks. They searched and found the mold in room 118.

They disposed of contaminated items and worked with the head of the maintenance staff to disinfect the inside of the sink unit.

The Assistant Principal in charge of the building was a woman who truly cared for the students and staff, but she'd been ignored and bypassed by the men in the Facilities and Environmental Services Departments. She was livid. She hauled the chief investigator for the Department of Environmental Resources into her office along with the Facilities manager who'd ordered the removal of the mildewed carpet. She laid down the law: there was to be no work of any kind done at Dumbarton unless she was briefed ahead of time. In addition, when vinyl floor tiles were installed in room 109 in place of the old carpet, the fumes had to be vented outside the building.

Later that evening, I phoned the teacher who'd left school early. Seeing my offered respirator made her realize that "they'd done it to me again." She'd seen her physician, who was stunned by her story. He put her on steroids (again) and told her to use her inhaler every two hours. She never recovered; a point-source illness if there'd ever been one. She was moved to another school, but she relapsed in March 2000 when the Dept. of Facilities again began painting inside the building (again) without informing the assistant principal in charge.

In June, 2000, I had my two-year evaluation conference with the school principal. She told me that the Director of Secondary Education had ordered that I be reprimanded. My handling of the occupational exposures was intolerable. The principal said she toned down the words that she'd been ordered to write. I was skewered.

On June 20, nine teachers gathered at a local restaurant to say farewell to our math colleague, who had to retire because of her illness. Her doctor said she had the lungs of a 75-year-old; she was in her mid-fifties.

TOOLS FOR SCHOOLS

In 2001, there was a notice that teachers interested in indoor air quality could attend a meeting with other interested people. The meeting led to the creation of the Citizen Advisory Board for Indoor Environment Quality in Schools (CAB). This grass-roots organization had fewer than fifteen members. What was unique about the organization was that the meetings were held in the office of County

Councilman Joseph Bartenfelder and were chaired by his legislative assistant Teresa Streb. Her interest in Indoor Air Quality (IAQ) issues began after some unhappy experiences with her children in a local school. Two members, Dr. Schaeffer, the chief industrial hygienist from Johns Hopkins University, and Dr. Hamilton, a Hopkins allergist, served as expert advisors on indoor air-quality issues. There were several teachers and parents in attendance. Also in attendance was Dr. Colleen Pietrowski, a physical therapist and health advisor, who had trained in Dr. Shoemaker's office. The meetings were both learning sessions and a chance to air grievances about how BCPS handled indoor air quality complaints. There were a lot of scary stories dealing with mold, mildew, lead paint dust and asbestos. The one common thread was that BCPS didn't take the IAQ complaints seriously and more often than not retaliated against the teachers who reported them.

After two years of meetings (justice moves slowly in Baltimore), the CAB requested Councilmen Bartenfelder's assistance in drafting legislation to "encourage" the Baltimore County Public School System to take indoor environment quality seriously. In the end, six out of seven councilmen co-sponsored his bill. It was the first I knew of in the country to establish an Indoor Air Bill of Rights.

COUNTY COUNCIL OF BALTIMORE COUNTY, MARYLAND
Legislative Session 2003, Legislative Day No. 23
Resolution No. 143-03 Councilmembers Bartenfelder, Oliver, Gardina, Moxley, Olszewski and McIntire

By the County Council, December 15, 2003

A RESOLUTION of the County Council urging the Baltimore County Board of Education to establish an environmental advisory committee to assist the Department of Education in its efforts to ensure the quality of the indoor environment in the Baltimore County Public Schools.

WHEREAS, the Baltimore County Public School System has an obligation to provide a learning and working environment that is free of recognized health hazards; and

WHEREAS, the quality of the indoor environment in the public schools can impact upon the health and effectiveness of students, teachers and staff; and

WHEREAS, students, teachers and staff deserve a healthy indoor environment that is conducive to learning; and

WHEREAS, environmental conditions in some County schools have led to complaints from teachers, students and parents and have raised the level of concern among all of these interest groups; and

WHEREAS, the Department of Education has established a program for the testing, evaluation and maintenance of the of the indoor environment in the County's 163 schools with a total square footage of approximately 15 million square feet; and

WHEREAS, the County Council believes that a multi-discipline group should be established, representative of all concerned interest groups, to assist the Department to identify potential environmental problems before they reach the level where occupant's health and safety are at risk, and to provide an avenue of communication so that teachers, students and parents can report potential environmental problems and work together toward a resolution of such problems; now, therefore

BE IT RESOLVED BY THE COUNTY COUNCIL OF BALTIMORE COUNTY, MARYLAND, that the Baltimore County Board of Education is urged to establish an environmental assessment advisory committee to assist the Department to evaluate its current building maintenance and testing procedures and to establish effective policies and practices to ensure the quality of the indoor environment in the Baltimore County Public Schools; and be it further

RESOLVED, that the committee shall be composed of representatives of the PTA council, parents, teachers, students, public school personnel who are responsible for building maintenance, the Executive and Legislative branches of County government, a recognized expert in the field of asthma and allergies, and a certified professional environmental auditor in the health and safety field; and be it further

RESOLVED, that the committee shall report annually to the

Board of Education and Baltimore County Council its findings and recommendations for effective testing procedures and effective policies and practices to ensure the quality of the school's indoor environment.

Baltimore County had its first IAQ Bill of Rights, with an order to create an environmental assessment advisory committee (EAAC). The County Council planned on monitoring the school system's compliance.

Rumor had it that members of the Facilities Department were angered by the legislation and that county personnel immediately started concocting ways to subvert the legislation's intent. But the BCPS Superintendent came on board, ordering creation of the required committee.

EAAC didn't start well. Drs. Schaefer and Hamilton volunteered to serve on the committee, but BCPS refused to accept them. BCPS immediately started stacking the committee with those who clearly had no intention of fulfilling the mandate of Resolution 143-03. No teacher was on the Committee. A member of the School Board said privately to a CAB member that the only reason the Committee was even meeting was because the Citizen Advisory Board was keeping the issue of Indoor Air Quality alive before the County Council.

There were a lot of unhappy faces on the EAAC when teachers and parents from the Citizen Advisory Board showed up at the next monthly meeting. The EAAC didn't want the CAB there, and for the next few months, there were a series of letters exchanged to establish that the EAAC had to follow Maryland's open meeting laws. We were allowed to sit and watch, but we weren't allowed to speak. We weren't even given copies of the agenda after the first few meetings. The message was clear: the CAB was not welcome. We sat in silence anyway and stared at them meeting after meeting.

After months of meaningless debate and posturing within the EAAC, the CAB wrote a letter of complaint to higher authorities. The CAB wasn't happy about the delays in fulfilling Resolution 143-03. EAAC responded by expelling CAB from the meeting room.

What had just happened? We weren't told why we were ejected or what the EAAC was going to discuss during the rest of the meeting. We had obeyed the no talking and questioning restriction. I contacted Teresa Streb and let her know what happened, and she communicated with Joe Bartenfelder.

No one recorded what transpired between the Superintendent of the Baltimore County Public School System and Councilman Bartenfelder. Bottom line? First, the County Council expected the EAAC to fulfill the terms of Resolution 143-03. Second, Councilman Bartenfelder was a member of the Citizen Advisory Board, and he expected his fellow CAB members to be treated with respect while they were attending the EAAC meetings.

The message from the office of the Superintendent to the EAAC must have been loud and clear. The next meeting of the EAAC had a completely new atmosphere. CAB members were greeted and people shook hands. There was food and drink available, and we were invited to partake. More importantly, the meetings became more productive when Teresa Streb introduced the Tools for Schools program for consideration by the committee. Tools for Schools is a proactive indoor air-quality program sponsored by the EPA and the American Lung Association. It stresses environmental input from teachers and parents, and yearly walkthrough inspections by trained Tools for Schools teams to spot IAQ issues before they become acute.

The EAAC opened up the meetings to questions from the CAB members present. We had plenty. On two occasions, I sat in for Teresa Streb, thus becoming the only teacher to serve on the EAAC while it was considering Tools for Schools. I sat next to the head of the Department of Environmental Services, the same man who addressed the Dumbarton faculty back in 1999. Did he remember me?

The good news is that after a year of meetings, the EAAC did fulfill the spirit that the County Council had intended. Even better, the Baltimore County Public School System formally adopted the Tools for Schools program. The School Board even hired two industrial hygienists to supplement the Dept of Environmental Services. Nancy Murray and I were chosen to be the teacher representatives on our school's Tools for Schools teams.

THE FLOOD

In October 2005, Dumbarton's ancient heating system was being replaced with a modern unit ventilation system. Instead of a unit that only produced heat, the unit ventilators would work year-round, cleaning the air of particulates and replenishing fresh air. Each unit had an easily changed filter. A technician told me that the unit could re-circulate the air in a classroom every 48 minutes. CO2 build-up (and the headaches that went with it) would be a thing of the past.

Unfortunately, there were problems with the plumbing. The new system was hooked up to the old copper piping, and leaks were common. In hindsight, the leaks seem like nuisance problems compared to what was to come. On October 25, the workmen performed a pressure test on the system. A cap on one of the main pipes in the second floor library workroom was loose. The pipe started gushing water during the pressure test, and since it was right above the school's rental copier, the copier was destroyed. Water flooded into the hallway, the bookroom and the library and started pouring through gaps on the second floor into the art room below and its adjoining storeroom, as well as into the adjacent art storeroom.

The workmen and evening maintenance staff did their best to clean up the mess with mops and buckets, but the dry-out was incomplete. No one even asked if all the water under the heavy wooden cabinets, bookcases and the library carpet might pose a health risk. That fall, I asked the librarian if my students could use the 3-volume set on *Notable American Women*. Unfortunately, the librarian had thrown it out; she said that it was moldy.

THE NEXT VICTIM

Christine Goldman taught art in room 121 since the early 1990s. She was honored with an Award for Excellence presented by the Baltimore Chamber of Commerce and the Board of Education in 1994. It was a prestigious award, and only a handful of people received it in a school system with more than five thousand teachers. There was more: her work with *Victims of Violence* and the homeless was a major contributing factor in Dumbarton's receipt of the National Blue Ribbon School award in the 1990s. She was in her fifties with thirty years of service on the books, but she wasn't yet ready to retire. She was also no stranger to life-threatening illnesses—she'd beaten breast cancer.

In mid-October 2006, Christine came to me complaining about the poor indoor air quality in her room. Sure enough, Room 121 had a sharp, musty odor. There was no airflow coming from the two unit ventilators. Without the unit ventilators, the air in the room wasn't being filtered, and fresh air wasn't entering the room. The art storeroom was even worse. This room had no air ventilation except for an inaccessible, sealed window. It made me feel sick to be in that room. I opened the unit ventilators; they'd simply been shut off. Once opened, they worked.

Restarting the unit ventilators came too late. Christine became seriously ill, with multiple health symptoms. Foods that she'd been eating all of her life now made her sick. She had brain fog and strange, stabbing pains. She worked at Dumbarton for just one more day.

I contacted my fellow Tools for Schools Team members; the administration notified the Department of Environmental Services. I pointed out the visible water tracks on the walls of the adjacent art storage room. The two new industrial hygienists came out and inspected room 121 and the storage room. They said nothing was wrong.

Christine's condition deteriorated during the next two months. She lost weight so fast because she couldn't eat more than a few foods.

She was constantly fatigued, she couldn't think straight. Toxigenic mold? Finding a physician who could successfully diagnose and treat the mold illness wasn't easy. She told me that she believed she was dying during that dreadful winter.

After suffering through several physicians who had no experience with her ailment and may have done more harm than good, she found Dr. Colleen Pietrowski. Dr. Pietrowski was one of the initial members of the Citizen Advisory Board and knew all about mold-related illness from her studies. She and her husband ran *The Center for Advanced Physical Health* in Timonium, Maryland. She specialized in physical therapy and nutrition, but many of her patients were there because she was one of the few health care providers in the Baltimore area with an up-to-date knowledge of biotoxin illnesses. Dr. Pietrowski ordered blood tests and conducted VCS (Visual Contrast Sensitivity) testing. The results were a good indication that Christine Goldman had been exposed to a water damaged building. Later, she began treatment with Dr. Shoemaker.

BCPS wasn't happy. She'd been away from work too long, now more than six months. You can predict what would happen next: the Dumbarton principal "excessed" her, a term used when a teacher's position was eliminated. But the position wasn't being eliminated, because a new art teacher was hired. It was the first step in a process designed to get rid of her. On May 22, 2007, she received a letter from BCPS noting that her position was eliminated. If she wanted to work for BCPS, she needed to complete a declaration of intent form and identify areas in the county where she was willing to work.

Her attorney, Howard S. Schulman wrote to Michelle Prumo, Risk Manager, on May 25, 2007:

> *Dear Ms. Prumo:*
>
> *On May 24, 2007, at approximately 4:30 p.m., Christine Goldman received a telephone call from Elaine Shaw in your office, who directed her to report to work on Tuesday, May 29, 2007 at the Dumbarton Middle School. Ms. Shaw also advised Ms. Goldman that because of the approaching Memorial Day*

holiday, there was insufficient time for your office to prepare a letter to her but that she would receive an e-mail, which she received this morning without copies of any medical reports.

It is our position that your office's directive places Christine Goldman in a position of harm. You are directing her to the very environment that has caused her medical condition.

On May 11, 2007, we sent you Dr. Shoemaker's report of May 8, 2007 regarding
Christine Goldman. Dr. Shoemaker has provided Mrs. Goldman with disability slips through the end of the academic year, which I understand she has furnished to you. Previously, Ms. Goldman has furnished your office disability slips from two other doctors. Dr. Shoemaker diagnosed Ms. Goldman's illness as being caused by mold exposure in her workplace. Dr. Shoemaker is a recognized expert and specialist in the area of mold toxicity and has been recognized by the courts as such. Dr. Shoemaker advised Ms. Goldman that if she is exposed to the spores in the school's water-damaged environment, she will sustain a worsening of her condition and place her overall health at grave risk. [Footnote:] *The school environment also appears contaminated by rodent feces.*

Because the directive places her health at risk, this is to advise that Ms. Goldman will not report to work on Tuesday, May 29, 2007.

We are also concerned about the timing of the directive and the cursory written communication from your office and the absence of any supporting medical report.

This ranks as one of the most frightening letters that I've ever read. A medical report, written by a specialist in his field (Dr. Shoemaker), was being discarded by the BCPS Risk Management Committee in favor of the report by their own physician, but the report by their physician wasn't going to be shared with Ms. Goldman or her attorney. The bias is even more explicit in a letter from the school on May 30, 2007:

Dear Ms. Goldman:

*Based on the reports of Dr. Siebert and Dr. Toney, **you are***

> ***able to return to work*** [emphasis mine]. *According to Ms. Michelle Prumo, Risk manager, you do not want to return to Dumbarton Middle School.*
>
> *Therefore, I am assigning you as a 1.0 FTE art teacher at Oakleigh Elementary School. You are directed to report to Oakleigh Elementary School on Monday, June 4, 2007. Failure to follow this directive may be considered insubordination and subject you to disciplinary action.*
>
> *Sincerely,*
> *Kim X. Whitehead, PhD.*

Christine Goldman had meanwhile sent in her "Reply Form – Involuntary/Excess Transfer." In it, she writes: "*I am requesting a position as an art teacher in any middle school in the central, northeast, or northwest areas. I am requesting that you make reasonable accommodations for my medical condition by placing me in a mold-free environment and a school without water damage. I would like to return to Dumbarton but cannot do so because of the mold and water damage there. Thank you – Christine Goldman, May 28, 2007*"

Ms. Goldman requested reasonable accommodations; the county refused.

Christine Goldman's attorney Schulman immediately replied to the hostile letter from the school on June 1.

> *Dear Dr. Whitehead,*
>
> *Christine Goldman has requested that I respond to your letter of May 30, 2007. Ms. Goldman must respectfully decline your directive, as her treating physician, Ritchie C. Shoemaker, M.D., has advised her not to return to work before June 30, 2007. Enclosed is a copy of my letter of May 25, 2007 to Ms. Prumo.*

The June 4, 2007 reply from the school was swift and furious. The bias against any treating physician except their BCPS/Concentra

(*Editor's note*: this outfit has long history in Maryland of providing "no-mold" opinions) non-treating physicians is apparent:

> *Dear Ms. Goldman:*
>
> *Based on documentation received by the Office of Risk Management, you have been unable to return to work since October 2006. Due to the length of time you have been on leave and in accordance with Superintendent's Rule 4153, you were directed by me (Dr. Kim X. Whitehead) to undergo an independent medical examination with Dr. Stephen Siebert on January 18, 2007. Dr. Siebert's report stated that "My opinion is that she can perform the essential functions of her position..." However, he recommended that you be evaluated by someone with expertise in occupational medicine. Based on this recommendation, you were directed by me to be examined by Dr. Robert Toney on May 8, 2007. As a result of that evaluation, Dr. Toney recommended that you not report to work until he had reviewed your medical records. On May 18 and 22, 2007, Dr. Toney advised Baltimore County Public Schools that there were "no contraindications to full duty."*
>
> *You were notified of Dr. Toney's recommendation by the Office of Risk Management on May 24, 2007. Since your position had not been released, you were told that there would be no contraindications for you to report to Dumbarton Middle School. However, you have related to the Office of Risk management that you would not report to Dumbarton Middle School due to your unsubstantiated concerns about the environment. On May 30, 2007, I directed you to report to Oakleigh Elementary on June 4, 2007. On Friday, June 1, 2007, I received a letter from your lawyer stating that* ***you refused to return to work for the remainder of the 2006-2007 school year*** [emphasis mine].
>
> *The last duty day of the school is June 19, 2007. In spite of your decision to disregard an administrative directive to return to work, you have the following options.*
>
> - *Report to work on June 8, 2007 in the Office of Art until June 19, 2007. You will be expected to report to work in August*

at a school determined by the Department of Human Resources.

- *Retire effective June 30, 2007. You must complete the enclosed resignation form and submit it to my office by June 8, 2007.*

Lastly, the Office of Risk Management is in receipt of your Sick Leave Bank form to cover your absence from June 8, 2007 to June 19, 2007. Because you have been medically approved to return to work, your Sick Leave Bank benefit will be discontinued as of June 8, 2007.

If you fail to choose either of the options presented, you will be considered insubordinate and recommended for disciplinary action up to and including termination.

Sincerely,
Kim X. Whitehead, Ph.D.

Forced to choose between her health and the career she loved, Christine Goldman retired. The loss of Sick Leave Bank benefits in June cost her more than $2,000 in salary. But one major question remained: was the school the source of the toxigenic mold that had ended the career of this gifted teacher? If so, where was the mold and who would be its next victim?

IT BEGINS AGAIN

There were a number of teachers with sinus infections in September 2007. Sinus infections aren't always an indicator of poor indoor air quality, especially in a school full of children. What was different about these infections was their severity. Multiple teachers needed more than one course of antibiotics to control the infection.

My sinus infection was so intense that it made my teeth hurt. By November, the pain was increasingly severe. My dentist examined me on November 5, noting how widespread my sinus infection was. Maybe I should have my loose filling fixed.

Meanwhile, I started becoming sensitized to air fresheners. My sinuses burned every time I entered the faculty room or one of the adult bathrooms. I knew from my training for Tools for Schools that air fresheners weren't allowed on school property, so I began to petition the administration to remove them. When faculty protests erupted about removing the cans, I copied all of the explicit warning labels from the air fresheners around the school and sent them to the principal and the Tools for Schools Team. By the end of November, all air fresheners were out of the building.

I saw Dr. W. Jeffrey Davis on November 29. He said that my throat was red and swollen. He put me on an antibiotic and a decongestant. The next day was sheer agony. I sat in my favorite chair in my family room in excruciating pain throughout most of the evening. Nothing I tried brought any relief. The next day, the pain had subsided. When I went to the dentist on December 18, he looked at the tooth and then he looked at me. The tooth had *split in two* from the root to the crown. He reached in and with ease pulled out a long shard of tooth. He said that he'd never seen that happen to a tooth like that before. He asked me if I was in pain; I said no. When the tooth split, the excruciating pain abated. He extracted the remains of the tooth.

There was a lot of circumstantial evidence: a student hospitalized, multiple sinus infections among staff members, multiple teachers complaining about dehydration, a teacher hospitalized under suspicious circumstances, and of course, my tooth. The sudden sensitization to air fresheners wasn't a random event. I asked Christine Goldman who had treated her illness.

My first meeting with Dr. Pietrowski was on February 12. She agreed that my symptoms were suspicious. She explained the concept behind VCS (Visual Contrast Sensitivity) testing, which I failed miserably. She ordered multiple lab tests. When I saw her on March 12, she said that my immune system was taking a major hit. I didn't have much Vitamin D in my body. C4a, MMP-9, MSH and other tests were all abnormal.

Both Dr. Davis and Dr. Pietrowski agreed that I should see Dr.

Shoemaker in Pocomoke City. I contacted his office and a huge packet of materials arrived at my house a few days later. I was impressed by its efficiency and organization, but it was also intimidating. There were some important preliminaries that had to be done, including a visit to Memorial Hospital in York for a suggested stress echo test.

I counted on getting better without ongoing exposures and I was pleased to find that this was the case. Still, I went back to Dumbarton for five days. I was hired to work with some other teachers on the "Thrill of the Skill" curriculum project. We met in the air-conditioned library. I couldn't believe it: there was water on the floor of the library and on the windowsills caused by two old leaky air conditioners. I felt bad by Friday.

On the final day of the workshop August 11, I went to see Dr. Pietrowski. The nasal swab had discovered something worse: my nasal cavity was being colonized by antibiotic resistant staph, though the kind that makes biofilm and not the kind that invades tissues. This was horrible news. With Dr. Pietrowski's report in his hands, Dr. Davis wrote his first ever prescription for "BEG" nasal spray to combat the staph. It contained antibiotics that I'd never even heard of before, like Bactroban and gentamicin. Dr. Davis insisted that I take erythromycin orally as well.

With the school year beginning on August 18, I was ready. I had a duty and the right as the Tools for Schools teacher representative to document visible mold. Within days the vigil began. I sent an email warning teachers not to use the sink in the library workroom where the drainpipe was leaking. I remembered the old maxim: *where there is moisture, there is mold*.

Later that week, I looked at that sink unit. Sure enough, there was already mold. Just to be safe, I took some photographs showing the extent of the obvious mold growth. On Tuesday, Sept 2, I notified the principal and the Tools for Schools Team of the mold problem. According to protocol, the school's industrial hygienists would come to the school, analyze a mold sample, and determine whether or not it was toxigenic. At least, that's what I thought the protocol called for.

That afternoon, the industrial hygienists spoke to me in the hallway outside my classroom. They couldn't find any mold under the sink in the library workroom. Their examination noted that I'd seen a chemical stain and I'd probably mistaken it for mold. I told them that I took a sample of the mold on a sterilized cotton swab and had it in a freezer bag. I offered to give it to them to analyze, but they refused.

I was dumbfounded! I ran up to the library workroom and looked under the sink. Sure enough, the mold was gone and all that was left was a five-inch chemical stain. I was floored when I learned that the school's chief custodian had been sent by the principal to deal with the problem before the industrial hygienists could arrive. For the next hour, she methodically scrubbed away all signs of the mold, so when the industrial hygienists arrived, it looked like I was crying wolf. I even got an email from the principal, which remarked innocently that the chief custodian hadn't been able to find any mold. The principal gleefully sent me a copy of the industrial hygienist's official report. *False alarm! There was no mold.*

Lesson learned: trust no one! Lies and deception are part of the game. (*Editor's note:* see 15 things a mold patient must learn.)

Lesson learned: document with photos.

I emailed the photographs to the principal, the Tools for Schools Team and the industrial hygienists, who, acting like dutiful Keystone Kops, promptly amended their report to read that there was mold, but it wasn't toxigenic, so there was still no need to worry about it. How did two trained industrial hygienists surmise that the black mold was not toxigenic by looking at a photograph? I held the freezer bag with the mold sample in my hand and sighed. The school system's new industrial hygienists weren't going to be of any use in this struggle.

Tools for Schools remains a wonderful program. There are elementary schools in Baltimore County that have embraced this program. Some schools used the program successfully to identify potential problems. Even our Department of Environmental Services has been quick to remediate potential problems with the program in place. In other areas of the county, however, especially at the middle and high school level, the Teams tired of the exhaustive walk-through process, usually done in June and often with no air conditioning. The walk-throughs at Dumbarton could take four to five hours. Schools began omitting or faking the walk-throughs, and some schools even neglected to complete the required yearly reports.

Environmental Services did nothing in 2008 to remedy this bastardization of the Tools for Schools idea. Worse, teachers who complained about the indoor air environment were quickly silenced. TABCO had some advice for me: create a paper trail. On September 17, I filed a formal complaint with TABCO detailing the deception that went on concerning the disappearing mold in the library sink unit. TABCO might not have the political power of a real union, but it had some sharp teeth when given the right tools to work with.

In late September, I received an alarming email from a language arts teacher. The refrigerator in the second floor bookroom was leaking, even after several weeks of attempted fixes. Some teachers had covered the floor with cardboard, but now the cardboard was growing mold. Sure enough, the floor was wet and I could see the colorful mold on the cardboard. It did not "look" very dangerous, but how did I know? I reported the leaky refrigerator and the moldy cardboard to the principal and the Tools for Schools Team.

The reply that I received back was a nasty email from the principal. Was I "touring" the school without being accompanied by the assistant principal in charge of Tools for Schools or the chief custodian? The message was loud and clear: if you find something wrong with the indoor air quality in the school, even by accident, then you'll get in trouble if you report it. One positive thing came out of this episode - on November 1, an email went out to the faculty to get their things out of the second floor refrigerator, because it was being trashed.

On September 24, Evan Krantz, the art teacher who'd replaced Christine Goldman, returned to school after visiting his physician. He stated that his doctors blamed his illness on the poor indoor air quality in the school. I'm not sure how the physician could tell, but TABCO needed to get involved. Their representative arrived at Dumbarton on October 1. He went into rooms 119, 121 and the storeroom. He wasn't happy about what he saw. Room 121 was mouse infested and Mr. Krantz was catching over a dozen mice on sticky traps every month. Mouse feces were everywhere. The next day, the principal unhappily notified me that the environmental people would be coming to the school in response to "my complaint."

I met with Dr. Pietrowski in October 14. My blood tests from August showed that my biomarkers hadn't returned to normal. C4a had increased to more than 32,000 - an incredibly high number. Was the library a source of toxigens and inflammagens (not just mold, mind you)? Could only five days in a library doing a curricular workshop really have ramped up my innate immune system to that degree? Dr. Pietrowski said that I needed to take some time off from school *immediately*.

The next day, a mouse got caught in my unit ventilator. The smell from the fried mouse and the burned ventilation parts was sickening. I felt ill and so did more than half of my students. Two of my students told the school nurse that they thought I was dying. I must've looked horrendous when I showed the C4a result to the principal that day and said that I needed to go on two weeks medical leave. It was the same week two years before when Christine Goldman got sick and had to leave school.

I had my first appointment with Dr. Shoemaker a few days later, on October 21. He wrote a prescription for cholestyramine (CSM) four times daily. There were more blood tests and even a magnetic resonance spectroscopy study showing abnormal levels of chemicals (lactate, glutamate and glutamine) in four areas of my brain. I had read *Mold Warriors* in preparation for the meeting, but as much as I liked the science, it was hard to memorize. He asked me if I knew which regulatory neuropeptide was affecting my heart's ability to pump blood efficiently to the lung during exercise. I replied VEGF. "Close," he replied, "It is VIP." I didn't have any. I was supposed to have some, as every cell in the body has some controls built in that are initiated by VIP.

Christine Goldman had given him permission to discuss her case, so we started to figure out why two healthy adults in classrooms only fifty feet apart had come down with the same biotoxin illness. He said that I had to get better microbial data. Technology had come a long way since the mold illnesses that had gripped my school in 1999 prior to the encapsulation. A DNA analysis of the mold spores in dust samples (called Environmental Relative Moldiness Index, ERMI) tells the tale. A good place to start would be room 121.

As my two-week leave of absence drew to a close, I met with Dr. Davis. He had no formal training in biotoxin illnesses, but he sincerely believed that toxigenic mold could cause serious problems in the human immune and respiratory systems. Dr. Pietrowski said that family physicians like Dr. Davis were rare. Most physicians that she's dealt with would never diagnose a mold-related disease, let alone treat one. It was now Dr. Davis' duty to write my return-to-work (discharge) paper. He wrote that there was no prognosis yet, and more testing was needed. He also wrote that he and the other treating physicians were waiting for ERMI data from the BCPS Dept of Environmental Resources.

The formal request from Dr. Davis for speciation data must have seemed like a gauntlet thrown into the lap of the Dept. of Environmental Resources. On October 11, the principal told me that one of the industrial hygienists would be arriving at Dumbarton on November 11 to meet with us. The principal took notes from the meeting, and they were revised several times. The first part of the November 13 notes begins as follows:

> *The principal began the meeting by thanking people in attendance and by directing the group's attention to Lee's return to work form, on which it was indicated that Lee's return was pending further testing, particularly ERMI testing. She asked what results the doctor wished to see with this testing. She also said that David was the scientific representative*

> *at the meeting and would have information from his environmental expertise.*
>
> *Lee presented information concerning his question, "What degree of proof is the Risk Management Committee looking for to determine the risk posed by hazardous airborne particulates at Dumbarton after experiencing water damage in the fall of 2005?" In addition to his involvement as shown in blood work, heart involvement, MRI, and VCS data that he shares with Christine Goldman, he suggested that the ERMI test was needed to show exactly what species of mold were in the air or in the storerooms, classrooms, and library.*

The industrial hygienist wasn't happy with this request, and he found excuse after excuse for why it wouldn't be done. He did agree to inspect rooms 119 and 121 and their storerooms, and he responded to my request for extended unit ventilation time in the classrooms (beyond 4 p.m.) by saying that he'd make a recommendation based on his inspection. The meeting notes added that *"the role of BCPS, therefore, is to find any problem, to clean it up, and to go forward."*

Paragraph five contained a summary of my views on the recent fried mouse in the unit ventilator incident. It was a paraphrase at best, but a year later, Bobette Watts-Hitchcock in the EEO Office would treat it as a direct quote. *"Lee indicated that he felt that the airborne spores were emanating from the designated areas and coming into the common areas into his room. He did not feel that an inspection of his room was needed since, as long as the ventilators are working, his room is fine now that the mouse has been removed and the electrical shorts corrected so the ventilator works properly."* Ms Watts-Hitchcock would edit this line to read, *"Mr. Thomassen did not feel that an inspection of his room was needed since, as long as the ventilators are working, his room is fine now...."*

The final paragraph in the notes reads: "*The principal thanked Lee for his work with Tools for Schools and for including her in the loop of emails so that she can be the point person with the people at Pulaski in correcting these needs. She did ask if Lee needed to adjust his time after school since the ventilators went off at 4 p.m. He indicated that with his disabled right hand, he needs to work until 6 p.m. Services of a parent volunteer were offered.*" What exactly would a parent volunteer do? Grade papers? Please – send me five!

True to his word, the two industrial hygienists did a visual inspection of rooms 119 and 121 on November 19. They found nothing in the visual inspection. But they did no scientific testing in either room. The day before, I collected dust samples for ERMI from rooms 116 and 121. The results showed that my room was not in bad shape. There were minor amounts of *Aspergillus penicillioides* in room 116. It's a very toxigenic mold on the EPA group 1 list of Mold Species, but the numbers indicated that these spores were probably entering into my room from somewhere else. It was a bit more problematic in room 121: ten percent of the mold spores in floor dust from the group one list in room 121 were *Aspergillus penicillioides*. Could this have caused the illnesses that were afflicting me and Christine Goldman? (*Editor's note*: this organism is one of nine target organisms Dr. Shoemaker relies on in assessment of WDB links to human illness). My gut feeling was that my search would have to expand to other areas of the school as well, given the distribution of which teachers were known to be ill. But would I be able to stay healthy enough to find it, or would my career be ruined like Christine's?

SAIIE
SEQUENTIAL ACTIVATION OF INNATE IMMUNE ELEMENTS
(CORRELATING REPETITIVE EXPOSURE DATA)

I had my first phone conference with Dr. Shoemaker on November 24. I was better on the CSM, no question about it. He started out by echoing the opinion of Dr. Pietrowski in October. I might not look that sick, but the lab results showed the exact opposite. He explained the abnormal osmolality result to me: I was dehydrated. C4a, MMP-9 and MSH were all abnormal. He explained a process called SAIIE (pronounced sigh-ya) that could help to establish whether or not the school was the source of the compounds that my innate immune system was reacting so negatively to. This prospective acquisition-trial type of evidence is used in Maryland courts. I decided to do the SAIIE protocol over Christmas break instead of Thanksgiving to give my body extra time away from the water-damaged building.

SAIIE testing includes five consecutive days of blood draws and each blood draw, while routine at Dr. Shoemaker's office, can appear intimidating to a lab tech. The directions are clearly written, in English, with all specimen requirements and all codes listed. Such a draw wasn't a problem if there are no phlebotomy lab errors. I tried to get the first of the five sets of blood tests for the SAIIE study on the 23rd of December, but the lab couldn't figure out Dr. Shoemaker's directions. I tried a different lab on the 24th, but the technician there struggled with the directions and ended up sending me home. Finally on the 26th, the lab successfully made the first blood draw, but the von Willebrand's vial arrived thawed, not frozen, and had to be discarded. The second blood draw went off without a hitch on New Years Eve.

Meanwhile, the dust samples that I'd gathered from the library and workroom the week before had a group one mold that didn't appear in the samples from the first floor: *Aspergillus niger* and lots of it. I mailed the results to Dr. Pietrowski and Dr. Shoemaker. There are twenty-six molds on the EPA's group one list, many of which wouldn't raise an eyebrow from either physician. *Aspergillus niger*, however, would.

When I returned to my water-damaged school on Monday January 5, 2009, I was deliberately not taking the CSM after the next blood draw. I felt good off CSM and away from the school, giving me hope that this illness wasn't permanent. I really wasn't prepared for what re-exposure to the school would do to me. I didn't think that a few days would make much difference in how I felt. It took me years to get as sick as I had been and Dr. Shoemaker said if the school truly made me ill, I'd be right back where I was with a multisystem illness in *three days*. I wasn't prepared for him to be right, but my goodness, was he ever.

The foul air in the school hit me hard. After just a few hours at work, I was struggling mightily. I've given the National Geographic Bee for the past fifteen years, but now I was looking at the materials as if they were written in Martian. I'm just on Day 1 of unprotected re-exposure and the Bee materials just didn't register. I had no idea what to do. Dr. Shoemaker had warned me about disorientating sensations like this. On Tuesday, January 6, I went straight to the lab for the third set of blood draws before going to school. I repeated the same procedure for the 4th and 5th blood draws on January 7 and 8. That week was marked by short stabbing pains; itchy skin, like bugs were crawling over me; an aching shoulder; and a strange feeling in the frontal lobe of the brain. I couldn't wait to get back on CSM.

I went to my appointment with Dr. Pietrowski on January 8. The receptionist told me that the appointment had been on the 6th. I was back on the CSM but now I was scared. I pulled out the appointment card. Sure enough, it had said January 6th, but I'd seen January 8th.

On Saturday, January 10, I was one of the assistant tournament directors at a chess tournament at Dumbarton. My duty was to supervise the chess players in the "cleaned-up" library. I was incredibly sick when I went home that evening. I was having massive leg joint pain. The next day, I was supposed to do the scripture reading at Stewartstown Presbyterian Church. My eyes kept going in and out of focus. When I returned to the pew and sat down, my mind focused on the Dumbarton Library—what was wrong up there on the second floor?

On Monday, January 12, I received congratulations for the third place trophy that our chess club had won, but I couldn't stop worrying about the horrible smell in the hallway outside my classroom. There were three rectangular shaped rooms between my room (116) and Christine Goldman's Room (121) fifty feet away. One was the art storeroom. The second was the old dark room, a leftover from when photography classes needed special equipment and chemicals to develop pictures. Now it was a storage room. In between was a custodial room. That one was a long room, just over twenty feet long and about eight feet wide. The left side had wooden shelving dating to 1957. There was an old heavy-duty sink with a leaky faucet. Here was the source of the stench: it smelled like dirt.

The math teacher in room 117 was new to Dumbarton during the 2008-2009 school year. Her eyes reacted to the air in the school on the first day of her arrival. Her room was directly opposite my room. Her sinuses had been giving her hell for five months. All three teachers closest to the custodial room were sick. The smell from the custodial room reinforced our growing suspicion that there was a serious mold problem there. Although the floor was too littered with equipment and debris to get a good dust sample, the shelves were easily accessible. The dust on these shelves and boxes was an eighth of an inch thick in places. My scientific team contacted Assuredbio. Would an ERMI canister filled with dust from the shelves be a viable test sample? Yes.

As if January 2009 wasn't busy enough with the SAIIE study and the growing concerns about the custodial closet, a ceiling panel fifteen feet in front of the library was obviously moldy. Water from a leak in a fire sprinkler pipe kept the tile wet, and the wet tile kept making so many of us sicker. The top of the panel was wet, but mold free. I took a sample with a sterilized swab and put it in a freezer bag.

The results of the SAIIE study began coming in during February. I was following the exact pattern that Dr. Shoemaker has shown to occur in hundreds of patients. The molds on the second floor were moisture lovers and toxin-formers, especially *Chaetomium.* Water-damaged buildings with *Chaetomium* often have a second toxigenic mold nearby—*Stachybotrys chartarum*. The result from the custodial closet was equally ominous: *Aspergillus fumigatus*. There were 3,514 spores per milligram of dust, which was a significant number. *Stachybotrys chartarum* and *Chaetomium globosum* were also present in minor quantities.

A the Tools for Schools teacher representative, it was my duty to inform the principal and the team about the visible *Chaetomium* bloom on the ceiling panel. The principal wasn't happy; she started asking questions that seemed designed to get me into trouble. Fortunately, the principal received an email from the head of Environmental Services. Don't worry about the moldy ceiling panel; mold on ceiling panels happens all the time. The panel was removed and replaced, and I kept my job.

I spoke with Dr. Shoemaker on the phone on February 19. The repetitive exposure protocol results were horrible. Three of my biomarkers had been abnormal in the December 26 labs, but had returned to normal between December 31 and January 5. Only twenty-four hours of exposure to the contaminated air at Dumbarton had caused TGF-beta 1 to skyrocket from 2,097 (normal) to 6,273. After 48 hours of exposure, C4a had jumped from 1,019 (normal) to 36,900. After 72-hours of exposure, MMP-9 had risen from 255 (normal) to 639. The message from the biomarkers was clear: get out of that school! The image of Christine Goldman flashed into my mind. This would have happened to her body during September and October 2006. How long could my immune system take this kind of punishment?

Dr. Shoemaker also had comments about the ERMI data. He echoed Dr. Pietrowski's warnings about the toxicity of *Chaetomium globosum*, but his attention was focused on others, including *Aspergillus fumigatus*. He called the ERMI results "a witch's brew of toxin-formers." I pointed out that Christine Goldman's hall duty assignment during the fall of 2006 had been ten feet from the door to the custodial room with the *Aspergillus fumigatus*. As the conversation wound down, Dr. Shoemaker wished me good luck. He said that I was brave.

THE TRUTH IS OUT THERE

This wasn't a good situation. I was in a classroom ten feet away from a room filled with several of the most toxigenic molds known to humankind. The Environmental Service personnel were more interested in denying the existence of mold in the school than actually remediating the problem. I could take my supplements like vitamin D and all the rest, I could mix and drink my CSM four times a day, and I could wear a facial respirator after school hours when the door to the custodial room was open, but those interventions were like Band-Aids. I needed to find a way to convince the school system that it was in their best interest to remediate the mold in the custodial room. It was now March, and the humidity levels were rising in the school with the start of spring. More molds were coming.

Friday the 13th was significant. First, I emailed Dr. Pietrowski about the feasibility of using Section 504 of the Rehabilitation Act of 1973 and the Americans with Disabilities Act to force BCPS to grant me accommodations that would deal with the burgeoning mold problem in the school. She was supportive. It turned out that any employee in a building that gets federal assistance can ask for 504 accommodations just like parents can ask for accommodations for a child.

Second, the school librarian was straightening books that morning. She was startled to find that the books on a shelf beneath the windows were covered with black mold. Some were contaminated from cover to cover. I rushed to the library to obtain a sample, but I was too late. She'd also notified the principal, who'd immediately phoned Environmental Services. One of the industrial hygienists had arrived at the school within an hour and whisked away all 31 books. The librarian herself had been ill. It appeared that the industrial hygienist had been in a hurry to make the evidence disappear. It would have taken several hours to inspect all of the books on the periphery of the main exposure. After years of collecting books, I knew to go to the cloth-backed books.

Within minutes, I had two books with three excellent samples from the dust jackets, all saved in freezer bags, but these were not the cover-to-cover books with black mold that the librarian had reported. Those were probably already in a dumpster at the Facilities building. I emailed the industrial hygienist, congratulating him on finding and removing 31 mold-contaminated books from the library. (*Editor's note*: these actions are called despoiling evidence. It's tantamount to hiding the murder weapon). I asked him to let me know which species of mold were present. He never responded. Two months later, I filed a complaint with the Faculty Council. The teachers hadn't received a copy of the written report that's required from Maryland Occupational Safety and Health (MOSH). Both industrial hygienists and the head of Environmental Services met with the Faculty Council and explained that a report was not needed for minor incidents like this. All of the molds on the books were harmless. I guessed we were supposed to trust them.

According to our Tools for Schools training, the teacher representatives were invited to forward the Environmental Services director any questions concerning the program or possible problems in the school. We were told that we were "the troops in the trenches," the front line in the battle against poor indoor air quality. My principal, however, was very controlling, and *nobody contacted anybody* unless it went through her. I had two protocol questions. (1) Who should be responsible under the Tools for Schools program for informing the principal that a fourth teacher in the school year has been put on a prescription for an inhaler? (2) Should it be the teacher representative or the school nurse (also a Tools for Schools member)?

You know you're in trouble when the principal notifies the area assistant superintendent about a personnel issue. The principal was posturing in her emails to paint me in the worst possible light with the assistant superintendent: "I've gotten in touch with Environmental every time you've raised an issue. [The industrial hygienists] have been out numerous times. They have tested and reported. I have shared everything with you in good faith."

I fired back, "It would have been my expectation that Environmental

Services would speciate the mold samples using DNA analysis to see if any of the fungi present were on the EPA group one list of toxigenic molds and to share that information with the teachers and staff of Dumbarton, as per Tools For Schools protocols. As a teacher with a documented biotoxin illness, it would be nice to know if the library is environmentally safe for me to participate in the Maryland State Chess Tournament Saturday, March 21 and 22, where I will be supervising the high school tournament matches." I then listed every possible element of a biotoxin illness that I was facing and noted, "...as the Baltimore County Public School system is a recipient of federal funding and has not acquiesced to even the most minor of accommodations in my workplace environment, I am hereby requesting a formal 504 hearing as specified in the Americans with Disabilities Act of 1990 and Section 504 of the Rehabilitation Act of 1973, with the goal of establishing a 504 plan with approximately one dozen reasonable modifications to the workplace environment, including communication rights not expressly forbidden in the Master Agreement that would be beneficial for a biotoxin-infected teacher with multiple disabilities working in a water-damaged building."

For years, I've complained about having to attend students' 504 meetings. I now look on those meetings as one of the most valuable things I've ever done in my educational career. The words "reasonable modifications to the workplace environment" were from Federal guidelines, and it must have stuck her like a dagger through the throat. After receiving that email, she snapped. She yelled at the secretaries and railed about teachers who thought they deserved bathroom breaks during the MSA testing that was going on. One eyewitness said that she retired to her office to play solitaire. Well, I said, let the cards fall where they may.

On March 21, I took my post at the Maryland State Chess Tournament. Just being in the library was making my throat burn. I walked over to the empty shelf that had once held the 31 mold-contaminated books. There was a shelf of books above it, and a leaky air conditioner with a drip pan above that. The librarian said that the industrial hygienist had been in a hurry to get the books out of the building. Was it possible in his haste that he didn't thoroughly check the books on the shelf above? If water from the air conditioner had nourished the mold on the bottom shelf, it certainly could have encouraged fungal growth on the shelf above. I started looking at a dozen books. They were all plastic hardbacks: not very appetizing for fungi, but then I spotted a single thick cloth volume. If any book was going to have mold, it would be this one. There was no apparent mold. I took my pocketknife and cut the tape holding the dust jacket to the book. The lower half of the cloth binding was covered in mold. Bingo!

Assuredbio contacted me about the samples from the three books. They'd never seen mold samples like this before. The counts of spores per milligram of dust were massive. From the group two list was *Aspergillus ustus*: 12,850,920 spores per milligram of dust. *Penicillium chrysogenum* was 1,712,919 spores per milligram of dust. The book was an asthmatic's worst nightmare. Ten different molds from the EPA group one list of toxigenic molds were present. Some of them had huge numbers, but Dr. Shoemaker's attention was on three molds known to have toxins that can cause illness: *Aspergillus penicilloides, Chaetomium globosum* and *Aspergillus niger.*

I had more proof, but in a school system that covers-up environmental problems and where an email can get you fired, you can't just say, "Look what I found!" I decided that the best way to present the mold data was in my 504 application. The 504 seemed like the best vehicle for getting the mold situation in the contaminated custodial room remediated as well. I wrote up twelve accommodations and then had Dr. Pietrowski make suggestions. I met with Dr. Shoemaker and we finalized the list.

MEDICALLY NECESSARY AND REQUIRED 504 ACCOMMODATIONS FOR PATIENT LEE THOMASSEN

[Revised June/July 2009 superseding previous copies]

1. There should be twenty-four hour unit ventilation in room 116 at Dumbarton (including weekends) maintaining a constant air flow of at least 15 cubic liters of air per second [*note that on or about June 4, 2009, ventilation time increased in room 116 beyond the 5:00 PM shutoff*]. A HEPA 0.3m filter should be placed in room 116 as a companion to the unit ventilator.
2. Airflow checks will be conducted yearly to confirm that the unit ventilator is operating at or exceeding the required air flow in #1 above.
3. Repair requests on the unit ventilator in room 116 will be considered a priority with 24-hour or less turn-around workplace time on repairs and needed maintenance.
4. Once monthly, the unit ventilator will be opened up and vacuumed by the custodial staff using a HEPA vacuum cleaner.
5. The custodial staff will use normal clean-up procedures in room 116, with the modification that the sweeper brooms will not be shaken out in the classroom or hallway within 50 feet of room 116.
6. It is required that one or two window air conditioners be installed in room 116 as part of the window renovation project during the summer and fall of 2009 with adequate BTU for the size of the room to maintain absence of fluctuation of room temperature.
7. Room 116 is ideally suited for a patient with a biotoxin illness with nearby restroom access and a sink for use in cholestyramine treatments. BCPS will keep the restroom and sink in good working order and provide a key to the frequently locked restroom.
8. BCPS will take proper precautions to make sure that room 116

is not damaged by water intrusion, rodents, and construction damage during the window renovation project during the summer and fall of 2009.

9. Mr. Thomassen will be permitted to wear a facial respirator as he deems appropriate for his condition.

10. Because of the highly scientific and technical nature of illnesses relating to the Indoor Air Quality in a water-damaged building, including but not limited to possible exposure to airborne toxigenic substances, toxigenic fungi, biotoxins, actinomycetes, potentially dangerous bacteria and mycobacteria (many having antibiotic resistant properties), inflammagens including, but not limited to, beta-glucans, mannans, proteinases, hemolysins, volatile organic compounds, and spirocyclic drimanes, dangerously high concentrations of inflammatory cytokines in the bloodstream (represented by MMP-9), elevated C4a and TGF-beta 1 readings, low VIP protein levels, low VEGF levels resulting in reduced oxygen delivery into capillary beds, elevated osmolality resulting in severe dehydration, low vitamin D levels, and a critical reduction of MSH, an important hormone that regulates inflammatory responses of the innate immune system and helps the immune system to keep commensal antibiotic resistant strains of staph from colonizing the deep nasal space, BCPS will allow Mr. Thomassen to monitor Indoor Air Quality and humidity levels in the school using appropriate modalities.

11. The BCPS Industrial Hygienists will supervise the remediation of the mold-contaminated dust and dirt in the first floor custodial room between rooms 116 (Mr. Lee Thomassen) and room 121 (Mr. Evan Krantz and former teacher Mrs. Christine Goldman), including but not limited to *Aspergillus fumigatus, Aspergillus unguis, Aspergillus penicillioides, Cladosporium sphaerospermum, Eurotium amstelodami, Penicillium purpurogenom, Stachybotrys chartarum, Aspergillus flavus, Aspergillus niger, Aspergillus sydowii, Aureobasidium pullulans, Chaetomium globosum, Paecilomyces variotii, Penicillium*

glabrum, Penicillium variable, Scopulariopsis brevicaulis, Stachybotrys chartarum and *Wallemia sebi.* Said remediation will follow professional standards (including isolation, containment and exhaust to the outside) and OSHA guidelines to prevent further contamination of the first floor of the school and to protect the health of those active in the remediation process and those who work in school.

12. The OSHA Act of 1970 section 5(a) (1) requires the employer to provide the employee with a workplace free of organized hazards likely to cause serious harm. Multiple incidents of visible mold and dirt and dust containing toxigenic fungi have endangered the health and productivity of BCPS employees at Dumbarton. BCPS is required to:

    - use modern scientific techniques including speciation to identify Indoor Air Quality problems and remediate them;
    - correct the problems in the custodial rooms that are encouraging fungi growth, including dust and dirt accumulation, inferior flooring, plumbing issues, lack of ventilation and lack of air filtration.

The BCPS EEO Office received the final list in mid-July – and they sat on it.

THE SCHOOL SYSTEM AT BAY

I knew that the 2009-2010 school year would be a tense one. I wanted to lay low and wait for the system to react, but then in September, when I was walking down the second floor corridor, there was a cluster of three new mold blooms on a ceiling tile outside room 213. Within twenty-four hours, I was walking toward the front door of the school and I saw a huge black mold bloom on a ceiling panel in the main lobby to the school only forty feet from the principal's office. I used my position on the Tools for Schools Team to report the mold appropriately. Any guesses on how my reports were received?

For the sixth time in less than two years, I was accused of touring the school without being accompanied by the assistant principal or the chief custodian. Oh no, I protested, I was just walking towards the front door to the school. It was a good thing that I did, too, or else that mold bloom in the lobby would've grown to twice its size over the next few months! The ceiling tile outside the principal's office was removed and replaced, but not the cluster of three smaller blooms. They're still there as I write this.

There were so many wet ceiling panels in the school lobby from leaking pipes that I decided to bring my digital camera to school to document them. I counted seventeen, with more in the gym lobby. It was a good thing that I brought the camera, because I was able to capture images of a flood in the custodial room. The sink had clogged up, and the leaky faucet had filled the custodial sink until it overflowed. Water went through the wall into the art storage room, ruining the art teachers' materials. The water then went through the walls and flooded the adjoining boy's bathroom. The Art Department was furious. Their storage room smelled like mildew and the asbestos floor tiles were coming up.

A few weeks later, the custodial room flooded again. All of the teachers on that end of the building knew it, but no one wanted to tell the principal because the policy in the school was always to shoot the messenger. So we waited until 8:15, when a sixth grade student did his daily chore of going to the custodial room to fill a pail of water for his teacher. The child came out of the room ashen-faced, saying that there was water everywhere. The teachers were ready for this and immediately called over the assistant principal to let her know that a 6th grader had discovered that the custodial room had flooded again. Who's smarter than a fifth grader? A sixth grader.

By October, I developed a serious limp from the numerous pains in my legs. Sometimes they were short stabbing pains; other times, it was aching joints. I kept getting Charlie horses and what Dr. Shoemaker referred to as "claw toe." It was very painful, and I knew that MMP-9 was jumping from my bloodstream into joints, muscles, nerves, and other places where it wasn't supposed to be.

I subbed for a math teacher's class one day, and when the children opened their books to begin their class work, I noticed black spots all over the book belonging to the child in front of me. I thumbed through the book, and the mold was on every page from front to back. I walked around the room. Two other students had moldy textbooks.

Tools for Schools gained its first parent representative, Karen Kruger. I'd been communicating with her since early spring and she'd seen my application for 504 accommodations. She was also a member of the Citizen Advisory Board. She wanted to know why I had not contacted MOSH (Maryland's branch of OSHA) about the contaminated custodial room.

On November 6, I received written notification from the EEO Office that BCPS would *not* grant me a 504 hearing and had unilaterally rejected most of my 504 accommodations. In particular, BCPS was formally refusing to remediate the contaminated custodial room. The document was both insulting and biased. Environmental Services and the industrial hygienists denied that there were any problems:

Accommodation 1(a): *"24-hour unit ventilation – request denied. There has been no evidence of any problems in classroom 116. Ventilators run all day while employees and students are in the building, providing ample fresh air. Notes from a meeting held November 13, [2008] state that 'if, after an inspection of the areas the employee thought were problematic, there was a need for long hours for ventilator usage,* [the industrial hygienist] *would work with the automation people to lengthen the hours." That inspection turned up no problems. In addition, by the employee's own admission at that meeting, 'he did not feel that an inspection of his room was needed since, as long as the ventilators are working, his room is fine now..."* [Note that the part about the dead fried mouse caught in the fan assembly was omitted, which changes the entire context of the quote. Many teachers in the school have turned off their unit ventilators because they are too noisy to allow instruction to take place.]

Accommodation 1(b): *"Installation of a HEPA room filter – request denied."* The explanation states that HEPA filters are designed for smaller rooms. Dr. Pietrowski read this and said, "That is why you put two HEPA air filters in the room!"

Accommodation 2: Annual air flow checks – *"Request granted."*

Accommodation 3: Repair requests on the unit ventilator with 24-hour service– *"Request granted."* [I requested that the unit ventilator be serviced on November 30 because the noise was unbearable. It was fixed in mid-February.]

Accommodation 4: Vacuum the air filter monthly with a HEPA vacuum – *"Request denied."* They did promise to clean the vents and wipe the area and change the filters every three months.

Accommodation 5: Normal cleaning procedures, but don't shake out the brooms in or near the classroom – *"Request granted."*

Accommodation 6: Installation of air conditioning in room 116 – *"request denied. Employee has not submitted documentation to support the need for air conditioning."* [Actually we did, but they ignored it.]

Accommodation 7: Keep the sink in room 116 in good working order; keep the nearby restroom in good working order, provide a key to the frequently locked bathroom – *"request granted."* [As I write this, the bathroom has been out of order because of flooding and tile damage for over a month. I never did get the key.]

Accommodation 8: Prevent damage to the room during the window renovation – *"request granted."* [As I write this, I'm staring at a gaping hole in my ceiling where water comes into my room from the newly installed leaky windows.]

Accommodation 9: Use a facial respirator as needed – *"request denied."* [I wear it after 3:00 anyway.] *"There has been no medical documentation submitted to support this request. Proper air quality is maintained in the school on a daily basis. If the employee is experiencing immune response complications, he should not report to the school as it would pose a safety issue for him to be around other people. In addition, if he does not have proper flow of air into his lungs, he should call 911."*

Accommodation 10: The employee will be allowed to monitor indoor air quality and humidity levels – *"request denied. If the employee has evidence that there is a problem in the school related to the indoor air quality or humidity levels, he should report it to the administration who will then report it to the experts who handle that area. The employee is responsible for carrying out the duties of a teacher and all time spent during school hours should be spent on those duties. There is no evidence that the indoor air quality at the school is problematic."* [The building is not air-conditioned. The heat and humidity are stifling in September-October and May-June. As for contacting the administration with indoor air quality concerns, been there, done that.]

Accommodation 11: The remediation of the mold-contaminated dirt and dust in the first floor custodial room – *"Request denied. There is no evidence of mold infestation in the room. The room will be cleaned of all dust and dirt."* [Big mistake! Never promise to clean something in a 504 document and then fail to do it!]

Accommodation 12: *"BCPS is required to assemble personnel to work together to establish a working written scientific IAQ plan to bring the school up to OSHA standards. This statement is not a request for an accommodation."* [Not only that, it proves that the EEO Office was not even reading the July accommodations application document. The actual accommodation request was quoted earlier. It required fixing the problems that were "encouraging fungi growth including dust and dirt accumulation..."

Then to add a further kick in my groin, the school sent letters announcing that BCPS and I had reached a 504 agreement. I had to write to the school board, TABCO and the Office of Personnel denying that any such agreement had occurred. I wanted my hearing!

MOSH

I bowed to pressure from my colleagues and the parent representative on Tools for Schools and contacted MOSH. Karen Kruger said that she'd seen MOSH sweep down on unsuspecting violators after complaints had been filed.

November 11, 2009

MOSH Compliance Unit

1100 North Eutaw Street

Baltimore, MD 21202

Dear Sir/Madam,

I am a teacher at Dumbarton Middle School in Towson. As the teacher representative on the Tools for Schools Team, I have brought a number of serious issues to the attention of BCPS (Baltimore County Public School System) including mice, roaches and mold. We have several teachers (including me) who have illnesses related to the poor indoor air quality in the building, and we think that it is related to fungal growth caused by flooding in the library and library workroom (2nd floor) and art rooms 119 and 121 and their adjoining storerooms (first floor), and a sink that overflows repeatedly in the first floor custodial room located between rooms 116 and 121. We also have major problems with humidity, which seems to be sustaining fungal growth.

I have reported IAQ-related illness to BCPS, as have others. There were four teachers (all within 50 feet of the flooded rooms) who were prescribed inhalers during the 2008/2009 school year. In 2006, the art teacher in room 121 got sick and is now partially disabled and 95% housebound due to exposure to toxigenic mold. In 2007, blood tests showed that I had the same biotoxin illness that she had. The replacement for the art teacher is now sick and his doctor stated that his illness is 100% caused by the poor indoor air environment in the school. The teacher across the hall from me suffered from sinus infections throughout the 2008/9 school year and now faces surgery. She was new to Dumbarton Middle School in 2008/9 and was ill within days of arriving at the school.

There's been a lot of visible mold in the school since August 2008. Visible mold at Dumbarton was rare before that. In the 13 months between August 2008 and Sept 2009, there have been the following instances of visible mold in the school:

- *Black mold in the library workroom sink unit;*
- *Black mold on a ceiling panel outside the library;*
- *Black mold on a second ceiling panel near the library outside room 213 (it is still there in fact, the affected area being about 3-4 inches in diameter);*
- *Black mold on 31 library books;*
- *Black mold on a ceiling panel in the lobby.*

There may have been other instances that I don't know about. I've recently noticed students walking around with mold in their math textbooks.

The Department of Environmental Services will usually come to the school and remove the visible mold, but they're refusing to speciate the visible mold to determine if it is on the EPA Group 1 and Group 2 lists. They're also refusing to remediate the contaminated dirt and dust in the affected areas. They assure us that all molds in the school are harmless. They're also refusing to keep written records of these visible mold removals (despite MOSH recordkeeping requirements) by claiming that they're minor inconsequential incidents not worthy of reports.

As the Teacher Representative on the Tools for Schools Team, it is with regret that I believe that the Department of Environmental Services has been lying to the faculty and administration about the toxicity of the visible mold and airborne mold spores in the school. One of my physicians is a noted scientist, the author of Mold Warriors and the Maryland Family Practice Doctor of the Year in 2000. He has had me collect dust and bulk mold samples for speciation. I am enclosing two of the lab reports for your inspection (please keep these lab reports confidential). The first is the lab report from a library book that Environmental Services missed when they were removing thirty-one mold contaminated books in the spring of 2009. Of particular alarm is the presence of ***Aspergillus niger*** *(22,882 spores/mg dust),* ***Chaetomium globosum*** *(10,331 spores/mg dust); and* ***Aspergillus penicillioides*** *(113 spores/mg dust). The exceptionally high spore counts from both the Group One and Group Two lists are also a major concern. [This book is still in my possession should you wish to inspect it]. The second lab report is from shelf dust in the custodial room between my room (116) and the art room with the sick teacher (121).* ***Aspergillus fumigatus*** *is present (3,514 spores/mg dust). A small amount of* ***Stachybotrys chartarum*** *was present in the dust sample. I have also enclosed a color picture of the disgusting and unsanitary conditions in that custodial room.*

On March 18, 2009, I requested a formal 504 hearing as specified in the Americans with Disabilities Act of 1990 and Section 504 of the Rehabilitation Act of 1973. In July, my medical team sent the EEO Office of BCPS twelve required accommodations to bring BCPS into compliance with ADA. Required Accommodation number eleven reads as follows: "The BCPS Industrial Hygienists will supervise the remediation of the mold-contaminated dust and dirt in the first floor custodial room between rooms 116 (Mr. Lee Thomassen) and room 121 (Mr. Evan Krantz and former teacher Mrs. Christine Goldman), including but not limited to Aspergillus fumigatus, Aspergillus unguis, Aspergillus penicillioides, Cladosporium sphaerospermum, Eurotium amstelodami, Penicillium purpurogenom, Stachybotrys chartarum, Aspergillus flavus, Aspergillus niger, Aspergillus sydowii, Aureobasidium pullulans, Chaetomium globosum, Paecilomyces variotii, Penicillium glabrum, Penicillium variable, Scopulariopsis brevicauisi and Wallemia sebi. Said remediation will follow professional standards (including isolation, containment and exhaust to the outside) and OSHA guidelines to prevent further contamination of the first floor of the school and to protect the health of those active in the remediation process and those who work in school." I received word this week that most of the accommodations that were required by my medical team have been rejected [by] the Dept of Environmental Services and the EEO Office including number eleven.

Another one of my physicians, Dr. Colleen Pietrowski, notified the EEO Office that I've been diagnosed with a health impairment that substantially limits the major life activities of respiratory function, vascular function, and immune function, which qualifies as a disability under Section 504 of the Rehabilitation Act of 1973. According to EPA protocols, anywhere that there is an immune compromised individual, you must use the highest level of containment and remediation procedures. A letter to me from the EEO office indicates that BCPS intends to use normal cleaning procedures on the contaminated custodial room in the near future (probably without even a facial respirator for the unsuspecting custodian), which will cause Aspergillus fumigatus to contaminate the first floor of the school. The custodial room is only ten feet from my classroom (room 116). I'm worried that exposure to Aspergillus fumigatus will further compromise my respiratory, vascular and immune systems.

It is for these reasons that I am notifying MOSH in a formal complaint about the unhealthy Indoor Air Quality conditions at Dumbarton Middle School and my good-faith effort to get the BCPS Dept of Environmental Services and the EEO Office to investigate the IAQ concerns of the faculty and my physicians using the most modern scientific methods available, including speciation.

While I waited for MOSH to sweep down like the cavalry from old and save the day, I attended my first ever meeting of the Tools for Schools Team at Dumbarton on November 24. The Tools for Schools Teams county-wide had been ordered by one of the industrial hygienists to do two walk-throughs per year instead of one. Three members of our team didn't want to do it, so we basically ignored a mandate from the Dept of Environmental Services to inspect the building. I brought three of the moldy math books to the meeting. We agreed that students shouldn't be carrying moldy textbooks, and I volunteered to write a letter to the principal and turn the books over to her to give to Environmental Services. We recommended collecting the math books for inspection and replacing any that had mold in them. It was now 6 p.m. on the day before Thanksgiving vacation, so I placed the books in a sealed bag with the cover letter and put them on the office counter for the Principal's inspection the following Monday.

Meanwhile, MOSH had completely blown me off. Not only did they say that they didn't investigate complaints, but as a courtesy, they wrote to BCPS to let them know that I had complained. They identified me by name to my employer. Their only recommendation to BCPS was to increase the unit ventilation.

As if that weren't enough, Karen Kruger emailed the Dumbarton head of the Tools for Schools Team on Sunday, November 29 about issues that had been brought up at the meeting on the 24th:

> *Jerry: This sounds like a needed fix for the boy's lavatory – thanks for the follow–up. While I was at school last week after our meeting, I took the opportunity to look around. I did notice those water stained tiles in the foyer, which I understand had been changed a few weeks earlier. I take it that this represents continuing damage from those old water pipes. I also peeked into that awful custodial closet near the darkroom in the 6th grade wing – boy that should be condemned! I hope there is some plan to remediate that mess – nobody should have to work in or near that. Also, I wonder if it would be prudent to request some radon testing in the basement in the spring. I've only recently learned about the basement since my daughter just started weight training for gym. Finally, did we learn anything new about those math books?*
> *Thanks for your work in this important area and for allowing me to participate as the parent representative.*
>
> *Karen J. Kruger*
> *Senior Assistant County Attorney*

The assistant principal immediately forwarded the email to the principal. Between the notice from MOSH, the three moldy math books, and the parent Tools For Schools representative complaining about the custodial room, the principal was fuming on the morning of Monday, November 30. She stormed into my classroom and shouted, "What's that noise?" I noted that it was my loud unit ventilator. Could it please be fixed? But that was not why the principal was there. I was accused of taking a parent on "a tour" of the building and I was removed from my position as the teacher representative on the school's Tools for Schools team.

In addition, I was formally reprimanded to the area assistant superintendent. The letter of reprimand states that I admitted my crimes to the principal. I answered back with a letter of my own and denied any such thing. I asked to receive a copy of the Karen Kruger email, but the principal refused to give it to me. The principal offered to meet with me or my representative, but when Karen Kruger tried to meet with her, she was rebuffed. The school board had decided ages ago that any parent can walk through and inspect any room in any BCPS public school.

RIGHTS VIOLATED

One day during the winter, it hit me: BCPS had granted five 504 accommodations. I contacted the EEO Office. While I was in the appeals process, were the accommodations that BCPS and I agreed on in effect or in hiatus? The school wrote back that they were in effect. Suddenly, the principal came alive and started reading the 504. On March 1, the principal contacted the facilities people about my request to have my unit ventilator fixed. My room was dusted for the first time that school year. My trash, which had not been emptied in three days, was emptied. To my amazement, however, the principal was not fulfilling two major accommodations: I had not been given a key to the frequently locked bathroom near my room and the unit ventilator filters had not been changed quarterly.

Teachers had been sick throughout the building since the beginning of the school year. This is normal in a school building. Principals are required to complete OSHA Logs and Summary forms yearly and post the Summaries for inspection. The Dumbarton Summary stated that not one teacher had a respiratory illness in 2009! And that was with four teachers on inhalers for the first time in their lives! I protested to the Faculty Council. The principal explained that her orders were to not record any respiratory illnesses unless the teacher had filed for Workers' Compensation. I contacted MOSH and got the answer that I wanted: the OSHA Summary forms weren't being completed accurately. MOSH contacted BCPS with their finding in May 2010 and announced an investigation.

After being rebuffed in a letter to the Baltimore County Administrative Office, I contacted the Office of Civil Rights, U.S. Education Department, Philadelphia Office, complaining that BCPS had denied my right to a 504 hearing. I noted that BCPS wasn't even following the accommodations that they'd accepted. The Office of Civil Rights agreed that my complaint had been filed within the required time period and merited an investigation. BCPS received notification.

A MOSH investigation and now an investigation by the Office of Civil Rights— I can only imagine the pressure was now building on the BCPS Dept of Facilities. The air filters hadn't been changed in six months in most rooms and more than a year in others. Late at night on Monday May 11, the unit ventilator filters were quietly changed when the teachers were out of the building, but things went wrong. The filter in room 210 was accidentally left on the floor, and it was dated April 9, 2009 (thirteen-months old!). Then on Friday May 14, I noticed the spent filters stacked on top of the dumpster as I pulled into the parking lot. The chief custodian saw me taking the pictures and stormed out of the building screaming that I wasn't allowed to do that. (*Editor's note*: there was no physical violence that followed from this screaming at Mr. Thomassen.) Her face was blood red and she was flailing her arms madly in the air.

I knew that pressure was building for Facilities to clean out the contaminated custodial room. I borrowed the video camera from the Tech Ed Dept and began shooting images of the room. Then I took more still pictures. Whatever level of remediation was being planned for the custodial room, I wanted to preserve images of what that room looked like. I showed the photos to Dr. Shoemaker during my appointment on May 12.

THE STEALTH REMEDIATION OF THE CUSTODIAL ROOM

On May 11, Dave Gilotty from the Department of Facilities signed in at 11:05. All Facilities personnel were required to sign in when visiting or working in a school. He entered the contaminated custodial room and took stock of the situation. People who saw him stated that he didn't look happy.

The chief custodian went to the chairman of the Art Department that week and asked if she had any paint available. She said no and cautioned the chief custodian that there were people in the school with sensitivities to paint.

On the morning of May 17, the chief custodian and Paul Hergenhahn entered my classroom. He hadn't signed in on the log and he wasn't wearing any visible identification or a visitor's pass. I introduced myself, but he wouldn't give me his name. He said that he needed to photograph my unit ventilator. They opened it up and the custodian wrote "May 11, 2010" in permanent marker on the filter and photographed it. I asked if he was from Facilities and he said yes.

They went into the now empty custodial room and began to scrub down the walls, shelves and floors. Then they started to paint with industrial grade enamel paints. The door was closed and there was no ventilation in the room – not even a window. The Art Department chairman protested to the chief custodian that you can't do this around children, but she ignored the complaint.

That evening, I entered the custodial room along with four curious members of the evening custodial staff. I began taking pictures.

The next day, the indoor air quality in my classroom was horrible. It was my worst nightmare – an improper remediation in the custodial room that would send *Aspergillus fumigatus* spores throughout the school and into my room. As I related at the beginning of this chapter, I passed out in my own classroom during my planning period. One of my eighth graders passed out and struck his head on a desk. Students complained about not feeling well all day long. I developed migraine headaches and had to take time off from work. Then I came down with bronchitis and Dr. Davis put me on an antibiotic. I notified MOSH in May with two letters and a photograph of the two enamel paint cans. On June 14, I filed a formal protest with TABCO.

> *I am a teacher at Dumbarton with documented sensitivities to industrial strength paint. I have also been diagnosed with a health impairment that substantially limits the major life activities of respiratory function, vascular function and immune function. This qualifies as a disability under Section 504 of the Rehabilitation Act of 1973. On October 28, 2009, the Baltimore County Public School System granted 504 accommodations for my illness.*
>
> *It is my sad duty to report that Baltimore County Public Schools began remediation procedures on Monday, May 17 in a water-damaged custodial room ten feet from my classroom without following EPA guidelines. According to EPA protocols, anywhere that there is an immune compromised individual, you must use the highest level of containment and remediation procedures. The custodial room in question is only ten feet from my classroom. BCPS facilities personnel painted the custodial room on Monday May 17 and Tuesday May 18 during school hours with two industrial gloss enamels in a stealth operation without notification of the school's administration and without properly ventilating the room to the outside of the building. This large unventilated custodial room (approximately 20 by 8 feet) vented fumes into the hallway and into my classroom, and these fumes are the probable cause of my passing out in my classroom on Tuesday May 18. An hour and twenty minutes after I passed out, an eighth grade student passed out in my classroom. He was injured when his head impacted on a desk as he fell and he was sent home.*

I would like to file a protest with Baltimore County Public Schools using the good offices of TABCO relating to the following issues:

13. The principal, building facilitator, school nurse, teachers and parents were not notified in advance or on the day in question that painting with industrial grade enamels would be occurring inside the building during school hours.

14. The man from Facilities who provided the two industrial strength enamels did not sign in when he entered the building. I believe that the man is Paul Hergenhahn. (see attached log – note the lack of a signature on May 17). When he entered my classroom on May 17, he was not wearing BCPS identification and he did not have a visitor's sticker. I introduced myself as Lee Thomassen, but he would not provide his name in return. He only stated that he was going to photograph my unit ventilator. I asked him if he was from Facilities and he said yes. I believe that BCPS regulations require Facilities personnel to sign in on the Maintenance/ Operations Personnel Log and wear visible identification at all times.

15. The school's chief custodian permitted and took part in the painting of the custodial room during school hours on May 17, despite a protest from Ann Summerson, the head of the Art Department. She also gave the chief custodian an explicit warning the previous week that there was a person in the school with sensitivities to paint.

16. BCPS failed to use full containment and remediation procedures despite a written requirement from me, Dr. Pietrowski and Dr. Shoemaker in my application for 504 Accommodations dated July 2009. The requirement for remediation in the custodial room demands that BCPS "... follow professional standards (including isolation, containment and exhaust to the outside) and OSHA guidelines to prevent further contamination of the first floor of the school and to protect the health of those active in the remediation process and those who work in school." The remediation on May 17 and 18 did not follow the requested professional standards.

17. BCPS facilities personnel failed to follow required EPA guidelines when remediating the custodial room. The EPA Technical Bulletin # 148, June 2001, page 2 states that “It is recommended that all persons with asthma, hypersensitivity pneumonitis, severe allergies, immune suppression, or other chronic inflammatory lung diseases be removed from the contaminated area until remediation is complete.”
18. BCPS facilities personnel exposed the children and staff of Dumbarton Middle School to fumes from an industrial-grade enamel paint that carries the following warning label on the front of the can: “WARNING! COMBUSTIBLE, VAPOR HARMFUL” (see attached photograph taken on May 17, 2010).
19. BCPS facilities personnel exposed the children and staff of Dumbarton Middle School to industrial grade enamels that carry the following information in their Material Safety Data Sheets:
    - *SHER-CRYL HPA High Performance Acrylic Semi-Gloss Coating, extra White/Tint Base; EFFECTS OF OVEREXPOSURE – EYES: Irritation; SKIN: Prolonged or repeated exposure may cause irritation; INHALATION: Irritation of the upper respiratory system. In a confined area vapors in high concentration may cause headache, nausea or dizziness. [MSDS provided by Dave Gilotty; May 31, 2010; see attached.]*
    - *TECHGARD Maintenance Gloss Enamel, Base 1; EFFECTS OF OVEREXPOSURE – EYES: Irritation; SKIN: Prolonged or repeated exposure may cause irritation; INHALATION: Irritation of the upper respiratory system. May cause nervous system depression. Extreme overexposure may result in unconsciousness and possibly death. Prolonged exposure to hazardous ingredients in Section 2 may cause chronic effects to the following organs or systems: the liver; the urinary system; the*

> *reproductive system; SIGNS AND SYMPTOMS OF OVEREXPOSURE – Headache, dizziness, nausea, and loss of coordination are indications of excessive exposure to vapors or spray mists. [MSDS provided by Duron, May 28, 2010; see attached.]*

I consider the actions of the Department of Facilities to be a violation of my rights under Section 504 of the Rehabilitation Act of 1973, the health and safety clause in the Master Agreement, and The OSHA Act of 1970 section 5(a) (1) which requires the employer to provide the employee with a workplace free of organized hazards likely to cause serious harm. Please discuss these serious concerns with BCPS about the questionable procedures employed by the Department of Facilities during school hours at Dumbarton Middle School on May 17 and 18, 2010.

THE END? NOT IN A WATER-DAMAGED BUILDING

I can offer no happy ending to this story. I have active protests being investigated by my union (TABCO) and MOSH. I've filed a formal complaint with the Office of Civil Rights, U.S. Department of Education. There are two new mold blooms that have formed on the second floor of the school. By the time this book is published, my health may have been compromised to the point that I'm no longer able to work.

Let me close with the words of Marc Danzon, the WHO Regional Director for Europe:

> *"Healthy indoor air is recognized as a basic right. People spend a large part of their time each day indoors: in homes, offices, schools, health care facilities, or other private or public buildings. The quality of the air they breathe in those buildings is an important determinant of their health and well-being"* [forward, pg xi, WHO Guidelines for Indoor Air Quality: Dampness and Mould].

WHO chapter 2, page 7 notes that at least 20% of the buildings in the United States, Canada, and several European countries had one or more signs of dampness. My school is one of those 20%. You see evidence of it in the buckled floor and loose tiles in the locked boy's bathroom. You see it in the wet ceiling tiles in the lobby. You see it on the floor of the custodial room, which has no floor drain. You feel it every May and June and again in September and October when the humidity in the school is often over 90%.

A healthy indoor air environment is a basic human right, and I intend to fight for that right for myself, my fellow teachers and my students for as long as my strength lasts or until the day that I succumb to my biotoxin illness. For those individuals who lie and minimize indoor air quality concerns and who refuse to use modern science to test for the toxicity of molds in schools, let me state that you injure not only your fellow employees in the school system but our precious children— the most innocent of victims— and the very reason why our public schools exist in the first place. Our children deserve better than this.

CHAPTER 21

A Tourists' Guide to Moldy Buildings in Washington DC: Where NOT to Visit

If you travel on airplanes a lot like I have to, there comes a time when doing more paper work or reading some immunology journal has just got to stop. Sleep might come, until the thunderheads at 35,000 feet over Kansas wake the dead and scare the living. At that time, I wish I were back on Kansas and not over it.

But after such abrupt awakenings it is hard to snooze again. What to do? Simple: read the travel magazines. Sure, the articles all feature wonderful places accessed by the airline you're on and the sales pitch to visit the Shangri-la after traveling on the airline is never-ending. It took me a while to learn that the writing in these pieces is pure marketing.

I couldn't help but share some nervous laughter (the turbulence awaited the next west-bound plane; we were in smooth air) reading a two-page spread recently that featured a large Midwest city, a city of shrinking population and financial woes shared by so many other big cities in these economic times. Still the writing was almost seductive: "Ample opportunity for new opportunities" really meant that the work force and consumers had fled. "A wide-open horizon for economic rebirth," meant that the infrastructure was so damaged that any change would only be coming from stimulus package programs. "Tawny hues of the fall skyline" were actually rusting hulks of unused manufacturing plants.

The article didn't talk about a 25% high school graduation rate, 33% unemployment and a 66% foreclosure rate for new mortgages within the metro area.

But other than that the article talked about elements that were:

unique, classy, sleek, pure enjoyment, indulgence in bliss and comfort,spacious,uncrowded(weknowwhatthatmeans),impeccably presented, ultimate, intimate surroundings, complimentary Wi-Fi, affordable (yes, we know what that means, too), explore. You get the idea. And how about, deep, *sugary* (what does that mean, sugary?) sand (I don't want that sand on me from a lake; what makes it so sticky? Something that kills birds, puts holes in fish and makes rivers burn?). Stunning, impossibly turquoise, inexpensive, snag the bargain, wrap your hands around a luxuriously succulent lemon-scented fabric (succulent fabrics?), elegant, rave reviews, were just a few of the literary sequins on this jeweled masterpiece of tourism.

Even if I trusted the airport to be safe (I don't; see Appendix 6), there is no way I would go visit that relic city from the industrial past. But that's just my spin on life. I'd rather be listening to the chorus of frogs in mid-March (they are actually called 'chorus frogs' and aside from the calls of tundra swans overhead, I think amphibian calls are most restful sounds of Nature) or planning to visit the Japanese cherries blooming along the Tidal Basin in our Nation's Capital beginning in two weeks. That is if the cross country skiers have left town. Imagine DC getting more snow in 2010, and a lot more snow at that, than Buffalo! (The Kansas storm clouds were bringing even more piles of more snow to the Mid-Atlantic.)

On second thought, driving all the way up to DC and back in one day is a bit much; that Beltway traffic will just frustrate anyone. As much as I like to travel, I value the serenity of the open road, without crowds all piling in the left lane when I wanted to be there with space ahead of me. At least in DC, there isn't too much worry about people driving too fast on the Beltway. In DC, it is so odd that people drive on the parkway, park on the driveway and then park on the Beltway, too.

So when one of my 'distance patients', one just devastated by her moldy and water-damaged home, but now doing so much better, called the other day wanting to go sightseeing in DC after her follow-up visit in Pocomoke, I might have been a bit negative about driving in the area. "Take the Metro (train) from Rt. 50 East and find a place to stay in town close enough so that you can walk to the attractions."

"What should we see?" her husband asked.

"Well, wherever you go, don't go to the Rayburn Building. We know that one is moldy. Just look at the ideas coming out of that group if you don't believe me. And don't go to the big courthouse on Constitution Avenue; the maintenance guys showed me some areas where they put buckets to catch the rain. (And was that "soot" way up on the ceiling? I had to testify in that building!). And it's OK to go to the Museum of Science and Industry, but stay out of the actual Smithsonian Castle. I've had two people go through repetitive exposure protocols there. Made them sick as can be.

"Come to think of it, I have a list of patients made ill by buildings in the DC area. I probably should write it up for you," I said.

The problem is that I have a roster of hundreds of people proven to be made ill by a repetitive exposure protocol and they aren't listed by building location. So there was the person from the FAA building and then there was the lady from the fancy hotel, and the Ritz Carlton condos, of course. Oh, and the USDA building and the Pentagon too. I needed some organization.

Maybe it would be easier to make up a list of the Federal buildings where I had NOT found mold illness.

The reality of mold illness is that if there is one person proven to be made ill, confirmed by the repetitive exposure protocol, the likelihood is overwhelming that when proper screening is done, and there is decent evidence that water intrusion wasn't dried out in 48 hours, we will find about 24% of people ill from the water-damaged building. I often will have a group of people come forward after some brave soul says she is ill from a Federal Building. Until the Detroit Metro Tower Air Traffic Controllers case (Haefner v. FAA, Appendix 6), the Federal Workers' Compensation Office (OWCP) didn't agree that mold exposure made people ill, but after that precedent setting decision, not only was "mold illness" identified as a source of disability but a diagnosis code was assigned to it as well. So some of these older cases I'm going to tell you about suffer from not having the benefit of that correcting precedent.

With all the argument about health care these days, I hear that there is a lot of money lost through fraud and waste, with Medicare being a frequent victim. But what if the Medicare claims were being handled by moldy workers who had trouble with recent memory, concentration, decreased ability to remember what they read, confusion and disorientation? That would be the case if there were water and microbes thriving in their work place, right? Let's go to the Center for Medicare (and Medicaid) Services Building at 7500 Woodlawn Boulevard. After the big flood from Hurricane Isabel in 2003, nothing was done about the soggy carpets on concrete floors. And the-water stained ceiling tiles under the condensation from the AC lines above the dropped ceiling say the building is not safe. With a sickened patient proven to be made ill by the building and nowhere else, the managers did commission an air quality study which (of course, this is the way to hide bad news about air quality) took but *six air samples* in this huge office building, each of which showed enrichment of Aspergillus and Penicillium compared to the great outdoors, but because so much mold was found in the outdoor air, the inspector said all was well indoors. We know now that such sampling is desultory at best and it is basically worthless to compare spores of outdoor fungus A to indoor spores B. Yet that was the repulsive attempt at logic used by the building managers. In 2010, no one without bias agrees that trivial air sampling, done for five minutes on one day with no recorded indoor conditions has any relevance to what the actual chronic conditions were.

Needless to say, I would really not want my Medicaid tax dollars to be allocated by someone with cognitive impairment. So don't go to that big building to see health claims for the entire country processed.

Imagine that you are a relative of a highly placed inspector for the USDA. You work in the USDA South Building at 1400 Independence Avenue SW and the nearby Cotton Annex (owned by General Services Administration). You are so sick you can barely get to work and because you are one of the few people trained in Food Safety, you continue to become sicker and sicker at work but also at home, too. The DNA testing done on dust from the South building shows massive elevation of total fungi and a predominance of organisms from the "Group I" roster of fungi found inside water-damaged buildings. Wouldn't you want the valuable employees, one of those few who are familiar with the dense jargon of the USDA, to be safe from the attack of the bioaerosols? Don't go visit the USDA South building.

Please recall that one of the first abstract functions that is impaired in those patients with mold illness who have cognitive impairment (and over 90% do) is the ability to perform simple arithmetic. I ask patients to divide 7 into 91 (no pencil and paper, just see the numbers) because most of them can't even see the numbers to do the division. Imagine if you were a trusted advance degree staffer working for the US Bureau of Labor Statistics at 2 Massachusetts Avenue NE. Would you be surprised to find the ERMI in your office was equal to that of a New Orleans house in St. Bernard's Parish after it had dried out from being under water for a week? Would you want the analysis of our economy and jobs to be performed by someone who had a TGF beta-1 over 15,000 and a C4a that was way too high as well?

I'll tell our visitors to stay way from the Bureau building, too.

Come to think of New Orleans, wasn't there concern about the work on the levees done by the US Army Corps of Engineers? Perhaps those involved in the New Orleans sites worked at one time in the US Army Corps of Engineers building at 441 G Street NW in DC, another building owned by GSA. I only have one patient from that building. She saw mold, smelled musty smells in most of the building, worse around the vent that blew onto her cubicle. High levels of Aspergillus/Penicillium were found. Her main duties were focused on reconstruction of Iraq. She was treated by transfer away from the building and then was compensated by an award for disability.

Don't go into the US Army Corps of Engineers building.

In a sad twist of perspective, one of the sources of illness in DC might be the FAA building at 600 Independence Avenue. Apparently, the same people who made such a bungle out of the job of remediating the Metro Airport Tower in Detroit are headquartered there. To be candid, this FAA building case has a lot of "baggage," but no patients can make up the lab results. Good science isn't naïve.

Still, stay away from the FAA building anyway.

Whatever you do, stay away from the US Department of Treasury Building at 3700 East-West Highway in Hyattsville. I have been given a thick stack of reports that condemned indoor air quality in that building. Patient after patient has come from this site. It's bad all right. Even NIOSH, investigating after 94 people signed a petition asking for immediate help, found something to worry about. NIOSH rarely reports out obvious illness. They missed the fact that the basement is a birdbath, subject to repeated flooding. The flat roof leaks and there are problems with pipes on the second floor of the seven-story building. I wonder what analysis of Federal contracts takes place in that moldy place.

But you can drive by the office. Park and look at the fixed windows of the building; look at the way the water from the land is funneled into the basement. Imagine where the air returns are for the building (hint: the basement). Marvel at the ability of the HVAC system for the entire building to be a universal distributor of bioaerosols from the moldy, wet basement. But don't go in.

This next one is scary to me. The J. Edgar Hoover Building is implicated by the repetitive exposure protocol performed by an employee there. Who knows what kind of National Security issues are discussed in that building. I wouldn't want anyone in a top secret role to be fatigued, in pain or brain-fogged, would you? Don't go in, even if you have the needed security clearances.

Maybe you'd like some outdoor recreation while you're in DC. I'm sure by now you have lots of concerns and some loss of confidence in the maintenance of the integrity of the indoors in our nationally owned buildings. How about a nice ride on the George Washington Memorial Parkway? The outdoors is fine. Molds found in piles of leaves, compost and rotting wood won't hurt you. They are subject to nature's evolutionary balance of competing organisms, unlike the man-made internal environment of water damaged buildings in which the most toxic molds out-compete the lower ranking members of their relatives. This National Park is part of the U.S. Department of the Interior National Park Service. Along the 18-mile road that runs from Mount Vernon, home of George Washington himself, to Key Bridge, extended several times to create a 7,600 acre national park along the Potomac, you'll find the Great Falls of the Potomac; Clara Barton National Historic Site; Arlington, home of Robert E. Lee; Dyke Marsh Wildlife Preserve and much more. Don't forget the Iwo Jima Memorial and Arlington National Cemetery. Just along the highway, along the Potomac River, is Ronald Reagan Washington National Airport. Enjoy the spectacular views of Washington DC, just across the river, but stay out of the buildings. Especially don't go into the boiler room area of the Clara Barton House and you will be fine. You might want to suggest that someone fix the roof on the house and correct the water intrusion in the boiler room area.

I've got a list of more places to stay away from, but I think the idea is clear. If there is a man-made structure, it will leak. When maintenance costs are cut to save money and poor building design is built into the Federal idea of where its workers should be, there will be sick people. If you want to work for the Feds and you need accommodations because the structures owned and maintained by the Feds are making you ill, just remember that the person in the big office who has control over your job is marking time off his clock to qualify for his pension. If he lets you have your way, politics being what it is, then his job will be toast. Besides, you've only got 12 more years to go to reach your pension. Then you can get out of the building and LIVE!

<u>Unless your illness is then too far advanced to fix.</u>

So maybe you would like to re-think going to DC. Tour the Eastern Shore instead. Just imagine all the wonderful marketing language we could use to lure you to stay a while on this beautiful part of Maryland. Let's see how the advertising spin might read.

The Pocomoke Nature Trail is a restoring walk through pristine cypress swamps. If the birds don't attract your attention then the unique vegetation of this boundary zone between North and South, all along the Atlantic Flyway will surely whisper secrets of Nature to your day. Crane fly orchids? Swamp alders? Possum haw? And maybe, if you watch closely you'll be one of the few to see the locally rare and endangered red-cockaded woodpecker. Maybe you would rather skip the 2-mile Trail to see how a waterman makes his living. Or visit our newly-opened museum, the Discovery Center, home to the live sturgeon in their aquarium and replicas of life along the Pocomoke River from times long forgotten elsewhere.

Make your way just a few miles south to see the descendants of Misty of Chincoteague. The ponies are still wild on the island of Assateague, living free like their ancestors who swam ashore from shipwrecked Spanish galleons. On your way off the island, don't pass up the Island Creamery because if their ice cream isn't the freshest and best you've had, it will be a close runner-up. Or maybe you'd like to take the open air sight-seeing boat trip to Tangier Island after your walk through our ancient forest. Experience the Chesapeake Bay first hand; take yourself back to the bay breezes and salty sprays, no different now from the days of Captain John Smith in 1607. The sea food is plentiful on Tangier! I hope you can make room for the famous Smith Island cake, now the state dessert of Maryland, after you finish the piled-high plate of fresh succulent seafood, cooked one meal at a time, just so, by experienced hands. Be sure to listen for the Elizabethan twang to the speech of the few remaining residents, all calling John Smith's crew their forbears. By the time your last scrumptious forkful of jumbo lump crab in the freshly made Maryland crab cake is a warm memory, you've got just enough sunshine left to follow the Bay to Beach Indian Trail to lead you to fun-filled Ocean City. Don't forget the French fries at Thrasher's on the boardwalk! Take a walk along the rolling breakers of the Atlantic. The sky is so clear, you can almost see to England!

Just do anything else besides going indoors in a Federal Building. Said another way – try to stay as far away as possible from the government.

CHAPTER 22

Black Blizzard: A Lesson From History

Can you tell me what the Black Lizard was? No, not the Texas-styled barbecued end pieces of a Gila monster (my first thought was "blackened Lizard"). I mean the Black Lizard of the dust storms in Oklahoma in the 1930's. I'm not sure why this environmental disaster from misguided soil practices of the Great Plains is called the Black Lizard: Blizzard sounds more appropriate.

We read about the 1930's Dust Bowl in grade school. The monster dust storms of that era weren't just dry winds. Those storms were filled with top-soils and sub-soils from hundreds of square miles of previously fertile farmland. I've talked with a few who survived what they used to call the "Black Blizzard." The rare photos on TV about the 1930's in the Central Great Plains don't portray even a small portion of the environmental disaster that compounded the Great Depression's economic destruction.

Oklahoma and Texas still suffer droughts, wildfires, dust storms and tornadoes, but the Black Blizzard hasn't returned.

The History Channel aired a show about the Dust Bowl in mid-April 2009. As one problem after another unfolded on the screen, I could see a natural parallel to innate immune responses in my patients. Imagine finding analogies of indoor environmental disasters in outdoor environmental disasters. Inflammation in my patients was out of control as a result of failed building management techniques, exacerbated by a deep lack of understanding from physicians about CIRS-WDB. That idea is no different, really, from the consequences that Great Plains farmers and townspeople faced when control of the soils was lost due to failed land management techniques, which were worsened by an ever-deepening drought.

Black Blizzards haven't returned partly because President Franklin Roosevelt pushed for the Emergency Farm Act, which formed the Soil Conservation Service (SCS) as a part of the U.S. Department of Agriculture in 1935. Before the drought years of the Depression, in times of ample rain, plowing the rich silky soils of the Plains yielded ever-increasing yields, so much so that any land that could be farmed was probably going to be farmed. Cutting down natural windbreaks along ridges and ravines provided even more land to plow and plant. When the rains disappeared, though, there was nothing left to hold the silts on the land, so when the winds began, tiny particles of soil blew away. Unchecked, the attacks of ongoing natural forces exerted devastating damage when the controlling factors that protected Oklahoma's ecology disappeared. Doesn't that sound just like the cytokine and complement attacks, the natural forces of our innate immune defense, going wild when the protection of MSH and VIP is removed?

The Dust Bowl is a story about desperate measures tried in desperate times. At first, the drought brought hunger to cattle and people alike. As cows died from thirst and starvation, people no longer had milk or meat. As the drought progressed, dust storms brought dust pneumonia, which first hurt the very young and very old.

Hopelessness and despair rode along with inhabitants as they fled, seeking a better life in places west, especially California. They had no crops to sell and titles to "worthless" land didn't help. Those who could buy land for a penny did, if only to hold on until some work appeared. The History Channel simply said that "so many grew old quickly."

Every day was slow torture as the grinding particles of dust tore at skin and clapboard. The fear that the wind would pick up soon tore at people's thoughts, too. When no home is safe, protected from damaging effects of what's in the air, no one feels secure. Each day without rain added to the massive soil erosion, which rose to levels not recorded before or since. Imagine the millions of years it took to build twelve inches of topsoil, which was blown away in just a few years, to the point that nothing grew where the soils eroded.

Rains were welcomed at first but the moisture they brought was costly. There was no top soil to hold the water. Even a brief thunderstorm washed away the overly dry soils, too light to stay on top of the ground and too dry to absorb water fast enough. More loss of soils from rains exposed more subsurface soils to the wind, which led to even more erosion. The next rain would cause even more damage carving chasms into hillsides such that the next run-off was even faster. Flash flooding was more common.

The modern homeless person, driven out by microbial growth in water-damaged buildings, suffers the same erosion of health, and each new episode of exposure brings a new level of illness. If the Dust Bowl represents one of the worst environmental disasters of all time, then human illness acquired from water-damaged buildings is an ecological health disaster on a scale no one has yet measured.

The worst of the Dust Storms happened on April 13, 1935, on a day that's come to be known as "Black Sunday." That storm dumped high-plains dust from Denver all the way to New York City. In one day, a storm created dust clouds 20,000 feet high and removed 300 million tons of top-soil. One day! Even counting the massive dust storms in the Sahara Desert that blow dust into the Gulf of Mexico every summer causing respiratory problems for a few weeks in Tortola (British Virgin Islands), Black Sunday was the biggest dust storm of all recorded time.

For those few gritty people who refused to leave their homes due to drought, storms, poverty and illness, there was a seemingly never ending supply of tortures ahead. I can't imagine living with constant static electrical shocks as the History Channel show portrayed. Filling a car with gasoline became no different than handling a Molotov cocktail with a lighted match in one hand. Mold patients know all about the zaps they get from any electrical ground, including light switches, car doors and people.

And then there were the plagues. Every creature suffered not just the livestock and people. Scorpions needed water, too, and when there was none left in their natural habitat, they were forced to leave their safe hiding places and go look for water elsewhere, which meant they threatened people in their homes. Grasshoppers devoured anything that was green, as if the parched land had any extra to share. Without predators, rabbit populations exploded, though the rabbits became a welcome source of protein for those who killed them. New illnesses from rabbits, like tularemia, took hold, and there are more new cases of rabbit fever today in the Central Plains than anywhere else in the U.S.

People with mold illness suffer from one attack after another once their defenses are destroyed by the relentless attack from uncontrolled innate immunity. Opportunistic organisms, like biofilm-forming coagulase negative staphylococci, move into hidden recesses and attack already weakened immunity, which lowers MSH even further. VEGF falls, adding more shortness of breath and brain fog. The list of predictable complications mold illness patients face daily isn't a short one. Don't forget the sting of the scorpion king, TGF beta-1.

The Black Blizzard's lesson was that we tried to outwit nature. We lost. Wrongheaded agricultural practices didn't cause the droughts but they surely guaranteed that the droughts would have a massively increased effect.

Even now, we try to outwit nature by thinking that our homes, schools and workplaces will be safe even if we site them foolishly, build them improperly or don't maintain them correctly. We build basements at the bottom of hills and wonder why they fill with water; we put ductwork in crawl spaces and wonder why dank crawl space air usually becomes laden with bioaerosols, which is then pumped by our HVAC system throughout our homes. We build high rise buildings as fast as we can, with time to completion being more important to the bottom line than any single other factor. We put another story on the high rise condo tower on the beach just as soon as the concrete from the story below has cured enough to support the upper story. The moisture in the concrete will be slowly released over many months, years actually, providing a constant supply of water to building materials, especially if vinyl wall covering was slapped on the inside walls of the concrete wall cavities. If you are HLA susceptible to mold illness, don't purchase a brand new high rise condo unit. Wait for the water to come out of the concrete walls and floors. Remember that mold needs 2 things to flourish – water and cellulose. The latter is almost everywhere – even in dust.

If we provide moisture and food for the modern microbial plagues flourishing in the water sources indoors, are those blooms different from the mass migrations indoors of scorpions that happened in the 30s? Is the uncontrolled reproduction of fungi and bacteria in a sheltered indoor ecosystem any different than a warren of rabbits whose population is growing by leaps and bounds?

As yet, I don't see mold illness resolved the way the Black Blizzard was. There's no Emergency Mold Act; there's no indoor air conservation service. Sadly we can't just plant trees and change farming techniques to conserve soil, slow down movement of water and block destructive winds. The fixes we need include educating physicians and patients, contractors and real estate brokers, judges and juries. Finding enlightened intelligent fixes from the Feds these days is almost an oxymoron. I just don't see the parallel now in the EPA or the CDC to the agricultural advisors that FDR had. Of course neither Big Tobacco nor Big Insurance had a vested interest in the Dust Bowl days and the Black Blizzard was visible for all to see.

Simply saying mold illness is real and exposure makes people sick, as the 2008 GAO report did, and as the 2009 WHO report did, are good steps. By April 2009 there were no federal or international agencies charged with protecting human health telling us that water-damaged buildings *aren't a threat*. That's another good step.

When I hear the unified public outcry to fix flat-roofed schools, which can prevent learning problems in children, then I'll know that Feds, state and local government people "get it." Until then, just like those who suffered from the Black Blizzard, there's no hope.

Still, I remain the ever-optimist. As much as I know the Big Tobacco-type war is being waged against the educated siting, design, construction and maintenance of buildings, there is an ever-increasing national awareness of what business can do to our environment. For every BP oil disaster in the Gulf of Mexico that angers and saddens us there are one hundred thousand buildings that would benefit from reduced energy use and increased use of green technology based on logic and science.

The lessons from the Black Blizzard must not be ignored. What we do to our environment stays with us forever. We know what we should do to safeguard our lands from erosion and our lungs and brains from WDB's. We know what we should do to protect our buildings from excessive moisture and if we simply guarded our buildings from interior environmental mistakes we wouldn't have health disasters to worry about later.

We need to put the Black Mold Blizzard in its new mold perspective. We all know that the Soil Conservation Service has done wonders for the U.S. farmer. What if that example were mirrored by the "U.S. Indoor Conservation Service?" Just imagine the textbook that high school students could read twenty years from now: "The people demanded that the Feds learn from their many misconceptions and tolerance of faulty building practices. They put the safety of their children's environment ahead of letting someone make every last dollar, while many suffered. It was the collective will of the people that finally demanded that indoor air quality be preserved for all, in all buildings. Health screening for mold illness became mandatory, as VCS testing is no more time consuming than a simple vision screening to obtain a driver's license. For those with VCS deficits, additional screening was mandated and then followed by effective treatment that was readily available. The Feds enacted medical procedures taught to them by those who actually knew what to do. They stopped the abuse of science by those who profited from loss of health. Genomics testing for treatable inflammatory illness became as useful and as widespread as fingerprinting.

"Court systems rose up as one to demand honesty in the courtroom. Conflicts of interest that damaged the life or health of individuals became grounds for the offenders to lose their income and university appointments. The Nation's productivity boomed when American talent and brains were no longer lost to American environmental illness."

Now that's a history book I'd like to live long enough to read. And when we can read that history book, then the Black Blizzard can finally rest in peace.

CHAPTER 23 Battling Mold: Policy and Politics

By Thomas D. Harblin, Ph.D.

ACKNOWLEDGEMENTS: The author is grateful to Arthur Jutton, Mary Mulvey Jacobson, Ritchie Shoemaker, and Dorothea Harblin for suggestions and comments on this chapter. However, the opinions and any errors are solely the author's.

An under-recognized global epidemic

Chronic human illness ("mold illness") acquired following exposure to buildings with a history of water entry and subsequent growth of illness-causing microbes is a real and present danger in the U.S. and around the world. This danger imposes often unrecognized profound health and economic effects on people and on communities.

In 2009, the United Nations World Health Organization issued a comprehensive report, "Guidelines for Indoor Air Quality - Dampness and Mould" that states:

"Indoor air pollution—such as from dampness and mould, chemicals and other biological agents—is a major cause of morbidity and mortality worldwide."

Further, it notes,

"The prevalence of indoor dampness varies widely within and among countries, continents, and climate zones. It is estimated to affect 10-50% of indoor environments in Europe, North America, Australia, India, and Japan."

The enormity of the resultant potential global health concerns cannot be overestimated. Data in this book demonstrate that 25% of exposed patients are at risk for a definable illness that occurs in people exposed to the growing number of buildings known to harbor pathogens. Using even the lower range of WHO estimates of WDBs, *the global population either already ill or at risk is 200 million people*.

Despite the fact that this mold epidemic has important human health consequences, which have been documented in countless peer-reviewed and published academic papers, the U.S. Public Health system hasn't recognized or responded to the problem. As a public health problem, mold illness is both an acute and a chronic condition. However, unlike influenza, syphilis, and gonorrhea, for which there is a standard public health alert process that routinely reports the cumulative extent of illness, there is no such reporting required for reporting mold contamination in communities, so the degree and impact of mold distribution is unknown. Mold activists must overcome that lack of awareness. This chapter suggests ways for activists to advocate for and to make change happen at a local level.

The purpose and role of new federal legislation

If all current laws, regulations, codes, and penalties were aggressively enforced, there would be little need for additional federal legislation. Slums are technically illegal. Shoddily constructed or natural disaster-vulnerable buildings are as well. Every municipality has codes for safe housing. But where's the enforcement?

Just so you know, whenever the word "mold" is used in this chapter, it's referring to the complex mixture of organisms and inflammation-causing chemicals found inside man-made environments (buildings) when excessive moisture is present. Curiously, the diversity of these harmful organisms and their simultaneous presence is essentially the same from San Diego to Manhattan and from Miami to Seattle. Wherever they occur, these organisms are linked to nearly identical associated metabolic problems in people. In a way that commonality of illness isn't surprising since we are talking about a similar source – an enclosed habitat where the indoor temperature is artificially maintained around 65-72 degrees year round.

Most materials found in a typical building contain or hold moisture. In the absence of highly efficient indoor air quality control, environments conducive to mold (i.e. containing cellulose) begin producing mold when the temperature changes and water intrudes due to structural deficiencies.

Thus, it's no surprise that we're awash with growing numbers of citizens, especially children, who are seriously ill from mold exposure. Unhealthy schools, water-damaged moldy work environments, and deteriorating large public housing complexes take a toll. If our current laws were enforced, these conditions wouldn't be allowed to continue. We urgently need new federal policy legislation.

The public health concern about mold is confounded by our federal agencies' delay in recognizing that WDBs hurt people. Natural and human-caused disasters like Katrina and floods in Nashville, Tennessee, or Fargo, North Dakota, have compelled many who not long ago routinely dismissed mold illness as insignificant to recognize that it's an epidemic. Further, a growing number of legal cases decided in favor of plaintiffs who claimed that moldy buildings caused their personal injuries have resulted in mold-caused illness gaining a higher profile. A growing number of community-based initiatives (usually involving schools that are WDBs) have further generated a wider public awareness of this serious health problem.

Federal legislation, and the process that creates it, stimulates enforcement efforts at all governmental levels by bringing to light expert testimony that informs and alerts the public. Legislation focuses attention on an issue, raises expectations about enforcement, directs resources to remediation and prevention, and encourages transparency and accountability for mandated outcomes. But federal legislation only produces such benefits when it's effectively designed and linked to local proactive code enforcement efforts.

Half-hearted legislative proposals filled with loopholes and enforcement exemptions are a waste of everyone's time. Any new legislation must mean business and it must rest on a foundation that includes public education. This latter is woefully absent due to the decades-long efforts of those with economic interests they see as adversely affected by the negative human health effects of WDB.

Desired legislative outcomes

Simply stated, the desired outcomes of new mold legislation are:

(1) prevent new construction that's inherently conducive to mold contamination;

(2) promptly and effectively remediate existing WDB to make them safer for human habitation;

(3) establish and make nationally available, easily accessible screening program to identify mold illness victims; and

(4) deliver prompt restorative healthcare to victims of mold illness.

Political challenges for containing mold

The three key strategic challenges are:

(1) advocate for new meaningful, enforceable federal legislation;

(2) engage in political action that strengthens enforcement at the local level; and

(3) disavow and debunk opponents' arguments and misinformation via local media and the web.

Ideally, new federal legislation would include funding to increase scientific understanding of the issue. The resulting understanding must be directed by an agency serious about its mission of enforcement, remediation, and prevention and must influence public health efforts. Unfortunately, an example of agency involvement that demonstrates the absence of sound logic was the 2007 HHS approval of $10 million to fund the National Toxicology Program (NTP), an effort dominated by veterinarians, which attempted to study mold exposure in rats at a time when the parameters of allergy that NTP chose to examine were already shown to be of no relevance in people with mold exposure.

New legislation must develop standards and offer solutions to remediate mold-affected structures; enhance methods to prevent water intrusion; and develop ways to rehabilitate and compensate mold patients for health and economic damages resulting from negligence and/or contractual liability on the part of those responsible for the mold conditions that made them ill.

It's time to reintroduce an expanded version of Toxic Mold Safety and Protection legislation. In the 109th Congress 1st Session, Rep. John Conyers, a Democrat from Michigan, introduced H.R. 1269, "The United States Toxic Mold Safety and Protection Act of 2005." Also known as "The Melina Bill," this legislation wasn't acted on and as of Fall 2010 hasn't been reintroduced. Congressman Conyers expects to reintroduce the bill at some point, but hasn't committed to a date.

Creating political change isn't fast or easy. Mold activists must become an organized and respected political force in their local communities, and must insist on honest and diligent law enforcement that protects the public from mold exposure. They must insist that code enforcement decisions be made transparent and routinely scrutinized. Rigorous code enforcement must become a compelling issue in local elections.

Using the media, activists must develop and lead a campaign to educate and mobilize mold victims and their advocates to counter the relentless efforts by those who deny or defend conditions that put people at risk for mold exposure. Some groups like GIHN provide an excellent example for others to follow. A comprehensive inventory of useful information sources is beyond the scope of this chapter, but in addition to the Action Committee for Health Effects from Mold, Microbes and Indoor Contaminants, noteworthy examples include Homeowners Against Deficient Dwellings (**www.HADD.com**), Policyholders of America (**policyholdersofamerica.org**) and **www.biotoxin.info**.

Any political change regarding science and health can only come from an educated populace. When facts defeat fear, and data, not dogma, are the basis for public opinion, the case for a ready resource of information about WDB and mold illness will be established. No single comprehensive resource inventory or website exists. The highest priority should be to create a site that links websites on mold, its health effects and treatment; the environment; the real estate and building industries, insurance, codes and inspection officials' qualifications and licensing; and policy and political action agendas. Such a "clearinghouse" would leverage the cumulative impact of currently fragmented activist groups. Creating such a site would be a great undertaking for a college or university-based community service project.

In short, new federal legislation, determined local political action, and a public education campaign are essential keys to improve the current situation.

Unintended consequences and the inadequacy of good intentions

In pursuing solutions to the above challenges, it's important to remember that all human actions, no matter how well intended, create unforeseen consequences. Some consequences are negligible, some profoundly threatening to human wellbeing. Human interventions to overcome serious problems often create more serious difficulties - consider the impact of levees constructed in New Orleans to prevent flooding from Lake Pontchartrian overflow and how their failure during Hurricane Katrina compounded the water damage in New Orleans.

In the legislative process, once desired outcomes are specified, proponents must focus on actions rather than intentions, particularly in the way regulations are drafted. Regulations provide form and meaning to ambiguous legislation concepts. Even H.R. 1269, a seemingly effective federal legislative response to mold prevention and management, includes sections that provide an exemption of liability absent proof of "intent to defraud." Gaining approval of legislation that doesn't contain such escape clauses is unlikely, but such language adds to the cost and difficulty of legal action often needed to enforce even the clearest of statutes. We must avoid legislation that provides aid and succor to those likely to back the status quo.

Converting adversaries and empowering legislative champions

In designing effective federal legislation, it's wise to anticipate the reactions of likely opponents. While anyone may draft a bill, only elected members of Congress can introduce legislation. In that process, the standing committee(s) to which the bill is referred seek input from affected parties. (*See* **Congress.org** *for a succinct summary of the legislative process.*) There are no ways for proponents of new mold legislation to avoid facing their adversaries during the legislative development process, so be prepared!

How many healthcare bills or environmental legislation initiatives have been blocked by those whose interests would be adversely affected by new laws? That's the Washington way, but early 2010 political election outcomes seem to suggest that the American public is demanding that their interests, however diverse, be given precedence over those of narrow special interests. In the case of mold issues, human health effects must be linked to health effects from mold in general rather than concentrating on WDBs alone. Focusing on the buildings alone will provide property owners who seek immunity from the health consequences of their defective structures with an easier target to defeat. It is the entire process of habitat creation that must be monitored.

Legislative proposals have maximum impact when they respect

all potentially affected parties' interests. Where healthy buildings legislation is concerned, the oversight Congressional committee(s) will invite comment from scientists, physicians, public health representatives, victim advocacy organizations, and litigators, among others. Representatives of insurance companies, architects, builders/construction associations, unions, building products companies, standards organizations, code administrators, chambers of commerce, and political officials from state, county, and local governments will be asked for their views as well. Many who testify will question the benefits, costs, administrative complexities, and the need for new laws and public funding of what some will call an "insufficiently documented alleged public health issue."

One of the top challenges for mold activists is how easy it is for credentialed defense "experts" to deny the link between mold and its health effects. Determining which particular microbes and biological agents cause which particular symptoms and illnesses is a problem because people are commonly exposed to multiple organisms at the same time when they come in contact with the "soup of microbes" commonly found together in "moldy" environments. In a legal case, if a judge requires a person who is ill from mold to prove which specific single organism from the "soup" caused him/her to be sick they are placed in a "Catch 22." If a plaintiff exposed to many illness-causing organisms cannot demonstrate that a particular one caused a particular symptom, a jury may be convinced that *none* of them did. Fortunately, both the GAO and WHO reports have specifically *decried ascription of causation to one single element* in a mixture of potentially causative agents. Precise, reliable measures of exposure vex epidemiologists, whose job is to establish causal connections between pathogens and illnesses. Further, individuals exposed to organisms causing illness don't all react in the same predictable ways. Some people seem to be more tolerant or resistant to extreme effects than are others.

This reality is no surprise to veteran mold activists. Proponents of the bill can't rely too heavily on testimony from victims, as important as this is. Instead, they need to bring to the table credible allies from each likely adversarial group. These groups invariably include educable people of conscience, "whistleblowers," individuals who've been victimized by mold or have family and friends who've had their lives destroyed from exposure. As a result, opposition can be muted from those who seek to subvert or weaken the bill.

A bill will reach the floor for a vote only if it has the support of a majority of the committee members. Each member must be able to defend his/her position by citing overwhelming evidence of the need for new legislation. Key to this is their ability to cite credible testimony from witnesses linked to opponents, who nonetheless believe the legislation to be essential and to be a solution to a real and present danger to public health.

Efforts to bring to the committees witnesses who come from opposing groups, but who have "seen the light," generates a better outcome than simply forging ahead with "the truth," opponents be damned. Failure to make that investment likely helps explain why H.R. 1269 stalled in 2005. Nonetheless, despite genuine efforts to forge a workable alliance, when the next round of hearings begins proponents need to be prepared to be humbled, yet undismayed, by the vehemence of opponents.

Getting legislation that works

In advocating for legislation that has the best chance to deliver desired outcomes:

- Specify the purpose of the bill and its intended outcomes. Insist on unambiguous language, especially in regulations. Ambiguity is the friend of power.
- Indicate resources needed to achieve outcomes. Avoid "unfunded mandates."
- Propose transparent oversight arrangements, clear chain of command authority, especially across agencies and political units, and identify who's accountable for failures and negligence. ("Heckuva job, Brownie!")

- Understand the consequences of compromises necessary to pass the bill. Do these create loopholes that will be routinely used by key players? Do they exempt specific actors from transparency and accountability? Do they excuse liability? Do they ultimately compromise or defeat the bill's very purpose?
- Beware exemption from liability based on absence of proof of "intent to defraud." Ideally, bill language should state that liability comes from negligent actions irrespective of intent. Physician malpractice isn't excused because the doctor intended no harm. So why should those who cause or enable toxic environments be exempted because they "intended no harm?"

Where advocacy action matters most: local enforcement

Once desired legislation becomes law, maintaining and assuring healthy buildings is largely about enforcement, especially at the local level where we live, work, study, and carry on our daily lives. Federal, state, county and city government laws, regulations, and codes mean little if they're not vigorously and consistently used to address, remediate, and prevent sick and moldy buildings.

Enforcing relevant codes is a highly political process. Most local officials are committed to the idea that growth is good, even essential to the economic wellbeing of their communities. Growth suggests prosperity, jobs and a growing tax base. New commercial buildings, conveniently available professional services, appealing residential developments, added or expanded public facilities such as schools and sports/recreation complexes, or enhanced environmental amenities attract more people and add to a community's diversity and quality of life. Most people believe that brand new buildings won't have microbial problems, but the opposite is more likely if the site was poorly chosen and /or not correctly drained, there was wet weather during construction, if construction techniques were haphazard, or if the builders used defective or contaminated building materials.

To achieve these community amenities, there's an attitude among local growth boosters that favors limiting constraining regulations. Anything that increases development costs as regulations invariably do, is often criticized as "anti-growth" or even "un-American." This sentiment is bolstered by a prevailing low tax ideology that's a frequent and popular theme in local politics. Whether such thinking is accepted as a given or viewed as extreme, it enables shortcuts and compromises in code and permits enforcement that leads to unhealthy buildings.

Even if every structure were perfectly built, the realities of entropy, natural disasters, the ceaseless physical dynamics of Mother Earth and human foibles all compromise living and working environments - at times suddenly - and creates settings dangerous to human health. How bad can it get? Sadly, recent examples such as Haiti and post-Katrina New Orleans suggest it can be catastrophically bad. How much human misery still occurring in these places might have been prevented?

Why enforcers don't enforce

Explanations as to why codes aren't enforced in any given case, especially involving toxic mold, can be complex or quite simple. The "culture of local enforcement" matters. As it's practiced now, code enforcement encourages discretion and trade-offs by administrators. In many jurisdictions, there are so many permit applications and "suspected violations" that overburdened and under-budgeted enforcement staff can't handle them all. If code enforcers are poorly qualified, if there are political pressures for quick approvals, or if routine public scrutiny of decisions is lacking, it results in lax enforcement. When you're dealing with mold, code ambiguities and ignorance of the health implications enhance the potential for mold damage outcomes.

It's ironic that water sprinklers in public buildings are encouraged as a counter to fire damage, since once activated, their health effects may be insidiously lethal to those who return to a saved building.

Post-fire and water damage cleanup and remodeling are uncertain remedies. Molds and microbes survive evolutionary threats because they can hide.

The above barriers to aggressive enforcement are compounded when growth investors and speculators, builders, and developers are courted by public officials. In many communities, these officials want their town to become the site of new corporate expansion or relocation, so they discount public health concerns and offer tax breaks to gain new jobs while quietly pledging to be "permit and regulation friendly" as an incentive. When it happens, corrupt elected officials further undermine code enforcement integrity.

"Overly" rigid and strict enforcement of building codes and permitting processes raise building costs and add incentives to evade or soften certain laws or codes. Public entities' predisposition to seek and accept the lowest bids on publicly funded projects adds pressure to discount regulations that increase construction costs.

Further, many builders believe the cost of "zero tolerance" would put them out of business. Public officials and their staffs are expected to exercise judgment to benefit the community. I haven't even mentioned the reciprocity based on friendship, the "I scratch your back, you scratch mine" thinking prevalent in many communities.

To counter this, it's helpful to recall Rene Dubos' mantra, "Think globally, act locally!" Hence, armed with compelling legislation and its attendant regulations, codes, and standards, citizen advocates can bring sustained, proactive local political action and pressure to enforce codes to minimize mold exposure risks for local residents.

Local action goals and resources

Most communities have local media and groups that pay attention to public life. They expect transparency about public officials' actions (or inaction), and they're on the lookout for things detrimental to community well being. These groups may be formal or informal and range from political parties seeking an edge for the next election (nothing helps more than the hint of corruption or dysfunction among officials in power) to activist "gray panther" retirees, book clubs, church and school groups, and service clubs, though the latter are often a bastion of advocacy for growth, limited government, and low taxes. However, most service clubs' guest speaker programs provide useful opportunities for informative presentations on important community issues. Expertise and volunteer energy can also be found among local health care providers, lawyers, writers, retired teachers, librarians, executives, and people with experience as a public official or public employee.

However, in this age of sensationalism, it's difficult to get people motivated about an issue that's not particularly visible or problematic in their community, nor yet a part of their own personal experience. A galvanizing event is needed, some personal connection that captures public attention and cries out for collective action. There are people who have confronted mold and some who've been sickened by it in every community. Mold victims may not be aware of others like themselves nearby. Unlike the many disease advocacy organizations that exist and sponsor annual events to benefit awareness and funding, who rallies mold victims and turns their personal troubles into a public issue?

Mold victims themselves are often overwhelmed with simply trying to get a healthcare provider to provide a diagnosis and treatment regimen. They may be battling health and property insurers who refuse to recognize the connections between their illness and its causes. They may be frustrated by efforts to hold accountable the building inspector or engineer who certified that the home they recently bought was sound and unlikely to harm them or lead to loss of resale value. They may be stymied by managers at their toxic workplace who refuse to see a link between their work environment and their illness. And they may feel let down by public officials whose bureaucratic offices deny or fumble their quests for justice. Some while healthy themselves, may be concerned about local school officials' inability to get the resources and action needed to stop the water damage to classrooms, damage which results from the ubiquitous flat-roof money-saving design characteristic of so many public schools. As an experienced builder once observed, "every building needs a hat!"

These very circumstances can generate the attention, momentum, and funding to compel new federal legislation that would authorize and fund research on health effects and effective treatment protocols for mold victims:

1. mandate and promulgate standards for codes to assure healthy buildings;
2. direct insurers to be responsive to their policy holders;
3. define necessary qualifications for those given a public trust to enforce the laws;
4. establish meaningful penalties for dereliction, harmful action, and inaction by those who enable mold environments to exist;
5. coalesce and streamline agencies that administer current and newly created mold programs.

These requirements and standards would arm citizens and advocates with the information and tools to hold local enforcement officials accountable. With such legislation enacted, community groups could identify and coalesce their constituencies and make healthy buildings a priority issue in local elections.

CHAPTER 24 Testing and Remediation

By Greg Weatherman

Mr. Weatherman is principal owner of Aerobiological, Inc, Arlington, Virginia. He has a series of thoughts regarding common problems with indoor moisture.

My goal with this chapter is to allow the reader to have a fighting chance to take care of their problems with mold whether they pay someone to fix it or do the work themselves. I can not write everything you need to know in one chapter but I can tell you where the problems occur when you use common mold remediation standards from the EPA, New York City and the IICRC. I have acted as contractor and/or consultant at different times over the past thirteen years. Perhaps my perspective has differences compared to others in my field: I view different parts of the process - from investigation of microbial contamination to its remediation-from many different sides. As a word of warning, I try to be evenhanded so I beat up on everyone without regard, be they contractor, consultant or regulator. I try to give credit when deserved. I only give verbal jousting to those who are known to have a good sense of humor or those who richly deserve it.

When I first started working with moldy environments in 1997, things were very different. We knew that WDB could hurt immune compromised people and make some allergic people ill, but that was about it. The only people who seemed to care about mold in buildings were healthcare facilities or government agencies where employee unions could file a grievance, though not much ever came of the complaints. Granted, there were some exceptions for large remodeling contracts with the GSA with no public mention of mold remediation in bid requirements. The FBI's HQ (J. Edgar Hoover Building) had water leaking into the basement from the large potted trees on the plaza at street level in the late 1990's. My first jobs were in federal office buildings in Washington DC. If mold were suspected, insurance carriers at the time expected (and paid for!) contractors to spray disinfectants everywhere to deodorize musty odors in an attempt to "kill" mold.

These were primitive days for professional air quality people. Environmental mold and bacteria were just social nuisances like dusty furniture and drafty air vents. Microbial growth was just a sign that the janitor was needed. HEPA filtered vacuums rarely were stocked in stores. It was common for CIHs to recommend to vacuum with a double bag in the vacuum rather than use a HEPA vacuum.

The best governmental guidance for indoor microbial contamination at the time came from Health Canada rather than the US EPA. The ACGIH (1989) and New York City Department of Health & Mental Hygiene (1993) had meager writings at the time. The first guidelines were those from New York City Department of Health & Mental Hygiene, Guidelines on Assessment & Remediation of Stachybotrys atra in Indoor Environments (1993). This guideline was for Stachybotrys atra only (later renamed Stachybotrys chartarum). The CDC was in an upheaval over a report on infant mortality and its "association" with Stachybotrys so it was a bit surprising that the US DOT's building in Washington DC was completely evacuated and renovated due to mold at a cost of several million dollars. This evacuation was preceded by similar actions in federal courthouses in the southeastern US (*Editor's note*: The case of the Martin County Courthouse in St. Lucie, Florida, established battlefields about health effects of WDB as dollars were awarded to plaintiffs and a new courthouse was built).

The federal government found that their stake in mold was very expensive.

Some may argue that the Federal Government engaged then in a conflict of interest to minimize the claims of mold hazards in order to reduce their expensive (landlord's) stake in the problem (*Editor's note:* The argument is still made today with even greater emphasis on absence of reasonable public health responses from agencies mandated to be proactive in protecting human health and not the business interests of political supporters). The federal government owns and/or operates office buildings, housing and public places where negligence may have been an issue. And the Feds would invariably decide to ask a blue ribbon committee to study and report on a problem.

How many blue ribbon committees have claimed there is no problem with just about anything? Do you remember that beryllium was once considered to be an essential mineral according to the Federal Government and found at trace levels in daily multivitamins until the Clinton administration investigations said beryllium could kill you? Government scientists can do good work but politicians and politically appointed bureaucrats (from either party) just about always will make a mess in the name of self-preservation (see Chapter 19). Just ask anyone at the US EPA how much they learned about indoor air quality while working at the original headquarters in the old Waterside Mall building (see Chapter 9). If you can get past government "deniability," just ask how many people they hurt.

Mold can be found everywhere; "ubiquitous" is the term I see used a lot. It can be found in desert sands and the frozen tundra (JI Pitt). It is abundant in soil and the food that comes from soil. Even if you have a new house and a new refrigerator, you can easily find Penicillium roquefortii in your refrigerator even though you have never allowed blue cheese to enter your premises. Outside of cheese production, subspecies of this organism have been shown to produce some nasty mycotoxins (JC Frisvad). If you have Chinese refried rice, you probably have a small amount of Bacillus cereus in your refrigerator. This organism could theoretically make the average person very ill if ingested in high enough levels ($>10^6$ organisms per gram) according to the CDC. This organism grows on protein and cellulose. It produces an enterotoxin. It is very common in the soil and WDBs (*Editor's note:* These Bacillus bacteria are related to the very lethal B. anthracis. So is Bacillus thuringiensis, the widely used insecticide and gardening product).

I want you to understand what "normal microbial flora" is versus "abnormal." Under normal conditions, mold and bacteria are essential to human life. Imagine a world with no fungi. There would be no biodeterioration of dead plant matter which is necessary for the process of converting organic matter into food for all living creatures.

The most important element in the mold story to understand is water activity. Water activity (or Aw) is the available water on the surface of the substrate. Water activity is expressed as a percentage; complete saturation has a value of 1, with lower percentages being fractions of 1. Available water is not to be confused with water at the molecular level available to the inner structure of the substrate. Simply stated: water activity drives the growth of mold and bacteria. A piece of wood with 0.5 Aw is not going to support mold growth. Xerophilic (dry-loving) molds such as various species of Eurotium, Eupenicillium, Aspergillus, Penicillium and Wallemia can grow at lower water activity levels from 0.82 and lower. Mesophilic (moderate moisture) mold such as various species of Cladosporium, Alternaria, Curvularia and Epicoccum grow at moderate water activity levels of around 0.84 and higher. These mesophilic organisms are also common in the outdoor environment at higher concentrations. This is also the range where Staphylococcus bacteria can grow (JI Pitt). Hydrophilic (water-loving) organisms require water activity levels of 0.9 or greater. Stachybotrys and Chaetomium are examples of hydrophilic organisms. This is where you also expect to find large amounts of Bacillus and other bacteria (gram-negative). That means there isn't much of a range for water saturation that causes mold growth when you consider colder temperatures create higher water activity levels at the colder base of walls in humid environments. Theoretically, if we want safe buildings, we just need to avoid water saturations over 0.65 Aw. This is the reason why you will see publication state relative humidity should not exceed 60% when measured with an inexpensive relative humidity (RH) tester that probably has an error rate of plus or minus

5% RH. Relative humidity is the available water in the air rather than a surface. 100% RH is necessary to reach saturation with higher levels leading to water droplet formation or rain.

The best analogy for water activity and fungal growth I can give comes from activated dry yeast used in baking. It has little or no odor when you buy it. Yeast is a fungus. It produces spores to survive periods of dryness so it may live again with the right mixture of food, water and temperature. Spores are capable of surviving dry environmental conditions for along time. But when you pour the dry yeast in some warm water with a pinch of sugar, almost immediately, those spores and yeasts wake up and grow! The characteristic strong odors that permeate the kitchen and every adjoining room mean that the yeasts are releasing secondary products of metabolism (ethyl alcohols and carbon dioxide) that only occurs with unrestricted growth. To be precise, the odor is coming from microbial volatile organic compounds, or MVOCs, made by the actively growing yeast. In a building, especially in a crawlspace or a basement, the musty odor, often likened to a dirty sock, is a sure sign that water and mold and/or bacteria have met. Out of new mold growth come secondary products of metabolism. And health effects for some. From this analogy, you can see why testing for carbon dioxide or CO2 is important if you want to know if the mold is actually growing. I have had cases where the CO2 is a few hundred PPMs higher in the basement area where the hidden mold is growing compared to other areas in the same structure with the same air system.

Water activity also helps predict potential toxin production if the organisms have the genetic ability, food source and temperature, for example. There have been many studies evaluating various species of mold in everything from food to building materials, with regard to the minimum water activity needed for growth versus higher water activity needed for mycotoxin production (JI Pitt, KF Nielsen). Remember, water activity is the available water on the surface of the material rather than the relative humidity or available water in the air. With the exception of some species of Eurotium, mycotoxin production on building materials generally occurs at water activity levels greater than 0.9, with 1.0 being the point of saturation (KF Nielsen). The exact amount of water activity necessary will depend on the food source and other factors such as pH, temperature, genetic ability to produce mycotoxins and microbial competition from other organisms such as Streptomyces (actinobacteria). There is no guarantee that mold with the ability to produce mycotoxins will actually produce mycotoxins every time the water activity exceeds known water activity levels for any given food source. One problem with proving mycotoxin existence in WDBs is the amount of sample material necessary for testing to register with the detection level of various test methods. Although there are ELISA test methods with lower detection limits, there are more potential mycotoxins than commercially available ELISA tests based on mycotoxins. (*Editor's note:* Yet every study that has looked carefully for mycotoxin production in WDB has found those products of secondary fugal metabolism.)

In my work, I pay attention to controlling water (or moisture) and finding the wet locations that allow microbial flora to grow in unintended ways (US EPA's Building Air Quality - 1995). I cite this U.S. EPA reference to show how long this issue has been on the radar of the US government. Actually, Dr. John Pitt (a.k.a. The Penicillium Man) has been saying for decades that we must control water activity to prevent food spoilage and mycotoxin production. We know it is critical to pay attention to water activity when assessing where microbial toxin production may occur as we investigate indoor environments. In 2004, Dr. Pitt once told a group of students (including me) attending his Penicillium species identification workshop in New Orleans to never quote him because he would deny it. Notwithstanding his commands, you can buy his invaluable book, titled Fungi & Food Spoilage, co-authored by his research assistant, Dr. Ailsa Hocking. This book contains rare physiological data such as growth temperatures and water activity levels for growth and/or mycotoxin production. Although the book is about mold growing on food rather than on building materials, research by others have shown the temperature and water activity levels to be similar though not exactly the same (K F Nielsen).

Soil

Look at locations where chronic wetting may occur. Soil is always wet, even when it appears to be powdery dry (*Editor's note:* ignoring this simple fact causes a lot of people to believe that their crawlspace/basement can't hurt them). Water from ground sources continuously moves through the particles of soil. Crawlspaces are the most common places I find overlooked by others when investigating. You should never have bare soil below a structure where you intend to live or work. Geosmin is a microbial volatile organic compounds or chemical produced by various species of Streptomyces. It will be there with other MVOCs like acetone. This is the same chemical you can buy from a hardware store or paint supplier. When I smell a familiar musty odor from I am fairly sure a crawlspace is (1) present; and (2) communicating with the indoor air, since dirt contains high levels of this organism which produces this chemical with its characteristic earthy smell. I had a mold remediation job where I could detect the musty odors from MVOCs. The two principals of Building Science & Engineering Associates verified my suspicion with onsite MVOC testing using a Z-Nose in order to justify some expensive demolition in a lawsuit-driven case. The engineers and the Z-Nose showed I have a nose like a mold-dog (*Editor's note:* No comments about size of the nose are permitted).

Mold samples may yield higher numbers of Aspergillus niger in rooms at the bottom of a structure rather than in its upper portions since this organism is extremely common in soil. The soil in some structures could be appear to be dry as a desert, but the microbial flora growing in the soil will be sucked into the air of the structure. Using a smoke pencil to evaluate air currents moving through a structure is an inexpensive way to find possible microbial pathways. This is why all plumbing and/or electrical lines running through a crawlspace to the overhead floor should be sealed airtight – especially a cold water pipe on the external wall that receives the prevailing wind. It would also be a good idea to install a sealed liner system such as by Basement Systems Inc. in the crawl space if allowed by your local building codes.

Kitchens

Kitchens are high risk for excess moisture and mold growth. Dishwashers, refrigerators and plumbing sources are locations to check. You can usually remove the bottom plate below the door of a dishwasher to view under it. You should use a flashlight to check the sides. Refrigerators develop problems under the unit and above the floor due to slight leaks or condensation. You must be very careful when interpreting test results from kitchen areas since food is grown in soil rich with many species of mold and bacteria. A good physical inspection is vital. The kitchen should show the greatest microbial diversity because that is where food is stored, prepared and thrown in the garbage to decompose.

High-Rise Buildings and Modern Chlorination

A new spin on an old problem is pinhole leaks in copper water supply lines. With the old chlorination methods, copper pipes lasted 25 to 30 years on average before erosion from the inside led to pinhole leaks. New chlorination methods designed to be "safer" led to faster corrosion with copper pipes lasting 10 to 15 years. It is not uncommon to find pinhole leaks in lateral pipes after the elbow or T-fitting, especially in lateral ½ inch copper pipes. Pinhole leaks can be very difficult to find behind kitchen cabinets or tiled bathroom walls. If you don't look, you'll never find the source. And you will be surprised just how wet a wall cavity can get in a short time from a pin hole in a pipe.

A word about equipment. Unfortunately, an infrared thermography camera may not detect this problem (depending on the layers of construction materials and the plumbing/electrical chases behind the walls). An ultrasonic leak detector may be helpful but not guaranteed when used for the same issues with infrared thermography cameras. It may be necessary to remove some building material and visually inspect suspect locations. Borescopes can be very helpful and be the least damaging since they can work with a 3/8th to ½ inch hole (which is easy to repair). No matter which technology is used, the investigator must have experience with the equipment for this type of inquiry. People who use cheap test equipment usually end up with invalid results. A good infrared thermography camera may cost at least $20,000.00 if you pay attention to sensitivity issues (such as every pixel having a separate temperature reading). If a consultant can not explain how the testing device works in general, this should be your sign to worry about paying for results.

Pests

Kitchens can also have rodent and insect infestations which allow more mold to be brought into the environment with the invading pests. Subterranean termites and carpenter ants can bring mold, bacteria and moisture continuously to locations where there is no plumbing (and therefore no plumbing problems) or other reason to assume a problem. Rodents can do severe damage to the outer structure while building nests and gathering food. Wasps and hornets can also sneak into tiny cracks or gaps without caulking to build massive nests in the external walls. Anything that allows warm, moist air to enter on the cool drywall side of a wall will lead to trouble with condensation. This is very hard to predict and very easy to overlook. A thermal camera can be helpful in this case. I have seen this problem overlooked by many. It also causes remediation costs to skyrocket because structural problems may be involved. This situation is a nightmare: many remediation firms are not qualified to deal with structural issues.

Carpenters are generally not mold remediators; no one asked them to do everything. Similarly, people who investigate mold may not be very knowledgeable about construction or pest control.

Bathrooms

Bathrooms and restrooms are also wet spots often underestimated or overlooked. Water may condense on the supply waterline to the toilet. The angle of the stop valve may slope downward so that water drips into the wall cavity, unseen for years. The wax seal below the toilet may leak due to age or a loose fit. Now you have pathogenic bacteria and viruses to add to the checklist of possible disease-producers. I actually am surprised that there aren't more problems seen with showers. I have yet to find a bathroom with a tub or shower enclosure built according to the Tile Council of North America standards (2005 or later). This is particularly important for construction after the mid 1960's when paper-faced gypsum board replaced mortar beds on wire lath under the wall tiles. Floors and walls next to the tub or shower enclosure are also affected. The biggest, best change in construction from the old standard is eliminating paper-faced gypsum board in any bathroom with a tub or shower enclosure. Another improvement is the application of a special waterproofing paint on walls before tile is installed on the floor and walls. Older houses with wood floors and plaster walls still have issues, just not as severe since they lack paper-faced gypsum board. Plaster walls can be tricky. The old way to build plaster walls was to apply three coats to wood lath strips. Plaster does not get as moldy as paper-faced gypsum board. The "curve" thrown at mold investigators is knowing construction methods for the period. Paper-faced gypsum board was introduced in the mid 1920's as a means to replace the wood lath strips and a couple of layers of plaster. This was a labor saving move for construction. You should not trust your ability to knock on walls and listen to the sound to decide plaster versus gypsum board walls. It is very common from just after WW2 to the mid 1960s to find paper-faced gypsum board below the plaster, supporting mold growth very well.

It is normal to find high numbers of Aureobasidium pullulans in bathrooms and kitchens (think of the refrigerator gasket on the door). This organism may appear on painted surfaces as pink growth or black growth. This organism is the source of the matrix or solid part of Listerine® Breath Strips. Please don't let some questionable person for hire claim this black mold will kill you. (*Editor's note:* This is the organism that prompted use of fungicides and mildewcides in paints to reduce the fungal growth on paint films.)

Basements

Basements are always problematic since the ground shifts and solid structural elements will also shift or crack. Block walls are worse than poured concrete walls since they more often develop cracks. Also, interior drain French drain systems can be a real nightmare. Some contractors will use drain tiles or flashing along perimeter walls inside basement. Not in my house! The problems occur when the drain tiles or flashing extends above the concrete floor surface along the wall. Contractors may claim this design captures water coming through the walls, diverts it to the French drain system and then finally to the sump pump. These contractors clearly have no clue about humidity control or they would correctly dig outside the structure and waterproof the wall down to the footer. Not only do water vapors evaporate from the moisture infiltrating the basement wall but, they have left a large chasm for mold, bacteria, microbial VOCs and insects to enter the home and wreak havoc. Those French drain systems with open drain tiles were not designed to control humidity or other things you don't want in a basement. Add a heating, ventilation and air conditioning (HVAC) fan coil unit (think of a universal distributor) to the basement and you now have a "whole house misery" unit. Poorly sealed sewage ejector pits and sump pumps near HVAC units is also a common problem that is not defensible from a design standpoint.

I had a customer in Virginia who had one of these unscrupulous waterproofing companies do a wonderful job in their residence. They installed open drain flashing made of galvanized metal along wood framing in one room and paper-faced gypsum board in another room. I called the waterproofing contractors before installation and warned them that open drain tiles or flashing above concrete flooring was a nightmare waiting to happen. I explained the lack of humidity control with this method. I was told the same thing I always hear from contractors who refuse to listen, "I've been doing it this way for 20 years or more."

Any contractor who uses this defense with no proof or solid scientific theory to back their use of that approach is a bad contractor. In this case, the contractor went ahead and installed the drain system described. You wonder how long it would take for mold to start growing on the paper-faced gypsum board or the wood next to the drain tile or flashing after periods of heavy rains. Not long! The story gets juicier since the house is owned by an attorney. I also referred the customer/attorney to a licensed professional engineer with a Masters of Science degree in geotechnical engineering who was familiar with the area. These sad stories would not happen if people thought with their heads instead of their wallets. Unfortunately, when customers have little technical knowledge a smooth talking salesman is hard to differentiate from an honest person.

An unscrupulous contractor can encourage a homeowner to allow installation of the interior drain by claiming it will save money to not dig around the foundation from the outside. The unscrupulous contractor will omit important details of other methods or admit to not actually solving the problem. The original reason for waterproofing was to control water and humidity to prevent mold growth below grade or below ground. You have to understand that water vapors in the form of humidity will condense on cool surfaces like basements walls below grade. Some contractors are only there to take the money and run.

HVAC

Heating, ventilation and air conditioning (HVAC) fan coil units can also be a common source of trouble. The mixing chamber where the cooling coil is located can be a constant threat when moisture accumulates on coils and near the coils. If the system is designed and operates correctly, water condenses on the cold coils and runs down the drain pan to the drain line so that water cannot get stagnant with growth from bacteria and mold. If it cools, it condenses. The condensate has got to go somewhere. One common problem for older HVACs is that coils are sized too big for energy efficiency so the conditioned air is never dehumidified. This means the coil cools the air but the surface area of the coil is so large the water never condenses to the point of draining to the drain pan. The cooling cycle stops because the temperature demand is met and the water on the coil is reintroduced into the air-stream instead of being removed by the drain system. A similar problem is seen when old compressors are sized too large for energy efficiency. The temperature demand is met quickly since the compressor is way too large and excess humidity is never removed. (*Editor's note:* this is very common problem causing human health effects.)

Another problem with HVACs is the use of whole house humidifiers. Water and organic material will ultimately produce microbial flora (mold and bacteria) in the humidifier, the air ducts and everywhere they blow without constant cleaning maintenance. Humidity in a room should be maintained between 40% and 55% relative humidity. Console humidifiers are more work to use but they do a better job without rusting your ductwork if you keep them clean and keep an eye on relative humidity (**www.emssales.net**). "Clean" console humidifiers are absolutely necessary if you live or work in a high-rise building. I will suffer nosebleeds in the winter with lower RH like 20% until I can get the RH at 40% or higher for my home. There should be no rust on the ductwork or the cooling coils. Rust in the HVAC allows organic debris and subsequent microbial flora to accumulate and cause misery. Rust in the HVAC is a sure sign of failed moisture control. A fiberglass lining in the coil area or mixing chamber of the HVAC can be a real problem if moisture and organic debris accumulate; microbial flora will grow and produce a characteristic fish or urine-like odor that is very noticeable when entering the premises. This seems to be very common in apartment and condominiums.

HVAC systems may not have been designed or installed properly to address humidity control for your structure. The HVAC system may have been designed only to meet the local mechanical code. This becomes more apparent when you have two or more zones (basement HVAC and attic HVAC) in large houses. I had a client who had slow growing mold everywhere in the basement. The mold was exclusively various xerophilic or lower moisture species of Eurotium, Aspergillus, Penicillium and Wallemia growing in unfinished basement rooms with no ventilation ducts. There was no mesophilic mold (requires moderate moisture to grow) such as Cladosporium growing on any surface. There was not hydrophilic mold (requires high moisture to grow) such as Stachybotrys growing on the same surfaces. This is a clear example of what you expect to find when there is chronic humidity without wetting of surfaces over a period of many months.

Other problems occur when HVAC contractors don't know what they are doing and damage the structure. I had a customer who endured the time and cost of remediation and rebuilding. He lost furniture and other belongings. He finally was able to live in his house after a year or more of agony. He decided to invest in a new HVAC system as I had recommended doing just that in hopes for maintaining a healthy environment. After one year he reported to me he felt as sick as he did when his problems first started. My assumption was the house was under a negative air pressure differential which could be caused from improper HVAC design and/or installation. I then referred my old client to the engineers to see if the HVAC contractor made mistakes. This negative air pressure differential created a vacuum effect where the house sucked contaminants inside faster than they could be pushed back outside. Over a period of time, microbes could infiltrate through the basement and disseminate throughout the house in unintended ways. The important part is handing the case to those better suited to address the customer's problems

The take home message of the problem is to make sure the HVAC contractor understands what is needed. Test or commission the system to verify correct operation; let the HVAC contractor know you will have a professional engineer check their work. Don't sign any work orders unless they have contractual language to protect your investment by holding them responsible for any design or installation flaws that creates problems with moisture, pressure differentials or thermal differences. There is no guarantee your local building code addresses all the problems from these issues. You may also want to have the HVAC contractor pull all fiberglass insulation from the inside of the fan coil unit and replace it with black, closed-cell insulation like Armaflex or Rubatex.

Finding Consultants

Who are you going to call? "Mold-Busters?" Be careful; look for some red flags. Insurance is very important. Ask to see the certificate! Consultants should have "professional" or "errors and omissions" insurance with a "pollution" policy specifically stating "microbial investigations" or words to that effect on their certificate of insurance. A "commercial general liability" (CGL) policy covers physical damage from accidents while on the jobsite. CGL polices usually have exclusionary language for "professional acts" which includes testing, taking samples, giving opinions and writing plans or designs. A "professional" insurance policy says the consultant's business has been assessed by an insurance risk manager or underwriter who reviewed standard operating procedures (SOPs), contracts, typical reports without personal information, and remediation plans. This risk management review is the reason why many consultants do not have insurance since they cannot find an insurer who feels safe betting a million dollars on a shaky person.

Some consultants will claim they don't get insurance because they "don't want a big litigation target painted on their back." Run away from this guy. They assume you will not sue a person or company whose assets are a couple of air pumps. I have "professional" insurance since my opinions may lead to someone paying tens of thousands to a million dollars in remediation, medical costs, litigation and other associated costs. A building owner, construction company or insurance carrier is not going to sue if they know the findings are factual and the consultant can cover legal fees defending his or her opinion.

Another excuse is "I would have to pass the cost to my customer and they would suffer by paying more money for my services." Think of the idiocy of this idea: are there many physicians who are going to keep costs down by skipping malpractice insurance? Why would you hire an uninsured consultant who is helping you identify the microbial growth areas that can cost you tens of thousands or more dollars. I pay approximately $5,000 a year for my insurance policies (CGL, pollution & professional). If a person has any kind of real business, they should be able to afford paying $100 per week since they charge $95 to $150 per hour.

Experience in environmental microbiology/mycology, microbial investigations, sampling techniques, moisture sources, building science (construction) and basic public health concepts are necessary to find the source of trouble in most buildings. Anyone who claims they can do it all is generally fluffing their résumé at your expense. An honest person knows their weaknesses and gets help in areas the customer needs. I am not good with mechanical engineering so I ask for help from consultants at the appropriate time for the good of my customers and "my liability." This is why I call the local engineers at Building Science & Engineering Associates when I know I am at my limit of knowledge. If you can't answer a question in deposition or court, you probably should avoid the same situation with a customer.

A quick way to gauge the knowledge base of a consultant is to ask who or what they use for scientific references when drawing conclusions. There are notable individuals, studies and texts which should be common knowledge to anyone who claims understanding of indoor microbial flora. I'm not going to list the all the people and writings, but I will say Dr. Chin Yang will cite references for his opinions rather than say, "Trust me, I'm an experienced PhD mycologist. I'm old as dirt from studying dirt." I could name others, but most don't have Dr. Yang's sense of humor or professional background discovering more than one genus of mold. The point is that nobody is "all knowing" and we must keep up with scientific literature to be intellectually relevant.

It is good to know an engineer who can work with indoor air quality and other HVAC problems. Usually mechanical engineers do this type of work. You may have a hard time finding a mechanical engineer for residential HVACs. I use a couple of local engineers: a materials scientist (master's degree) with a long history of government IAQ work, and a Ph.D. environmental engineer who specializes in inhalation toxicology modeling. They have no conflicts of interest with homebuilders. You need to know that talented consultants can be found outside of strict professional titles. Some certified industrial hygienists (CIHs) are competent in HVAC system diagnostics. They should be able to run load calculations if they know what they are doing with mechanical systems.

The same consultant may be able to do moisture investigations and look for microbial growth. Traditionally, certified industrial hygienists (CIH) are trained in workplace safety, environmental exposures, taking samples and doing environmental risk assessments for a broad range of pollutants and irritants. The certification is extremely hard to get compared to many professions. Some CIHs will not work in residential settings. Some would argue their training is meant for the workplace with 8-hour exposures. Another issue with CIHs is that their field of knowledge is very wide for workplace safety, but it may not include environmental microbiology and construction knowledge. I do know some who have learned the differences in construction methods between most homes and office buildings or factories. The CIH certification by itself actually is meaningless for your needs unless the CIH is strong in microbiology and building science.

Building science is the core of knowledge for professional engineers (PE). You may find a materials scientist or environmental engineer. These guys are really strong in physics (ignored by many consultants). PE's are generally civil or mechanical engineers who are excellent with construction or heating, ventilation and air conditioning (HVAC) systems but may not be familiar with environmental health and safety or microbiology. Most PE's are there to get the structure built or find the construction defect. This is their mindset. This is why you may see some PE's who are also CIHs. If they also understand environmental microbiology and architecture, you have found a rare person.

You may be thinking a microbiologist or mycologist is the way to go. The trap is to think all microbiologists and mycologists will understand your problem. They may not have done many physical inspections of properties. Subsequently, they rarely get to see the conditions where the organisms were found. Also, environmental microbiology is not easy. Ask any medical microbiologist who changed to the environmental field. They will tell you they had growing pains. Unless a microbiologist/mycologist has specific experience with indoor environments, they will also be of limited use.

The same can be said for toxicologists. A toxicologist is supposed to be an expert in general health effects from toxic substances. (*Editor's note:* finding a toxicologist who actually understands the immunologic basis for human illness acquired from WDB is not a common event. Few testifying toxicologists have ANY reasonable knowledge of the diverse immune problems cause by inflammatory triggers found inside WDB. Face it: immunology is what makes people sick in WDB, not toxicology. As soon as some toxicology guy with a PhD (Note: if he is paid by an insurance company, I can guarantee you he isn't your friend) starts talking about dose-response in mold cases, run away from that nonsense opinion as fast as you can). At the end of the day, they still need to know about sampling, building science, and microbiology/mycology. Biochemists also might wear the toxicology hat depending on their focus. (*Editor's note:* Don't forget that if the toxicologist doesn't understand the role innate immune responses in human health, his opinion about human illness from WDB is of no value).

The American Counsel of Accredited Certifications (**www.acac.org**) has designations for various consultants ranging from general IAQ (CIAQC) to microbial (CMC) certification. The former name of the organization was the American IAQ Council starting in 1995. Their website lists the qualifications needed to receive the certifications based on verified independent proof of experience over eight years or more and passing a rigorous, closed-book examination. The test is separate from attending a training session or obtaining membership in an accrediting association (certification mill). The ACAC has other certifications which require less experience. The best qualified consultant has PE, CIH, CIAQC or CMC certification rolled into one. I know some of them and they are not quick to use generalized "cookie cutter" opinions. Investigations take time and time is money. These folks have endured a lot of reading, tests, continuing education courses along with relevant field experience. You can't expect them to charge handyman rates. The ACAC website also has a listing for "insured" companies. I personally love this concept since this forces honesty if consumers would just use the information provided free of charge.

Conflicts of Interests

A person or company engaged in microbial remediation should never be doing the post-remediation verification testing. The ACAC has a code of ethics that prohibits this type of conflict of interest for anyone holding an ACAC certification. Some remediation contractors may look at a job site to size-up the problems for a good estimate, but they cross the line if they take samples, interpret the results and do the work. I have been placed in this position and I insist that someone financially independent from my company must take the samples if they want me to do the remediation. I don't mind giving my interpretation of results, but I always tell clients to get a second opinion. I have taken samples for remediation jobs where others have done the investigation. This is done to keep the investigators honest or as internal quality analysis to make sure I'm performing like I think I'm performing.

In Virginia, we have licensing of contractors including mold remediation through the Department of Professional and Occupational Regulation (DPOR). Soon, licensed contractors will be required to hold a specialty license for mold remediation, just like asbestos abatement contractors. The "Asbestos, Lead & Home Inspectors Board" will oversee the licensing. They currently have a "conflicts of interest" document where asbestos abatement contractors must have a customer sign the form stating they were told the abatement contractor has a financial interest with the person or firm performing inspections or post-testing. Virginia has a similar and more stringent form for lead abatement contractors. This will probably be the case with mold in the near future. It would be nice if the rest of the country had the same oversight! The Virginia DPOR doesn't get in the way but they do pay attention. They also don't charge outrageous licensing fees. If you don't live in Virginia, just demand the same standard in writing from the remediation contractor and consultants. You should also pay the consultant directly so the consultant has a legal responsibility to you to protect your interests.

I had a customer who was a first time home buyer. She was suffering migraine headaches and suspected mold as the culprit. I was hired as a remediation contractor after a prior company did some investigating and air sampling. This company tore apart moldy gypsum board in the basement and laid it against furniture with no attempt to erect containment or protect the furniture. After this amazing feat of negligence, they took air samples with a spore trap essentially sampling what had recently been disseminated into the air at much higher levels than normal occurrence. My client was told the "Stachybotrys is off the charts." They told her they could have a crew there the following week to fix the problem. At some point, this company will get caught in court, a licensing board or the ACAC. The latter is important since they will notify insurance carriers if someone's certification has been revoked which would lead to the inability to get insurance coverage. Side note: Insurance is very important for many reasons.

Short History of Testing

I started with the standard air sampling technique with an Andersen impactor and different nutrient agars in 1998. At the time, the common practice was to use malt extract agar (MEA) for general fungi. Dichloran glycerol agar (DG-18) was used for xerophilic or low moisture organisms like various species of Aspergillus, Penicillium and Wallemia. The presence of organisms on the MEA like fast-growing species of Trichoderma or Rhizopus caused the true level of Penicillium to be under-reported. The protection from this false negative was to run a DG-18 sample which is unfavorable for hydrophilic or high moisture organisms such as Stachybotrys, Fusarium, Trichoderma and Rhizopus, among others. As an additional safeguard, I would also use an additional nutrient agar such as cellulose agar or cornmeal agar for hydrophilic or high moisture organisms such as Stachybotrys and Fusarium. Even with the use of three different nutrient agars, the results were not always representative of the environment since the nutrient plates could be overloaded, some organisms under reported due to competition and only "viable" organisms could be cultured.

Tryptic soy agar (TSA) was used as nutrient media for general bacteria. Although bacteria sampling is generally done by swabbing a surface, air samples with TSA would show higher levels of various species of Bacillus and Streptomyces in chronic wetting locations. Both of these genera are spore-formers and common in the soil. Surfaces with crawlspaces and paper-faced gypsum board were the most common sources.

Next came the spore traps in 2000. They supposedly captured everything if you are looking for whole spores or conidia (a.k.a. "boulders in the air") without microbial competition getting in the way. This study is the reason for the switch: **http://www.springerlink.com/content/q5430r0167246k44/**

Spore traps did a much better job of finding Stachybotrys and Chaetomium. They also showed much higher levels of mold in air samples. Even better, the lab fees are at least half the cost of basic fungal culturing with a single nutrient agar without going into a lot of species details for Aspergillus, Cladosporium and Penicillium. The problem with spore traps is they do not tell you what you need to know other than existence of Stachybotrys and Chaetomium. Except in winter, the Cladosporium count should outnumber the Penicillium/Aspergillus group count in most indoor environments. You also need to know what species of Aspergillus, Cladosporium and Penicillium you are encountering. The biggest problem is the under the microscope, many spores or conidia look clear, round to slightly oval and sized 2 to 5 microns in diameter. This includes species of Aspergillus, Penicillium, Wallemia and other fungal organisms. Aspergillus niger is very common; its role as a toxin-former is inconsistently reported. Aspergillus versicolor is listed in literature as producing high levels of mycotoxins. Penicillium chrysogenum is extremely common outdoors. You would expect to find high levels of this organism indoors. High levels of Wallemia sebi may suggest a problem from humidity and not necessarily a plumbing problem, for example. How would you know if you have a problem if you can not differentiate these basic organisms? To this day, the spore trap air sample is the universal test method even though it gives you the least useful information in many circumstances. It is universal because the test method is cheap and quick. I know Dr. Shoemaker doesn't agree with spore counting since he says the information will not report what levels of fragments of spores are (*Editor's note:* Please read the next few paragraphs for relevant studies).

Ultimately, I graduated to using liquid impingers (SKC BioSamplers) since I could have the liquid (10% phosphate buffer solution) separated for different types of analyses and/or applied to different nutrient agars. This was the only sampling method where it was literally impossible to overload a sample. This test method allowed for better capture rates for bacteria in air samples. I could take 8-hour samples by stopping the sampler every 30 minutes to top-off the sampling liquid to maintain the capture efficiency. This test method also allowed me to get the liquid analyzed for macrocyclic trichothecene mycotoxins by an ELISA test method at Texas Tech University. The drawbacks were moving around with sterile water and pipettes. You also have to use heavy duty Gast air pumps. The liquid impingers are made of glass so, extra care had to be taken. The liquid impingers also had to be washed and sterilized in an autoclave with each use (extra labor nobody will ever pay for). If I were to sample for microbial flora today, this would be my preferred method if cost was no consideration for the customer. This is still the best collection device for air sampling.

I went to PCR testing in 2004 because I was not satisfied with the spore traps for air sampling while dealing with extremely sensitive people. A study by Gorny (U of Cincinnati) on particulates that are fragments of spores gave weight to this move: **http://aem.asm.org/cgi/content/abstract/68/7/3522**

I went to surface PCR testing (ERMI) in 2005 because air sampling could never explain the age-old argument of dose response via inhalation. Test environments are greatly impacted by pressure systems from weather and other activities in the indoor environment. Dr. David Strauss from Texas Tech did a study showing small particles of Stachybotrys contained mycotoxins. **http://aem.asm.org/cgi/content/abstract/71/1/114**

For some reason, everybody just ignored the importance and kept using spore traps.

The smaller the particle, the greater the chance of deep inhalation into the blood stream not to mention the olfactory bulb route into the brain, especially if someone is breathing heavy or fast (think steps indoors). Dr. Shoemaker's patients are the reason I have been looking at how to control small inhalable particles by understanding particle behaviour. We are blessed to have to really intelligent microbiologist-mycologists telling us how microbes grow. Unfortunately, the best scientific literature for environmental microbiology is lacking in "aerosol physics." For all test methods and research conclusions, it is important to understand the limits of our knowledge before you can accurately understand what is known.

The importance of PCR and the ERMI test is simple. The 36-species list is useful for most situations where you can not see an obvious mold problem. If you compare the cost of culturing mold samples to the species level, PCR and ERMI are very competitive in price. You really need all 36 species since different environments may have different organisms that grow under the same conditions such as Wallemia sebi and Aspergillus penicilloides. The first organism will culture in common MEA media within 1 week. The second organism will probably not be seen unless a second culture in DG – 18 media for species identification is done for 2 weeks after the first culture rather than the commercially customary 1 week for species identification in DG-18 after the first week in culture with the other test organisms. There are other organisms on the same list that you will rarely find outdoors in air samples unless there is rotten building material or paper (Aspergillus versicolor, Chaetomium globosum, Cladosporium sphaerospermum and Stachybotrys chartarum). You should be careful when investigating and ask if any recent renovations or demolition has occurred since man made building materials are rich with these four organisms and there may not have been water damage in the test area where low levels were detected. It is uncanny how 24 hour PCR air samples will back-up known test results from Health Canada and a large study by Brian Shelton of Pathcon laboratory: **http://aem.asm.org/cgi/reprint/68/4/1743.**

I have been working with amosite asbestos and soot in the last 2 years in a process to clean the indoor air of small particles. I have been paying attention to Aerosol Technology by William C. Hinds to explain why the air-cleaning method works in scientific terms accepted by the scientific community. Depending on the specific gravity and particle size, particles can be small enough to start behaving like gases in the air rather than particles moving by gravity. Gas molecules have Brownian motion. Usually, 0.3 micrometers particles (3 times the size of an influenza virus) or smaller have this property of physics in the air depending on the aerodynamics. These fine and ultra fine particles tumble in the air until they agglomerate with other particles or coagulate with vapors in the air. Humidity plays a very crucial role in the ability of very small particles to float in the air.

A microbial particle will increase in weight and size as the particle absorbs moisture and swells. The same particle will shrink in size and weight as the water within the particle evaporates. At this point, the microbial particle will be readily able to re-aerosolize so some unlucky person can inhale the microbial particle deeply into the lung. I think the reason this has been overlooked is simple. Microbiologists/ mycologists have historically looked at various microbes as "whole" organisms that create illness or allergy. There have been some brilliant exceptions with research from the University of Cincinnati (Gorny) and Texas Tech University (Strauss).

This means, you may sample according to classic industrial hygiene methods for air sampling between 3 feet and 5 feet from the floor or the level where people inhale if they are sitting or standing. You may find nothing and conclude the complainant is crazy or getting exposure somewhere else. Maybe you sampled at a time when the microbial particles were dryer and higher or wetter and lower depending on relative humidity patterns. I cannot find any research on humidity effects and altitude for recovering microbial organisms or the fragments of microbial organisms. I have approached a government agency about this unknown area for research. Still waiting.

There are creditable data to show this idea is active. An air sampling study from the University of Cincinnati found Streptomyces spores expanded in aerodynamic diameter from 0.85 micrometers at 30% RH to 1.07 micrometers at 100% RH. The length of time for moisture condensing and evaporating is obviously a concern in this great study: **http://aem.asm.org/cgi/content/abstract/64/10/3807**

Look at environmental microbiology/biochemistry testing. Outside of medical locations for nosocomial infections, bacteria are largely ignored. There are many species of vegetative bacteria that produce exotoxins. The bacteria reproduce and die but the fragments bound with toxins and inflammagens are still with us. These chemicals aren't alive or dead: they just are. We really do not know what is growing and disseminating from the coils and drains of HVAC systems. Other than polymerase chain reaction (PCR) analysis, how else will you answer this question with today's commercially available technology? The EPA's PCR technology has not addressed these types of organisms due to the rapid change in genetics in bacteria. There is a PCR technology developed by LSU where you have to culture a sample before you can run a PCR analysis. It obviously requires more labor and material than the PCR methodology developed by the US EPA, so it will cost more.

Some people will claim particles have electrostatic properties that make them stick to surfaces in the dry air. This is only partially true. People create the necessary energy to release the particles when they walk across the floor. At an IAQA event, Dr. Richard Shaughnessy, University of Tulsa IAQ Program, stated open weave carpet put 4 times more dust in the air compared to closed back carpet with foot traffic. The study was done for schools. Cleaning agents are generally based on cationic surfactants like quaternary ammoniums. This means that cleaning tile floors with a dirty mop head and a positively charged mopping solution will counter the negatively charged particles, so that they become airborne after the mopping liquid dries and air currents are created. The best solution is to use wet disposable mops followed by dry static mops which can be thrown away. Air currents created by ventilation, people or wind can dislodge the particles into the air. If something is true all the time, you are looking at a constant of physics. Since this is rare, it is best to avoid "all the time," "never" or "the exact same."

What Can Go Wrong with Remediation?

In a word: everything.

I have dealt with many clients who found their particular mold remediation project did not go as planned from a human health standpoint. To understand why mold remediation fails, you have to understand why testing fails so you can understand what was overlooked. Further complications come from technology transfers where essential components are ignored. Another problem is that cleaning methods are often validated by means too superficial for sensitive populations such as Dr. Shoemaker's patients. Most remediation contractors follow whatever everyone else does, no matter the circumstances.

There are guidelines and standards for mold remediation. Guidelines are "suggested" practices for "common" situations involving "average" people. Always read the disclaimers before you cling too tightly to the guidelines or standards. You or your situation may be different. You may have different sensitivities or your situation may not be described by any guideline or standard such as an attic, crawlspace or other pollutants. You may have weather extremes like the humid summer or frozen winter.

Another problem with guidelines and standards is the democratic process used to make the standards does not always lead to the best outcome. The outcome is what the group voted by majority to approve as the guideline or standard. In the beginning, the

IICRC's S520 Standard and Reference Guide for Professional Mold Remediation (1st edition, 2003) stated to keep everything dry for removal since water leads to more mold growth. The second edition (2008) changed course and said some dust suppression is good. The second IICRC group that voted on the second edition changed their minds about the virtues of dust suppression over time. The ACGIH's Bioaerosol: Assessment and Control (1999) and New York City Department of Health and Mental Hygiene's Guidelines on Assessment and Remediation of Fungi in Indoor Environments (2000) had stated the need for dust suppression years before the IICRC decided to write the S520 Standard and Reference Guide for Professional Mold Remediation (1st edition). This does not mean the people involved with the first edition of the IICRC's S520 Standard and Reference Guide for Professional Mold Remediation were wrong. It just shows how some people think differently and may change their minds at a later date depending on new information or insight.

Guidelines and standards also may be the product of greatly differing groups as far as background and education. Some groups may have more input from the microbiologist/mycologist and CIH members who deal with building owners and make their money by testing and giving professional opinions. Some groups may be loaded with government officials with little knowledge of business operations (*Editor's note:* and a lot of knowledge of political spin).

Idealism about remediation often fades when confronted with economic survival. Some early guidelines stated a need to use excessive plastic sheeting like asbestos abatement with three-stage decontamination chambers. Instead of a shower in the middle decontamination chamber, the ACGIH recommended use of a HEPA vacuum while removing PPE. I'm guessing you could quiz 100 remediation contractors who work for insurance-covered jobs or commercial building owners if they go to this effort. You will find few who follow this guidance simply for economic reasons. I found it was not necessary (by internal sampling) to use more than a single chamber decontamination unit. I used science to justify my economic reason.

Government regulators and CIHs tend to over-regulate the methods and dumb-it-down for contractors they perceive to not have the capacity to think. Proof of this can be found in EPA and OSHA regulations for asbestos abatement – don't confuse the regulators with physics like particle adhesion principles or van der Waals forces with too much water. Politics can make any effort get derailed whether it is a bureaucrat protecting turf or politicians putting pressure on bureaucrats to follow their political ideology (chasing votes or contributors) rather than science or moral conscience. Look at lead paint requirements over the years. Residential lead paint waste was allowed in public landfills starting under the Bush Administration. The Obama administration has decided to micromanage renovation contractors rather than just enforce the much tougher OSHA regulations that have rarely been enforced by any administration. If the EPA needs 1 to 2 years to approve a new disinfectant, where will they ever find the time or staff resources to micromanage a lead paint licensing program for every handyman in America?

Some groups may be weighted towards the mold remediation contractors who deal with property insurance carriers and property owners. Contractors are basing their decision on an economic model that maximizes profit and limits their liability. Their standards for cleanliness and testing will be less particular since they do not make money on testing. Customers are generally not going to pay for the level of detail if you raise the technical bar of acceptable remediation for sensitive populations. The contractor is not generally there to give professional opinions beyond how to remediate or find what the investigator missed due to inaccessibility of building components. The contractor did their job if a HEPA vacuum was used per square foot of invoiced work even if the HEPA vacuum did nothing beneficial. This is why it is hard for many mold remediation contractors to obtain professional or "errors and omissions" insurance coverage to take samples, test, write plans/designs or give professional opinions. This latter wording can usually be found under Exclusions in a commercial general liability (CGL) insurance policy.

(*Editor's note:* the following text will require careful reading. It is not directed to the beginner.)

Many mold remediation procedures came from asbestos abatement procedures. The use of negative air pressure (differential) enclosures, decontamination rooms and personal protective equipment and work practices can be found in the OSHA Asbestos in Construction Standard (29 CFR 1926.1101). Negative air pressure differential means a room or area has less air pressure measure with a manometer compared to an adjacent room or area. This means there is a vacuum effect where particles are pulled into the negative air pressure differential enclosure rather than pushed outside the enclosure if you use a smoke pencil. The major difference between asbestos abatement and mold remediation is not wetting surfaces prior to removal of moldy building material to the extent done in asbestos abatement.

Some asbestos abatement methods came from clean room concepts from medical, pharmaceutical operations and nuclear power. The difference is (1) use of smooth surfaces to limit the chances of particle adhesion; (2) laminar air flow to push contaminants to filtration; and (3) positive air pressure gradients that decrease as you approach a hood exhaust under negative air pressure differential or vacuum. They kept the personal protective equipment and HEPA vacuums. The original federal guidance came from the US EPA report number 560/5-83-002 Guidance for Controlling Friable Asbestos-containing Materials in Buildings (March 1983). It is interesting asbestos and mold industries are using technology discounted in the 1950's to control infection rates in operating rooms if you read Clean room Technology by W. Whyte – University of Glasgow.

Originally, asbestos training taught us to move the portable HEPA filtered device or negative air machine (NAM) next to the area where you were doing abatement work. The device was vented outside the contained area just as they do it today. The regulations are vague with stating the NAM should to be next to the area of abatement demolition.

29 CFR 1926.1101(g)(4)(vi)

For all Class I jobs where the employer cannot produce a negative exposure assessment, or where exposure monitoring shows that a PEL is exceeded, the employer shall ventilate the regulated area to move contaminated air away from the breathing zone of employees toward a HEPA filtration or collection device.

Asbestos trainers may not mention this part of the directions for use since it poorly worded by federal regulation. This leads to NAMs being located at the opposite side of the room or contained area to minimize dust loads that lead to costly filter changes. This also means fine and ultra fine particles were pulled across the room or contained area, thereby not permitting prompt removal when the NAM is next to the demolition area. This makes no sense (see page 67 in Clean Room Technology by W. Whyte (2001).

SOLUTION 1: Use NAMs the way they were originally intended if you are creating a heavy airborne dust load from demolition activities. Place the machine near your demolition area but not in a manner where you are between the HEPA filtered NAM and the demolition location like a moldy wall. Move the NAM as necessary to capture dust and protect workers. Prefilters are cheap and should be changed as often as necessary while the NAM is operating.

Another problem with relying on HEPA-filtered NAMs exhausted to the exterior is the belief all the particles will be obedient and head straight for the NAM for complete removal in the HEPA filter. In clean rooms, there is laminar air flow from fans blowing filtered air at the HEPA filters so particles can only go in one direction or drop to the smooth, seamless floor for removal by HEPA vacuuming or wiping. This is the purpose of laminar airflow. We sure don't have several fans to create laminar air flow. Cleaning several fans would drive the cost higher since the decontamination would take significantly more time. We rarely get a smooth floor for mold remediation or asbestos abatement. Even if you put plastic on the floor, it will not be smooth once is has been used by many feet, wheels and other abrasives.

SOLUTION 2: This problem can be solved by using PVC shower liner or pond liner material. The PVC liner material is highly durable and reusable. The liner material can be cleaned by damp wiping or using dry Swiffer® cloths. The liner material can be moved as work progresses in larger areas especially where ladders or scaffolding may tear 6 mil plastic sheeting normally used by abatement/remediation contractors.

A bigger problem is HEPA filtered devices (P-100 respirators, HEPA air cleaners/ negative air machines and HEPA vacuums) may not filter as well as they claim or 99.97% efficiency with 0.3 micrometer particles. A 2007 study by Tina Reponen at the University of Cincinnati showed failure with N-95 respirator cartridges. **http://annhyg.oxfordjournals.org/cgi/content/abstract/50/3/259**

I have had clients swear they get sick entering a moldy home wearing a P-100 respirator which is the equivalent to HEPA filtration. An N-95 has 95% rather than 99.97% efficiency. The last guy to complain to me was SCBA certified and showed me he knew how to check the fit of this respirator by donning it in front of me. The poor guy thought he was going crazy but he was right. I could not reliably tell a medical referral patient which respirator to buy next. Manufacturers are generally not going to test each filter as they are manufactured. They test some filters and "hope" the results are representative of the lot of products manufactured by the same methods in the same production run.

Finally, HEPA filtered devices may have filter-bypass issues. The location where the HEPA filter meets the housing in the device (P-100 respirators, HEPA air cleaners/ negative air machines and HEPA vacuums) may allow particles to pass though that gap due to poor sealing of gasket material. Sometimes, there is no gasket material. It is common for the US Army Corp of Engineers to demand all HEPA filtered negative air machines (NAMs) be certified by an independent testing laboratory for asbestos abatement projects. That certification will add a few dollars to the cost but without the proof of efficacy, the money spent on the filter is 100% wasted.

Testing filters with DOP or Di-(2-ethylhexyl) phthalate, a chemical that forms droplets with a uniform size of 0.3 micrometers, is common. Another test method involves the use of latex beads sized 0.3 micrometers. I have a philosophical problem with the accepted test protocols for HEPA certification. I am removing particles that will break-up into smaller fine or ultra-fine particles as they desiccate over time. These particles are inhaled more deeply into the lungs and airways. These industry-accepted test methods tell me what happens with high particle loads over a short time course. What happens with mold and bacteria over a 6 month to 1 year period with light particle loading each time?

SOLUTION 3: Make sure you vent the NAMs to the exterior unless you can test them onsite with a laser particle counter. It is dangerous to assume HEPA filtration is all you need to be successful. The Indoor Environmental Standards Organization (IESO **www.iestandards.org**) is working on a standard for testing HEPA filtered devices like NAMs, air cleaners and vacuums: IESO 4310/ Portable High Efficiency Air Filtration Device Field Testing and Validation Standard.

Some structures are built in a manner where ducting to the exterior is infeasible due to window designs that do not open to the exterior such as high-rise buildings and low-rise commercial properties. It is not a good idea to run ducting from a NAM into hallways or stairwells since emergency egress may be blocked.

SOLUTION 4: In asbestos abatement projects, it is common to use a water baffle system to duct the NAMs to the exterior of the contained area but within the structure of the building. You can take a large leak proof container, cut a hole to allow metal HVAC ducting parts like an airtight vents (flat not saddle) attached to round ducts to match your flexible ducting between the NAM and the water baffle. The round, metal ducting must terminate ½ inch to no more than 1 inch above the water line without sloshing water to the exterior. Exhaust holes can be positioned above the water line near the top of the container. All inner ducting seams in the metal ducting must be sealed with foil tape common to HVAC ducting. 55 gallon drums made of plastic work well allowing as much as 1,000 cubic feet per minute (CFMs) of exhaust air with 12 inch ducting terminating approximately 29 inches from the bottom of the drum. The water must have a surfactant or particles may bounce from the surface of the water and become airborne. The surfactant or detergent must produce low suds or soap bubbles or you will make millions of soap particles in the area outside the containment. It is very important to build water baffles correctly and check to make sure they are working as planned by using sampling methods while they are in operation. I use 25 mm TEM cassettes operating at 10 liters per minute around the water baffle. You can take the air samples just outside the water baffle but, you may get some contaminants from the air outside the water baffle. It is better to take the air samples just inside the water baffle where exhaust air is forced to the outside of the water baffle. Laser particle counters are helpful but, tap water has natural salts or minerals that create will appear under a microscope at high magnification or create "noise" for particle counters. Direct microscopy lets you see the particles down to 0.3 micrometers even if the images are not as sharp as an electron microscope.

Negative air pressure differential containments can cause unintended problems. This technique came most recently from asbestos abatement. It replaced neutral air pressure differential containments which had problems with cross-contamination and worker protection. Usually, a HEPA filtered machine is exhausted to the exterior of the containment so the volume of air leaving is greater than the volume of make-up air entering the contained area to produce a vacuum effect. Manometers are used to measure the air pressure differential. Most guidelines and standards describing negative air pressure differential containments state a measurement of -0.02 inches of water column or water gauge (w.g.). This is a technology transfer from asbestos abatement (29 CFR 1926.1101). If the NAM is located near the entry/exit point which is opposite the area where a wall is removed, the pressure differential may be lower since the vacuum effect is higher as you get closer to the NAM and weaker as you get farther from the NAM. Another problem is high winds blowing on the external walls which are part of the contained area. This means the negative air pressure differential may in fact be a positive pressure differential since you cannot control all air leakage all the time. You can only go so far with duct tape, spray poly (glue) and 6 mil plastic sheeting. This becomes very important if the external wall is the site of demolition for microbial growth. Inner walls towards the center have less risk since they are not affected by driving winds from the outdoors.

SOLUTION 5: In reality, the negative air pressure differential will be higher to overcome the fact pressure differentials are not the same measurement around the perimeter of the contained work area. A negative air pressure differential between -0.03 w.g. and -0.04 w.g. can overcome most problems except high winds (20 mph or greater). You will need to raise the negative air pressure differential and reinforce the containment for high wind situations. Plastic sheeting may be pulled from the walls unless you used strapping material like wooden fir strips (1" x 2" or 1" x 3" boards) or Smart Seal's Fast Clip (**www.smartseal.cc**). The latter example is made of PVC and can be cleaned and reused on other projects if the pieces don't get too small.

Negative air pressure differential containment problems occur when make-up air is introduced with cross-contaminants or moisture. You should check your assumptions with an air current tester or smoke pencil. This is a glass tube connected to a rubber bulb like a turkey baster. The glass tube has granules that produce synthetic white smoke when exposed to the air. Sometimes the same granules are placed in a plastic bottle for the same effect when you squeeze to produce a puff of smoke to see where air currents are going or arriving depending on the structure. This easy and inexpensive test method has been championed by Joe Lstiburek (PhD engineer), Terry Brennan (MS Physics) and every EPA certified asbestos training program in America for good reason. Side note: Dr. Joe Lstiburek tells everyone to lick their hand, place it over an opening like the edge of a closed door and feel which direction or side of the door dries the hand from air currents and repeat to find pressure differentials between rooms. This may look strange to a person paying for professional services even though Dr. Lstiburek is absolutely correct. I wonder if his clients shake his hand afterward. Please read Dr. Lstiburek's website (**www.buildingscience.com**) for more useful information. There are also some highly technical books for sale if you want to understand proper construction versus the "crackerjack boxes" some homebuilders would like to call your future home.

Putting a ground floor room under a negative air pressure differential with the NAM located on the ground floor or above the ground floor will be a bad idea if there is a crawlspace below the ground floor. You will suck the dirt, mold, bacteria and other assorted pollutants into the area you think you are remediating (*Editor's note:* this is the reason why patients with mold illness shouldn't use an attic whole house fan if they have a crawlspace). In other words, you are heading for inflammatory disaster. I have seen it happen too often. Fixing the mistakes of another contractor should never be required! The Aspergillus and Penicillium counts from that contractor's errors were in the thousands per cubic meter of air.

SOLUTION 6: It is better to place the crawlspace under a negative pressure differential that is greater (between -0.04 and -0.05 inches of water gauge) than the ground floor (between -0.02 and -0.03 inches of water gauge) above the crawlspace or do not place the ground floor under a negative air pressure differential. You will need to check your containment system with the air current tester that can be purchased from The Energy Conservancy, Lab Safety Supplies or other sources.

A similar problem may occur when you are working in a moldy basement with window wells near ground level. You will suck the contaminants through the basement window wells since they are near the soil level. Some basements have open drain tiles along the perimeter walls. This is a really illogical design by basement waterproofing companies engaged in commercial hit and run. You may also have expansion joint material made from dense paperboard material soaked in tar which makes for really good place to grow with the moisture constantly flowing around the footer below ground. This material is fine for sidewalks and driveways but just a plain bad idea for basements hidden behind walls.

SOLUTION 7: Create airlocks or single decontamination chambers placed under a "positive" air pressure differential sealed at the entrance/exit so air cannot enter around the airlocks. HEPA filtered air will enter your contained work area through the airlock. You are just reversing the flow the NAM by turning it around so the HEPA filtered air is entering the airlock at a greater volume than the volume of air leaving the chamber. The frame can be made of wood or PVC pipe with self adhesive plastic on the two sides that do not open, the top and bottom. A peelable tack mat should be placed on the floor. The feet are the worst offenders for cross-contamination. These types of products can be found at **www.surfaceshields.com**, your local asbestos abatement supply house or paint shops like Sherwin Williams or Duron. It is also a good idea to wear plastic, non-skid shoe covers that can be discarded every time you leave the contained area. Don't rely on the shoe covering of disposable overalls like Tyvek or Kimberly Clark, for example.

A test with the air current tester or smoke pencil will show it is not possible for airborne particles to enter the airlock under a positive air pressure differential since air is always pushing outwards at all openings including unintended imperfections. This concept works for submarines under water and pharmaceutical/medical clean rooms. It is very important to make sure the work area which is not under a negative air pressure differential stays as clean as possible. You may want to hold a HEPA vacuum within inches of the wall as you are cutting the gypsum board or dry scrubbing solid wood with a plastic scouring pad. All water-damaged material must be double-bagged just before entering the airlock without dragging the bags on the ground. All personnel must disrobe and vacuum themselves before exiting into the airlock from the work area. This method works well in office buildings. You place areas outside the remediation work area under a positive air pressure differential so air currents are always pushing into the work area rather than escaping into the occupied areas outside the work area.

Negative air pressure containments can also bring hot humid air into a cooler environment to condense on the surfaces of the point of entry such as around window frames that probably already have moisture fatigue if they are on the side of the structure that receives the prevailing wind. The lower corners are especially at risk since water drops after it condenses from a vapor to a liquid. You can also cause the upper floors of a two-story house with a full basement develop mold on the upper floor even though your NAMs are in the basement if you have a large project in the summer with heat and humidity. I had a job in Fredericksburg where the insurance carrier had brought in a remediation firm which failed post remediation testing six times before I took the job. The upper floor with no prior mold problems had a thin layer of surface mold even though the water damage was an ice-maker leaking into the basement two floors below the top floor. The industrial hygiene firm failed to note the high number of basidiospores and ascospores on the post-remediation air sampling. This was proof the negative air pressure differential was pulling the outdoor air into the structure at a very high rate. In the absence of wood rot, there was no other explanation for the basidiospores and ascospore counts inside. Note: after a rain, these counts can easily reach 50,000 spores per cubic meter of outdoor air.

Weather extremes also create problems for HVAC thermostats located in the remediation work area under a negative air pressure differential. The thermostat may think it needs to keep cooling the hot humid air sucked in by NAMs. Ultimately, this load will make the coil freeze into a solid sheet of ice. I wish I could say I only see this once or twice but people just don't seem to think about their remediation plans beyond their narrow focus of removing the known problem. The opposite situation is cold winter air causing a thermometer to regulate a furnace to roast the house or worse.

SOLUTION 8: Do not operate the HVAC system while working in large areas or exposing the thermostat to extreme temperatures. As long as you don't cross-contaminate, you can operate the HVAC overnight to keep pipes from freezing or use portable electric heaters with a ground fault circuit interrupter (GFCI). You may have to pump HEPA filtered air to the return ducting leading into the HVAC fan-coil unit if it is near the work area like a basement. The filtered air should originate within the structure and outside the contained work area so you do not cause unintended negative air pressure differential problems noted in prior paragraphs. In commercial high-rise buildings, you may want to consult a professional engineer or CIH since the air systems can be more complex and present more problems in occupied areas like office buildings.

You can use your NAMs as air scrubbers in a contained work area. Two locations are helpful as long as the exhaust is not pointed in manner that pushes contaminants from the entry/exit point within the structure to areas outside the contained areas to be occupied. The location should be near the demolition surfaces described in Solution 1 and near the entry/exit. The NAM used like an air scrubber near the entry/exit point should be placed so it pulls air into the room while pointing the exhaust away from the entry/exit. This placement should also allow personnel to enter and exit the contained work area. You may want to use a water baffle described in Solution 8 to keep the exhaust air flow from pushing in any direction too hard. Remember plastic, non-skid, disposable shoe covers are important and use a tack mat.

These air scrubbers may be HEPA filtered but they will not clean all the air in the contained area. The only particles removed will be the particles that are pulled into the HEPA filtration (please read Chapter 5 of Clean Room Technology by W. Whyte – 2001). Many particles will bounce around the containment due to swirls and eddies in the air currents or temporarily adhere to surfaces. This is true whether you use negative air pressure differentials with laminar airflow in the work area with a NAM or use NAMs as air scrubbers. At some point, the particles on surfaces will release and become airborne – especially when the air becomes drier and/or hotter.

Negative air pressure differential containments are not the bullet-proof method everyone would like you to believe. I have purposely gone this far into the chapter to let you know this fact after leading you through all the problems that can be verified. In some cases they do work but, only in small areas without too many obstructions like walls. Clean rooms operate at ten air changes per hour at a minimum.

Negative air pressure differential containments for asbestos or mold generally operate at four to six air changes per hour. Manufacturing of products is done cleaner than your home will be remediated by contractors following the industry standards and guidelines. You can predict failure by using the methods used in clean rooms to correct the problems or find another way to address the failures other than HEPA filtration. You can use an inexpensive anemometer (**www.emssales.net**) to make sure the air in the contained area is moving at 60 – 100 feet per minute from every vantage point like clean rooms operate with laminar airflow. The corners of the room will be the areas where the pressure is not working if you use this clean room test principle. You can also use a smoke pencil to visualize the areas where laminar flow is poor. A laser particles counter can take it further to mirror clean room techniques. I try to reach ISO Class 3 clean room levels with particles the same size or larger than 0.5 micrometers at 35 particles per cubic meter of air (W. Whyte – Clean Room Technology; Table 3.3). There will be people who say this is unnecessary but, if you are remediating mold or bacteria, you have to address all particles. It is not humanly possible to discriminate which particles are allowed to enter the HEPA filter like a children's tree house. I have been able to reach these levels of cleanliness with Solution 9. A conscientious remediation contractor can take several samples with a laser particle counter before spending time and money with microbiology sampling. This state of cleanliness will be gone within days to 1 week depending on season, wind and other factors that re-introduce outdoor particles and mold to the indoor environment.

Large remediation areas may also be a problem since the CFMs of air movement for each machine has a range of distance where the minimum wind speed (60 ft/min) and the maximum wind speed (100 ft/min) may create spacing problems. (W. Whyte – Clean Room Technology - 2001) Note for post-testing: test corners instead of the middle of the room. This will teach contractors not to use one large NAM to place several rooms under a negative air pressure differential containment – if they put their complete faith in HEPA filtration to clean the air. Clean room engineers are very knowledgeable about these problems. I am giving this information for guys like Bob Brandys of Occupational & Environmental Health Consulting Services who has many years of experience in clean room manufacturing to go with his many degrees and certifications. He also has extensive experience with asbestos abatement and mold remediation for a very unique perspective.

SOLUTION 9: Fogging for dust suppression to clean the air of respirable particulates. This concept is not entirely new but, you have to understand how and why or it may fail. Thunder storms scavenge pollution gases and particles from the atmosphere more efficiently than most of our pollution control technologies. Instead of waiting for airborne particles to enter the NAM or HEPA filtered air scrubber, you can capture the particles in the air with billions of tiny liquid droplets and push them to the surfaces where they can be removed by HEPA vacuuming and/or wiping with disposable towelettes.

Commercial electric foggers commonly used were originally made to fog oil based mosquito pesticides in the late 1940s (B and G Equipment Company and Fogmaster). They are commonly used to apply deodorizers and disinfectants to air ducts (Please read the section on disinfectants and duct cleaning). These machines have pressurized tanks and send water droplets to the tip of the orifice or hose where a fan sends a plume of liquid droplets ranging from 20 microns to 75 microns depending on the Venturi valve settings. Depending on the model of fogger machine, the plume can be 10 to 25 feet long with the thickest part of the plume ranging from 5 to 8 feet. These machines are commonly used to disperse water based pesticides, disinfectants and deodorizers without changing the temperature like thermal foggers. The retail cost is approximately $400 to $550 depending on manufacturer and model. There are more expensive models used in commercial greenhouses (Dramm Corporation) with higher gallons per minute operation.

We tested our theory to clean the air with fogging by removing thermal surface insulation (TSI) with greater than 60% amosite asbestos on HVAC air shafts in a high-rise building. This is extremely dangerous asbestos since the asbestos fibers easily become airborne with very little hand pressure (friable) and they repel water. I personally removed the asbestos material without negative air pressure differential containment in the work area. At the sealed entrance/exit of the work area, there was a 3-stage decontamination unit with a shower in the middle required by OSHA regulations. A small HEPA filtered NAM was attached to the equipment or dirty room of the 3-stage decontamination unit sealed to the work area. This is where personnel remove their personal protective equipment before entering the shower. Make-up air entered the decontamination unit through the clean room where clean suits are donned before entering the shower and dirty room. Weighted flaps were installed at each section to limit airflow especially from the work area.

I allowed the airborne fiber count to reach high levels during removal and wetted the waste material once it was in the bag. My personal air samples were largely overloaded and impossible to count with a phase contrast microscope. A commercial lab was able to count one sample with more than 1,000 structures per square millimeter with a transmission electron microscope (TEM). For the trained (and usually licensed) eye, this is an abominably high level that would require more respiratory protection than the powered air personal respirator (PAPR) I wore. There were no issues outside containment. My post testing results were well below the 70 structures per square millimeter read by a transmission electron microscope (TEM). The work area did not have the air cleaned with a HEPA filtered air cleaner. After removal, I simply fogged my product, let it settle and removed with HEPA vacuuming and wiping of surfaces. The industrial hygiene firm who did the testing is well known in the area and has held large industrial hygiene contracts with the Pentagon, NIH and more.

I repeated the same scenario at a later date at a different location and used an employee to fog or spray as I removed the building material with asbestos. Again, there were no issues outside the containment and the post-testing was beautiful. The fiber counts for my personal air samples were below the 10 fibers per cubic centimeter required by OSHA regulations for PAPR. The second project was overseen by an IH who was previously on Virginia's DPOR Asbestos Board. In both instances, we vented the NAM of the 3-stage decontamination unit into the water baffles described in Solution 4 rather than outdoors.

In another experiment, we built a containment tent of plastic framed with PVC pipes. The containment tent had a footprint of 80 square feet with an 8 foot ceiling at the corners peaking to 9 feet at the center of the ceiling. We attached a duct system with an inline fan connected to a fireplace where we burned newspaper soaked in motor oil to produce black carbon soot. The entry/exit point had a flap and located on the side away from the prevailing wind. We only ran the tests on days when the winds were 10 miles per hour or less. The test location was also off the beaten path so automotive pollution would not interfere as much possible.

Mold is easy to capture compared to soot since it can easily absorb water. Black carbon soot is the perfect lab rat for studying air cleaning with dust suppression without any filtration. Black carbon soot is very light and small. Soot has a tendency to float in the air until it sticks to a surface, agglomerates to particle sizes too large to float or coagulates with moisture in the atmosphere until it cannot float. Soot from hydrocarbon combustion is also very water repellent. Soot is a major pollution problem because of these properties. Soot particles range in size from 0.02 to 0.08 micrometers or smaller than a flu virus (0.01 micrometer). Soot particles tend to agglomerate in chains or clusters like grapes that can be viewed on a light microscope at high magnification.

We ran samples at different time junctures with TEM cassettes (0.45 micrometer pores) with a small piece of clear tape measuring approximately 9 mm^2 to 16 mm^2 in the center so the mixed cellulose ester (MCE) filter would stay visually white for comparison purposes. This sampling method allowed us to visually assess the sample until we had samples that required a microscope to view. The MCE filters were mounted on glass slides, cleared with triacetin and topped with a glass cover slip. The samples were viewed on an Olympus CX-41 light microscope with a camera coupler lens connected to a Minolta

DiMage digital camera linked to a TV monitor. This set-up allowed us to use the digital camera zoom lens to so magnification could be increased to approximately 1,800 X. We used a stage micrometer to verify the measurements for magnification or particle sizes with the aid of a computer printer. Dr. Jim Tucker at Environmental Monitoring Systems, Inc. (**www.emssales.net**) was very helpful with sampling strategy and microscopy knowledge.

We sampled after filling the containment tent with soot, two hours after fogging the air, four hours after fogging the air and the next morning after fogging the air. We made comparisons with no fogging treatment, fogging 100% water, fogging water with surfactants (asbestos wetting compound), and fogging our slow evaporating solution (water, surfactants and semi-volatile organic compounds such as 0.32% buffered phenol) to slow evaporation of the water droplets in the fogging treatment so the solution clings. The basis of our solution is capture the particles longer to weigh them down to surfaces rather than float in the air for long periods of time. We followed with a "before" and "after" sample once we found the lowest levels of SVOCs to combine with water and surfactants. We found it takes four hours to allow the particles and droplets to settle in the best circumstances. In the real world, it would be best to allow overnight for the particles to settle on the horizontal surfaces. These samples were sent to a lab for combustion byproducts analysis by using various electron microscopy, polarized light microscopy and X-ray Fluorescence (XRF) techniques. We took these samples since a light microscope can tell you have particles but it cannot tell you difference between soot, ash or just carbonaceous material. These later samples were used to verify results with a different test method. The lab results showed a 90% capture rate with greater than 95% of the particles identified as black carbon soot.

This result is very hard to achieve by any other means. You could run several HEPA filtered air scrubbers for long periods of time if you have the machines and the electrical output. Never mind the clean room industry found this method does not work too well just after WW2 with the first laminar flow clean room in 1961 at Sandia, New

Mexico. This method is called conventional or turbulently ventilated (Clean Room Design; Second Edition – W. Whyte, 1999). Today's clean rooms are based on Unidirectional flow or laminar air flow.

We have since tried air cleaning by fogging with concrete dust and gypsum particles from drywall demolition. These particles are easier to capture since they absorb liquids if the water droplets don't evaporate before reaching the particles. Part of the problem with this approach can be addressed with aerosol physics. Another term for dust suppression of particles in the air is coagulation. This is covered in Chapter of 12 Aerosol Technology (WC Hinds - 1992). Coagulation is the attachment of airborne particles to gases or liquid vapors in the air. Thermal coagulation is gases and vapors bouncing back and forth in Brownian motion until the molecules collide and attach with particles. This process happens in the atmosphere with black carbon soot over a long period of time. Kinematic coagulation is liquid droplets (oil or water) colliding and attaching with airborne particles without the benefit of wind or air currents like rain drops capturing pollen spores as the rain is falling to the ground. Kinematic coagulation will not capture particles smaller than the droplets if the droplets are settling downward like rain without wind. A physics equation to support this theory can be found in Aerosol Technology (WC Hinds – 1992). This is why smoke particles (soot, ash and char) seem to pass through the rain or a fireman's spaying hose. Gradient or shear coagulation occurs when liquid droplets are aided by wind currents with different laminar layers or gradients traveling at different velocities. Theoretically, particles would never collide with liquid droplets if the droplets and particles were all moving at the same speed. The electric, cold foggers (B & G, Fogmaster, etc.) provide the gradient currents with a small fan at the tip of the nozzle where the tiny droplets are dispensed. The inner plume is moving faster than the outer plume of fog dispensed by the machine. It is important to use the fogger with a slow and gentle motion (in a manner similar to airbrushing paint to avoid runs). This creates more turbulence in the air to capture microscopic particles.

Caution: I cannot recommend the use a gas powered or thermal fogger in an indoor space to be occupied. Gas powered foggers are based on the Nazi V1 rocket to create very high pressure. They create carbon monoxide which can be life-threateningly hazardous to health. They also create carbon dioxide (CO_2). Common respirators with cartridges for purification cannot filter carbon monoxide and carbon dioxide. This can create a situation called immediately dangerous to life or health (IDLH) referenced by OSHA regulations. Thermal foggers that heat the solution were designed for oil based pesticides to create very small droplets between 2 and 10 micrometers. The alveolar pores in the lungs are generally 5 micrometers in diameter. This is where the air/CO_2 exchange occurs between the lungs and the bloodstream of the body. This means a person accidentally walking into the area with a fog will inhale droplets that pass the lungs and go straight to the bloodstream for a much larger dose than intended when risk assessments were made on many products. Also, the heat and/or pressure of these foggers may alter the chemical formulation to create unintended formulations or cause product ineffectiveness. The Ideal Gas Law has temperature, pressure and time as variables for liquid vapors and gasses. Small water droplets produced by these machines may cause water to be predominantly in a vapor form rather than the larger droplets that fall to the horizontal surfaces faster.

Finally, the average size of the droplets in the fog is important. Water based droplets will evaporate faster as the droplets are smaller because the ratio of the surface area to the volume of the droplet increases. A physics equation for this scenario can be found in Chapter 13 of Aerosol Technology (WC Hinds - 1992). Our experience with amosite asbestos, soot and gypsum board particles indicates the reported 35-50 micrometer size of a B & G electric fogger set at three to four full turns open is highly effective for average temperature (65F – 85F) and relative humidity (50% RH and higher). The evaporation increases as temperature increases or relative humidity decreases. This means we had to use more product to be successful with soot when the temperature was 85 F and a 20% RH. In fact we doubled the fogging time at the same fogging rate. This means what you do in the Washington DC area in the summer is very different than Arizona in the summer or the winter heating season just about anywhere in America.

The average droplet size is also important since you want the droplets to settle after they collide with the particles. In a study by Riemer and Wexler (Droplets to Drops by Turbulent Coagulation; Journal of Atmospheric Sciences; Volume 62; pages 1962 – 1974), they found droplets 40 microns and greater have settling coagulation. Droplets of 10 microns or less do not drop like fog (page 1967). It is not just a question of evaporation of the liquid droplet. Size and gravity matter. I put this study in this chapter for the most intellectually challenging types to read. This is the stuff necessary for professionals such as Wane Baker who is a mechanical engineer and CIH at Michael's Engineering. Wane is heavily involved with ASHRAE, AIHA and IAQA. Since he is a rare individual as far as professional pedigree, I am giving a really technical citation for evidence guaranteed to work that calculator. People like Wane keep us honest.

Another important point to understand is super-saturation. This is how rain occurs. If you read Chapter 13 Condensation and Evaporation in Aerosol Technology by William C. Hinds, you will find that water droplets continually shed water molecules as they attract water molecules from the air depending on the humidity levels. Rain occurs at super-saturation or a relative humidity (RH) greater than 100% RH. This means the amount of water vapors in the air exceeds the capacity of water that can be held at that given temperature and droplets grow into drops to fall. When you are wet fogging to clean the air, you are using super-saturation for a very short time period like 30 minutes to allow the scavenged particles to drop as fast as possible.

As far as contributing to water activity that may lead to microbial growth, every decent contractor and consultant has a moisture meter. It is important to remember organisms need a certain amount of time to grow under moist conditions. Paper-faced gypsum board needs at least two to three days of moisture before mold growth occurs. Solid wood takes much longer. Don't take someone's word for the Gospel: test it with commonly known technology.

We filed for a patent and started a dust suppression product for the process, AeroSolver. We did what any honest company does. We checked with the EPA's Antimicrobial Division in July 2009. They balked at any mention of mold, bacteria or microbial fragments on the label since they felt it was "mitigating" (in your spare time, if you read the definition of "pesticide" in FIFRA, you'll see my concern). Our company wrote a letter to be sent to the EPA Office of General Counsel – Philip Ross. We again raised the concern our product was a "cleaning product" with no claims to affect the life of any organisms which should be governed by Part 152.10 which exempts cleaning products from FIFRA. We used different examples such as contractors following the EPA's Mold Remediation in Schools and Public Buildings where it is recommended to use detergent and water to clean moldy surfaces to be salvaged. We raised the issue of pest control contractors using integrated pest control techniques in the EPA's Ariel Rios Building in Washington DC to clean odors and food residues that attract pests rather than spraying insecticide everywhere.

To date, the EPA Antimicrobial Division has sent a statement to Indoor Environment Connections stating the use of cleaning products to prevent insect infestation is illegal under FIFRA. I have a letter from Dennis Edwards, second ranking EPA AD official who has been at the EPA since Nixon in the 1970's, stating no product can claim to be used for mold (even for cleaning) without an EPA registration. I have posted the correspondence with the EPA at **www.aerosolver.com** under the dropdown menu for industries served. The tab is called "EPA Correspondence" if anyone wonders how this bizarre story is possible and we have to explain why mold and bacteria is not on the product label. The story is too bizarre to believe so we offer the evidence. If you have questions about EPA registered pesticides or using cleaning agents to perform mold remediation, I recommend contacting your state pesticide control officials. The can be easily located at the Association of American Pesticide Control Officials website (**http://aapco.ceris.purdue.edu/htm/directories.htm**).

These state officials are the individuals who actually do the footwork for pesticide enforcement with very little funding or staff. It is sad to think some states may have enforcement against contractors following the EPA's remediation guidance to use detergent and water to clean moldy surfaces that can be salvaged. Some states require a pesticide applicator license to use an EPA registered product for mold remediation.

I would be happy to assist any attorney representing a poor mold remediation contractor who follows the EPA mold remediation guidance from the EPA Office of Indoor Air. Unfortunately, the folks at the EPA Antimicrobial Division do not understand mold is physically removed in remediation even though a different group has spent years educating the general public. From what I see, it does not matter which political party is in control if bureaucrats are allowed to make or enforce regulations selectively from behind their administrative desks. The EPA has a long history of spending taxpayer money to defend this type of behavior, only to lose in court. Dealing with the EPA now makes me especially miss Frank Sanders who ran the EPA Antimicrobial Division for years while being reasonable and timely.

At the risk of being too detailed, I fear some of my comments might be too focused on professional issues in remediation and mitigation. Yet the air you breathe will determine your Survival! When you read a mold standard, remember green beans are canned in facilities cleaner than the common mold remediation job.

Moreover one must recognize that each testing methodology has its limitations. Know what these are before drawing conclusions. To do otherwise is to join the Ass^2 Club.

Just as in understanding real estate, to understand WDB is to be site specific. To paraphrase Conrad Hilton – remediation is all about location, location, location.

CHAPTER 25 David's Bubble: How He Plans to Keep Susan Safe

I don't often have the chance to talk to a statistician who is also the spouse of a patient. David Lasater Ph.D. is that person and he pays close attention to every detail, a personality trait that one expects from top-flight statisticians. He recently retired to join his wife in their "final home," one he'd make sure was safe for her. He'd watched her suffer for too long from the Lyme disease that was her first blow, and then, before she could regain her health, from exposure to moldy homes. She's been severely injured by her inflammatory illnesses, in part due to a heart problem that limits her some, and from increased susceptibility to bioaerosols found inside water-damaged buildings.

And yet, now she's active, attentive and witty. Now no one would look at her and think that anything was wrong. She's walked many a mile in the shoes of someone with chronic fatigue, and she won't go back to that old life willingly.

It was David's attention to detail, and his investment in a home that would be safe for his wife that keeps her healthy today. She was extremely sick from exposure to moldy buildings, and as a real estate person, she knew of countless upscale properties that she couldn't even enter, much less show to prospective buyers. "Buyer beware," is good advice in this country now, especially with previously unoccupied, foreclosed homes available as "bargains" all over the U.S.

She and David could have stayed in their safe house but the grandchildren were growing up near Reading, Pennsylvania, and they knew several years ago that soon it would be time for David to retire from his demanding international consulting work, which meant it was time to move.

Ever the organized, logical mathematician, David answered the questions that so many people ask. Where can we go? What do we have to do? How do we know if the next place is safe? His process is a model for what others could do, understanding his project might be bigger than what others might choose, but it's the attention to detail and the insistence that the job be done right that served to meet David's ends. At the end of the job, he still has a vibrant wife. And without David's work, she couldn't have survived living in their new home as it was. He took control of the remediation of the home, and now it's a showpiece of safe, healthy living.

While Susan regained her health, I'd see wonderful Excel presentations that David made of the sequential data I'd collected in her case. Yet he wasn't just crunching numbers, he was looking for hidden patterns and trends. The reality of her illness lay in the numbers: David trusts numbers. David knew that the time would come when he'd be responsible for Susan's ongoing safety and it would be his responsibility to make sure that what was done next was done right. He never tired of looking for simple, correct answers. As we've seen throughout *Surviving Mold*, the data shows us the way. "Trust, but verify" one of our Presidents was fond of saying. Trust people but verify what they say. Trust numbers after verifying that the collection methods are sound.

I've learned over my career that arguing with statisticians about numbers is a losing proposition. David is precise, and while polite, he doesn't suffer <u>any</u> fools. So when it came time for Susan to leave their "safe house" for the great unknown, David's approach became a problem-solving one. He identified concepts and fleshed out the details. He had an overall idea of what his next dream house would be and where he wanted to live. They'd search and find their jewel, and life would go on.

They soon found that if they wanted a dream house in which Susan could be safe, they'd have to either build it from scratch or see a property with potential and fix it. Either way, they were looking at spending a lot of cash. As David remarked one day, just look at what this illness usually does to people: it makes them tired, in pain, overweight (not Susan!) and broke.

Here's David's summary. I can tell you that Susan is doing quite well now. She crashes if she's exposed to a water-damaged building for less than ten minutes, but thanks to David's process of problem solving, Susan doesn't crash at home.

Learning prior to finding a solution

David will tell you that mold needs moisture and cellulose to grow. Warmth usually speeds the process. Any kind of leak or high humidity can support mold growth in a home. Where does water enter? Roof leaks, basement leaks, plumbing problems, HVAC that isn't installed or maintained correctly, condensation from cold water against warm surfaces; and the list goes on. Brick homes without weep holes to prevent condensation from draining to the outside world, fake stucco with a requirement for constant caulking, drip edges that aren't correctly installed, ice dams that quietly fill attics with moisture, not venting plumbing vents to the outside world or not using normal boots to keep the pipes from leaking back into the attic, insulation that closes off soffits in the attic.

David kept siding in his mind, as siding of any kind invariably led to water penetration and mold growth somewhere, even though it's not always visible to the eye. And if the home uses oriented-strand board (OSB), a conglomeration of wood chips and glue, look out. OSB is a lot cheaper than plywood, but it's a fabulous growth medium for microbes. OSB is particularly bad as material used under siding and in underlayment below carpeting. While the OSB might be dry (less than 10% moisture), it's a sponge if it sits outdoors in a stack under a flimsy plastic tarp during construction. If you've got OSB around windows or chimneys, it's also a sponge. What *was* dry becomes wet during construction or construction storage. I'm told there are now some waterproof OSB products for sale. I hope the advertisements aren't misleading, as all OSB I've seen just can't handle being damp.

Searching for a new home

By David Lasater

In two years of searching for another safe house, we learned that there are critical points that must be considered before building a new home or purchasing an existing one. We looked for a brick or stone exterior, one with weep holes and no problems with the pointing. The restriction of no OSB eliminated most new construction and manufactured homes. We remembered what Dr. Shoemaker advised in his book: be at the top of the hill and avoid the ever popular walk-out (or totally enclosed) basement, thereby eliminating most of the building sites we considered. I can't tell you how many people willingly buy a home, complete with an in-ground basement that sits at the bottom of a hill. There's no way that such a basement will ever NOT be wet. *Taking water out after it's gotten in* just won't protect Susan. I hope those people checked their genes before they moved into their birdbaths. Of course, we were endeavoring to move to a mountainous area along the Reading Plateau, where we also knew we'd have to remediate radon gas.

Custom building costs too much, plain and simple, even if a contractor could stick to a construction schedule. Most builders said nine months to a year even if we could find a suitable lot with no builder tie-in (*Editor's note*: tie-in means a restriction as to who's allowed to build on the property after it's purchased). Many "custom" builders have their basic models and the "custom" feature means giving you three kitchen ceiling fixtures from which to choose!

We also wanted to avoid any noxious odors in the surrounding area, such as landfills, factories, or mushroom houses. Dr. Shoemaker's office is in a rural area dominated by agriculture and chicken houses. Sure, there was land for sale as the farmers are being squeezed by taxes, but if the chicken house smelled on hot summer days when the wind came from the West before we build, it would smell after we build, too. Fertilizer doesn't always smell great.

We were also leery of overhead high-tension electrical lines. If I recall correctly, the biggest risk for heart attack in the Framingham,

Massachusetts, study of heart disease was density of telephone poles with high voltage lines just behind. In other words, we knew we were looking for a needle in the proverbial haystack while under time pressure and budget constraints. Susan was finally feeling well enough where we were before, after five years on the sofa. Now she was healthy enough to care a lot where she lived. She wanted her "bubble."

We felt that if we didn't get the job of relocating done easily, then we might have to work a little harder.

Through Dr. Shoemaker we'd met Greg Weatherman; we called him when we found a 22-year-old contemporary, well-built 4,000 square foot home on top of a hill, surrounded by woods and other high-quality homes. The entire living area is contained on a single level. It was in disastrous condition and reeked of mold so badly that Susan couldn't last even five minutes in it. She said "absolutely no way!!!!" but I said, and Greg agreed, "This is all fixable; it's a great house that has suffered deferred maintenance and prolonged water penetration."

The accountant in me looked at the cost of finding a house of our dreams versus the cost of buying this home and fixing it. The balance sheet said fix up the home. Up front, we had to pay the costs of rigorous inspections, with documentation of the remediation requirements. Susan finally agreed to consider the home after these numerous, lengthy inspections, expert opinions, and a reduction in the acquisition cost after the Seller received our volume of reports and photographs. It helped that the grandkids were only eight miles away, plus we were following the old real estate adage, "buy the worst house in the best neighborhood." Even with the extreme cost of the remediation, we've not out-priced the neighborhood.

Looking back on our house-hunting, we took so many buying trips further away, we had to wonder how we missed finding this place so close to where we wanted to be. Sure, it wasn't for sale, but we thought we'd surveyed the landscape well. All I can say is, searching a square mile takes time and motivation. If Susan hadn't been feeling better, she couldn't have done the searching as I was still working full time.

In addition to the beautiful setting and the other considerations, the house was brick with 1-inch by 6-inch tongue-and-groove sub flooring and all else was 5/8- or 3/4-inch plywood; no OSB anywhere! Most of the floors were hardwood with stunning tile floors in the solarium and baths. Importantly, there was also a large basement with dry walls and floors. Additionally, two crawl spaces were constructed like the basement – concrete block walls with concrete floors. If they hadn't been done right, I'm sure I would've insisted on such a measure as crawl spaces are killers for people who are sickened by molds and other microbes. To my relief, there were no ducts and flexible HVAC lines getting soggy from condensation from the warmed air passing in a chilly crawl space, though I have to admit that I was disheartened to find that the insulation around the ductwork itself was soggy from condensation.

Remediation planning and execution

Greg Weatherman authored the remediation plan and I implemented it via a myriad of contractors (each interviewed before signing any papers). There were three critical components of the work to be done – first, rip out all mold-damaged materials and replace them with mold-resistant materials; second, stop all sources of the water damage the home had received; and third, clean everything with physical scrubbing, HEPA vacuuming and finally, anti-microbial treatment of all surfaces. This cleaning included all surfaces – basement walls and floors, ceilings and walls in the main house, plus all cabinetry. (*Editor's note*: the three most important elements of remediation are the ABCs: after **Abating** water intrusion; containment and isolation before removal of damaged **Building** materials; then comes **Cleaning, Cleaning, Cleaning**.) After the very expensive rigorous cleaning including fogging with chemicals that Greg uses, I removed small particles and mold spores (*Editor's note*: whether or not the spore is alive means nothing for those who react to the chemicals on the spores) using a wet Swiffer cloth and thick knee pads. Finally, all hardwood floors were scrubbed with Murphy's Oil Soap and treated with a preserving liquid wax. Ductwork was vigorously cleaned while being scanned with video camera and then sprayed throughout with an anti-microbial compound.

Repairs required before cleaning

Initial steps were relatively easy, and included gutter cleaning, gutter and downspout repairs or extensions and the installation of a gutter guard product to prevent any future water penetration. Half of the downspouts didn't have underground piping to move rainwater away from the house; now all downspouts empty into the woods at least thirty feet away from the foundation. Carpeting in the four bedrooms that'd been installed over the 1–inch by 6-inch tongue-and-groove sub floor covered with an underlayment of 1/ 4-inch Luan plywood was removed. I see no value in carpeting much living space: I call that idea "reservoiring," meaning that a reservoir of particulates and microbial products too small to be removed by vacuuming and even steam cleaning is created in the carpet even before the volatile organic compounds have evaporated out of the new carpet.

Basement vents were *closed* with concrete blocks to prevent high-humidity outside air from entering the cool air of the basement and condensing – a major source of moisture in the house's lower level. The crawl space is treated as a battlefield in the War on Mold. All insulation in the basement and crawl spaces was removed, from under floor joists and especially the soggy insulation from around ductwork. It was all wet from condensation due to cool air from the air conditioning system inside and hot humid air outside before the wall vents were sealed.

Major repairs took longer. Damaged floor joists from a leaking shower were completely removed. Imagine my feelings when I learned that thanks to the defective shower, I now had to rebuild the entire master suite bathroom, including new tile and new cabinetry. It really doesn't take much time to check plumbing and grout tile properly and install green board water protection in a shower, but *not* doing that costs health and lots of dollars. Damaged roof sections were completely removed and re-roofed. The skylights in the solarium and elsewhere were both re-flashed and re-caulked, even if they appeared to be intact, or they were replaced if there was any hint of leaking. Only then were we ready to completely rebuild the large solarium. The two chimneys were completely re-flashed, with counter-flashing on the step of the gable roof, with contact points for the chimney, and with the side of the house thoroughly inspected with meticulous attention to detail.

The HVAC system needed lots of attention. All insulation was removed from ductwork. Vents were added to heat / cool the basement the same as the rest of the house to prevent any condensation possibilities. A new air conditioning condenser with a Seasonal Energy Efficiency Rating (SEER) rating of **20.5** was added. (*Editor's note*: the higher the SEER rating, the greater efficiency of prevention of energy consumption there is. A SEER of 12 is considered standard for the newer, high-efficiency units; a SEER of 6 is commonly seen from the energy-efficient units of the 1990s). In an attempt to increase efficiency, the whole house air-handling system, all three zones of it, and the basement were equipped with a humidifier for use in winter and a new de-humidifier for use in summer. Just to be sure, we installed a UV light-based air cleaning system with a new Minimum Efficiency Reporting Value (MERV) 16 filter. This filter is so efficient, it removes just about every aerosol from the air (a HEPA unit with a 0.3 micron filter will give a 16 MERV), with an "arrestance" value of over 99%. (*Editor's note*: cleanrooms in computer production facilities aren't usually this intensely cleaned). The addition of a new outdoor air intake ensures that the home has positive pressure in the whole house at all times. That means that any moisture we create from cooking, breathing, bathing, washing dishes or clothes or any aerosols we make are pushed outside.

All new construction materials included new paperless insulation and mold-resistant drywall that was Dens-Glass fiberglass bound. All paints were low- or no-VOC paint. We replaced carpet with hardwood floors in three bedrooms. The master suite has a water-heated porcelain tile floor. A humidistat-controlled venting system was installed in the spa area of the solarium to prevent the hot tub from ever increasing the humidity in that room above 55%. (*Editor's note*: as a general rule, wherever you see indoor spas, especially those installed without much regard to residual humidity, look out for moldy conditions).

We didn't forget landscaping and neither should you. We removed plants that were too close to the house, trimming bushes and trees away from house, so the house siding could breathe freely. The landscaping also included the use of river rock around the entire perimeter of the house – not mold-spore producing mulch. I know that Dr. Shoemaker says that outdoors mold is part of an entirely different ecosystem from indoor mold, and I'm not arguing with him. This is our house and I'll decide what mulch we have (or don't), as I'm going to do everything to protect my wife. As you might expect, we found the radon we were looking for. We fixed this problem by installing two independent radon remediation systems. A smaller house might only need one.

Preparation for the move

How do you say goodbye to your old friends, the mold holders? Goodbye will do and cross-contamination won't. We bought new leather furniture and beds. All our furniture that we kept was cleaned with an anti-microbial. All old bedding was trash. All our books from an extensive library were individually wiped carefully with an anti-microbial before packing. The only carpets in the entire house are now area rugs and oriental carpets that were thoroughly cleaned on both sides before being moved into the house. I couldn't part with my collection of walnut, cherry and teak lumber, so my entire wood collection was moved directly into a new out building. Susan can stay out; sawdust is a killer for her. Just for fun, I put up a playful sign, "No girls allowed."

Testing considerations

I wish the ERMI labs had discounts for quantity! Before we settled on this project, I asked sellers if they objected to my performing an ERMI test. If the seller balked, I walked. I'm not unobservant, but I'm telling you that most homes with dangerous levels of ERMI look just fine. As an aside, any ERMI over 2 is a hazard for Susan, according to Dr. Shoemaker (and based on what I saw in her reactions to exposure, I believe that even a 1 might be too high). Just look for the risk factors for microbial growth before you assume the freshly painted den with the chimney, older carpet and the sliding glass doors is safe. I can't tell you how many conversations I had with mold remediation firms about ERMI testing. Most weren't even familiar with the method. Some didn't even want to learn, so they were easily eliminated from consideration. Most use the common "air quality testing" that identifies only a few classes of mold spores. If all they know is air testing, I know where the Yellow Pages are.

ERMI is new. I understand that it's being enhanced even as I write this, with a study done with Dr. Steve Vesper of the U.S. EPA, Dr. King teh Lin of Mycometrics and Dr. Shoemaker. As a statistician, I'll tell you that ERMI needs improvement in documentation. The logarithm-based metric doesn't clearly identify harmful mold spores from harmless mold spores, only those from Group I and those from Group II. Further, the few experts like Dr. Lin and Dr. Ritchie Shoemaker can quickly point to one DNA identified mold spore count from Group I of "6" and tell you that it's "high" and another with a mold spore count of "3000" is harmless. This doesn't help lay people and victims of sick building syndrome to understand the metric. Standard air quality test results will never fully quantify the risks of a specific environment. Because I understand that the ERMI is a first attempt at a building index, I'd say that assessment of health risk is more important than a building index. The ERMI helps rule out a home as absolutely not safe, so don't even think about an exposure. But for marginal ERMI results, the proof of safety is re-exposure of the patient. The ERMI method can continue to be developed to quantify health risks with more focus on harmful spore types and documentation can be improved to better explain the basis and calculations to all – professional remediation firms and victims alike.

Conclusions

Our focus was to immediately remove all badly damaged materials but make repairs in such a way that the same damage can never recur. Best advice: educate yourself. Ask pointed questions of all inspectors and contractors; demand complete answers. It's your dime; pick the brains of all the various experts and keep learning. Compare bids. Reject vendors who are evasive or act superior – like the "mold expert" who told Susan: (i) he didn't smell anything in the house; (ii) he never heard of ERMI; and (iii) furthermore, he didn't care. He wasn't hired; he was actually lucky to survive (my cane is dangerous for those who would hurt my wife)! Susan now has her "bubble" and breathes quite normally in it, feeling the best she has since 2004. As I sit in the sunshine in winter, I'm warmed by the solar energy and the knowledge that my beloved Susan will enjoy her new healthy life as a grandmother and as a wife. I will never forget that I lost her for five years and we can never regain that precious time.

CHAPTER 26 Legalized Medical History

In Ass^2 Medicine (see Chapter 3), I spared little criticism for medical thinking based on assumptions. Guessing is a really bad idea when the basis for therapy relies on secure data points. The concept of "not-guessing" is particularly important when you're treating enormously complex inflammatory diseases, with treatment dependent on a series of individual steps taken one at a time.

As unusual as it seems, the people who most often expose "medicine by assumption" are attorneys. From where I sit, the basis of cross examination is to take an idea as presented and to distill for analysis from as many aspects as possible, attacking each piece or idea to see if it holds up to criticism. And again, from where I sit in personal injury litigation, the plaintiff and defendant each have their attorneys, who have their consulting experts, who have their opinions and documents they rely on to support their opinions and to refute the other side's consultant. It's true that a defense attorney has a bias, in that he wants to remove the credibility of an expert witness, and the plaintiff's expert must similarly destroy the defense expert's criticism. It's no surprise that truth isn't the point here, merely who will be able to stand up well enough to the attorneys' attacks. Truth becomes irrelevant if it can be twisted and distorted.

You'd think that assumptions wouldn't last too long in the legal crossfire, but some attorneys are better verbal jousters than others; some think faster on their 'feet' than others; and some simply know their material better than others.

What I see attorneys doing that physicians don't, is asking the same question from several angles, comparing the answers, and then putting the summarized answers into a construct that they can attack. Yes, there will be some twisting of words and ideas, too, if one lets an attorney sum up what one just said.

Applying that lesson to taking a medical history means that each point that's recorded is checked and double-checked. That means an unbiased "legalized" medical history takes more time than a history guided by guesses and assumptions. Sadly, I see wholly biased medical histories recorded by defense consultants in litigation. That's just a way of life for defense-hires, as if their bogus recording of events wouldn't go unnoticed and un-rebutted. Some of the defense docs for hire just don't know how to get away with misstatements; the real pros think that they can.

Very few patients are initially prepared for an unbiased legalized medical history. I require that prospective new patients send a detailed timeline of their important medical events and a complete medical history before their first office visit. Then I compare the timeline to what the medical records document. Somewhere between the records and the timeline is the truth. If I have to ask a question, "When did you first get ill?" three different ways to get a consistent answer, I'll do that comparison. Patients don't always associate, say, moving into the musty smelling apartment three months before they became ill, especially if the event was 15 years ago. It's easier for me to see the overarching summary of someone's illness because it didn't happen to me. Because I've seen so many of these kinds of cases, I can view and summarize a patient's files fairly easily, if not always quickly.

When people have little recollection of what happened, and time has preserved little actual evidence, one can reasonably question the history's accuracy. In medicine, absolute perfection in diagnosis isn't required; all we need is reasonable medical certainty. So when I see notes that recite "allergies" or "chronic fatigue syndrome" or "fibromyalgia," I know that the patient's symptoms might justifiably be considered other diagnoses as well. Often the alternative diagnoses are ones I'm very familiar with.

Sometimes the truth about environmental exposures that might be linked to illness is never determined. People can't remember, the medical record is silent or records were lost. When that happens, legal certainty basically disappears unless a family member or trusted onlooker provides additional information. If a defense attorney begins by asking, "Isn't it true that you can't tell me when Mr. Jones became ill?" and if I don't have verification of when the plaintiff's illness began, the opportunity to confirm when the illness began is lost.

Fortunately, most patients know exactly when they became ill, where they were, and what the likely environmental exposures were. In these cases, the legalized medical history is able to rule out multiple sources of exposures. Finding just one exposure strengthens the causal link to human illness. Finding multiple sources of exposures demands a lot more work and a lot more time, because you have to be sure that Building A made the patient ill and not Building B and C. In these cases, the legalized medical history provides the only opportunity to sort out who did what to whom and when.

The problem with our current medical practice paradigm is that physicians caring for people usually don't get paid for solving complex medical problems. Outpatient consultations that last more than an hour are rare these days, especially if the physician is being paid by an insurance company. Reimbursement for a 15-minute surgical procedure usually is far greater then payment for 90 minutes of thoughtful history-taking and decision-making. There's no reimbursement for time spent organizing two thousand pages of previous medical records.

As a result, look what we commonly see in outpatient workups: use of prior histories as proven fact; use of prior diagnoses as confirmed findings; and the perpetuation of countless medical assumptions (See Chapter 3).

It's the quality of the history and the careful study of prior medical records that provides a basis to answer the questions: "Who is the patient?" and "What happened to the patient?" These questions underlie "What's wrong with the patient?" When all three of these questions are answered thoroughly, then we're ready to obtain the database of physiologic parameters that will tell us what to do for the CIRS-WDB patient, and in which order we need to do it.

Where we are now in CIRS-WDB litigation

Until September 2009, defense attorneys could usually expect that their consultants would be able to say a lot of really silly, unsupported things about CIRS-WDB. These consultants' lack of experience in treatment really made no difference in court. After the U.S. GAO report of September 2008, suddenly the tables were turned on the defense consultants. All of their pet foils, like (i) ingestion caused illness; (ii) only mycotoxins made people sick but the levels were so low illness couldn't occur; (iii) specific causation of one factor in the diverse chemical mixture found inside WDB made sense; (iv) linear dose response made sense; (v) inflammation wasn't a factor; (vi) no one else showed exposure to WDB made people ill; all of those garbage ideas went "*poof*." Gone.

Those were the defenses used for years. But in just a few minutes after the Mold Expert position statement from the Policyholders of America was published in July 2010, those hoary defense ideas died. The defenses were shown to be untrue. This report came out after the both the 2008 US GAO study and the July 2009 WHO study were published. These latest and greatest studies reporting on chronic inflammatory response syndrome (CIRS)-WDB showed that what had been used in defense was just plain wrong. All of us who argued against such ridiculous statements from the defense were heartened.

Never underestimate the cleverness of the defense in mold litigation. Suddenly after the General Accounting Office (GAO) report, we started seeing the consultants being touted as treating physicians. *You've got to be kidding me*, I thought. Well, just let's see their published data. No, they don't have any. So how can they try to pass themselves off as treating physicians? Where are their legalized medical histories? There are none.

As an example of passing off book-learning for real learning, let's consider the person who has read lots of academic papers on the proper methods used by welders to piece together two sections of angle iron. He's never even picked up an acetylene torch, but because he's read about it, he's going to testify to a jury how the angle iron *should* be welded. The analogy here is clear: any expert has to have experience in welding far beyond simply parroting the literature on welding. Any *real expert* in CIRS-WDB would never cite the 2002 American College of Occupational and Environmental Medicine (ACOEM) opinion and the 2006 American Academy of Allergy, Asthma and Immunology (AAAAI) opinion alone as reliable documents. Only those who had no experience with treatment would use those unscientific papers to support their screwy idea that molds couldn't hurt anyone. They'd have to cite a series of patient data that they'd published, having assessed those patients before and after treatment, and they'd have to include a thorough, rigorous and rational defense of their treatment methods.

But no government agency had stated such facts plainly until the U.S. GAO report was finally published. Adding to the power of the U.S. GAO's sea change in perspective was the 2009 World Health Organization (WHO) report that really scuttled the old-time consultants as they tried to make a jury into a ship of the fooled.

The GAO laid out three cardinal elements for determining causation:

1. epidemiologic similarity of findings of a given person to those findings already published (documented).
2. epidemiologic similarity of lab findings to those seen in experimental studies on humans and animal (documented).
3. response to therapy (documented).

In other words, to determine that WDB caused your illness, a doctor had to be able to show that a given patient's history was similar to those of other WDB-patients already in the literature; that their lab findings were similar; and that a response to therapy was documented in the medical records.

Given that there's now a rich literature on epidemiologic findings and our group has published six papers on treatment using protocols employed by literally hundreds of physicians, you'd think the defense attorneys would simply roll over and hide. Logically, they probably should, since their medical position is without basis, but they don't. And since defense attorneys are paid to develop arguments, they do.

If we apply the lessons of legalized medical histories to those recorded by defense consultants, they fail. If we apply the lessons of legalized medical practice, using transparent, thorough and rigorous assessments with interventions confirmed to show benefit in the literature, the defense consultants have an incredible mountain to climb.

So what do I credit attorneys for in my practice? First, elimination of guessing. A legalized medical history is unlikely to be flawed. Second, a healthy dose of cynicism regarding what will be held as truth in court. I must document, document and document! Third, the actual scientific basis for what I have been doing for 13 years is published and peer reviewed. The latest and greatest science is rehashing some of what our group has published for years.

Now the fun begins in deposition: defense attorneys need to be prepared for a medicalized legal history!

CHAPTER 27 Lawyers, Lawyers, Lawyers

You'd Think That Truth Was the Judicial Standard

In *Mold Warriors*, published in 2005, plaintiff attorneys Dan Bryson and Dodd Fisher helped me try to explain the idea that a testifying expert must be qualified by the Court in order to sit in the witness box, talking with a jury. In the years since then, much of my time in court on behalf of my patients has been spent answering so-called Frye or Daubert challenges.[4] Every state has their own version of which criteria will be applied to the prospective expert but the basic idea is the same: the trial judge has tremendous power - and responsibility - to determine the outcome of cases simply by agreeing whether to allow an expert's testimony or not.

---

[4] The Expert Witness and the Daubert Challenge: For more than 80 years, the U.S. courts have used the Frye rule to make determinations about the admissibility of expert witness testimony in the area of science or medicine. The Frye rule (Frye v. United States, 293 F 1013 D.C. Cir, 1923) asks, as its two primary concerns, whether the findings presented by experts are generally accepted within the field to which they belong; and whether they are beyond the general knowledge of the jurors. In 1993, a critical ruling was handed down that was destined to redefine the landscape of medico-legal testimony of experts (Daubert v. Merrell Dow Pharmaceuticals, Inc 509 U.S. 579 [1993]). As a result of this watershed case, the Supreme Court effectively made trial judges the gatekeepers of scientific expert testimony on the basis of four criteria:

1. Whether the theory used by the expert can be and has been tested.
2. Whether the theory or technique has been subjected to peer review.
3. The known or potential rate of error of the method used.
4. The degree of the methods or conclusion's acceptance within the relevant scientific community.

If the plaintiff's expert is kicked out, so too is the likelihood that the plaintiff will receive any personal injury compensation.

For this discussion, let's assume the trial judge isn't biased. That means we won't be using one decision made by a Circuit Court judge who cited in his opinion items from the defense motion 39 times and only once for the plaintiff. Naturally, the defense won that one.

The power that judges have is muted somewhat by the appeals process. If a decision is made that one side doesn't like, they can try to take it to a higher level, but the reason for the appeal must be based on an error in law or its application made by the trial judge. In a world that's moving as fast as the science of mold illness (or any field for that matter), there's no way that a given judge will be *always* right. Maybe that explains why the higher courts always include a panel of judges to decide appeals cases. It's a built-in control against one judge making a mistake, and the idea that mistakes made by a group are less likely to be due to bias or to the influence of special interests.

And the idea is consensus of a majority rules, as we see in the U.S. Supreme Court. There's no difference in a 5-4 vote compared to a 9-0 vote, as both are total wins. Coming close counts in pin the tail on the donkey, hand grenades and horseshoes, but not in Court decisions. Still perhaps 5-4 decisions provide more hope for the losing side being able to change the outcome in the future than 9-0 decisions.

In sports, losing is a fact of life. Perfect seasons are rare, but losing once doesn't necessarily influence the impact on the results of the next game. However, that isn't the case in court decisions. They call it precedent. Every black mark placed by a judicial pen is made by indelible ink. When you read a motion in a legal case, there will always be discussion of what others have said and done before. But look out: here in the motion written by a defense attorney is where we see some real travesties of justice, because now we have a biased interest interpreting prior court decisions for a judge who might not have time to read multiple twelve inch stacks of documents. The danger is that the judge might *believe the biased slant* on prior judicial opinion or his/her clerk might not have read the cases which were cited in a slanted fashion.

Both the Daubert and Frye decisions are classic ones in how often what was actually said in the decisions is bastardized by the defense. Winning or losing counts, not how you play the game. Hitting below the belt and gouging eyes happens all the time in litigation and everyone involved knows that. It's all about *winning*. Our justice system is called the "adversary system." It isn't about truth, it is all about winning. In medicine we can only win if we follow the truth.

Still, one feature of the game (sad, isn't it, that such an important part of the legal process is an elaborate game?) of getting the expert into court is the reality that if I testify, the defense usually loses. Is it any surprise then that the defense will try anything they can to convince a judge to exclude my testimony? So far, their strategy hasn't worked very well, because I've been allowed to testify thirty-five times (here's a New York case, now 36; still counting) compared to six times when I wasn't. Yet when the defense boys get working on their spin machines making the motion to exclude, you'd think I'd never been accepted and no one believes me anywhere in the world. And yes, the defense is known to twist the truth. One can only wonder about plaintiff attorneys, but I bet they have the same proclivity to delete, distort and misrepresent the truth as well.

Still, a 36-6 record is pretty good. If you are rooting for your college basketball team to get a high seed in the NCAA tournament, that record will turn a lot of eyes. Or if you are a baseball pitcher for San Diego, that record will bring a free agent contract worth many millions.

But excluding my testimony isn't a strategy devised just for mold cases; it's a long-standing aspect of all litigation. The idea is that legal decisions about who is an expert should be based on methods of science and technology that's tried and true. Anything that isn't widely accepted by the relevant scientific community might just be too new or "novel." Considering that, then the chances of that novel idea being accepted as valid might be impacted adversely if the defense brought in a string of consultants who all say "This is not good science." Who will the judge listen to? The six well-heeled defense boys (usually wearing color-coordinated suits) who've never treated a mold illness patient or the one plaintiff's expert who's treated thousands?

The defense has lots of other weapons they use besides gaming the system regarding expert testimony. Rarely is a plaintiff able to afford paying expert fees for a string of experts, especially after the injury from exposure to WDB took away his or her health, job, savings, home, marriage and maybe more.

As you can imagine, close behind trying to exclude the plaintiff's expert is the tactic of spending the plaintiff to death. I don't know too many attorneys who'll fund a plaintiff's case, though there are a few.

And then there's the time-honored practice of delay. It is true that justice delayed is often justice denied. Imagine surviving the legal challenges to testimony and having adequate money to invest in your case but suffering from postponement after postponement.

In ten years of doing legal work and after more than one hundred depositions, I've met some decent human beings who are also attorneys. Let's see, I can think of at least five, and well, I'm sure that out of the hundreds of attorneys I know, I'm just forgetting the rest of the good ones. None of these attorneys take the Insurance dollar, none are defense attorneys. Maybe I forgot a decent human being who played defense games.

Still, it's the legal profession that has the weapons to bring about social change, protect human rights and bring justice to bear on critical problems in our world. The problem is that for every attorney who has idealistic thoughts, there are opposing counselors who will fight to prevent changes requested by the plaintiff. One might think the system is one of checks and balances, but from my point of view, *it's all about the checks*. And often truth is sacrificed in the balance.

Most people going through a deposition don't regard that time as enjoyable. The other side has a right to ask whatever they want, all designed to make a record or sworn statement that they can use later. And if you let them ask away without any interruption, the process becomes really one-sided. But since the attorney is usually asking me about things I know far better than he does, often the advantage swings my way.

It's usually pretty easy to see which attorneys have done their homework and which ones haven't by the questions they ask. Experts are allowed to make the attorney look stupid during the deposition, but most of the ones I've talked to do that all by themselves.

Some attorneys seem to think that the deposition is time for verbal World War III, and that the fate of the world hangs in the balance of this one case. It doesn't. Florida has a code of conduct for attorneys; perhaps all states do. The mere presence of this code suggests that others have violated the terms of the code before.

"Conduct that may be characterized as uncivil, abusive, hostile or obstructive impeding the fundamental goal of resolving cases fairly and efficiently will not be tolerated. Such conduct tends to delay and deny justice. Accordingly, in addition to the standards imposed on all attorneys by the *Florida Code of Responsibility,* the following standards will apply:

a. Counsel will treat all other counsel, parties and witnesses in a civil and courteous manner, not only in court, but at depositions and in all written and oral communications.
b. Counsel will not even when called upon by a client to do so, abuse or indulge in offensive conduct directed to other counsel, parties or witnesses. Counsel will abstain for disparaging personal remarks or acrimony towards other counsel, parties or witnesses. Adverse witnesses and parties will be treated with fair consideration.
c. Counsel will make good faith efforts to resolve by agreement and objections to matters contained in

pleading and discovery requests or objections.

d. Counsel will not, absent good cause, attribute bad motives or improper conduct to other counsel or bring the legal profession into disrepute by unfounded accusations of impropriety."

I know some attorneys who basically ignore most of these simple concepts. How they're allowed to keep practicing law is unknown to me. Perhaps attorneys say that about physicians, too.

What's fascinating is that the same attorney who was so obnoxious in deposition can become the epitome of politeness when a judge is right there listening. Being exposed as a jerk in a courtroom surely doesn't contribute to a better won/loss ratio.

And then we see the same two attorneys, bitter enemies in court proceedings, with all kinds of accusations and posturing going on, sitting down at the end of the day to share a beverage and small talk. Sounds like politicians, doesn't it? As long as the defense does a decent job, there's always a billable hour (or one hundred billable hours) ahead, especially if the plaintiff loses by a slim enough margin to make him think that he can win the next case. One wonders about the symbiotic relationship between plaintiff and defense attorneys and billable hours.

The posturing of the defense in mold cases is changing radically now that the reports from GAO, the WHO and the independent assessment of CIRS-WDB (published by Policyholders of America - POA - in July 2010). No longer can the little gambits the defense has used successfully in the past have any hope of surviving the counter-attack of the well-informed plaintiff's expert. For years, the defense could hide behind ACOEM and when that report got old and tattered, they trotted out a new version of the same tired and wrong ideas, the AAAAI paper published in February 2006. That combo used to be persuasive in a few cases, especially when an intelligent defense attorney like Lane Webb in San Diego helped a judge decide (erroneously) that mycotoxins must be measured and identified for there to be any human illness. (*Editor's note*: the case was called Geffcken, and it was heard on appeal February 26, 2006 in California; 137Cal.App4th 1298).

Now that we have clear academic support that illness can be acquired by inhalation following exposure to a chemical mixture of chemicals and organisms found inside WDB, the idea of ingestion as source of illness is laid to an ignoble rest. Lying face down beside that flawed idea is the toxicological concept of dose response. Indeed, now that the illness is shown to be immunological and not toxicological at all, perhaps the dark forces of highly paid toxicologists will be expunged from the defense side, too.

The best news however, is that the position paper from POA, itself including work done by experts in the Action Committee on the Health Effects of Mold, Microbes and Indoor Contaminants (ACHEMMIC), now named the Global Indoor Health Network (GIHN), lays out the arguments in cold black and white, from research to agency opinion, all written from the perspective of the treating physician. We're seeing the end of the Mold Wars in the near future. The science is clear; the agency opinion is becoming responsive to public demands and the attorneys who fight against mold plaintiffs are running out of ammunition.

Still, despite the ongoing explosion of solid data and incontrovertible proof, the forces of defense are resourceful and well funded. They'll find a way to make arguments, and though there's absolutely no science behind their current arguments now, they aren't going away any time soon. Take a look at Big Tobacco: their cover-up was exposed and their cause lost for years until the final verdicts costing billions of dollars came down, yet that industry survived and is possibly more profitable now than ever.

The defense consultants whose lives of undercutting truth began with Big Tobacco are getting older. A few new players have been recruited to the dark side, but the new consultants have nothing new to offer, just a different face spouting nonsense.

There are more than a few government employees who've stonewalled the advance of public knowledge and public acceptance of the reality that WDB are dangerous buildings. Their deeds might be hidden from view now but those deeds are recorded silently and inexorably by one Madame DeFarge[5] after another. The POA position paper has the potential to sound the death knell on their careers of manipulation and intentional sabotage of the truth. As the court decisions grow in number and consistency to set a new precedent, and as the public demands for integrity from government pile up on this issue, the government types will have a restricted list of career choices. Retirement and subsequent employment by the mold defense industry is one option for a few.

Or they might decide to wait out their time before pension by saying that yes they always knew that human health was caused by exposure to WDB.

Don't forget what Frank Fuzzell told us. The new government theme is predictable: they'll say, "We knew it all along and we've always acted on the side of truth."

That will be an outright lie, of course.

What Tom Harblin tells us will likely come to pass, too, and sooner rather than later, perhaps driven by "green ideas" about building. The target will be prevention of water intrusion and lower carbon footprints, but the new building ideas will also be driven by a desire to prevent mold litigation.

When the Golden Age of Truth about WDB emerges, I wonder what will become of the sickest mold activists, including many who were injured by exposure to WDB. It's not likely that they'll ever be made whole, not by medication *and certainly not* by the legal system.

What about the teachers injured by school administrators who

[5] Madame Thérèse Defarge is a fictional character in the book A Tale of Two Cities by Charles Dickens. She is arguably the book's main villain, seeking revenge against the Evrèmondes for crimes a prior generation of the Evrèmonde family had committed. These crimes include the deaths of her sister, brother, and father. (Wikipedia)

intentionally deceived staff and parents about the safety of the school? Will they have to contribute their life savings and retirement as compensation for those who they deliberately blocked from obtaining relief? Even now, as I see noble fighters for truth at Lanier Middle School in Fairfax City, Virginia, and Paul Taylor fighting to expose the microbial contamination at Goddard High School in Roswell, New Mexico, I wonder if they'll ever receive reimbursement for doing the right thing, for making their schools safe for students?

What about the peer-to peer physicians who have no qualifications to discuss CIRS? And what will happen to the government-employed physicians? Won't they have to get a real job?

I can guarantee you that lawyers will be involved with the answers to all these questions. Whenever there's an argument that involves money in some way, the attorney is bound to be there, taking a percentage or piling up billable hours. Sure, I feel that attorneys need to be held to a standard of behavior where lying or deceiving on behalf of someone paying for the opinion become grounds for permanent expulsion from the trade.

But if that were the case, we wouldn't have lobbyists anymore. We wouldn't have very many politicians either. A lot of government spin managers would be looking for jobs too.

Perhaps I'm getting too optimistic and the end of the Mold Wars will simply bring truth to one part of our society. There's a long way to go regarding the rest of the work that needs to be done.

CHAPTER 28

Darwin and Us: Is Unnatural Selection Changing Our Adversaries Faster Than We Can Adapt?

Charles Darwin's classic book, *Origin of Species*, was published in 1859, 35 years after his brief stop-over in the Galapagos Islands. Darwin is rightfully described as the Father of Evolution. His ideas that gradual genetic changes occurring over a long time permitted certain species to out-compete others for food and survival have themselves survived countless scientific and religious challenges. Environments change: inhabitants can either adapt or perish. What makes one organism give rise to a new group of descendants is called Natural Selection.

We think of Natural Selection as coming from natural forces. If one doesn't eat, one won't breed successfully. If one doesn't breed successfully, one won't continue the genetic lineage. But the habitat change we see from natural forces, be it volcanic activity, UV light injury or cell injury won't cause a change in *one or two generations*. Natural Selection demands a long time to act on small changes in DNA in order to result in a major change, for instance having a big beak and eating seeds found high up in a tree if one is a finch on the Galapagos.

We know that people sickened by WDB must change their lifestyles or suffer the inflammatory effects of re-exposure to other WDB. If they don't adapt, will they be able to out-compete others for survival? *Surviving Mold* warns that they won't.

In Darwin's construct, the species that *could* adapt better to a changing environment survived, while others, who couldn't keep up, faded into the fossil records. The changing environmental conditions seen over time created the basis for Darwin's idea of natural selection: Nature would select those organisms best suited for competition in the real world. If the particular organism (species or individual) was more intelligent, stronger, faster, more virile / fecund (you name the quality), compared to its competitors, then let the games of life and death begin.

Darwin didn't know much about Man's chemical use back in the mid-1800's. Pollution wasn't much of a factor in daily life. And who cared that the thousands of chimney pots in the United Kingdom spewed forth combustion products of a variety of woods and coals. Staying warm in the face of cold winds coursing through drafty houses was more important than worrying about warming the globe with the by-products of burning fuels in the smoke (*Author's note*: drafty homes rarely make people sick from water intrusion).

It is only in the last 15 years that we have known we must re-evaluate Darwin's ideas of Natural Selection, ones that held sway for nearly 150 years. We now know that large chunks of DNA can be sent from one organism to another, for example, either from mutations from fungicides (benomyl is one actor here, described in Chapter 11); or from ionizing radiation resulting from massive exposure (*Author's note*: think Hiroshima, Nagasaki and Chernobyl); or from bacteria playing "share the flag" a "game" in which large chunks of DNA, found on structures called plasmids situated outside of the main DNA repository in the cell, can be siphoned from one bacterium to another. Add to this idea the process sometimes called "kleptochloroplasty," common in one-celled creatures in which large segments of functioning DNA in cell structures are engulfed by another creature, providing energy or some other advantage to the engulfer if the newly acquired DNA is kept functional. If the engulfee provides advantageous genetic information, taking that information and assimilating it into the "new me" is a strategy that one-celled organisms use every day. Said another way, single gene changes in DNA like the ones Darwin talked about are chump change. Big chunks of DNA are sea changes in evolution. And that sea change is going on all the time. Look at lateral gene transfer. Moran and Jarvik have written in the April 2010 edition of Science how pea aphids carry genes that are derived from fungi (wonder if they have beta tubulin-1 too?) that make carotenoids that in turn help resist injury from oxidation. For the pea aphids, we can certainly say, "There is a fungus amongst us."

And how about the lake sturgeon? DeWoody and Hale, Genetica 2010, found those unusual fish have at least 15 genes from a fish parasite, a worm called *Schistosoma* that lives closely with the big fish. Lateral gene transfer again. Looks like in the microbial world, DNA is just a big bag of jelly with genes floating in and out, some transcribed and some not. The capability of an organism to change is thus related to its *plasticity*, or how it changes from day to day.

Darwin talked about a universal common ancestry (UCA) for all species, an idea that seemingly is derived from life beginning from one thing, occurring for unknown reasons and from which all life evolved. The Creationist believers might be asking Darwin where the original organism came from. Recently, Douglas Theobald of Brandeis University published in Nature (May 2010) his findings on UCA. He looked at the three main types of living creatures: many-celled creatures (eukaryotes); bacteria; and the oldest living organisms (Archea) - assaying for 23 proteins that all the organisms have. He says that the 23 proteins clearly show that life comes from one kind of creature and that we can ignore lateral gene transfer and all the rest. I wonder if he looked for the effects of benomyl on the modern versions of ancient organisms.

These sea changes aren't found in the sum of small changes accrued over time that Darwin thought was natural. I call this change in large amounts of DNA "**unnatural selection.**"

In the 21st century, we must deal with unnatural selection, as now genetic change can be so pronounced that its effects show up in *just one or two* generations and not one or two geologic eras with each spanning millions of years.

Such an unnatural change has made buildings dangerous since the 1970's. I am reminded of the warning about WDB in Leviticus, from the Old Testament (Leviticus 14:34-47: The Bible, King James version,

Oxford 1888), yet the literature on multisystem, multisymptom illnesses acquired from buildings is remarkably scarce before 1980. I guess it is possible that more people read Leviticus in the old days compared to now. If those readers of the Old Testament heeded the advice to abandon moldy homes, thereby not becoming ill, and that Biblical idea held sway in this country now, maybe we wouldn't have so many lives destroyed by WDB. Note that I am not saying that failure to read the Bible is the reason why so many people are sickened by WDB's in our era.

But if something had changed, just as Frank Fuzzell said, after the mutagen benomyl was introduced into our species and if the biochemical disaster it brought continues to occur, then such change is not natural. This fundamental occurrence, changes in segments of DNA, not just one base pair or one nucleotide, is what calls Darwin's ideas into question. He just didn't have clue on what would happen when a big chunk of DNA, created by chemical exposure, made a sudden move. Who can blame Darwin for not knowing what would be going on in Post-World War II in the US where better living from chemistry became a way of life?

There are many examples that mutated organisms are at risk to become toxin-formers. From the Buruli ulcer patients injured by newly emerging mycolactone (toxin) forming *Mycobacteria invadens* to the weaponized anthrax cases in the Washington D.C. Post Office in 2002, genetic manipulation of organisms by Nature or by Man isn't a good idea.

Seemingly minor genetic changes can create linkages to toxin formation. Consider a pathogen found in cereal grains called Fusarium graminearum (Matz S, et al, Fungal Genetics and Biology 2005; 42: 420-433). It normally makes a red pigment called aurofusarin, manufactured by a gene cluster. The mutated form of this fungus is albino with no red pigment. The mutant involves a chunk of DNA affecting the entire gene cluster. It has markedly increased growth rate and a 10-fold increased production of fungal fruiting bodies. They all make an increased amount of the mycotoxin zearoleneone. Whenever I hear of this toxin, one that has estrogen-like activity, I wonder about the new studies showing precocious puberty in our young girls, with onset of secondary sex characteristics as young as age seven in over 20% of our kids. Are any of our children getting zearolenone in their cereal? Or how about in their soybeans?

As an example of what I am saying about sea changes in one-celled microbes, where did the new bacteria from New Delhi (NDM-1) come from? Bacteria with the NDM-1 gene are resistant even to the antibiotics called carbapenems, used as a last resort when common antibiotics have failed. The mutation has been found in E. coli and in Klebsiella pneumoniae, frequent culprits in respiratory and urinary infections. They are resistant to every antibiotic known to man. All their resistance factors are on plasmids. It is not too far-fetched to think about bacteria living in sheltered environments, like biofilms, sharing plasmids like baseball cards ("I'll trade you quinolone resistance for a combo of macrolide and tetracycline resistance"). The New Delhi traders appear to like to have every card in the deck.

If we live in a world of pollution with chemical additives found everywhere, where can we find sanctuaries that adhere to the laws of Natural Selection? When we find such pristine areas, will we learn about our goal of *Surviving Mold*, that is "can we live without suffering inflammation hits from exposures day in and day out?" The reality for mold survivors is that in modern America we are essentially guaranteed to find unnatural selection everywhere so we need to learn to avoid those exposures that create excessive inflammation.

Galapagos, 2007

I had never heard of a Shellback, yet here we were crossing the Equator near the Galapagos Islands and there was an elaborate Shellback ceremony about to start. For those like me who had never crossed the Equator on board a ship, we had to be initiated into the Shellback fraternity. Shipboard traditions must carry on!

There are many reasons to visit the Galapagos. I wanted to experience the wonderful world of the World's newest lands, with its finches, hawks and mockingbirds, boobies, tortoises and penguins. I was also looking for the *absence of use of fungicides* and then the absence of indoor dwelling, toxin-forming fungi.

Owned by Ecuador, these island treasures of the Pacific are now closely guarded and protected. Thankfully, preservation of the wilds of the Galapagos is now mandated by the government. Tourism is the biggest source of income for the Galapagos, yet the number of annual visitors is regulated and restricted. There are many islands that any visitor would like to see, but the itineraries of boats visiting the islands are strictly enforced by issuance of permits. When permitted to see selected islands, visitors must be accompanied by a native-born naturalist who in turn must keep visitors on the designated paths.

Chemical use is also regulated. No benomyl; no dithiocarbamates. Maybe no resistant organisms?

And that wonderful cone shell I found on a beach: leave it there. As an avid shell collector, that one was a prize I'll never see again. But since I left the treasure there, maybe another visitor will have a chance to hold it like I did. I guess that is the point of the prohibition against taking any natural item home from the Galapagos. If I wanted a souvenir (and I did), there were plenty of t-shirt shops, book stores and museum vendors to help me.

I looked for homes with active water intrusion, and there were some, but nowhere did I experience any health symptoms in *any indoor* environment that warned me to leave right away. I asked as many people as I could about use of chemicals to protect their plants. No, such use was not permitted.

Scotland, the North-West Highlands, 2010

I am sitting on three billion year old rocks looking at the majesty of Beinn Eighe, the tallest mountain in the first National Park of Scotland. We don't see anyone else visiting here at the end of July. Just below is the 20-mile long Loch Maree, with the fishing village of Gairloch disappearing into the mists of rain coming off the Sea of the Hebrides. The Gulf Stream is just west of the Outer Hebrides, bringing rain and warm weather to this land well north of Ireland.

This is temperate rain forest, no different really from the Inner Passage of Alaska. I was surprised at first to see the majestic cedars towering over the untouched forests, but of course they live here as this northern Gulf Stream climate isn't too different from that of the Pacific Northwest. What is different from the Pacific rain forests are the moors. And sheep. Lots of sheep. We saw mountainsides with stone fences that climbed nearly vertically. Sheep wouldn't leave that pasture. For miles though, from the Cairngorms (mountains) on to the coast, the vegetation was low and flat, dominated by heathers and gorse. Until we reached the rain forest, few trees were growing unless recently planted by the Scots for a commercial source of softwood.

Still, with just a few settlements inland, what we saw were some of the most pristine environments, including the Galapagos, in the world. Oldest and most pristine. Yet this wasn't Antarctica or the Himalayas.

What about the chemical use? Not for the sheep and not for the corn that wasn't grown and not for the soybeans not planted.

What about the moldy buildings that would make me ill? Nope. No fungicides, no toxigenic fungi indoors. It looked to me that the magnificent old buildings we saw in Scotland don't grow toxin-formers in the absence of fungicides. Yes, there were WDB and yes, there were evidences of fungal colonization, but no illness? Amazing.

The geology of Scotland is complex. We are looking at rocks from diverse origins that collided in the Paleozoic era. Major fault lines are readily visible, including the Great Glen Fault that separates the North West Highlands from the Grampian Mountains. From the Firth of Moray near Inverness to Loch Ness, Loch Lochy and Loch Linnhe leading to the Firth of Lorne, the land is further complicated by volcanic activity that occurred in years gone by. Geologists tell us that the land masses that make up Scotland and Britain were forced together back in the dawn of time, as Scotland drifted up from well below the Equator, nearer to Antarctica than to the Arctic. The Galapagos serve as an example of moving land masses that we can see; the land mass of Scotland was well-traveled before it arrived in the UK.

I wondered if the tectonic plates moving Scotland across the Equator held a Shellback ceremony for the drifting land mass. I am looking for the primordial forces of time that are no different in the world's oldest rocks at a maximum age of three billion years (*Author's note*: recently geologists have confirmed that rocks on Baffin Island, just across the big ditch from Scotland, are 4.5 billion years old, now the world's oldest) compared to the Galapagos with the world's newest rocks.

If Charles Darwin is right then natural selection will not bring about unnatural toxin formers in pristine environments. If Charles Darwin is right then people like me and my patients with 4-3-53 and other susceptible haplotypes must be well adjusted to old mold. It survived and so did our direct ancestors and theirs before them. We co-existed. The mold multiplied and so did we.

Princess Anne, Maryland 1985

Here is the William Johnston House, largely built in 1815, with older nails serving as evidence that some structure was on the site in 1805. With over 5000 square feet and so many fireplaces, this "side hall" Federal home, an elegant old home in an elegant old town, is one of the best examples of the transition from Federal to Greek Revival architecture in the Lower Eastern Shore. The large in-roof gutters take the rainfall off the steep roofs, funneling the water into dual downspouts on both sides of the home. Heavy rains overwhelm the ability of the soils to absorb water so the basement suffers water intrusion frequently. It probably has for over two centuries! Yet the massive beams and sill plates of the basement are still functional, without evidence of the wood rot I see so often in modern wet basements. And despite the seasonal floods in the basement, I didn't find terrible fungi destroying my innate immune responses when we were there. We bought it.

When I put insulation in the ten-foot high walls of the first and second floors and then the attic joist spaces, I thought I was reducing energy loss. It never occurred to me that I was reducing the very draftiness that may have kept the house from making its inhabitants ill. And it never occurred to me that I was putting modern-day food stuffs right where moisture-loving fungi, bacteria and actinomycetes were waiting for dinner. Still, we had no problems with illness from that old house, even though the food supply for toxin-forming microbes was ample and the basement was wet.

My colleague has very recently had the same experience with a Victorian home in Oakville, Ontario. Built in 1886 and augmented in 1940, then doubled in 2002, it has the original 1886 basement. It sits a few hundred yards from Lake Ontario and both sump pumps run pretty much constantly. He is a 4-3-53 and his other haplotype 17-2-52A is mold-susceptible. His MSH is unmeasurable and when he is well, his C4a runs about 200% of normal. He has had no adverse reaction to this home to date nor has his son – who is a homozygous 17-2-52A and thus also mold-susceptible. He asks whether the organisms in our 19th century basements, especially fungi, had established squatters' rights to the territory and would defend it against any young whipper-snapper toxin-formers that floated in? Of course they do, that is the prime reason some fungi produce and deploy compounds (fungistats, not toxins) that inhibit the growth of other fungi. Is old mold really gold? It lives with us, not agin us, and it defeats the new mutated forms that make some of us so ill. Does our HLA glove fit the (antigens from) old mold perfectly and so our immune systems do what they should? Remember that the old mold and our ancestors had eons to learn to live together. It is called Natural Selection.

Face it: fungi are ubiquitous. The new toxin-formers might not be able to drive out the ancient cellar dwellers. We see this all the time in biology. Here is a colonizing methicillin resistant staph aureus

(MRSA) growing in someone's nose (like an ICU nurse). Why doesn't she become ill? Does her own native bacterial flora suppress the invasiveness of the MRSA? Given that there are over 300 different kinds of bacteria growing on our skin, why don't we have more skin infections?

What I am asking is whether there can there be equilibrium of microbes in a given econiche, a balance that keeps all living harmoniously until the tranquility is disrupted by something external? If that is so, then natural selection involves changes not just in a given organism but in the entire system of organisms as they interact with each other.

Darwin didn't talk about the effects on selection of ecosystem change, but that process is everywhere around us. Unnatural indeed when, for example, coal smoke from tall smokestacks falls to Earth, large scale spreading of sewage for fertilizer on fields happens daily, and clear cutting of rain forest occurs. And that is every day.

The Galapagos are volcanic islands located at the junction of so many Pacific Ocean currents that the ecology is unique. Look at how the new land, acted upon by constant forces of wind and current, garnished by ongoing volcanic activity, is changed inexorably over time. While we were there, the Ciera Azul volcano was active, so more new land for colonization was being formed right as we watched, spellbound and awestruck. The Earth's tectonic plates are moving the islands ever so slowly, like a conveyor belt, towards the mainland. Eventually, I guess, the Galapagos island of Floriana will become part of the back yard for the three-million people who now live in the city of Guayaquil. The newer islands are located to the West and the older islands, eroded over time (like Floriana), have been moved along to the East.

The islands have been visited by all sorts of people from many countries. Situated as they are near the shipping lanes to Easter Island, Tahiti and Japan, the Galapagos provided a welcome supply of fresh water and giant tortoises for visitors in sailing ships (including Darwin's). The tortoises required almost no food after they were taken into captivity, so their maintenance costs aboard a ship were trivial. They just needed a place to stay, probably not far from the galley. The fresh meat of the slaughtered tortoises surely was a welcome source of protein. Each island had its own variation of giant tortoise; each with a complex life style that didn't account for slaughter by man. Fortunately, the Darwin Research Center is collecting tortoise eggs and re-populating each island with its own indigenous tortoise. No longer are the tortoises a walking meal ticket for many.

Within the individual islands of the Galapagos are seven climate zones, making a truly remarkable compression of ecosystems in a small area. Such compression creates the possibility for establishment of unique grouping of plants, birds and animals found in a limited area, but it also made the ecology much more vulnerable to damage following introduction of "exotic" species like rats, pigs, goats, dogs - and Man.

Exotics are non-indigenous species. Most conservationists I listen to agree that prevention of spread of exotics is a basic principle of protection of the integrity of a given land form. When I see phragmites invading and taking over the wetlands of our tidal wetlands garden in Pocomoke City, I want to dig them out while others say "use a herbicide." We don't want any of those "come-here" plants around here. This garden is not big enough for the mallows, wild rice, arrow arum *and* the phragmites. Personally I don't trust the manufacturer of the poison used to kill phragmites to know what will happen when I put the killer chemical into the soil and therefore, inexorably, into the Pocomoke River. Dig we shall!

How do I know I am NOT looking at a phragmites version of "Son of benomyl?" I didn't poison the exotics. And I didn't get rid of them quickly either.

Yet, look at the colonization of land emerging from a volcanic eruption. *Everything* is an exotic! The lesson from Mount St. Helens in Washington State was that initial colonization came from airborne organisms, followed by emergence of bacteria tolerant of adverse soil conditions. The desolation of destroyed forests was only short-lived. Now nearly thirty years later the volcanic eruption has given birth to a newly populated area that will look much like the areas not destroyed by the eruption - *in our life time*. Nature finds a way! Nature can move quickly – especially relative to geologic time.

So how long does it take for a new land mass in the Galapagos to develop its own ecosystems? Actually, from the perspective of geologic time, not very long. From insects and spiders blown aloft to 20,000 feet to seeds carried by birds and winds, colonization starts fast from above. What we haven't spent much time discussing is the time it takes for *microbes to start to colonize* new land. Algae and bacteria are so diverse in their tolerances that they will be growing in some of the most adverse areas on Earth. As an example, just pick up a microbiology journal and read about the bacteria and algae (extremophiles) growing in the boiling water of Morning Glory Pool in Yellowstone National Park. How they got to the steaming volcanic areas of the Galapagos isn't clear to me. Yet there they are.

When we see the devastation of all that crude oil washed into wetlands along the Gulf of Mexico, we need to remember that there are bacteria (think of *Pseudomonas fluorescens*, especially) living there now that will say 'thank you for the new food source.' Will they out-compete other indigenous bacteria that can't eat oil for a living? For a while, sure, but not for long. One might ask how come these non-toxic bacteria weren't seeded onto booms designed to protect the shorelines and marshes. There were people who had the bacteria for sale, proven to be safe and effective. And then too, what about the fungi living in the marshes, digesting plant life that has died? What looks like massive decay to us on the nightly news from the oil might be re-interpreted to be simply liberation of stored foodstuffs for other creatures. Out of the wasteland will come vibrant new life.

But will the new life be the same as the old life? Is the oil spill unnatural selection? You bet. Is the benefit of oil-food for the *Pseudomonas* bacteria natural selection? Sure it is, just like the upwelling of nutrients into the cool currents off Alaska is natural. Killing the bacteria with quinolone antibiotics from our waste water treatment facilities would be unnatural selection but movement of "new bacteria" into the empty niche created by die-off of the Pseudomonads is natural selection.

How long will it take oysters in the Gulf to re-populate? And what about shrimp? The truth is a lot less time than it will take to replenish sea turtle populations. Even now in August 2010, the shrimp harvest in the Gulf looks clean and robust from the TV news cameras. Small creatures reproduce faster than bigger creatures, yet just look at the conifers growing just thirty years later on the previously desolated land near Mount St. Helens.

When Darwin arrived in the Galapagos in 1831, he didn't spend much time on the islands, visiting on land for about 19 days. His powers of observation must have been truly remarkable as he saw enough differences among animals that he was able to leap to an entire idea of evolution based on differential rates of reproduction. If an organism didn't reproduce fast enough compared to its competitors, it died off as the reduced numbers of offspring couldn't maintain enough food intake to support a population. Changes in genetics let some populations take advantage of available food better than others, thereby breeding more and passing helpful genes along to the next generation. Darwin called that idea Natural Selection.

Darwin used the different habitats of the Galapagos finches as one of his examples of how Natural Selection would have driven finches with different beak sizes into different habitats. I have to wonder if that example wasn't a stretch of imagination, as during the short time we were watching those same kinds of finches that Darwin did, we saw plenty of ground finches eating in trees and tree finches eating close to the ground. The beaks of these birds weren't so one-dimensional after all. Imagine the hungry finch thinking, "Wings do your thing!"

Still, the generation time for changes in birds is limited by the number of broods the bird pair makes, the number of eggs that are hatched and baby birds that are kept alive long enough for the babies to reach the age when they can reproduce. Compare a successful pair of finches boosting their population by ten birds a year to a successful fungus dividing every 30 minutes. How many billion fungi babies is that in a year? The number is more than 10, for sure. And if a genetic change (maybe a beak that is one one-thousandth of a millimeter thicker on the bottom side) is passed on to ten new birds a year, how many years will it take for that changed bird to impact on the diversity of the finch population? Compare that to genetic change in a microbe that is reproducing so much faster than any bird, rabbit or elephant population.

Could Darwin have missed something?

His Galapagos model for Natural Selection had so many variables that Darwin ignored. He was there for only a short time. Seasonal changes weren't noted as he was there for a month! He didn't have much time to collect many examples of many species, in part because there aren't many species there and the ability to collect a lot of finches is tempered by the small finch populations. Add to that problem the trouble of shooting little birds without blasting their beaks to different parts of the Equator. Maybe if he had looked at the huge frigate birds his killing wouldn't have ruined such a proportionally large percentage of the bird. Handling cumbersome nets to catch birds would mean less destruction of specimens but less time would be left to catch multiple birds in multiple diverse areas of 15 islands in only 19 days.

If beak size differences in finches generated a lot of "signal," then the variation in size and shape aspects of tortoises on different islands would have generated a lot of "noise" if Darwin had looked at them. Trying to show changes from gradual genetic change is much harder in species that live 150 years compared to those like fungi that might live 150 hours. How can observation of change be relied upon if only one generation of an animal is available? At least Austrian geneticist Gregor Mendel could look at peas that reproduce more quickly than tortoises.

Darwin didn't control for availability of fresh water in an environment either. Face it; organisms do a lot better in moist conditions than in arid conditions. Some finches might have a difference in their beaks that let them drink water from crevices in a rock, for example. It wasn't food as a force of Natural Selection that changed their shape, it was water.

Now add salinity of water to the mix. Who can forget the famous marine iguanas warming up in the sun spitting out the salty water they ingested when they ate algae in the cold shallows of the Ocean. Imagine a finch trying to develop a niche on the shoreline by eating algae but NOT being able to unload the extra salt it took in. Dead finch is what that organism would be. Any exotic pig with a penchant for algae would become country ham pretty quickly.

Move on to soil fertility. New arid volcanic soils don't give up plant nutrients easily. Birds living in those environments will struggle to feed their babies compared to those atop the aged rainy volcanic forests. Arid volcanic soils grow plants richly only after eons of time have released food for plants.

Then add heat tolerance to the mix. If a finch had a great beak to crunch the poisonous little apple seeds but couldn't handle the heat of the interior of the particular island, sorry, no baby finches. Even though the chilly ocean currents of the Galapagos support populations of cold-loving penguins and sea lions, the interiors of the islands are really hot!

Don't forget distance. Isolation doesn't always bring biological diversity. The Galapagos are 600 miles to the West of South America. Traveling that far might not be a problem for birds or whales but getting a new colony of a mainland species will be hard if only a few adults make the journey. Who knows what will happen to them? Being a pioneer on a desert island might sound like a romantic idea, but just imagine landing with no food, no water (and no Internet); how would *you* survive?

The isolation idea also applies to inter-island exchange of DNA. Land tortoises might walk across a land bridge but they sure won't swim. So even if a hypothetical land bridge connected islands in the Galapagos, after the bridge collapsed or eroded, whatever animals were trapped on one island would stay there.

One more element missing from Darwin's analysis was the use of selective poisons to kill just about all members of a race of pests. Benomyl was the best killer of fungi. DDT was our best insecticide. What could one consider to be natural about these selective mass produced killers? And then what would be natural about the survivors of such dealers of death?

From the newest land to the oldest

Next to part of the Monie Thrust, an ancient geologic formation close to Loch Assynt, north of Ullapool on the North West coast of Scotland, three billion years young, we walked a bit in the moors, looking for signs of gradual change in geology that paralleled the gradual changes in evolution discussed by Darwin. This is an ideal location to think about Survival of the Fittest. The environmental conditions could be described as forbidding, yet I kept imagining Sir Charles Lyell, a Scottish geologist from Angus, writing his Principles of Geology. Who knew that his ideas would have such enormous impact on Darwin, as he read the encyclopedic geologic text on board the HMS Beagle? How much did Lyell's idea that "geologic change was the steady accumulation of minute changes over enormously long spans of time" become transmuted into Darwin's natural selection causing biologic change as the steady accumulation of minute changes over enormous periods of time?

Maybe it didn't matter. By the time we could see any changes from time we would be several hundred thousand years old. We won't leave any marks of who we were or what we thought. I laughed a bit, as I thought these Scottish rocks were *built to last*. There weren't many people here; the life must be tough with the rain, winds and chill of winter. Warm summers? Scottish Highland oxymoron. The nearest large city, Glasgow, is well to the south near the Clyde River and the Firth of Clyde and it seemed a long way away today. Matthew Hudson reminded me that *Clyde-built meant durable*, as ships like the Queen Elizabeth I and II were built in the Port of Clyde. He also noted that the Titanic was built in Ireland.

Still, if rocks are durable, then the ability of biology to adapt is the most durable aspect that I work with. The only thing stable about changing organisms is the change itself. With Mount St. Helens as a back drop, we can ask what happened at Chernobyl after the nuclear disaster 30 years ago. Would you expect to see new organisms appearing in less than a generation, organisms that are adapted to environments with a reduced availability of carbohydrates? That is exactly what the Russian investigators showed at the International Mycological Congress held August 2010 in Edinburgh. Fungi around Chernobyl now grow well in the face of increased oxidative and radioactive stress, elevated levels of hydrogen peroxide and activation of anti-oxidant enzymes. Talk about an unnatural selection! Wipe out the ecosystem with radiation and then see what comes out of the hot muck in a few years (this wholesale set of mutations took a ***single*** human generation). Creatures from the sci-fi movies like giant ants, giant spiders and giant rabbits (remember Night of the Lepus)? Nope, but lots of fungi! Surprised? We are seeing the result of the evolutionary processes accelerated by unnatural selection and extreme conditions.

Frank Fuzzell showed me documents that confirmed that Pittsburgh Paint and Glass (PPG) began adding benomyl to indoor paints in 1970 to control the growth of *Aureobasidium pullulans*, a common contaminant growing on paint films. Frank blamed DuPont for the application of benomyl worldwide as the reason for the biggest agricultural disaster the world had ever seen and it was introduced to the indoor world as well. He believed that the new organisms, new toxin-forming fungi living indoors were simply the consequence of the selective pressure of benomyl – introduced by Man at his own cost. I don't see any refutation of Frank's claim anywhere but he isn't here to prove his claim either. As far as proving Frank's ideas, genetic change won't be found in the fossil record, especially not when the change was initiated just forty years ago.

Does the idea that a mutated group of genes can survive passage to multiple generations of other invertebrate species have any implications for people? The reality is that we know so little about the fate of genes that are inserted into our DNA, whether the genes are from a herpes virus, an XMRV virus or maybe, some *Schistosoma* worm. Does benomyl have implications for vertebrates? Just ask the plant pathologists at Virginia Tech. Oh, yes, they know all about benomyl.

Epigenetics

Darwin was involved in an 'argument' with Jean Lamarck who, writing nearly 100 years earlier, argued that evolution could occur in a generation or two. Think back to your high school biology class discussion of giraffes. Lamarck said longer necks helped them eat acacia leaves better so the giraffes grew longer necks rapidly. Darwin said the longer necks started as a random mutation but survived better and therefore bred better, just like the finches bred more successfully.

Unnatural selection would seem to side with Lamarck's ideas. A 'modern' variation of Lamarck's ideas, that environmental influences can affect immediate inheritance, comes from the rapidly growing field of epigenetics (environmental factors affect gene activity). No less than the esteemed scientific journal, Time Magazine, weighed in on 1/18/10 (pp. 44-53) on the debate. Darwin's supporters might be screaming heresy, but the basic idea that, "whatever choices we make *during our lives* might hurt us but they *won't harm* our offspring by altering our genes," central to Darwinism, just doesn't hold water based on new data. Amazing. If Darwin could turn circles in his grave this new thread of science would be a good circle-starter.

A Swedish researcher, Lars Bygren, published data in 2001 that clearly showed that grandchildren whose grandparents survived famine in Norrbotten County (north of the Arctic Circle) lived far longer than grandchildren of grandparents who lived during normal and feast years. One might ask how longevity of a succeeding (+2) generation is caused by something that Darwin could explain?

Basically, the idea of epigenetics is turning the idea of Nature versus nurture upside down. There is a lot going on here. What the epigenetics advocates say is that gradual genetic change is bypassed by external factors that control DNA activity and the result is changed gene activation seen in a *single generation*. So we ask "Given there are changes in a single generation and so they are not traditional natural selection, which of these non-Darwinian changes are caused by a chunk of DNA being damaged versus external factors modifying the gene activation of the DNA – the epigenetic change?"

The answer probably won't surprise readers who have read this far in this book. Inflammation gone wild means loss of control of inflammation. Genes gone wild means loss of control of genes. Epigenetics means that genes are being told to turn on or off by outside influences and that change is passed on to future generations. Loss of control of genes from epigenetics? You bet!

Here is a crucial concept to understanding where Darwin is being re-evaluated. We are looking at changes in gene activity based on *what we are doing to ourselves now*. If those differences in gene activity are part of our germ cells (sperm and ova), then our offspring *will also have* the differences in gene activation. It makes no difference what you do once your offspring are born but what you do before they are born may have profound effects on their lives. Epigenetics isn't traditional evolution; it is simply a response to environmental stressors that is durable over time in our children. So, should we be screening our third grade students for uncontrolled innate immune inflammatory responses? You bet!

Inflammation gone wild certainly could be one of the stressors. Does this idea make one think differently about illnesses that are seemingly exploding in our modern world? How about autism and cystic fibrosis?

I hope you've heard about the inheritance of mitochondria that all (in the beginning) come from a woman in the ancient Rift Valley in Africa. (*Author's note*: ova have mitochondria, sperm do not. All mitochondria, therefore, come from women. The original mother is now thought to come from early hominids in Africa.) So if we are looking for differences in children then think about what Grand-Pa was doing before he was eleven years old. Think about it. Do we need to look for epigenetic changes in sperm around the age of puberty?

Does this idea mean that a prospective mother needs to explore the boyhood health of her prospective partner before deciding on him as the father of her children?

Maybe the lessons for us from the Galapagos and the Scottish Highlands are that the pristine environments provide for pure natural selection. Any environment less pristine, altered by pollutants, behavior or pestilence, becomes the host for adverse effects on our offspring. If we return to the link between environment, host defenses and inflammatory responses (TGF beta-1, HLA DR, for example and hypermobility) as affecting the quality of life in parents, it is not much of a leap to wonder about their children as well. Should we move the parents out of one environment to another before they become parents? Should those with certain haplotypes identified as high risk for epigenetic phenomena not be permitted to be parents? Oh my; I worried about Big Brother in the VIP chapter. I'm not ready for this Big Brother.

May I suggest that questions of ethics and human rights aren't likely to be processed well by government – which has become the lowest common denominator?

Summary

Medicine involves the interaction of Hygeia and Panacea. Lifestyle and medications. Some ideas for survival are clear:

- If you see a WDB, clean it up
- If you see WDB-illness, correct it
- Don't go into a WDB to become sick a second time, much less for the first time
- If you see prospective father with bad habits, fix them

before he is a Dad
- Pick your parents well; insist that they weren't moldy when they made you (not serious – alive and susceptible still beats 'not in existence')
- Make sure there is love and respect between parents and children

Whether or not epigenetics truly is a source of helpful interventions is unknown. We are opening doors to such therapies by identifying the genes that are active when we are ill and that should be silenced as they are when we are well. We have learned much about the genetic aspects of hyperacute illness acquisition by observing prospective re-exposure trials. We have learned much about chronic illness and resolved chronic illness by following the genomic aspects of patients. Inflammation tells us much when we look at its basis!

If inflammation is the unleashed beast of genetic change in our environmental community then for the first time we now have the potential to tame the beast. We know what genes are hurting us; let us use genetically active drugs like Actos, VIP and erythropoietin to cage the modern beast.

The days of finding no genetic monsters among microbes are not for us. Too late. We can adapt and survive - or not. Fungi can evolve much faster than we can – remember a generation can take 30 minutes – but we can think. We have technology and science. Let's use them. If you are a 4-3-53 or any of the susceptible haplotypes, you were well adapted due to gradual evolution over the centuries – gradual fungal evolution, gradual human evolution. There was no external genetic interference. Now there has been and is. Act accordingly – you are at risk! Your children may well be at risk. If you are a double susceptible, your children are at risk.

We can build safe buildings. We can identify children at risk for illness acquired following exposures. We can safeguard partners before they become parents. We must do so because we can't live in the North West Highlands and then go to work in Pocomoke the next morning. Yet we can make it our duty to provide shelter and safety for our families in the world away from the safety of the Highlands.

Who shall be saved from the slings and arrows of unnatural selection? Maybe those who learn to read the signs that say danger is present. *Surviving Mold* will help those people.

Those who choose not to learn, not to pay attention or simply assume they are right (without basis) aren't likely to be allies in the search for safety in the Era of Dangerous Buildings. Perhaps they will change their opinions through education, understanding and example.

More likely, those who have built their reputations based on wrong ideas won't change at all. Change will only come when the new generation of thinkers emerges from the suppression of truth that has been experienced by every WDB-illness patient. So it is to those who come along after us that will bear the final burden of integrity and honesty. Our understanding of mold illness will make our children more likely to survive and breed, making effective application of information part of Natural Selection. The final judgment on evolutionary fitness is in the future. We cannot know the results now. Now that we know what to do about human health effects acquired following exposure to the interior environments of WDB, let us not be reticent to act to defend our children, as our children and their children will benefit.

Such benefit is close at hand. The time to reach for and hold that benefit is now. There is no person, no government agency, and no insurance company that can stop the truth. This book is part of that battle for the dissemination of this essential knowledge.

Survival begins with knowledge. It continues with understanding. It succeeds with compassion. Since we are all one under the color of our skin regardless of the faiths we hold, then let us join as one to make our homes, our schools, our workplaces and our selves safe.

CHAPTER 29

Six Years On: A Patient's Personal Perspective

My name is Matthew Hudson. I am 4-3-53 and 17-2-52A. Not lucky in the HLA DR lottery. Each of my children is thus guaranteed at least one susceptible allele. My eldest son is 17-2-52A and 17-2-52A. He was fortunate he did not inherit my 4-3-53. Interesting to note that his mother has at least one susceptible allele which is identical to one of mine. See Dr. Shoemaker's comments on spousal choices. The mother of my other children is like me, a 4-3-53 (my other allele) and like me she is a 401 (the subgroup of "4" that is the worst to have). Another reminder of this book's comments on the counter-intuitive prevalence of genetic spousal choices. A mystery for another day perhaps.

My own story was told in the final chapter of *Mold Warriors*. Suffice to say that my senior staff operated in an old, extremely water-damaged building on a waterfront in Portland Maine. After buying this business (Scotia Prince Cruises) and spending a year working to improve the ship, the business and senior management, it became clear to me that something was wrong with senior management. They were slow, they were forgetful and their ability to assimilate the new information I was providing was far below par. I happened to read a magazine article during a flight and the recitation of mold symptoms was a "hallelujah" for me. We brought in a testing company, using as our lab the same lab used by the EPA and CDC and hired perhaps the leading toxicologist then active in this area - Dr. Richard Lipsey. It quickly became clear that the building was infested; indeed later work by NIOSH not only showed that the building was "unsafe" but also the worst building them (or Dr. Lipsey) had ever tested until Katrina. I immediately pulled my staff and our customers out of the building and we operated from the parking lot for the balance of the summer and autumn season. My research online and through Dr. Lipsey brought me to Dr. Shoemaker. Naturally I was skeptical and rather hoped that someone would convince me that the building was safe as my family had paid several tens of millions of dollars for the business and there was no other facility that would accommodate our special maritime requirements. As well as being a businessman (with almost 45 years of experience by 2004), I'm also an attorney. I like to think that I'm not easily fooled.

I subjected Ritchie Shoemaker to a comprehensive test of what he had said, what he did, what our lab results might show. Here are just two of my 2004 questions and his answers.

Me – "Your "first identified" biotoxin patient led to the serendipitous use of CSM for her GI symptoms and then her other symptoms also abated – actually it was an inadvertent n=1 experiment. When was that?

RS – "1997"

Me – "CSM, as we know, is inert and can only remove things from the body and was thus approved many years ago for cholesterol reduction. Was this the reason you deduced the presence of a poison which the CSM helped to eliminate?"

RS – "Yes, it could not be adding anything to her body. It can only work by taking something out"

As a result of the information Dr. Shoemaker provided I determined to go ahead and recommended to my colleagues that they all travel to Pocomoke from Portland. Expenses for everything were paid by me, of course, including any lab tests our company health scheme would not cover. When the results of the HLA DR assays were back I asked RS which of us he predicted would make a full recovery. I should explain now that our small group of 14 people who worked in the building and one senior executive who worked elsewhere (Canada) turned out to be statistically unusual. Except for one young man, each of us had one or two susceptible alleles whereas the prevalence in the normal population would be 25% thus 3 or 4 of our group. The one person who did not work in the building or even in Maine had one susceptible allele but he was not a case. He was not ill. Thus the finger of suspicion pointed more squarely at the Portland Marine Terminal. The only person who did work in the building and who had two normal alleles, was initially ill because of his daily exposure to huge amounts of Stachybotrys and more than 30 other toxic molds. Once removed from exposure and treated with 30 days of cholestyramine (CSM) this young man made a complete recovery both subjectively (he felt fine, better than he had in years) and objectively – all of his labs resolved to normal. The number of symptoms in the average non-case population is 2.7. Our 14 people with exposure to the Portland Marine Terminal had an average of 20 symptoms.

In my own case I waited a number of weeks before going to see Dr. Shoemaker by which time I had received the HLA DR and initial labs for each of my employees. Dr. Shoemaker had explained his theories to me in advance and the lab results meant I was able to see that he had not misstated anything. The only one of us with 2 normal alleles did make a full recovery. No one else did. Our blood work showed we were all objectively sick, not just feeling badly with lots of symptoms. So – I needed to know if it was the building. Two brave people, our Chief Operating Officer and our Corporate Counsel, volunteered to be guinea pigs. Both had begun to feel better out of the building and on CSM.

We followed the Shoemaker repetitive exposure protocol. Their cognitive abilities were assessed by an industrial psychologist and bloods were drawn before they reentered the building for 24 hours over a weekend. Then they took the same battery of mental tests and blood was drawn again. Previously high flyers and well known to me before I brought them to Portland, the MBA and the JD were not high achievers on the tests beforehand. I was not surprised. They were much worse after the renewed exposure. They were sicker, quicker – 5 minutes for Bob and less than 30 for John. They quickly resumed therapy. Their Before and After labs came back exactly as predicted by Dr. Shoemaker. That was enough for me. We were not going back into the building and we could not carry on unless the City of Portland replaced the old structurally unsafe building. By the way they didn't. They pretended to remediate and they lied about the structure.

As for myself, on examination and interview I had 23 symptoms out of 37 on the standard roster. After treatment with CSM for 3 weeks, 12 symptoms were the same, 8 had improved and 3 were gone. Dr. Shoemaker then treated me for MARCoNS - Doxycycline and Rifampin orally with Bactroban applied topically to my nasal passages. This treatment continued for another 30 days during which the CSM was reduced to two times a day from four. At the end of that time only 2 of my symptoms were the same (both had preceded my Scotia Prince Cruises exposure to the Marine terminal), 12 symptoms had improved and 9 were gone. With "only" 14 symptoms I had greatly improved my quality of life but had hardly joined the normal population. Still it was clear that Dr. Shoemaker knew what he was talking about. I am by nature suspicious of "new" knowledge brought by strangers, but my own understanding of this genetic susceptibility - borne out by my own experiences, those of my staff, my family and friends - makes clear that Dr. Shoemaker is the clarifying beacon of knowledge and truth in this essential area of human health. Those who stand against his results appear to be at best gossips with no hands-on knowledge, or worse, are paid to give "opinions" (perjury rules do not apply) in court in order to help the insurance industry or some employers avoid its / their proper obligations. Think EPA. Think Jerry Koppel.

I note that no physicians who have repeatedly and successfully treated a large body of patients producing smiling faces and probative lab results have yet appeared in the lists of those who (i) chatter to "colleagues" about CFS or Fibromyalgia at conferences (travel and lodging tax deductible); or (ii) perform their Iago interpretations for large hourly fees. I also note that as opposed to employers – sadly even those government departments charged with safeguarding public health – who deny safe working conditions and fair compensation to their employees, my working class family forfeited many millions of dollars in order to protect the health of 14 individuals and the members of the public who would spend up to two hours in the dangerous building.

Is there a lesson here? I have no axe to grind but, as is portrayed in the personal stories you have just read, it does seem clear that the balance between individuals and large organizations has tipped decisively against individuals. It is sad to note that this great nation, built upon a foundation of citizens rebelling against distant rulers, now finds itself in that self-same position of subservience, one in which too many of our political rulers and their lackeys stuff their pockets and carryalls with the bounty laid out by the citizenry, such that the citizens themselves now go hungry.

During the writing of this chapter I asked Dr. Shoemaker a few other questions. Here is a sample.

Me – "How many biotoxin patients have you treated?"

RS – "Now nearly 8,300 with 6,100 being mold illness patients."

Me – "Of those who followed your advice how many were <u>not</u> helped to better health?"

RS – "Remember that people must be removed from exposure and follow my protocols faithfully. Before we could correct C4a with epo we used to define "better health" as a greater than 75% improvement in symptoms but now that I have VIP we define it as higher than 95% improvement. In fact for the very great majority of patients (95% +) their symptoms resolve to the point that they resemble controls.

Me – "So, that is a 95% reduction in symptoms by most of your patients since your discoveries of the role of epo and VIP, plus your work to gain approval for the use of VIP in the USA. May I say in simple terms that the no improvement rate has almost zeroed out?"

RS – "Yes."

Me – "I understand that for those with unhelpful HLA DR it is not possible to talk of a "cure" as we will always be susceptible, however now that your research has led you to discovery of the roles of VIP, MSH, TGF beta-1, C4a and VEGF – is it unwise to discuss complete recovery?

RS – "Compete resolution will mean removal of the adverse effects of illness on HLA. I don't think we can say that yet but as it stands, once the inflammatory process is converted back to its normal self, patients can have normal lives again."

Turning back to the science. Neither color, creed, culture, education, intelligence nor neighborhood defines us. We are defined by our innate immune system and we do not have the technology to change it. Some recent work suggests that random genetic events may change it enough to make a difference to our children and their children. However we are prisoners of our HLA DR. Most of the time most of us are thoroughly pampered prisoners. Sometimes some of us are not. This book is of course about the latter group, the victims. First WE need to know what is wrong with us. Interveners who are vague or subjective or self-serving – none of those folks can help. If you don't actually know, if you haven't objectively helped most of the patients you have treated or, worse, if you talk but don't treat, then move out of the way – you're blocking the light. Let the truth in.

CFS – Is it real? Erik Johnson's battle to be heard. He is inconvenient for the inventors of the syndrome and their franchisees. Perhaps he is an inconvenient truth. No-one likes to see their pet theory discredited. But like snake oil – where is the data? Where are the lab results showing successful treatment? XMRV has been "discovered." It may be associated with CFS although if CFS is just a name for inflammatory disease caused or triggered by inhaling air shared with the pathogens inhabiting water-damaged interior spaces, then all talk of CFS is as much use to patients as a vanity license plate. Where is the data?

In 2004 Dr. Shoemaker gave a talk to the IACFS. He pointed out that in his experience "two uncommon haplotypes (DRB1-4, DQ-3, DRB4-53 and DRB1-11, DQ-3-DRB3-52B) found in 4% of the population actually represent 88% of cases of refractory CFS." Now I call that data to wrestle with; or to hug and take home for much closer examination. Nope. The CFS club continues to deny. Just ask Erik.

On the other hand understanding innate immunity provides verifiable insight into a vast array of medical problems. Host defenses, antigen processing and inflammation that is ubiquitous. It hurts us daily - from asthma and fibrotic lung disease to cirrhosis and autoimmune processes, not to mention the fundamental concerns regarding cancer and vascular disease. The fact driven part of me asked Dr. Shoemaker - "Do you think that some of the patients with Chronic Fatigue (CFS) or Fibromyalgia might also have problems arising from WDB?"

His response - "On the contrary, I think biotoxin patients are the vast majority of those called CFS and Fibro. When we look at the final common pathway of illness, even if we can stretch to think of XMRV being a player (another unproven theory), it is prior immune injury coupled with day to day exposure that leaves an individual vulnerable. Once MSH and/or VIP fall, then the patient becomes vulnerable to inflammation introduced by WDB exposure - they become sicker, quicker. In addition there are those with an original, not repeat exposure, causing their illness. No, you will find a huge number of biotoxin people if you look at those labeled as CFS or fibromyalgia."

My personal involvement in this effort now spans some six years and I am gratified to see that the general body of knowledge is beginning to catch up with the detective from Pocomoke. Since I began to edit this book several weeks ago we have learned that there is a set of genes which predisposes certain people to meningitis. An international team compared DNA from 1,400 people with bacterial meningitis and 6,000 healthy individuals. They found differences in a family of genes involved in the immune response that seem to make people more or less susceptible to the infection. Sound familiar? **http://www.bbc.co.uk/news/health-10893055.**

10 days ago US researchers announced that in a 20 year study of 4,000 people, half with Parkinson's disease, the team found an association between genes controlling immunity and the condition. It was reported that "During their search, they discovered that groups of genes collectively known as HLA genes are associated with the condition. These genes are key for the immune system to differentiate between foreign invaders and the body's own tissues. In theory, that enables the immune system to attack infectious organisms without turning on itself - but it is not always an infallible system." **http://www.bbc.co.uk/news/health-10956490**

These same researchers also said that "Multiple sclerosis has already been shown to be associated with the same HLA genetic variant seen in the latest study in Parkinson's disease." I am reminded of Ritchie Shoemaker's recent success with a small number of patients who were said to have MS. Note that when Ritchie has "success" this means objective success as indicated by lab results. Of course it also means happier patients.

I turn to Ritchie's chapter on Frank Fuzzell and Benomyl. Sometimes when things are very scary we feel they ought not to be totally credible. For my part I find it more comforting to approach things in that way. Yet this month we read in the New York Times that "A dangerous new mutation that makes some bacteria resistant to almost all antibiotics has become increasingly common in India and Pakistan and is being found in patients in Britain and the United States who got medical care in those countries, according to new studies." Moreover "Experts in antibiotic resistance called the gene mutation, named NDM-1, "worrying" and "ominous," and they said they feared it would spread globally." **http://www.nytimes.com/2010/08/12/world/asia/12bug.html?ref=world**

This new and unusual bacterium is plasmid-mediated. Remembering Ritchie's chapter on benomyl and plasmids, I wrote to him to say "what does mediated mean in this context?" He responded that "it is a jargon term for 'being the effector.' The gene comes from another source. The donated gene, a chunk of DNA is an evolutionary force. Our only hope is that the super bug will not compete as well with the bacteria that live here already. War of the World's idea is that the bacteria we have on earth killed the aliens. Where could the antibiotic resistance genes have come from in the case of NDM-1? I bet on a biofilm former, as within that nice little protective igloo, bacteria do all kinds of gene transfers."

Having harbored a biofilm in my own sinuses (MARCoNS) and having seen the positive results attending its removal under Ritchie's protocols, the idea of a gene transfer factory 2 inches away from my brain was a powerful and unwelcome image.

Finally I would like to comment on Ritchie's work using magnetic resonance spectroscopy scans to help diagnose and track the treatment of biotoxin patients with executive cognitive dysfunction. This may seem unusual to many mainstream doctors but it should not. According to the BBC news this month, a brain scan will show a distinctive pattern associated with autism. "Experts at King's College London ... looked at 20 non-autistic adults and 20 adults with Autism Spectrum Disorder (ASD). They were initially diagnosed using traditional methods, and then given a 15 minute brain MRI scan. The images were reconstructed into 3D and were fed into a computer, which looked for tiny but significant differences. The researchers detected autism with over 90% accuracy, the Journal of Neuroscience reports."

What I found particularly interesting about this report in the Journal of Neuroscience were the following comments: "The findings have been welcomed by the National Autistic Society, who say they add to the understanding of the condition. They say adults can find it very difficult to get a diagnosis of autism, and this may help.

"However, they say without more awareness among doctors, it may be of limited use.

"There's still a woeful lack of awareness in GPs' knowledge of autism," said NAS centre director, Carol Povey.

"People with autism are often dismissed when they go to their GPs for help, so we have to make sure front-line professionals have awareness of autism so they can make appropriate referrals."

I wonder if any of these comments sound familiar to any of you who have been dismissed by front-line medical personnel, perhaps being told that you are malingering or depressed, suffering from allergies or the dreaded but incredibly vague CFS with its outrider Fibromyalgia. Substitute biotoxin illness or if you will, CIRS-WDB for autism in the above and you have our problem - a "woeful lack of awareness in GPs."

My eldest son and I are both WDB victims along with 11 of our former Scotia Prince Cruises colleagues. As with the other guest authors in this book we are victims of our immune systems compounded by (i) the ignorance and self-serving chicanery by government and building owners – in the Scotia Prince case one and the same – and; (ii) the inadequacies of the dispute resolution system, controlled as it is by lawyers whose knowledge base is narrow and skill-sets are self-centered. It is after all the "adversary" system of justice. It has nothing to do with truth. Winning is all that counts.

Sadly this mindset is irrational when the paradigm has to do with water-damaged buildings and innate immune systems.

In the case of our nation's schools, Lee Thomassen has had to resort to the Federal Office of Civil Rights (OCR) and he updates his chapter today with this - "There were a few positive results from my August labs after being out of my WDB for four weeks. C4a dropped from 16,406 in May to 4,494. Osmo showed the biggest improvement: 287. Other results: TGF beta-1 dropped from 6,720 to 4,840. VEGF is 38. VIP is not in yet and the lab apparently ignored the instruction to do MSH. MMP-9 is at 488. I had a good summer and got a lot of my energy back. The pain in my legs went away, although it has come back with a vengeance now that I am back in my WDB.

"BCPS turned down my request for a transfer for the second year in a row. When the federal mediator asked in July why BCPS didn't just move me to another school, the response was that all of the vacancies had been filled. The case-workers at OCR have taken my case and they want to see all of my data right away."

It may be that Lee's crusade will create a precedent in which the OCR will intervene on behalf of children, parents and teachers. We can but wait. Often Big Government means Big Election Donors – so we shall see.

You will not fail to have been moved as I was by these horrendous and heartfelt stories that must have been very difficult to write for each of the patients and/or their loved ones. In each case great determination and courage was required to move through and past the ignorance and often, the uncaring attitudes, of the "caregivers". The institutional attitudes of our government bureaucrats are perhaps the most shocking of all. Obviously they no longer work for us. You can do a great deal to rectify this situation. Give this book to your friends and perhaps more importantly give it to your local family physician and insist that he read it. Ritchie Shoemaker's protocols are based on solid data and have been repeatedly proven to work. They do not require Ritchie Shoemaker in order to be effective. All they require is intelligent use. You can help to make that happen by encouraging physicians to learn and by encouraging patients to be knowledgeable about their own health and to always insist that their caregivers educate themselves about water-damaged buildings, biotoxins and genetic susceptibility.

That susceptibility may well have become more important in the last several decades, not because buildings have become more airtight but because fungi, mankind's ancient fellow traveler on spaceship Earth, have changed. After eons of living in stasis and harmony with humans whose HLA DR, following Darwinian theory, had weeded out those unable to cope with the ubiquitous fungi, suddenly fungi are changing and they are doing it incredibly rapidly. Whether it is benomyl and the plasmid-mediated cascade of its aftereffects or it is Chernobyl or it is any number of other chemicals which we have introduced into the same pool of air that we share with fungi, fungi appear to be changing, leaving some of us without a chair each time the music stops. We are the ones now susceptible to the toxic stew that inhabits water-damaged buildings.

Genetics determine destiny. What does that mean for those of us susceptible to WDB now and to our children? My own view is that Pandora's Box simply contained a go-faster time machine that would speed up the evolution of our enemies large and small – but not change the tempo of our own evolutionary responses. Result – extinction for us. Survival for the microbes, the beasts and the insects.

But mankind can think, so perhaps not all is lost. In 1913 Dr. Charles Richet won the Nobel prize for his work on anaphylaxis in which he injected sea urchin toxin into his subjects and noted that a second small injection could lead to a much more rapid and more dangerous reaction, sometimes life-threatening. In 2010 Ritchie Shoemaker calls this biotoxin phenomenon "sicker, quicker." As Louis Pasteur, whose own career involved serendipity when he accidentally discovered that attenuated microbes can be used for immunization, wisely noted, "In the fields of observation, chance favors only the prepared mind." I concur. Fortunately Ritchie Shoemaker was that prepared mind. Thousands of us are the better for it.

See Appendix 6 for the March 2009 letter from the US Dept. of Labor - proof that we can win this fight. Indeed I can recommend all of the Appendices for those of scientific bent.

I can also recommend all of the Appendices if you think that you, a member of your family or a friend may be ill. Appendix 1 provides the step by step road back to full health. Please remember that biotoxin patients often suffer from brain fog and this condition may be worse than they perceive, since it is pernicious. I know this from watching my colleagues in Portland. Biotoxin patients can be difficult to help but it is essential to persevere. Be patient, be supportive and don't give up.

Please introduce Dr. Shoemaker's work and these personal battle histories to your friends, your colleagues and your physicians. A letter faxed to his office on a physician's letterhead will almost certainly result in a personal call from Dr. S who will gladly share his protocols with any other physician. 1-410-957-3930. If this becomes unwieldy his office will follow up and make sure the information is provided. Just remember that this is not Chinese food with choices from columns A, B, C. It is a step-by-step treatment process guided by lab results at every step of the way.

If you do this you will save many people from devastation. That is the sole purpose of this book.

EPILOGUE: ROSWELL, NEW MEXICO 9/21/10

"I expect everyone to have the common goal...of what is best for the young people of the community not just their education but also their health and that should be the focus of what we're trying to accomplish... Our duty to our students, to our young people is a continuous duty." Judge Thomas Rutledge, 9/21/10

Paul Taylor led the way across the stretch of sandy scrub between the two parking lots. He is walking fast and I'm not used to the nearly 4000 foot elevation. Besides, I've got this suit on, and a tie, not the jeans and T-shirt I'd normally wear. A few years ago (and a few pounds ago too), I would have kept up better with this ramrod-straight West Point graduate. His eyes set on the modern Courthouse in downtown Roswell, New Mexico, Paul is finally having his day in court. Today, General Patton himself would stand aside as Paul marched toward his destiny.

I don't think he knew then that what was going to happen to him, his wife Dr. Terri, and his daughter Paige would make such a difference forever in the "mold movement." I didn't either.

"Paul, please take a look. You've got rattlesnakes around here, right? Are these sidewinder tracks?"

"Oh, sure, Doc. Sorry, I wasn't thinking about snakes. We've got about 30 minutes to get to the court room. I was thinking about the motions we've got to survive in the morning session. I know you'll do well in the Daubert. The defense has only a bunch of bull to use. I can't believe their malarkey is something they think a judge will swallow."

I heard him, but I couldn't stop looking at what I thought was the mark of a snake. "Paul, what is that track?"

"Ritch, these are tracks of a different kind of sidewinder. No question these tracks are from defense attorneys. I can't tell which ones, but no doubt about it. Look here, they were too weak to pull their file trolleys through the sand. They had to weave back and forth, making those "S" tracks.

The "S" tracks were all over this case. All aspects of the well-deserved contempt for the defense interests in mold litigation were present here. In spades. And the senior 'managers' of the school system too, acting as if the sick teens didn't matter. We had a collection of data on sickened high school students that should have made every parent in town rise up in anger. But they hadn't because the school system administrators said everything was just fine. They paid a lot of public money earmarked for education to some consultants, who then said everything was just fine. Under oath, no less. I guess the consultants could ignore the wet wall that fell in because of the moisture and they could ignore the countless ERMI tests showing pathogenic fungi everywhere.

Oh, the defense finally agreed that Paige was sick, but "not from my building." Since you have read this far in *Surviving Mold,* you know what will happen next. The defense started making up things to escape justice. Are you surprised the defense used Frank Fuzzell's smokescreen concept? Sure, just *look away from the truth* long enough and the truth might get lost in the smoke and mirrors that provided the foundation of the defense here. They blamed Paul's house for Paige's illness (not true); they said she had chronic allergies (not true) because she had taken some sinus medications for ten days as a fourth grader; they said she was just making up the illness (a lie). They said that mold exposure doesn't cause inflammation (not true). They hid behind the ACOEM report and its clone, the AAAAI paper of 2006 (both lies based on lies and more lies). They said whatever they thought they could get away with.

And of course, "they" had never examined or treated Paige. Paige's case was obvious to anyone who actually worked with sickened kids. Paul had spent many thousands of dollars on testing and experts to show that just about every area in the school was contaminated with fungi, bacteria and asbestos. This was not a safe place for children and was certainly not a place to learn. What was it the World Health Organization said? That ***everyone had a right to breathe air that was safe.***

It wasn't safe in the high school, no question about it.

What Paul wanted was a safe school that his daughter could attend so she could graduate with her friends. Doesn't sound like too much to ask, is it? So he found a legal firm that suggested that he could sue for an injunction to make the school safe. Paul wasn't suing for money. He didn't demand the school be shut down (*Editor's note*: it should be, just like countless schools in this country that make kids and staff sick). Paul was suing just for the right for Paige to re-enter the school and not be sickened.

My job was to correct the illness of Paige. And now I was in Roswell to tell the Judge all about her illness. The defense knew I wouldn't lie and I would tell the Judge that the defense and their hired guns were not telling the truth. The defense had to try to exclude my testimony. If they couldn't do that with facts or data (they couldn't, ***they don't have any data***), here came their consultants in color-coordinated suits to mouth nonsense as truth. And I see the same sickening games played in litigation from Washington State to Florida and from Maine to Roswell. You'd think that such garbage would rot from its own weight. It doesn't. Not yet.

Paige was sick alright; she had a pair of the worst HLA genes. You already know that if even one of the "dreaded genotypes" meets a moldy building the bearer of those genes will be ill. That was Paige. Due to accommodations provided by the other high school in town (since she couldn't attend the one she was supposed to attend), she was still a straight A student taking AP everything. Sports? Soccer, basketball, tennis. Dance? Sure enough, there wasn't much Paige couldn't do. Except it was overdoing; when she came home, she crashed for several days. Yet she'd get up from her bed and push herself even harder when she felt bad. It is fair to say that if Paige didn't have an anvil of innate immunity holding her shoulders down she would have done even greater things. She had been to Dallas and Denver and so many other places to find an answer to her illness. Countless thousands of dollars later, Paige was still sick. Yes, she had been treated with cholestyramine, but that is just the first step (see Appendix 1) as readers already know. We needed to get her all the way up the treatment pyramid, not simply stopping on the first floor.

Paul called me. I simply said, "Bring your princess here and let's get her fixed." With Actos and other interventions, Paige started getting better, but the no-amylose diet is tough for anyone, not to mention a teenager. The anaerobic threshold exercises she started require that patients abandon thoughts of doing too much exercise. Simply stated, that isn't Paige. Not when the soccer team needed her in the State championships and she had dance competition too.

Dr. Terri, her Mom, knew she needed VIP. But Paige was too young so we didn't know if VIP would survive a clinical trial that was testing safety and efficacy. I *wouldn't* prescribe my miracle drug for Paige. Dr. Terri said little about my refusal, but it was clear to me that she didn't agree.

Time passed. Paige was another year older and I had my data about safety of VIP. OK, if she has the *expected abnormal pulmonary artery response to exercise*, she can take VIP.

I think Dr. Terri would have walked the three hundred miles to Albuquerque to have the required stress echo testing, as that was the nearest facility that could do the test (*Author's note*: she drove). As expected - Paige flunked. VIP, here we come.

The drug worked, beginning with the first dose. And as you might expect, race horse Paige pushed herself further and harder than what the meds would let her do. She had to dial back on what she wanted to do, waiting for the hormonal and inflammatory remodeling to occur. It did.

Paige is now getting ready for college. Her friend from Arizona wants Paige at Arizona State; a relative wanted her at University of New Mexico. I wanted her at Duke, as another student there, Josh Sommers, had started to raise awareness of mold illness on a national scale. Not to mention the fact that Duke women's soccer has been dominated by UNC for a long time. Let's put an end to that right now! (*Author's note:* In fairness, UNC has dominated a lot of schools in women's soccer for many years.)

The hopes for Paige were almost giddying. At long last, accomplishment was stepping on disappointment and confidence was brimming over the top of disillusionment.

But the high school wasn't safe for Paige. Her journey wasn't completed.

In my mind the best answer to any argument about the health effects of a water damaged building are answered by a human health survey and a repetitive re-exposure trial (after informed consent, of course). The school said, "Where are the other sick students?" Fifteen came forward, many treated by Dr. Scott McMahon, a Duke residency-trained pediatrician. The data were impressive in that the innate immune inflammatory problems we would expect to see were right there in black and white. In fact, the levels of TGF beta-1 in this group were some of the highest in any data set I have reviewed. The kids with the worst genes had terrible illness, and that was expected, but so too did some of the kids with less malevolent haplotypes. This school was stealing life from these students. As the list of students I could see were proven to be sickened by the school mounted, the Judge would have no choice but to close the school. Just let me show him the data. One student did a re-exposure trial right after school opened for the fall 2010 semester. His symptoms exploded, his C4a and TGF beta-1 doubled, his VCS crashed. No way that school is safe.

Both sides challenged the other's experts. The depositions of the other side's 'experts,' were <u>our</u> best weapons. They were really consultants and let's be clear, neither of the two MDs had ever treated their first mold patient. Those guys were caught making up factoids about simple things they should have known. I had prepared for hours: bring it on, buddy.

Before the hearings on admissibility of experts there was to be a perfunctory hearing on summary judgment. No way was that going to be supported I thought. Wrong.

Judge Thomas Rutledge of the Fifth Judicial District listened carefully to both sides. All parties agreed: Yes, the case was just about Paige (which meant goodbye to the rosters of all the other sick kids). The injunction would run out when Paige graduated from high school which would be May 2011. Let's see, now. The earliest that testing could be done was Christmas break, as the school would have to be vacated for the destructive testing Paul wanted. There the school district was arguing that the destructive testing posed a health risk to the students (Paul, they just admitted the school hosts the organisms that make people sick!). Then the results would have to be produced, with discussion among both sides about the findings. More depositions too. And then the school would have to put out bids for the scope of work, unless an emergency was called, and even then the earliest that remediation could begin would be the end of April, lasting until about the day Paige graduated.

This timeline didn't look good for Paul.

After the Judge retired to chambers to ponder his ruling, the attorneys for both sides were summoned back to meet with the Judge. The attorneys came out, their body language almost telling the tale: a smirk from the defense guy and stone faces from Paul's lawyers.

And that was it. Since the injunction would expire before Paige could go back to the school, the matter was moot. Paul's application for injunctive relief would not be heard.

I didn't pretend to know why an injunction was requested, but it just felt odd. Wouldn't the injunction apply to all the sick kids we had documented? No, the case was about Paige and an application, injunctive relief. Wasn't the case all about all those nasty things the defense tried and the hiding/destruction of evidence by the school? Nope, the case was about Paige.

Judge Rutledge wasn't done. The school district won, but the judge had seen our data and read all our briefs. He knew there was a problem and he knew other kids were sick. His comments came as he said he was putting the school district on notice that they had better clean up the moldy school and prove it was safe for the students.

"The one thing I would remind everyone, counsel, parties and in the courtroom that I expect everyone to have the common goal even though you are on opposite sides of this issue, and the common goal is what is best for the young people of the community not just their education but also their health and that should be the focus of what we're trying to accomplish.

"In this particular case, as the Court has indicated, there are significant problems going forward as to this injunctive action. But that doesn't mean that efforts should not continue to go forward.

"And I appreciate (counsel), your comments on behalf of the school district that they are going to go forward and not walk out of this courtroom going, yay; it's over, because it is not over. Our duty to our students, to our young people is a continuous duty. With that, we will be in recess."

I asked Paul if he had won. The school had to admit that it was not safe and that clean up was mandated. Wasn't that what he wanted? His reply was "What I wanted was for the school to be held accountable. The newspaper will say tomorrow that the school was vindicated - when it wasn't."

I live in a small town. Roswell, with nearly 50,000 inhabitants dwarfs the 3,000 people who live in Pocomoke. But the small town aspects were true for Roswell too. Seeing someone in the box chain grocery store might prompt a discussion about the court decision. Such talk wouldn't focus on the unidentified sick kids or the duty that the school district was supposed to uphold to make the building safe. The chit-chat would be about Paul losing.

Before the Roswell decision and thus before I wrote this epilogue my editor and friend wrote in the last Chapter of this book about

"...the inadequacies of the dispute resolution system, controlled as it is by lawyers whose knowledge base is narrow and skill-sets are self-centered. It is after all the "adversary" system of justice. It has nothing to do with truth. Winning is all that counts." He also said, "Sadly this mindset is irrational when the paradigm has to do with water-damaged buildings and innate immune systems."

This is what happened in Paige's case. Judge Rutledge knew the truth but he could only rule on matters properly before him and those were only two - Paige and injunctive relief relating to her. Even if he granted the injunctive relief Paul sought it would not have had a practical result for Paige. So the Judge was powerless to act.

If the administrators and the schools insurers had acted in good faith then the finding of such incredible mold and sickened students should have prompted a careful school health survey. And we aren't talking about a bunch of untrained personnel masquerading as experienced health care providers reading a check list to satisfy their responsibility for a comprehensive health survey. And we aren't talking about substituting some federal personnel as "responsible investigators" to do the survey when they have a track record of ***not seeing mold that was all over the flooded crawlspace and sickened people who already had massive data bases of information*** when they were in a school in Virginia. NIH is a government term for "Not Invented Here." If they don't think of it, it doesn't exist. If they deny it, it doesn't exist. Facts don't matter when the organizational mandate rules.

As an aside, in October 2010 the CDC released the NIOSH paper on the investigation of the Fortier High School near New Orleans in 2005. The NIOSH team cited our groups two early papers; duplicated and therefore validated our VCS testing results; and validated ERMI, all the while adding the weight of the CDC to the diffuse list of symptoms NIOSH reported the exposed patients had. These included all the symptoms that the CDC had fought for years like fatigue, cognitive problems, headache, abdominal pain and unusual rheumatologic problems. These symptoms, of course, are always found in all groups of mold patients: the only ones who didn't know that fact were the consultants who never looked.

There sure was a lot of shame that the nay-sayers earned in the Roswell case. Liars? You bet. Intentional deceivers of the truth? Paul commissioned extensive ERMI testing; the school was a rat's nest of water intrusion and mold growth. The school ordered a few air samples and said everything was fine. Fabrication? You bet. Were the school's representatives actually *intentionally and knowingly* exposing some of Roswell's children to toxic bioaerosols in the school? You bet.

Jodi Ashcraft has worked with Paul in his business for years. After the legal process failed Paige, Jodi and Paul agreed that no one should ever have to suffer the way Paige and her classmates did.

Paul and Jodi are as savvy about use of the Internet as anyone I've met. Jodi didn't mince words: "We can get the word out about diagnosis, testing, treatment, resources, legal cases, access to data bases, references and much more. We can become a one-stop resource center for anyone injured by moldy buildings. We have 2,000,000 names in our data banks. You need help with genomics testing? We can help. You need to share information; it will just be a click away. There hasn't been a coordinated approach to solving the problems in schools; we can help. There hasn't been a clearinghouse for information for plaintiffs like the defense bar has. Now there will be. You have my word on that."

The sacrifice of the Taylors will pave the way for safety and justice in schools in Roswell, in Fairfax, Virginia, in Pawley's Island, South Carolina and in Baltimore, Maryland. And from their work, the Taylors will save thousands of students like Paige from the losses; social, physical and emotional, that Paige suffered.

Let Judge Rutledge's words be heeded for all of us living in the Era of Dangerous Buildings.

Frequently Asked Questions

1. Do I literally have to clean everything in my house (i.e. plastic storage containers, books, toiletries, candles, knickknacks, canned food etc) or just the major things like clothing and furniture?

   Yes, everything.

2. Is it possible to clean electronic devices that may have mold spores inside of them, seeing as I have no way to clean the inside of these devices (i.e., DVD players, computers)? Is it good enough to simply clean the outside of them?

   Clean the outside and vacuum the openings.

3. Do pictures and documents have to be scanned/copied or can the originals be kept?

   Photos can be washed. Porous frames are trash. Documents need to be copied.

4. What about books, journal? How do you clean them? Or do they have to be thrown out?

   Books are difficult. If the books are on a shelf and not opened, HEPA vacuum each one. Opened books are trash.

5. Do I need to use a new cloth for every item that I clean (one clean cloth per moldy item) or can I use the same cloth for several items?

   One cloth can be re-used. You will be vacuuming, then wiping and vacuuming again anyway.

6. Can leather furniture be cleaned since it is not porous like

cloth?

Use quaternary cleaners on leather. Vacuum the nooks and crannies of the piece of furniture and do the same thing again. I know some people will say throw away the leather, but I have seen good results if the leather is finished and not rough.

7. Is it likely that I have cross-contaminated my car seeing as I drive in it, in my moldy clothes and with other items from my home on a daily basis? If that is the case, do I need to sell my car as well or can it be cleaned? The interior of my car is cloth not leather, if that makes a difference.

   The trouble with cleaning for mold is where do you stop? There are many chemically sensitive people who have to get new cars. For most people the car isn't the problem. Be careful about the antifreeze system in a car as that system's glycol ethers are a real problem.

8. Is it likely that I have contaminated my friends' homes just by being there as well? If so, is that going to be problematic for me when I go to visit with them? I plan to move in with one of my friends for several months while I look for a new place to buy. I have slept at this person's house already with my moldy clothes and suitcase quite a few times... will I need to clean her floors and table tops etc, or is any possible cross contamination from my simply having slept there too minimal to worry about? Have I cross-contaminated this person's bed by sleeping in it?

   Clothes are much less of a problem if they are laundered before wearing to someone else's home. I get mold hits from families that come from moldy homes, but the risk of illness and cross contamination from clothes is less if the clothes are removed from the home, cleaned or dry-cleaned.

9. I plan to purchase a HEPA vacuum cleaner. After I clean things with it will it need to be thrown out, or is replacing the filter sufficient?

   Replace the filter. A good HEPA will last a long time. Better yet, invest in a central vacuum.

10. I assume that plates, dishes, silver ware etc. can simply be run through the dishwasher. If so, where do I clean them? My current moldy dishwasher before I move out or the dishwasher in my new place?

    I have not seen people made ill by a cleaned dishwasher. Rub down the outside and vacuum it but the wash cycle is enough to safeguard your silver and china.

11. I am a little worried that work could be moldy too. None of my co-workers are sick and I have asked about water damage. I am told there has not been any. But I have noticed tiles that you can tell have been wet. Any way to determine if the building is moldy short of doing a test?

    ERMI is the key concept here. I don't trust any building to be safe just because a landlord said so. Spend the $300 to know for sure.

12. Would it be better to clean everything myself or to hire a mold expert?

    As far as abatement and removal of materials, you need to learn the tricks of the trade. But there is no reason you can't do the work yourself if you have carpentry skills. The cleaning you probably should do.

What Is Generally Involved in Remediation?

Understanding what remediators are supposed to do may save you lots of money and failed work. Don't forget the ABCs of mold survival: **A**bate the water intrusion first; remove or encapsulate all **B**uilding materials that are contaminated; **C**lean, Clean and clean. If your remediator doesn't know about ERMI (appendix 3) and human health effects, get another one. If your remediator tries to tell you that air samples have value in assessment of efficacy of remediation, get another one.

But let's assume you have found someone who will (i) give you a quality job for your money and will (ii) give you confirmation that the process of remediation is done properly.
So what process do you look for? Be certain that the *potential* for water intrusion has been taken seriously. Basements and crawlspaces are the source of over 75% of the moisture intrusion that results in the illnesses I see. Don't guess. Don't assume that dry dirt means safe dirt. Seal it! If there have been leaky roofs or windows, check wall cavities with reliable moisture meters (see Chapter 24). Make sure HVAC isn't acting as the universal distributor of bioaerosols.

So what about the actual work? First, make sure the area is contained, sealed off, isolated from the rest of the world. That means that no bioaerosols from the vortex created by remediation will be able to spin into your nose and lungs. Containment means no cross-contamination. Look for plastic sheeting to be sealed with tape at the top and bottom around the work space. Before you start work, make sure that no air is blowing through any parts of the isolated work site (check for air leaks with a candle or a moistened hand, for example).

I can't tell you how often I see remediators forgetting to seal off all HVAC vents. Unbelievable. And don't turn on the HVAC!

You might have to make a "vapor lock," a chamber that will seal off the sealed off area that is opened by going in and out. The logic gets a bit difficult here as the chance for cross contamination is increased every time anyone enters the work chamber. A back-up is to exhaust room air outside the plastic room to the outside using a strong fan. If you can exhaust the work site too, that's even better.

If the work site is large or there are multiple sites, and this is especially true in the fake stucco homes with bad caulking, the logistics of sealing everything is nightmarish. Make sure an air scrubber with HEPA is being used constantly.

If you are going into the work site, use a respirator. A lot of publicity goes into the use of N95 masks, but I'm not so sure those devices are adequate, so scrub the air you will breathe before you breathe it. Some studies show that (i) N95 is no good; and other studies show that (ii) N95 is great. Yikes.

Wet drywall and wet insulation are trash. Bag it, seal it and when you take it out of the building don't let any dust from the work site follow along. Weight-bearing structural elements are more difficult to analyze. You've got to re-surface the exposed wood, so sanding or some destruction of the uneven surface of the wood has to occur. Vacuum thoroughly, using HEPA vacuum. Wipe it down with something (at one time Greg Weatherman recommended Sporicidin) that removes spores, fragments of spores and the tiny particulates that no one can see. Vacuum it again and wipe it again. Then cover with Bin or Kilz to seal the wood in a lacquer. If the wood isn't bearing weight, take it out and put in a new stud as it is much cheaper than trying to re-finish the old wood.

Cleaning includes all surfaces and contents. This is maddeningly slow work! All non-porous items should be wiped with something like Sporocidin or quaternary ammonium compounds (even Fantastic is better in my mind than diluted bleach, but I have heard a lot of heated arguments here). Throw away the wipes. Then vacuum with a HEPA vacuum, repeating the wipe and vacuum process. Porous items basically are trash. All the walls and non-porous flooring (tiles, carpet, etc.) should be vacuumed with a HEPA vacuum, wiped down with a mop and re-vacuumed after they are dry.

All of the trashed materials need to be handled as if the contents are going to kill you. Double bag in plastic, tape the opening shut, don't throw the bags or slide them on concrete slabs. You don't want known contaminated materials making you ill by having them open to the air you breathe.

Appendix 1

What Do I Do?

I receive the following phone call at my office a lot: "Dr. Smith is on the line and he wants to know what he should do for Mr. Jones."

The protocol I use is straightforward, but there are a series of steps to be taken, each in its proper order. The idea is that these treatment steps are similar to how one would climb a pyramid. If the patient feels fine with simple removal from exposure, no more steps are needed. There are a few people who are fixed with this step alone and thus, the number remaining to be helped is reduced by those fixed at this first step. CSM is the next step. Some patients are finished with therapy at this step, with fewer of the initial total of ill patients therefore needing more steps.

Ascending the pyramid step by step leaves a steadily smaller number of people with illness refractory to intervention. At one time, the top of the pyramid was treatment of high C4a. After that step was taken, still a few were left needing help, with those with high TGF beta-1 being identified as one group we didn't know about before. Along came losartan to treat TGF beta-1, and some at the top of the pyramid got better, but a few others were still sick. The pyramid was higher than we thought, with the top still containing some people who still needed help! Now we treat this group with VIP. We want to know what steps are left to be taken after VIP has done its magic. Until we've taken the "final" step, we don't know that it *is* final.

So the road to the top of the pyramid begins at its base. You can't get to the top by leaping; you've got to plod uphill step by step. Some of these steps are high.

I start with differential diagnosis. Plug the patient into the Biotoxin Pathway; follow the order of treatment that our group has published without being creative or substituting favorite unproven ideas in place of those proven to work. Monitor the "prognosis labs" compulsively (MMP9, VEGF, C4a and TGF beta-1) and visual contrast sensitivity (VCS) at each step. Continue to enact the next intervention that fixes each uncorrected item until you get to the "top" of the treatment pyramid. Monitor exposure like a hawk each step of the way. If there's a new WDB exposure, then like Sisyphus pushing a rock up a hill in Hades only to have it roll back down from the top, you've got to go back to the bottom of the pyramid and start over. Then make sure the patient stays well by avoiding re-exposure.

Yes, avoiding re-exposure means the patient **and the doctor** both need to learn *Surviving Mold*.

When you are done pyramid-climbing, have a nice life, Mr. Jones. Good job, Dr. Smith.

The sequence of events that Dr. Smith will need to learn have an order of treatment and an overarching structure, similar in form to a traditional Japanese *kaiseki,* a sumptuous 14-course feast. Each step must be "served" in a certain order to achieve the desired effect. Sure, there's some freedom provided for the chef to substitute fresh ingredients in each course, providing different colors and textures by season, but the meal, said to be designed by Zen monks and fed to samurai and shoguns back through the mists of time, maintains its structure despite the delightful wobbles in presentation. The structure of the meal is designed to foment contemplation. Likewise the structure of the treatment protocol demands paying attention to symptoms and the changing variables of Visual Contrast Sensitivity and lab parameters. Skipping steps is strictly forbidden.

So the individual courses of our treatment *kaiseki* involve:

(1) Sequential differential diagnosis begins with a compulsively obtained data base, i.e. what could be wrong? What that means is the labs needed to show inflammatory abnormalities are collected and the labs that are always normal in WDB-illness are also collected. (*Editor's note:* at the end of every step, the physician must re-visit his differential diagnosis);
(2) Performing ERMI testing (contact **www.mycometrics.com**) to ensure there's no exposure to a building with an ERMI *greater than 2 if* the patient's MSH is less than 35 and C4a is less than 20,000; or no exposure to ERMI *greater than negative 1 if* MSH is less than 35 and C4a is greater than 20,000;

(3) Removal from prior exposure (this step can variously mean antibiotics for Post-Lyme; no more consumption of reef fish for those with ciguatera; and no more working, schooling or living in a moldy environment for WDB-illness patients;

(4) Correcting toxin carriage with CSM and/or Welchol, using VCS monitoring to assess endpoints;

(5) Eradicating biofilm-forming MARCoNS;

(6) Eliminating gluten for those with anti-gliadin positivity as shown by a positive blood test, with celiac disease ruled out;

(7) Correcting elevated MMP9;

(8) Correcting ADH/osmolality;

(9) Correcting low VEGF;

(10) Correcting elevated C3a;

(11) Correcting elevated C4a;

(12) Reducing elevated TGF beta-1 and

(13) Replacing low VIP; and finally

(14) With a final check to verify stability off meds.
Each of these steps using FDA-approved medications is available to practicing physicians.

Thus the protocol I use is a process, one that monitors progress transparently. The protocol is one that's based on peer-reviewed, published material, using differential diagnosis continuously. Note that there's nothing in this protocol that says the illness is psychological. This illness isn't somatoform disorder or malingering, *it is physiology*. The logic goes something like this: "If disease X is present, then abnormal physiology Y will be there as well as a necessary marker for X. Removal of Y will lead to the remaining abnormalities Z. Once Y and Z are removed and there are no more remaining physiologic abnormalities, then X is gone as well. <u>Further, if Y is not present then X can't be the diagnosis."</u>

We have a case definition for mold illness that our group published in 2003. The earlier case definition contains two tiers as follows below. Given the September 2008 case definition of mold illness from the General Accountability Office following (in time) that from our group, I certainly defer to that body's panel regarding what's causation and what isn't. Basically, they say the illness is caused by mold if all epidemiologic findings of the individual are consistent with that of others reported; that the physiology has been shown to be present in laboratory animals and/or people previously; and that treatment is successful. Until September 2008, however, according to our data gathered in thousands of cases any diagnosis of environmentally acquired biotoxin illness, including that from mold, must also include:

(1) The potential for exposure;
(2) The presence of a distinctive grouping of symptoms; and
(3) The absence of confounding diagnoses and exposures.

This first tier of the case definition is adopted from the initial CDC case definition of *Pfiesteria* cases from 1998.
The second tier of objective factors demands presence of three of six of the following:

(1) HLA DR by PCR showing susceptibility;
(2) Reduced levels of melanocyte stimulating hormone (MSH) in a properly performed /prepared specimen;
(3) Elevated levels in matrix metalloproteinase-9 (MMP9) in a properly prepared serum specimen;
(4) Deficits in visual contrast sensitivity (VCS);
(5) Dysregulation of ACTH/cortisol in simultaneously obtained specimens;
(6) Dysregulation of ADH/osmolality in simultaneously obtained specimens.

We welcome the GAO report but our protocol, while more demanding, has the research and science to support it. The second tier is adapted from similar use of different parameters in illnesses such as systemic lupus erythematosis and rheumatic fever, among others. The case definition is derived from looking at what thousands of mold illness patients demonstrated that none of the control patients demonstrated.

When physicians start saving money (They don't! Avoid being penny wise and pound foolish) by skipping testing for elements of the case definition, they lose the ability to (i) know what's wrong; (ii) know where they are in the therapeutic process; (iii) know when they're on the right track; and (iv) know when they *aren't* on the right track. Similarly, when elements of the case definition aren't assessed, the clear diagnosis of mold illness will fade from its proper place in the front row. If the patient doesn't have mold illness, then prove that, too. Very important.

Understanding that physicians aren't often ready to use every element of *someone else's* treatment protocol, especially since the terms might be new and the concepts not clear, errors arise when the physician becomes creative without basis. In our 14-course feast analogy, the individual course may be altered some, but not the structure.

I used to teach new medical providers much more than I do now. I felt that students and residents needed to read about their patients. If a student saw someone with chronic ulcerative colitis (CUC) at noon, I recommended that the student read about it that night, including doing a literature search on PubMed. We'd talk about CUC again the following day. The student was able to observe the illness, see the patient in his environment, review published literature, and then assimilate all into a patient-specific treatment protocol. We call that evidence-based medicine. At the end of a six-week rotation, the student would usually have 40 or 50 literature searches to back up careful observation, perhaps new techniques in history taking, combined with treatment tricks from the old country doctor who'd seen 300,000 patients in his career. When the student then compared what he saw in Pocomoke, Maryland to what he saw in the tertiary institutions in Baltimore or College Park, he could make his own criticisms and comparisons. To me, that kind of evidence-based medicine (EBM) fulfilled the modern criteria for what EBM should be.

As time has passed, however, it appears to me that each succeeding generation of students (the residents were worse by far than the med students) seemingly did fewer literature searches, even though they had become easier to do, and read less about their patients. Maybe the teaching career of the old country doctor is over and the problem is with him rather than the students!

The old country doctor says, "If you don't know everything about something in medicine and you're being called upon to be responsible to your patients and help them with that something, find out more about that something. And don't go to sleep until you do."

I guess I'm getting old because my ideas of responsibility for patient care now are old-fashioned. Reading about illness is tough when you're trying to play slow pitch softball for the hospital team, romance the unit clerk floating from ward to ward but often on the 5th floor, or moonlighting, trying to pay the huge med school tuition bills. Maybe the problem is that these days a primary health care provider can simply refer unfamiliar problems to a specialty provider, hoping that the specialty provider will know the answer.

So if the physician is approached with the 14-step biotoxin treatment protocol, one applicable to just about every chronic fatiguing illness, what does the physician do? Read about the protocol? Call someone, asking for their insight?

Do a literature search? All the above? Or does the physician guess, making errors of assumptions, omission and commission? Or simply refer the patient out, not wanting to be bothered learning something new.

I reviewed one chart today that demonstrated some typical errors. A physician in Kansas was given a copy of *Mold Warriors* to read by a patient who knew she was exposed to a WDB. She wanted the physician to treat her. The physician admitted he had no idea what some of the lab tests were but because “they were new” (wrong!) and “might not be covered by insurance” (wrong again!), he decided to give the patient some CSM to see what it would do. But instead of using the protocol for CSM published in *Mold Warriors*, the physician felt that a just little bit would be enough, and besides, there were too many side effects from CSM, anyway. He rationalized that if the patient did well, he would’ve saved the patient potential expense and side effects. The real kicker was that the physician wanted the patient to see an allergist in Wichita first, as *that physician* might know what the labs meant.

I work with so many good physicians who don’t guess; who don’t assume and who ask for advice when in doubt that the attitudes of the above Kansas doc have got to be in the minority? Well. Maybe not.

Jumping into a complex case investigation by guessing isn’t good medicine. Imagine if the Crime Scene investigators in Las Vegas just took a look at the dead body and certified the cause of death according to a guess. What happened to the forensic evidence? What happened to the process of investigation?

The process isn’t much different looking at Chronic Fatiguing Illnesses.

I had another example today. This time the physician prescribed CSM for a month; wouldn’t do a nasal culture looking for biofilm-forming staphylococci; and decided that since the patient wasn’t better, to give the person Ampligen instead. I don’t know what the cost of the IV treatment is, but I’ve heard it’s very expensive. It was almost palpable. This doc said, “Don’t bother doing what is published to work: do something different.”

The mantra of assumptions and guesses had a new theme several months ago: use VIP even before the workplace had been assessed for mold (by the way-the ERMI was over 20 when I had it tested!). “Let’s use VIP because it has helped a lot of people in Dr. Shoemaker’s practice. We don’t need to follow the protocol he uses.”

I remember hearing a clinician from Alabama who used this approach say, almost mournfully, “We have a lower rate of success with VIP than Dr. Shoemaker.”

Clearly the VIP didn't work, right? Actually, *no*. The VIP won't fix people with ongoing exposures (see the Chapter 16, VIP) and the initial rise in TGF beta-1 measured fifteen minutes after the test dose of VIP told the patient and physician alike that *there was an ongoing exposure*. Did someone skip that step?

The list goes on. The worst offenders for patients in the Chronic Fatigue Syndrome and mold worlds are providers who discount mold exposure. And then there are those who guess about exposures (I've heard all the excuses not to spend $300 on ERMI: the crawl space was dry; the remediation guy said the work was successful; the air sample ruled out any problem; the sump pump never runs; the home is brand new). It's all guessing. Just do an ERMI. The fungal DNA results are very helpful. Make life easy and objective; analyze data.

Or how about the guy who decides that CSM is too much bother and, even though the VCS remains positive (showing incomplete treatment of toxin carriage), decides that it's time to move on to something new.

From where I sit, the importance of the nasal culture is the step most often under-used by docs. Maybe some just aren't oriented to microbiology. All of us are pushed to read one thing or another every day just to stay current. Assimilating new information into clinical practice takes time away from home life and life outside the office. I read three microbiology journals each month but none from Family Practice and only one from Internal Medicine. Physicians are allowed to make different choices about what they read; my choices lean towards bacteria and microbes. In the microbiology world, biofilm formation is a giant topic now, though it was much less important in 1998 when I started doing the cultures for biofilm-formers. If the physician doesn't look for the slow-growing, fastidious coagulase negative staphs, he won't know the biofilm-formers are there. And without eradication, the patient will not show much improvement with CSM alone. Without a culture or (worse) with a routine nasal culture, (*Editor's note*: the technique must be one that looks specifically and selectively for the coagulase negative staphs, like the API-STAPH), there's no way the physician can be sure he knows where he is in the therapeutic process. If the lab won't use API-STAPH or some similar technique, then use another lab. Cambridge Biomedical (617-787-8998) does an excellent selective culture for the biofilm-formers.

Poor MMP9 gets short shrift from some physicians, too. Lowering MMP9 usually means both using pioglitazone at full dose (45 mg daily in one dose) and using the no-amylose diet (see *Lose the Weight You Hate*). I often see three errors routinely made here: no use of pioglitazone; use of inadequate-dose pioglitazone (full dose is 45 mg daily); no use of the required combo of full dose pioglitazone and no-amylose diet. For those with leptin levels lower than seven, I *never* use pioglitazone as it lowers leptin too much and will reduce the stimulus to make MSH. For these slender patients, high dose Omega-3 fatty acids will do the job, but only in high doses (2.4 grams of EPA and 1.8 grams of DHA daily).

I don't have too many docs consulting me about using low-dose erythropoietin (epo) any more. The black box warnings that the FDA put into the package insert scares people, as it should. Having said that, low dose epo, when used according to an IRB-reviewed protocol is quite safe and effective. The molecular biology of what epo does is stunning to say the least, from stopping over-production of C4a to stabilizing harmful remodeling in brain, nerve and heart. Epo is a drug that saves lives. For those with C4a that goes over 20,000 rapidly, it truly restores health. Using VIP in the face of uncorrected C4a (or C3a or VEGF or MMP9, for example) is an error in logic.

The use of VIP in an IRB-approved protocol has revolutionized my practice. For the very worst patients, ones who just don't get better with anything short of VIP, this new approach is a life-returning protocol. Many who've gone off VIP after their illness is corrected haven't needed anything more providing they aren't re-exposed to a moldy environment, so I dare to use the "cure" word. VIP is only used for patients that have been moved up the treatment pyramid with persistent illness noted following each previous step. When docs ignore or skip steps or make assumptions about therapies, the patient is guaranteed to suffer. The doc must insist that an ERMI be done looking at the potential for exposure to WDB before starting the VIP. If the ERMI at home or at work is over 2.0, don't start VIP.

Maybe if one hasn't been introduced to a kaiseki, then that person won't know what the experience would be like. But just because one doesn't know what kaiseki entails, don't make assumptions about its existence or assign value judgments to its worth. We have the hard data and we share it with anyone. No cost. Except the cost of this book but maybe you borrowed it?

Appendix 2

Rosetta Stone

	DRB1	DQ	DRB3	DRB4	DRB5	
Multisusceptible	4	3		53		
	11/12	3	52B			
	14	5	53B			
Mold	7	2/3		53		
	13	6	52A,B,C			
	17	2	52A			
	18*	4	52A			
Borrelia, post Lyme Syndrome	15	6			51	
	16	5			51	
Dinoflagellates	4	7/8		53		
Multiple Antibiotic Resistant Staph epidermidis (MARCoNS)	11	7	52B			
Low MSH	1	5				
No recognized significance	8	3,4,6				
Low-risk Mold	7	9		53		
	12	7	52B			
	9	9		53		

1. Look at the LabCorp report. There are five categories of line entries: DRB1, DQ, DRB3, DRB4, and DRB5. Each individual will have two sets of three alleles, unless the DRB1 is 1, 8 or 10. Those patients will only have a DQ and won't have DRB 3, 4, 5. Each individual with a DRB1 other than 1, 8 or 10 will have a DQ and one other allele from DRB 3, 4, 5. If you are expecting to find two entries in, say DR 3, 4, 5 but only find one, the patient is homozygous for that allele and only one allele will appear on the PCR. Don't be worried that the numbers and letters look impossible to understand. They just have to be translated.

2. You will translate these categories into B1, DQ, 52 (A, B, C), 53 and 51 respectively. If the translation were easy and made sense, you wouldn't need this!

3. The numbers and letters in each of the 5 categories are given to an excessive amount of detail. Write down only the first two numbers in each line.

4. When you see 03 as one of the two genes, an allele, for DRB1, rewrite it as 17.

5. Record the entries in B3 by converting the 01 to A, the 02 to B and the 03 to C; this will give you 52A, 52B and 52C, respectively:

6. B4 is 53.

7. B5 is 51.

8. Record the genotypes in two columns, one each representing each parent, using the templates in the appendix.

9. When the international Language on HLA DR changes again, we might have to change the Rosetta Stone again. I hope not.

HLA DR Haplotypes

1-3	9-3-53	13-7-52C
1-4	9-7-53	14-3-52B
1-5	9-9-53	14-5-52B
1-8	10-5	14-7-52B
4-3-53	11-3-52B	15-5-51
4-4-53	11-4-52B	15-6-51
4-7-53	11-5-52B	16-5-51
4-8-53	11-7-52B	16-6-51
7-2-53	11-8-52B	17-2-52A
7-3-53	12-3-52B	17-2-52B
7-4-53	12-5-52B	17-2-52C
7-9-53	13-3-52A	17-3-52A
8-3	13-3-52B	17-2-52B
8-4	13-6-52A	17-3-52C
8-5	13-6-52B	17-4-52A
8-6	13-6-52C	17-4-52B
8-8	13-7-52A	18-4-52A
9-2-53	13-7-52B	103-5

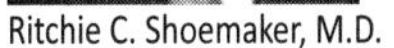
Ritchie C. Shoemaker, M.D.

King Teh Lin, PhD

Appendix 3
ERMI

INSIDE INDOOR AIR QUALITY

Environmental Relative Moldiness Index (ERMI)SM

By King-Teh Lin, PhD, Ritchie C. Shoemaker MD

Concerns about health effects caused by molds growing in the indoor environment of water damaged buildings (WDB) affect many people. Just picture the questions that accompany thinking about occupying a new living space. Is this musty smell a warning? Can I trust the joints of the flexible duct work attached to the air handler in the crawlspace to be airtight? And what about that bubbling of the paint in the living room ceiling by the chimney? Is this basement play room next to the dirt crawlspace safe for my children?

Mold illness comes from any indoor environment that is damaged by water intrusion and not just by natural disasters. Yet there are been no standardized, objective methods available to quantify the indoor mold burden in homes.

What would you do if you faced the concerns of three actual patients? (1) You are a new home buyer. You have a history of unusual fatigue, cognitive problems and chronic respiratory problems. Your doctor says indoor mold makes you sick. How can you tell if the beautiful home across town is safe? (2) Now make yourself a 55 year old secretary at a large manufacturing site. Your office had visible mold growth; you were proven to be made ill by re-exposure to the office. Your employer assures you the office has been cleaned thoroughly.(3) Now have three sick kids in a riverfront town in Massachusetts.Your children were told they had Lyme disease, but they didn't get better with tons of antibiotics. Another physician says your kids are sick from exposure to WDB.

How do you know if toxigenic molds are in your indoors? Spend a chunk of cash to bring in an industrial hygienist who takes a few air samples? Spend more money on more samples? When do you pay big bucks for mycotoxin testing?

Face it: Human illness that follows exposure to WDB has moved into daily medical practice, in part because confirmation of causation of human illness is backed by intense scientific research (1, 2, 3). Now that physicians can diagnose mold illness using simple tests and treat mold illness effectively, prospective inhabitants of dwellings all want to know: How can I be assured of safety?

Research has come a long way from earlier thoughts that exposure to WDB wasn't confirmed to be dangerous. A recent paper from the CDC on molds in New Orleans states, "Molds, endotoxins and fungal glucans were detected in the environment after Hurricanes Katrina and Rita in New Orleans at concentrations that have been previously associated with health effects (4)." And in the paper's acknowledgments, "We are indebted... to the US Department of Health and Human Services for ensuring the safety of the sampling teams (4)."

We're glad the CDC has caught up with current research on mold illness. Thankfully, that research gives us answers for the questions posed by our three patients. They can do home sampling for fungal DNA. For less than $500 and in less than 10 days, prospective occupiers of new building spaces have a chance to avoid inhabiting risky interior environments by first using the Environmental Relative Mold Index (ERMI). We know that the DNA testing, will not replace either industrial hygienists or careful home inspection as the best way to ensure safety but now no one interested in safety of a building can skip doing an ERMI.

WHY DEVELOP ERMI?

The tests we have used for years to assess mold contamination are flawed. Air samples taken for a few minutes were just a snapshot in time; they didn't actually represent a complete picture of ongoing health risks for occupants. We compared levels of organisms found indoors to outdoors, not distinguishing between genera found. "Mold is mold" was the underlying concept here; we all know that some genera of molds won't cause illness and others do. We tried to establish thresholds for levels of indoor molds but there are so many variables that impact on sampling that reproduction of results is difficult. Spore counts? Not when NIOSH told us that there were toxins on 500 tiny fragments of molds we missed for every spore we found (5). Why not test for mycotoxins alone? Mycotoxin testing has to be thorough, with multiple samples for multiple compounds. Talk about costs!

Thanks to the pioneering work of Dr. Stephen Vesper (6,10) and scientists at the Microbial Exposure Laboratories of the EPA, Cincinnati, we believe the problems involved with indoor testing may be solved. Just look for the DNA! The development of Mold Specific Quantitative Polymerase Chain Reaction (MSQPCR) and its application called the Environmental Relative Moldiness Index (ERMI) has brought the light of illuminating science into the darkness of indoor mold testing. ERMI is an objective, standardized DNA based method that will identify and quantify molds. The science behind this breakthrough that led to MSQPCR is now patented (US Patent No.6,387,652). In 2006, the Department of Housing and Urban Development (HUD) used this technology to complete the American Healthy Homes Survey (AHHS). Based on this national survey and MSQPCR analysis of the settled dust in these homes, a national Environmental Relative Moldiness Index (ERMI) was developed.

In the American Healthy Homes Survey, dust was collected in a nationally representative sampling of 1096 homes byvacuuming an area three feet by six feet in the living room and bedroom for 5 minutes, each with a dust sampler-fitted vacuum (Figure 1). The settled dust is collected in a special in-hose device that is sent to a reference laboratory. At the lab, the individual samples are evaluated for quality and reliability against internal standards. Each satisfactory sample is then mixed and sieved through a 300 micron screen. The samples are each analyzed for DNA of 36 species of molds that can distinguish between molds found in WDB from molds found in non-WDB. ERMI doesn't measure DNA of all fungi, just those that describe the "relative mold burden" that has validity anywhere in the country.

WHAT IS THE ERMI

These 36 species were divided into 26 species/clusters associated with WDB (Group 1) and 10 common species/clusters not associated with WDB, called Group 2. The number calculated as the ERMI is actually the sum of the logs of the concentrations of the DNA of the different species. The "mold index" is the difference between Group 1 and Group 2. The laboratory will report the Concentration of the 36 species in your Sample (Table 1).

The computed ERMI values are graphed From lowest to highest (Figure 2). The scale ranges from -10 to 20. On the y axis, the percentage of homes that fall into different ERMI percentages is shown. For example, an ERMI of 14 is in the top 25 % of homes for relative mold burden. An ERMI of -6 would be in the lowest 25% of homes. Each value is plus or minus three.

USING THE ERMI

The ERMI scale was derived from the analysis of the settled dust in the common living room plus bedroom of a home. Even if most of a water-intrusion problem in a home comes from the basement, we won't suggest sampling the molds in the basement first, as all the national standards are derived from sleeping areas and living areas.

So what should our patients do? The ERMI costs several hundred dollars, providing information that is potentially far better than limited testing done by an expert whose time can be a large part of the bill. Each of our patients did home samples. The beautiful home across town had a bargain price tag because of its multiple problems with the roof flashing by the chimney. An ERMI of 18 saved the patient a mountain of trouble. The secretary found an ERMI of 0.02. She has done well after remediation. The Massachusetts Mom found that her home was terribly contaminated, even

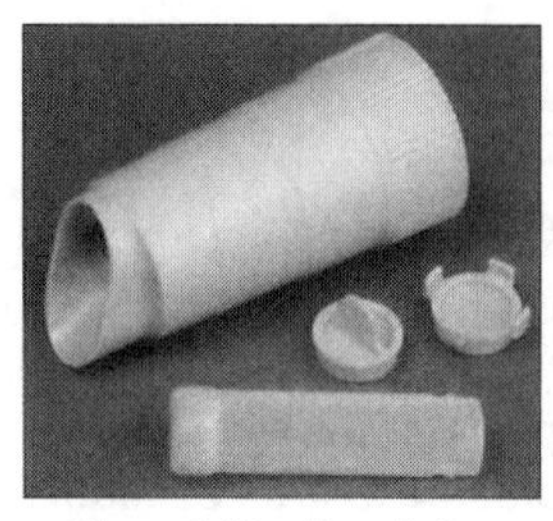

Figure 1. The Dust Collector contains a main holder, its caps on both ends & a filter insert.

without visible mold, musty smells or abnormal air sampling from two prior mold inspectors. She says to this day that ERMI saved her children's lives. Maybe that is too much credit, but the truth is that her family only now is well.

Make no mistake; presence of health effects shown by a protocol that evaluates health will always trump an ERMI. ERMI is a mold index, not a health index. If the ERMI is elevated, you have mold trouble. If the ERMI is low and there are people in the home with a typical mold illness, consider repeating the ERMI in different areas. If the ERMI is low and no one is ill, your sense of security increases. If you are not ill, an ERMI helps determine if your home is safe for visitors and loved ones who might have a different genetic susceptibility to mold exposure than you do. If the ERMI value suggests the home is in the upper 25% of the scale (i.e. ERMI above 5), then an investigation for water damage could be health-saving.

Group 1		
Fungal ID \ Unit	House A	House B
	Spore E./mg	Spore E./mg
Aspergillus flavus/oryzae	ND	<1
Aspergillus fumigatus	<1	1
Aspergillus niger	<1	ND
Aspergillus ochraceus	ND	11
Aspergillus penicillioides	81	4600
*Aspergillus restrictus**	ND	ND
Aspergillus sclerotiorum	ND	13
Aspergillus sydowii	ND	ND
Aspergillus unguis	ND	ND
Aspergillus versicolor	ND	56
Aureobasidium pullulans	610	450
Chaetomium globosum	1	5
Cladosporium sphaerospermum	<1	24
*Eurotium (Asp.) amstelodami**	16	3600
Paecilomyces variotii	ND	ND
Penicillium brevicompactum	19	34
Penicillium corylophilum	ND	ND
*Penicillium crustosum**	ND	ND
Penicillium purpurogenum	ND	1
*Penicillium spinulosum**	ND	ND
Penicillium variabile	ND	6
Scopulariopsis brevicaulis/fusca	3	43
Scopulariopsis chartarum	ND	3
Stachybotrys chartarum	ND	1
*Trichoderma viride**	ND	3
Wallemia sebi	8	2400
Sum of Logs (Group 1):	8.56	24.13

Group 2		
Fungal ID \ Unit	House A	House B
	Spore E./mg	Spore E./mg
Acremonium strictum	ND	1
Alternaria alternata	ND	ND
Aspergillus ustus	ND	1
Cladosporium cladosporioides 1	31	140
Cladosporium cladosporioides 2	1	4
Cladosporium herbarum	87	13
Epicoccum nigrum	37	570
*Mucor amphibiorum**	2	22
Penicillium chrysogenum	1	ND
Rhizopus stolonifer	<1	<1
Sum of Logs (Group 2):	5.3	7.96
ERMI (Group 1 - Group 2):	3.26	16.17

The Institute of Medicine's report (8) on dampness and health expressed the opinion that there was scientific evidence linking molds and damp environments

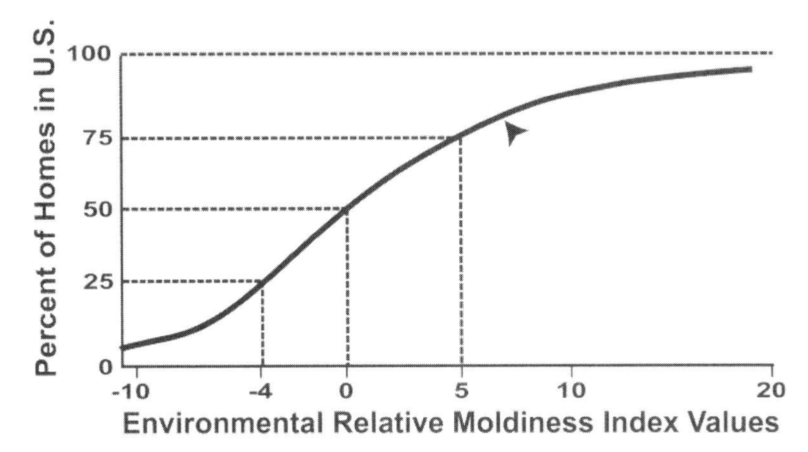

with respiratory symptoms. The cut-off for literature to be considered by the IOM was 2003; the pace of mold illness research has long ago outstripped the earlier IOM recommendations. ERMI isn't discussed in the IOM.

ERMI is useful in clinical studies

In a recent paper (7), ERMI values were correlated with laboratory assays, symptoms, neurotoxicological studies and measurement of brain metabolites, lactate (indicating capillary hypoperfusion) and ratios of glutamate to glutamine (indicating the balance of excitation versus inhibition of neurotransmission). There was a clear association between an elevated ERMI and elevated levels of lactate measured by magnetic resonance spectroscopy (MRS), in hippocampus (memory) and frontal lobes (acquisition), together with reduction of normal ratios of glutamate to glutamine. An elevated ERMI was closely linked to brain fog, memory deficits and abnormalities in executive cognitive function.

Do high levels of mold, therefore, translate in genetically susceptible patients into inflammation that reduces blood flow in particular parts of the brain such that the brain doesn't work? Yes! Even better, (i) following treatment abnormal brain metabolites are reduced and (ii) the benefit of treatment maintained with re-occupancy of the home provided the post-remediation ERMI is less than 2. Relapse occurs if the ERMI is higher.

In a study conducted on homes of asthmatic children by Case Western Reserve, remediating water-damaged, moldy homes significantly reduced the asth- matic child's need for medical intervention (9). In a prospective study of atopic infants (6), measuring the mold burden with MSQPCR was a better predictor for development of wheeze/rhinitis than the home inspection.

Air samples can be useful to pin-point the location of a hidden mold problem. In order to take air samples for MSQPCR analysis, the polycarbonate filter is useful with either 0.45 or 0.8 micron pore size. The flow rates range from 2 to 16 liter/minute. The holder for the filter can be a button sampler or cassette. In MSQPCR analysis, the filter cannot be overloaded, meaning air samples can be taken for prolonged periods. However, there is no ERMI scale for air samples; dust is preferred.

What do we do with the ERMI kit?

Help is always just a few clicks away at **www.mycometrics.com**. First, locate the most commonly used area in the living room. Using a tape measure and masking tape, mark a 3-foot by 6-foot sampling area on the floor. Record what these dimensions were and where you took them for later comparison. Next, do the same in the main bedroom.

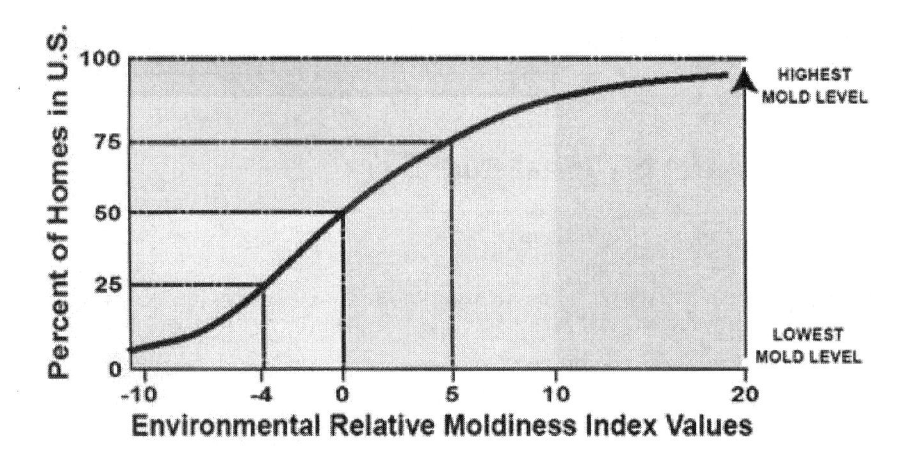

Take off the protective caps of the sampler. Insert the filter into the dust sampler and place the sampler inside the vacuum hose. Use a separate dust sampler for each area sampled. Vacuum for 5 minutes, pull out the sampler and cap it. Send in a sealed bag for an ERMI analysis at an EPA-licensed ERMI laboratory. You should ask for a repeat ERMI, taken in the same spots as before remediation to assure clearance.

No sampling can replace the skill of the experienced mold inspector in investigating mold problems. ERMI is a helpful tool. As further research refines the use and application of ERMI we will have greater ability to direct use of ERMI testing.

Summary:

Identification and accurate quantitation of indoor molds to the species level is now available, using DNA analysis, the MSQPCR. This automated analysis provides for rapid, reproducible results that can be reliably interpreted. For patients, prospective home-buyers, industrial hygienists and remediators alike, ERMI shows great promise for the future.

Conflicts of interest: Dr. Lin is an employee of Mycometrics. Dr. Shoemaker has none.

References

1. Shoemaker R, Rash JM, Simon E. In: Bioaerosols, Fungi, Bacteria, Mycotoxins and Human Health; ed. by E Johanning MD 2005. Toxicology and Health Effects pp 66-77.

2. Shoemaker R, House D. Neurotoxicology and Teratology 2005; 27; 29-46.

3. Shoemaker R, House D. Neurotoxicology and Teratology 2006; 28; 573-588.

4. Rao C, Brown C, et.al. Applied and Environmental Micro 2007; 73(5); 1630-1634.

5. Gorny R, Schemchel D, et.al. Applied and Environmental Micro 2002; 68(7); 3522-3531.

6. Vesper SJ et al., J. Occup. Environ. Med. 2006; 48, 852-858.

7. Shoemaker R, Maizel M. IACFS meetings 1/14/07, Fort Lauderdale, Florida.

8. Institute of Medicine, National Academies of Science. Damp Indoor Spaces and Health. The National Academies Press. 2004; p. 355.

9. C.M. Kercsmar et al., Environ. Health Perspect. 114,1574 (2006.

10. Vesper SJ et al., J Exposure Anal. Environ. Epidemiol. 2007; 17; 88-94.

Ritchie C. Shoemaker, MD is a Family Practice physician from Pocomoke, MD. He writes from his experience of diagnosing and treating the world's largest series of 5500 patients with acute and chronic illness caused by exposure to biotoxins made by molds, spirochetes, dinoflagellates, and blue-green algae. His practice experience is the foundation that enables him to challenge much of what we hear about chronic illnesses from agencies and academics that don't treat the illnesses. For more information, contact Ritchie C. Shoemaker, MD, 500 Market St, Suite 102 Pocomoke, Md 21851.

King-Teh Lin, PhD is Laboratory Director at Mycometrics, LLC, Monmouth Junction, NJ. He earned his PhD in Molecular Genetic and Microbiology from Robert Wood Johnson Medical School. Soon after his postdoctoral fellowship, he continued on at the same University as a faculty, up till recruited by PandK

Microbiology Services as a Director of RandD, where he was the first to commercialize the EPA-licensed for wood decay fungi. He has analyzed more than 10,000 PCR samples for both fungi and Legionella. His works have been published in many leading peer-reviewed Journals. In 2005, he established Mycometrics, LLC. For more information, contact King-Teh Lin at kingteh@mycometrics.com

Appendix 4

References for Benomyl

1. Desperation Medicine, by Ritchie C. Shoemaker, MD, published by Gateway Press, 460 pp. 2001
2. Famine on the Wind; Man's Battle Against Plant Disease, by G. L Carefoot, published by Rand McNally and Co., 231 pp. 1967
3. Pfiesteria: Crossing Dark Water, by Ritchie C. Shoemaker MD, published by Gateway Press, 350 pp. 1998
4. **www.sptimes.com/2006/09/24/news_pf/Business/**Benlate®_s_bitter_lega.shtmlPhone call with Frank Fuzzell, 9/15/06
5. Provided by Frank Fuzzell
6. Mills HA, Sasseville DN, Kremer RJ. Effects of Benlate on leatherleaf fern growth, root morphology and rhizosphere bacteria. J Plant Nutrition 1996; 19(6) 917-937.
7. Florida Department of Environmental Protection Site Investigation Section, Microbial Assessment and Work Plan for the Flatwoods and Casteen Roads Site, Leesburg, Florida. 7/24/00
8. Spradling A, Ganetsky B, Hieter P, Johnston M, Olson M, Orr-Weaver T, Rossant J, Sanchez A, Waterston R. New roles for model genetic organisms in understanding and treating human disease: report from 2006 Genetics Society of America meeting. Genetics 2006; 172(4):2025-2032.
9. Hyland KM, Kingsbury J, Koshland D, Hieter P. Ctf19p: a novel kinetochore protein in Saccharomyces cerevisiae and a potential link between the kinetochore and mitotic spindle. J Cell Biol 1999; 145(1): 15-28.
10. Rathinasamy K, Panda D. Suppression of microtubule dynamics by benomyl decreases tension across kinetochore pairs and induces apoptosis in cancer cells. FEBS J 2006 (Epub ahead of print)
11. Fink G, Steinberg G. Dynein-dependent motility of microtubules and nucleation sites supports polarization of the tubulin array in the fungus Ustilago maydis. Mol Biol Cell 2006;17(7): 3242-3253.

12. Kim JM, Lu L, Shao R, Liu B. Isolation of mutations that bypass the requirement of the septation initiation network for septum formation and conidiation in Aspergillus nidulans. Genetics 2006; 173(2): 685-696.

13. Tanaka K, Mukae N, Dewar H, van Breugel M, James EK, Prescott AR, Antony C, Tanaka TU. Molecular mechanisms of kinetochore capture by spindle microtubules. Nature 2005; 434(7036): 987-994.

14. He X, Rines DR, Espelin CW, Sorger PK. Molecular analysis of kinetochore-microtubule attachment in budding yeast. Cell 2001;106(2): 195-206.

15. Gupta K, Bishop J, Peck A, Brown J, Wilson L, Panda D. Antimitotic antifungal compound benomyl inhibits brain microtubule polymerization and dynamics and cancer cell proliferation at mitosis, by binding to a novel site in tubulin. Biochemistry 2004; 43(21): 6645-6655.

16. Ovechkina Y, Maddox P, Oakley Ce, Xiang X, Osmani SA, Salmon ED, Oakley BR. Spindle formation in Aspergillus is coupled to tubulin movement into the nucleus. Mol Biol Cell 2003; 14(5): 2192-2200.

17. McCarroll NE, Protzel A, Ioannou Y, Frank Stack HF, Jackson MA, Waters MD, Dearfield KL. A survey of EPA/OPP and open literature on selected pesticide chemicals. III. Mutagenicity and carcinogenicity of benomyl and carbendazim. Mutat Res 2002; 512(1): 1-35.

18. Bentley KS, Kirkland D, Murphy M, Marshall R. Evaluation of thresholds for benomyl- and carbendazim-induced aneuploidy in cultured human lymphocytes using fluorescence in situ hybridization. Mutat Res 2000; 464(1): 41-51.

19. Hess Ra, Nakai M. Histopathology of the male reproductive system induced by the fungicide benomyl. Histol Histopathol 2000; 15(1): 207-224.

20. Suwalsky M, Benites M, Norris B, Sotomayor P. Toxic effects of the fungicide benomyl on cell membranes. Comp Biochem Physiol C Toxicol Pharmacol 2000; 125(1): 111-119.

21. Hatvani L, Manczinger L, Kredics L, Szekeres A, Antal Z, Vagvolgyi C. Production of Trichoderma strains with pesticide-polyresistance by mutagenesis and protoplast fusion. Antonie Van Leeuwenhoek 2006; 89(3-4): 387-393.

22. Ogawa K, Yoshida N, Gesnara W, Omumasaba CA, Chamuswarng C. Hybridization and breeding of the benomyl resistant mutant, Trichoderma harziantum antagonized to phytopathogenic fungi by protoplast fusion. Biosci Biotechnol Biochem 2000; 64(4): 833-836.

23. Ma Z, Yoshimura MA, Holtz BA, Michailides TJ. Characterization and PCR-based detection of benzimidazole-resistant isolates of Monilinia laxa in California. Pest Manag Sci 2005; 61(5): 449-457.

24. McMahan G, Yeh W, Marshall MN, Olsen M, Sananikone S, Wu JY, Block DE, VanderGheynst JS. Characterizing the production of a wild-type and benomyl-resistant Fusarium lateritium for biocontrol of Eutypa lata on grapevine. J Ind Microbiol Biotechnol 2001; 26(3): 151-155.

25. Broco N, Tenreiro S, Viegas CA, Sa-Correia I. FLR1 gene (ORF YBR008c) is required for benomyl and methotrexate resistance in Saccharomyces cerevisiae and its benomyl-induced expression is dependent on pdr3 transcriptional regulator. Yeast 1999; 15(15): 1595-1608.

26. Mukherjee PK, Sherkhane PD, Murthy NB. Induction of stable benomyl-tolerant phenotypic mutants of Trichoderma pseudokoningii MTCC 3011, and their evaluation for antagonistic and biocontrol potential. Indian J Exp Biol 1999; 37(7): 710-712.

27. Jung MK, May GS, Oakley BR. Mitosis in wild-type and beta-tubulin mutant strains of Aspergillus nidulans. Fungal Genet Biol 1998; 24(1-2): 146-160.

28. Bello VA, Paccola-Meirelles LD. Localization of auxotrophic and benomyl resistance markers through the parasexual cycle in the beauveria bassiana (Bals.) vuill entomopathogen. J Invertebr Pathol 1998; 72(2): 119-125.

29. Park SY, Jung OJ, Chung YR, Lee CW. Isolation and characterization of a benomyl-resistant form of beta-tubulin-encoding gene from the phytopathogenic fungus Botryotinia fuckeliana. Mol Cells 1997; 7(1): 104-109.

30. Wu TS, Skory CD, Horng JS, Linz JE. Cloning and functional analysis of a beta-tubulin gene from a benomyl resistant mutant of Aspergillus parasiticus. Gene 1996; 182(1-2): 7-12.

31. Bogo MR, Vainstein MH, Aragao FJ, Rech E, Schrank A. High frequency gene conversion among benomyl resistant transformants in the entomopathogenic fungus Metarhizium anisopliae. FEMS Microbiol Lett 1996; 142(1): 123-127.

32. Yan K, Dickman MB. Isolation of a beta-tubulin gene from Fusarium moniliforme that confers cold-sensitive benomyl resistance. Appl Environ Microbiol 1996; 62(8): 3053-3056.

33. Tikhomirova VL, Inge-Vechtomov SG. Sensitivity of sup35 and sup45 suppressor mutants in Saccharomyces cerevisiae to the anti-microtubule drug benomyl. Curr Genet 1996; 30(1): 44-49.

34. Huang KX, Iwakami N, Fujii I, Ebizuka Y, Sankawa U. Transformations of Penicillium islandicum and Penicillium frequentans that produce anthraquinone-related compounds. Curr Genet 1995; 28(6): 580-584.

35. Nara F, Watanabe I, Serizawa N. Development of a transformation system for the filamentous, ML-236B (compactin) producing fungus Penicillium citrinum. Curr Genet 1993; 23(1): 28-32.

36. Picknett TM, Saunders G. Transformation of Penicillium chrysogenum with selection for increased resistance to benomyl. FEMS Microbiol Lett 1989; 51(1):165-168.

37. Goldway M, Teff D, Schmidt R, Oppenheim AB, Koltin Y. Multidrug resistance in Candida albicans: disruption of the BENr gene. Antimicrob Agents Chemother 1995; 39(2): 422-426.

38. Horio T, Oakley BR. Human gamma-tubulin functions in fission yeast. J Cell Biol 1994;126(6): 1465-1473.

39. Annas, GJ. Intelligent judging- Evolution in the classroom and the courtroom. NEJM 2006; 354: 2277-2281.

40. Bottaro A, Inlay MA, Matzke NJ. Immunology in the spotlight at the Dover 'Intelligent Design' trial. Nature Immunology 2006: 7(5): 433-435.

41. Blwise RJ. In defense of Darwin. Duke Magazine March-April 2006, pg 29-35.

42. Yamada T, Furukawa K, Hara S, Mizoguchi H. Isolation of copper-tolerant mutants of sake yeast with defective peptide uptake. J Biosci Bioeng 2005; 100(4): 460-465.

43. Berg J, Tom-Petersen A, Nybroe O. Copper amendment of agricultural soil selects for bacterial antibiotic in the field. Lett Appl Microbiol 2005; 40(2): 146-151.

44. van Bakel H, Strengman E, Wijmenga C, Holstege FC. Gene expression profiling and phenotype analyses of S. cerevisiae in response to changing copper reveals six genes with new roles in copper and iron metabolism. Physiol Genomics 2005; 22(3): 356-367.

45. Griffiths BS, Kuan HL, Ritz K, Glover LA, McCraig AE, Fenwick C. The relationship between microbial community structure and functional stability, tested experimentally in an upland pasture soil. Microb Ecol 2004; 47(1): 104-113.

46. Garcia-Villada L, Rico M, Altamirano MM, Sanchez-Martin L, Lopez-Rodas V, Costas E. Occurrence of copper resistant mutants in the toxic cyanobacteria Microcystis aeruginosa: characterization and future implications in the use of copper sulphate as algaecide. Water Res 2004; 38(8): 2207-2213.

47. Konstantinidis KT, Isaacs N, Fett J, Simpson S, Long DT, Marsh TL. Microbial diversity and resistance to copper in metal-contaminated lake sediment. Microb Ecol 2003; 45(2): 191-202.

48. Pfaller MA, et al. Journal of Clinical Microbiology, 2006; 44(10) 3578-3582. Candida rugosa, an emerging fungal pathogen with resistance to azoles: Geographic and temporal trends from the ARTEMIS DISK antifungal surveillance program

49. Nowak C, Kuck U. Development of an homologous transformation system for Acremonium chrysogenum based on the beta-tubulin gene. Curr Genet 1994;25(1): 34-40.

50. Seip ER, Woloshuk CP, Payne GA, Curtis SE. Isolation and sequence analysis of a beta-tubulin gene from Aspergillus flavus and its use as a selectable marker. Appl Environ Microbiol 1990;56(12): 3686-3692.

51. Diaz Borras A, Vila Aguilar R, Hernandez Gimenez E. Synergistic effect of fungicides on resistant strains of Penicillium italicum and Penicillium digitatum. Int J Food Microbiol 1988; 7(1): 79-85.

52. Praitis V, Katz WS, Solomon F. A codon change in beta-tubulin which drastically affects microtubule structure in Drosophila melanogaster fails to produce a significant phenotype in Saccharomyces cervisiae. Mol Cell Biol 1991; 11(9): 4726-4731.

53. Cameron LE, Hutsul JA, Thorlacius L, LeJohn HB. Cloning and analysis of beta-tubulin gene from a protoctist. J Biol Chem 1990; 265(25): 15245-15252.

54. Delves CJ, Ridley RG, Goman M, Holloway Sp, Hyde JE, Scaife JG. Cloning of a beta-tubulin gene from Plasmodium falciparum. Mol Microbiol 1989; 3(11): 1511-1519.

55. Sarcina M, Mullineaux CW. Effects of tubulin assembly inhibitors on cell division in prokaryotes in vivo. FEMS Microbiol Lett 2000; 191(1): 25-29.

56. Zhang X, Smith CD. Microtubule effects of welwistatin, a cyanobacterial indolinone that circumvents multiple drug resistance. Mol Pharmacol 1996; 49(2): 288-294.

57. Khan SA, Wickstrom ML, Haschek WM, Schaeffer DJ, Ghosh S, Beasley VR. Microcystin-LR and Kinetics of cytoskeletal reorganization in hepatocytes, kidney cells, and fibroblasts. Nat Toxins 1996; 4(5): 206-214.

58. Ben-Yaacov R, Knoller S, Caldwell GA, Becker JM, Koltin Y. Candida albicans gene encoding resistance to benomyl and methotrexate is a multidrug resistance gene. Antimicrob Agents Chemother 1994; 38(4): 648-652.

59. Sanglard D, Kuchler K, Ischer F, Pagani JL, Monod M, Bille J. Mechanisms of resistance to azole antifungal agents in Candida albicans isolates from AIDS patients involve specific multidrug transporters. Antimicrob Agents Chemother 1995; 39(11): 2378-2386.

60. Summerbell RC. The benomyl test as a fundamental diagnostic method for medical mycology. J Clin Microbiol 1993; 31(3): 572-577.

61. Supreme Court of Florida, John Castillo v. E.I. DuPont de Nemours and Co, Inc, No SC00-490. July 10, 2003. Re-hearing denied Sept. 4, 2003

62. Rull RP, Ritz B, Shaw GM. Neural tube defects and maternal residential proximity to agricultural pesticide applications. Am J Epidemiol 2006; 163(8): 743-753.

63. Hewitt MJ, Mutch P, Pratten Mk. Potential teratogenic effects of benomyl in rats embryos cultured in vitro. Reprod Toxicol 2005; 20(2): 271-280.

64. Amdam GV, Hartfelder K, Norberg K, Hagen A, Omholt SW. Altered physiology in worker honey bees (Hymenoptera: Apidae) infested with the mite Varroa destructor (Acari: Varroidae): a factor in colony loss during overwintering? J Econ Entomol 2004; 97(3): 741-747.

65. Sato ME, Raga A, Ceravolo LC, De Souza Filho MF, Rossi AC, De Moraes GJ. Effect of insecticides and fungicides on the interaction between members of the mite families Phytoseiidae and Stigmaeidae on citrus. Exp Appl Acarol 2001; 25(10-11): 809-818.

66. Childers CC, Villanueva R, Aguilar H, Chewning R, Michaud JP. Comparative residual toxicities of pesticides to the predator Agistemus industani (Acari: Stigmaeidae) on citrus in Florida. Exp Appl Acarol 2001; 25(6): 461-474.

67. Alston DG, Thomson SV. Effects of fungicide residues on the survival, fecundity, and predation of the mites Tetranychus urticae (Acari: Tetranychidae) and Galendromus occidentalis (Acari: Phytoseiidae). J Econ Entomol 2004; 97(3): 950-956.

68. Alston DG, Thomson SV. Effects of fungicide residues on the survival, fecundity, and predation of the mites Tetranychus urticae (Acari: Tetranychidae) and Galendromus occidentalis (Acari: Phytoseiidae). J Econ Entomol 2004; 97(3): 950-956.

69. Miozes-Koch R, Slabezki Y, Efrat H, Kalev H, Kamer Y, Yakobson, Dag A. First detection in Israel of fluvalinate resistance in the varroa mite using bioassay and biochemical methods. Exp Appl Acarol 2000; 24(1):35-43.

70. Martin SJ, Elzen PJ, Rubink WR. Effect of acaricide resistance on reproductive ability of the honey bee mite Varroa destructor. Exp Appl Acarol 2002; 27(3): 195-207.

71. Adamczyk S, Lazaro R, Perez-Arquillue C, Conchello P, Herrera A. Evaluation of residues of essential oil components in honey after different anti-varroa treatments. J Agric Food Chem 2005; 53(26): 10085-10095.

72. Peng CY, Zhou X, Kaya HK. Virulence and site of infection of the fungus, Hirsutella thompsonii, to the honey bee ectoparasitic mite, Varroa destructor. Jinvertebr Pathol 2002; 81(3): 185-195.

73. Kanga LH, James RR, Boucias DG. Hirsutella thompsonii and metarhizium anisopliae as potential microbial control agents of Varroa destructor, a honey bee parasite. J Invertebr Pathol 2002; 81(3): 175-184.

74. Peng CY, Zhou X, Kaya HK. Virulence and site of infection of the fungus, Hirsutella thompsonii, to the honey bee ectoparasitic mite, Varro destructor. J Invertebr Pathol 2002; 81(3): 185-195.

75. Omoto C, McCoy CW. Toxicity of purified fungal toxin hirsutellin A to the citrus rust mite phyllocoptruta oleivora (Ash). J Invertebr Pathol 1998; 72(3): 319-322.

76. Maimala S, Tartar A, Boucias D, Chandrapatya A. Detection of the toxin Hirsutellin A from Hirsutella thompsonii. J Invertebr Pathol 2002; 80(2): 112-126.

77. Liu JC, Boucias DG, Pendland JC, Liu WZ, Maruniak J. The mode of action of hirsutellin A on eukaryotic cells. J Invertebr Pathol 1996; 67(3): 224-228.

78. Aardema MJ, MacGregor JT. Toxicology and genetic toxicology in the new era of "toxicogenomics": impact of "-omics" technologies. Mutat Res 2002; 499(1): 13-25.

Appendix 5

Guide to Pronunciation

1 BEN-o-mill (benomyl)
2 die-thie-o-CARB-a-mate (dithiocarbamate)
3 fis-STEER-ee-uh pis-kuh-SEED-uh (*Pfiesteria piscicida)*
4 fue-SAR-ee-um SOL-an-nee *(Fusarium solani)*
5 SUE-da-MOAN-us FLOR-uh-senz (Pseudomonas fluorescens)
6 sul-fun-nil-URE-ee-us (sulfonylureas)
7 mike-crow-opp-THAL-mee-uh (microophthalmia)
8 an-in-SEF-uh-lee (anencephaly)
9 aa-ZOLE (azole)
10 as-ko-kite-a (Ascochyta)
11 AS-per-JILL-us (Aspergillus)
12 boh-TRY-tis (Botrytis)
13 sir-RATE-oh-SIS-tee-uh (Ceratocystia)
14 sir-co-SPORE-uh (Cercospora)
15 sir-co-spore-RELL-uh (Cercosporella)
16 sir-co-spore-RID-ee-uhm (Cercosporidium)
17 clah-doh-spore-RID-ee-uhm (Cladosporium)
18 COLE-luh-TOE-tree-cum (Collatotrichum)
19 core-RINE-uh-SPORE-uh (Corynespora)
20 die-plo-CAR-pon (Diplocarpon)
21 fue-SARE-ee-um (Fusarium)
22 fue-si-CLAD-ee-uhm (Fusicladium)
23 moan-i-LINN-ee-uh (Monilinia)
24 MY-ko-SPHERE-ell-uh (Mycosphaerella)
25 NEU-ro-SPORE-uh (Neurospora)
26 pen-uh-SILL-ee-um (Penicillium)
27 FEE-uh-LOFF-fur-uh (Phialophora)
28 RIE-zoc-TONE-ee-uh (Rhizoctonia)
29 scler-oh-TIN-ee-uh (Sclerotinia)
30 sept-TORE-ee-uh (Septoria)

31 YOU-still-LAH-go (Ustilago)
32 vert-tuh-SILL-ee-um (Verticillium)
33 VEN-ture-REE-uh (Venturia)
34 car-BEN-duh-zeem (carbendazim)
35 meth-uh-SILL-in ree-ZIS-tent co-AGG-you-laze NEG-uh-tiv STAFF-uh-luh-cokey (methicillin-resistant coagulase-negative staphylococci)
36 meth-a-SILL-in re-ZIS-tent staff OR-ee-us (methicillin-resistant Staph aureus)
37 IN-doh-lee-uh-SET-ick (indoleacetic)
38 fue-SARE-ee-um OX-ee-SPOOR-um shlect (*Fusarium oxysporum schlecht)*
39 SUE-da-MOAN-us FLOR-uh-senz (Pseudomonas fluorescens)
40 SUE-da-MOAN-ads (pseudomonads)
41 my-TOSE-is (mitosis)
42 my-crow-TUBE-yules (microtubules)
43 kin-NET-oh-chore (kinetochore)
44 SEN-tro-meer (centromere)
45 ak-tin-NO-muh-SET-eez (actinomycetes)
46 TRI-koe-DERM-uh (*Trichoderma)*
47 all-ter-NARE-ee-uh *(Alternaria)*_
48 gee-OH-trick-uhm (*Geotrichum)*
49 chuh-TOME-mee-um *(Chaetomium)*
50 ACK-ree-MOAN-ee-um (*Acremonium)*
51 THIE-oh-BEN-da-zole (thiobendazole)
52 BENZ-a-MID-a-zole (benzimidazole)
53 KLOR-oh-THAL-uh-nil (chlorothalonil)
54 SILL-in-drow-spur-MOP-sis (*Cylindrospermopsis)*
55 mike-crow-SIS-tus (*Microcystis)*
56 sy-AN-oh-faj (cyanophage)
57 sy-AN-oh-back-TEER-ee-uh (cyanobacteria)
58 QWAT-tur-na-ree uh-MOAN-ee-um (quaternary-ammonium)
59 AS-per-JILL-us FYUM-uh-GOT-us (*Aspergillus fumigatus)*
60 EE-nil-CON-uh-zole (enilconazole)
61 my-CON-uh-zole (miconazole)

62 key-toe-CON-uh-zole ketoconazole)
63 flu-CON-uh-zole (fluconazole)
64 OR-ee-oh-bae sid-ee-um PULL-you-lans *(Aureobasidium pullulans)*
65 AS-per-JILL-us NID-yu-lans
66 AS-per-JILL-us par-a-SIT-a-cus *(Aspergillus parasiticus)*
67 pen-a-SILL-ee-um si-TRIN-um (*Penicillium citrinum)*
68 pen-a-SILL-ee-um cri-so-GEN-um (*Penicillium chrysogenum)*
69 ACK-ree-MOAN-ee-um cri-so-GEN-um (*Acremonium chrysogenum)*
70 AS-per-JILL-us FLAY-vus (*Aspergillus flavus)*
71 pen-a-SILL-ee-um eye-TAL-uh-cum (*Penicillium italicum*)
72 pee dij-a-TATE-um (*P. digitatum)*
73 bo-VARE-ee-uh BASS-ee-AN-uh (*Beauveria bassiana)*
74 en-tuh-mo-PATH-uh-gin (entomopathogen)
75 DRO-so-FIL-uh mill-LAN-o-gas-tur (*Drosophila melanogaster)*
76 pro-TOC-tists (protoctists)
77 plaz-MODE-ee-um fal-SIP-ar-um *(Plasmodium falciparum)*
78 pro-CARE-ee-otes (prokaryotes)
79 SAC-crow-MY-set-eez *(Sacchromycetes)*
80 meth-oh-TREX-ate (methotrexate)
81 CAN-deed-uh AL-buh-kanz (*Candida albicans)*
82 CAN-deed-uh ruh-GOSE-uh (*Candida rugosa)*
83 CAN-deed-uh glob-RAHT-uh (*Candida glabrata)*
84 CAN-deed-uh CREW-sigh (*Candida krusei)*
85 MIKE-row-THAL-mee-uh (microphthalmia)
86 METH-oh-mill (methomyl)
87 AY-pis MELL-a-FER-uh(*Apis mellifera*)
88 var-O-uh dee-STRUCK-ter *(Varroa destructor)*fl
89 flu-VAL-in-ate (fluvalinate)
90 PYE-reth-royds (pyrethroids)
91 HERE-suh-TELL-uh TOM-sun-eye *(Hirsutella thompsonii)*
92 HERE-suh-TELL-in A (hirsutellin A)

Appendix 6

Haefner/OWCP Decision

File Number: 092085516
reconvacate-D-

U.S. DEPARTMENT OF LABOR

EMPLOYMENT STANDARDS ADMINISTRATION
OFFICE OF WORKERS' COMP PROGRAMS
PO BOX 8300 DISTRICT 9 CLE
LONDON, KY 40742-8300
Phone: (216) 357-5100

March 30, 2009

Date of Injury: 06/20/2007
Employee: ROBERT D. HAEFNER

ROBERT DANIEL HAEFNER
50464 ELMWOOD CT
PLYMOUTH, MI 48170

Dear Mr. HAEFNER:

This concerns your compensation case and your request for reconsideration received 02/17/2009.

We have evaluated the evidence submitted in support of your request for review. Your case has been reviewed on its merits under Title 5, United States Code, Section 8128, in relation to your application including supporting evidence. It is determined that sufficient evidence of file now exists to accept your claim. The reasons for this decision are discussed in the attached notice.

Therefore, the decision dated 02/12/2008 is vacated and your case is accepted for chronic inflammatory illness due to mold exposure in the Detroit Airport Tower in November 2005, December 2006, January 2007, and February 2008.

Sincerely,

Melissa Myers
Senior Claims Examiner

DEPARTMENT OF TRANSPORTATION
FEDERAL AVIATION ADMINISTRATION
WORKERS' COMP DIVISION-AHL-100
800 INDEPENDENCE AVENUE,SW,RM 521
WASHINGTON, DC 20591

File Number: 092085516
reconvacate-D-

NOTICE OF DECISION

IN THE CASE OF ROBERT HAEFNER, 0920855156

March 30, 2009

ISSUE:

The issued for determination is whether the evidence of file is sufficient to vacate the prior decision of 02/12/08.

BACKGROUND:

You are employed as an Air Traffic Controller for the Department of Transportation, Federal Aviation Administration, at the Detroit Metropolitan Airport. On 09/10/07, you filed timely Notice of Occupational Disease, Form CA-2, for chronic inflammatory illness with associated symptoms as a result of exposure to black mold in the workplace.

Upon review of the initial evidence submitted, specifically only the Form CA-2, insufficient factual and medical evidence was received to support that you were exposed to mold, as alleged, and that you had a definitive diagnosis related to that exposure. Subsequently, on 09/25/07, a development letter was issued to you outlining the required factual and medical evidence needed to substantiate your claim. Specifically, you were asked to submit a factual statement outlining where and how often you were exposed to black mold in the Detroit Tower, any exposure to mold outside your federal employment, a description of your symptoms and what makes them better or worse, and a discussion of any history of similar illness. We further outlined the necessary medical documentation to support your claim, specifically a comprehensive medical report from your treating physician that discussed your symptoms, results of diagnostic studies, diagnosis, treatment provided and its effects, and your physician's opinion, with medical reasoning, as to the cause of your condition.

No response was received. Thus, by formal decision dated 10/29/07, your claim was denied. It is noted that you did, in fact, submit additional evidence for review, but it was not viewable at the time the 10/29/07 decision was issued.

You disagreed with the 10/29/07 decision and requested reconsideration by completion of the appeals request form on 11/21/07. As a large volume of factual and medical evidence had been received, a merit review of your case was performed. Additional development was undertaken with you and your employing agency concerning the facts of your exposure. The entire record was extensively reviewed and by reconsideration decision dated 02/12/08, the 10/29/07 decision was affirmed but modified in part. Specifically, it was accepted from a factual standpoint and a determination was made that you were exposed to some level of mold in November 2005 and to elevated dust and airborne particulates, Fusarium mold, and several types of fungi in December 2006 and January 2007(it was noted that this claim, 092085516, only considered mold exposure beyond May of 2005 as two other claims were of record pertaining to exposure prior to May 2005 under case files 092056731 and 092060424).

However, from a medical standpoint, the claim remained denied for fact of injury as your treating physician had not provided a thorough discussion of your findings upon examination and diagnostic studies objectively establishing the ill defined diagnosis of chronic inflammatory illness. It was explained

File Number: 092085516
reconvacate-D-

that although Dr. Shoemaker indicated that you met all criteria for the diagnosis, he neglected to outline what that criteria was and how your diagnostic testing, laboratory results, and clinical findings upon examination supported the diagnosis of chronic inflammatory disease as opposed to a more definitive diagnosis. It was further pointed out that Dr. Shoemaker did not indicate knowledge of your positive antinuclear antibody and how this correlates with your medical condition. Thus, your claim remained denied as the medical aspect of fact of injury had not been met. You were also advised that the medical evidence of file was devoid of a well rationalized medical explanation as to whether and how your diagnosis was attributed to work related mold exposure of November 2005, December 2006, and January 2007.

You disagreed with the 02/12/08 decision and requested reconsideration by completion of the appeals request form on 02/03/09, postmarked, 02/11/09. A large volume (more than 100 pages) of documentation was received with your request including: letters to this office from Attorney Jennifer J. Kukac dated 02/11/09 and 02/12/09; a 51 page typed letter to an attorney, Dodd Fisher, from Ritchie C.. Shoemaker, MD dated 05/14/08 (marked Exhibit 2); a two page typed letter to Registered Nurse Faith Widders from Dr. Shoemaker dated 06/06/07 (marked Exhibit 1); an 11 page typed letter to you from Dr. Shoemaker dated 08/31/08 (marked attachment D); a six page typed letter to Dr. Shoemaker from Ernest P. Chiodo, MD dated 11/19/08 (marked attachment E); a letter from Dr. Chiodo to Dr. Shoemaker dated 12/08/08; 2 pages of Biological Bulk Sample Results from Wonder Makers Environmental, date sampled unknown; a document that is five pages typed entitled causality (marked Exhibit 5); a 14 page typed document entitled Cognitive deficits and inflammation (marked Exhibit 3); Fungul Culuture Report from Wonder Makers Inc date sampled 02/02/08; Fungul Spore Information; a letter to Vincent Sugent from Michael A. Pinto, CEO of Wonder Makers Inc dated 02/20/08 (marked attachment C); a17 page typed document entitled Table of Contents for WDB References II (marked Exhibit 6); Diagnostic Results for C3a and C4a dated 05/18/07 (marked Attachment 4); Memorandums to all Employees from Joseph Figliuolo and Shirlee Coppo dated 01/06/09, 01/07/09; 01/08/09, 01/09/09, 01/13/09, 01/20/09, 01/21/09; and a 14 page typed medical report from Mary Ann Gudice, MD, dated 02/09/09; a Memorandum to all tower employees from John Guth dated 01/06/09; a Memorandum to Everyone from Gary Aniner (sic) dated 01/20/09; and an excerpt from The Weekly Newspaper of the Motown District dated 01/30/09

DISCUSSION OF EVIDENCE:

A complete review of all the above additional evidence was reviewed, as was your entire case file. Your case has already been accepted from a factual standpoint. It is not disputed that you were exposed to mold in the Detroit Tower in November 2005, December 2006, and January 2007. That being said, the new factual information received, specifically the 02/20/08 letter to Vincent Sugent from Michael A. Pinto, substantiates that in February 2008 additional swab and tape samples were taken from various faces of the sandwiched gypsum board, wall cavity insulation, insulated beams, and interior greenboard finish material of the Detroit Tower and were shown to have high levels of fungal spores from 18% to 50% of the sample constituents. This included Stachybotrys, Ulocaldium, Aspergillus, Pithomyces, Alternaria, Cladosporium, Penicillium, and Acremonium. Mr. Pinto ended his letter stating that the samples support active colonies of various fungus and their "presence in the building supports the contention that the health effects suffered by many of the controllers are likely related to mold exposures." He further noted that the samples further demonstrate that the fungul contamination had not yet been properly identified and quantified and that previous remediation had not been effective in resolving the fungal contamination issues in the tower over the past three years. As you were still working in 2008, and exposed to these fungal spores, your exposure has been expanded to include February of 2008 in addition to November 2005, December 2006, and January 2007.

Turning to the medical evidence, in a letter dated 06/06/07 to Registered Nurse Faith Widders from Dr. Shoemaker, he stated that you were seen in his office on 04/24/07 and that he believed you to have the

File Number: 092085516
reconvacate-D-

diagnosis of chronic multi system illness, ICD-9 code 989.7, caused by exposure to toxigenic organisms, including but not limited to, fungi present in the Detroit Tower. He opined that your prognosis without treatment is poor, but after follow up labs are performed, he can reasonably predict it will be fair. He expanded stating that your diagnosis will not be upgraded from fair until proper remediation takes place in the workplace, and reports that ongoing exposure to a water damaged building can progress to total disability. Dr. Shoemaker reported that you had been orally administered cholestyramine four times a day and that follow up lab work to that treatment still needed to be discussed with you. Dr. Shoemaker advised that he was attaching the lab studies and an MRI report and noted that you have evidence of abnormalities in lactate and ratio of glutamate to glutamine seen on magnetic resonance spectroscopy. He summarizes that the enclosed objective test data results in your meeting the criteria of a published, peer reviewed case definition for a chronic inflammatory illness caused by exposure to a water damaged building. He further supported his opinion with his ten years experience in the field with over 6400 patients with illness caused by exposure to biologically produced neurotoxins and over 4300 patients with illness acquired following exposure to water damaged buildings. He then states that his research group has published three papers on this topic

The 51 page letter to Mr. Fisher from Dr. Shoemaker, dated 05/14/08 discussed you as a cohort member amongst 14 others exposed in the Detroit Metro Tower. This letter is extremely comprehensive and notes that each of the cohorts have been treated successfully with cholestyramine, resulting in a reduction of symptoms, improvement of laboratory testing and correction of deficits as per Visual Contrast Sensitivity. Dr. Shoemaker notes that success in this manner can only arise from treatment of a biotoxin-associated illness following exposure to the interior of a water damaged building. He definitively notes that no other illness improves with such treatment.

Dr. Shoemaker continues by expressing his dissatisfaction with the lack of action taken by the FAA in acknowledging or resolving the issues concerned with mold exposure in the Detroit Tower. He outlines his presentation of baseline lab findings for all 15 cohorts and finds them consistent with similar patients who have illnesses caused by exposure to water damaged buildings. He advises that the health parameters of the cohorts "are singularly distinct from any other group of patients other than those exposed to WDB or those who have a systemic biotoxin-associated illness" and reports that the lab results are objective. He states that each member of the cohort (including you) meets the peer reviewed published case definition of mold illness and each have permanent effects of the illness. He further reports that the time line in this case mirrors the time line "in thousands of others similarly sickened by exposure to the indoor air of a WDB."

Dr. Shoemaker provides an in depth connection of you and the other cohorts to various studies and data pertaining to mold exposure and provides information regarding his extensive involvement in this area of medicine. He advises that there is a genetic susceptibility in almost a quarter of the population He then states that it is his opinion that the cohort (you) suffered a chronic disease as a result of your exposure to the Detroit Tower as a result of "a gradually developing injury caused by exposure to toxigenic organisms, including, but not limited to, toxigenic fungi."

He then advises that certain HLA DR haplotypes are found in "significantly higher percentages in cases of biotoxin illnesses when compared to control populations and those with exposure but no illness. Dr. Shoemaker indicates that these "dreaded" haplotypes are found in at most 4% of the population." He states that 33% of the members of your cohort have this haploytype and 93% of the cohort has the mold susceptible haplotype. He notes that the cohorts have this gene by chance but when exposed to the interior environment of the Detroit Tower, became quite ill.

He further outlines that when he saw each individual cohort for evaluation, he reviewed your history, performed a physical exam, and conducted visual contract sensitivity (VCS) testing, which is a most useful tool in diagnosing biotioxic illness. He advised that results from those evaluations demonstrated that you each met the preliminary criteria for mold illness. He then indicates that his office performed

File Number: 092085516
reconvacate-D-

blood draws and a battery of laboratory tests for each of the cohorts whose results "unequivocally supported" his preliminary opinion regarding the etiology of your illness. He reports that the cohort had a 66% positive VCS compared to less than 1% found in control populations; 93% HLA DR positivity compared to 24% of the normal population; MSH deficiency in 93% of the cohorts versus 10% of controls; Too high MMP9 in 33% of cohorts compared to 5% controls; dysregulation of ACTH and simultaneous cortisol in 53% of the cohorts compared to 5% controls; abnormal ADH/osmolality in 93% of the cohorts versus 5% controls; C4a above control levels in 93% of the cohorts but in less than 5% controls; Anticardiolipin antibodies of the IgG class were 30% positive in the cohort and less than 1% in controls; lipid profiles and nasal cultures for biofilm forming organisms were more common in cohorts than controls; and VIP was undetectably low in 93% of cohort versus 5% of controls.

Dr. Shoemaker then states that the cohorts' symptoms and diagnostic testing unequivocally support the diagnosis of a biotoxin illness caused by exposure to the workplace. He stated that causality is confirmed by application of the Causality model to the symptoms, VCS scores, laboratory findings, and diagnostic studies including EKG, pulmonary function test, and EKGs. He further supports his opinion with his over 30 year experience as a treating physician, his treating over 4,700 patients with chronic illness following exposure to a water damaged building, and his treating over 6,900 patients with chronic illness acquired following exposure to biologically produced neurotoxins, including those from water damaged buildings. He ends his summary stating:

> Given all the above it is my opinion to a reasonable degree of medical certainty that the multiple symptoms, profound cognitive impairments and multiple physiologic and neurotoxicological abnormalities present in the cohort were solely caused by exposure to the complex mixture of biologically active compounds, more likely than not present in the water-damaged indoor environment of the workplace. There is no evidence of exposure coming from any other sources; this statement is made with full recognition of the two positive Lyme tests. There is no evidence that any other medical condition could cause the multitude of laboratory abnormalities in the cohort that respond rapidly to use of targeted therapies beginning with CSM.

He further reports that you will be at risk for recurrence of illness without the use of medications and with exposure to molds and other toxigenic organisms. He reports that this includes exposure to the Detroit Tower in its current state.

In the letter from Dr. Shoemaker to you, dated 08/31/08, he begins by advising that he reviewed a sworn affidavit of a phone call between you and FAA surgeon, Dr. Jackson (note: this document is not on file), and acknowledged that you were medically disqualified from working as an Air Traffic Controller. He indicates that your cognitive impairment is due to "an illness caused by breathing air in the water-damaged building (WDB) that is your workplace." Dr. Shoemaker goes on to outline why he emphatically disputes Dr. Jackson's opinion and why the FAA should be held responsible for your illness. He notes that you have metabolic problems including the presence of peripheral inflammation that leads to the central nervous inflammation demonstrated on magnetic resonance spectroscopy. He further outlines that you do not have lupus, as all confirmatory tests have come back negative. He states that there is not doubt as to the cause of your illness as you were ill before treatment, improved markedly after cholestyramine treatment, didn't become ill with exposure to all environments except the workplace while off medication, rapidly became ill after you re-entered the workplace off medication, and improved again with removal from the workplace and retreatment with the same medication. He summarizes that the proof is in the time of exposure variable, i.e. going into the Detroit Tower, and objective testing done afterwards which show "profound decline" and match your symptoms. He then outlines his professional experience in similar type cases in the field and his influence on research, data, and treatment in this regard.

Dr. Shoemaker further states that your health parameters are "singularly distinct" from any other group of patients other than those exposed to water damaged buildings or who have a biotoxin-associated

File Number: 092085516
reconvacate-D-

illness. He notes that the features of the illness are "reproducibly reliable laboratory testing results; these results are objective and cannot be altered by any particular spin." He ends his letter reiterating his opinion that your "multiple symptoms, profound cognitive impairments and multiple physiologic and neurotoxicological abnormalities" were "solely caused" by the mixture of biologically active compounds present in the Detroit Tower. He states:

> There is no evidence of exposure coming from any other source. There is no evidence that any other medical condition present could cause the multitude of laboratory abnormalities in you that respond rapidly to use of targeted therapies beginning with CSM but that relapse over time with ongoing exposure."

He further explains that you are genetically susceptible and that even with successful acute phase treatment, patients with susceptibility are subject to a reacquisition of illness following any additional exposure (i.e. "sicker, quicker").

In the letter from Dr. Chiodo to Dr. Shoemaker dated 11/19/08, Dr. Chiodo begins by noting that you have been medically disqualified from your Air Traffic Controller position and summarizes your history of symptoms including short term memory problems, chronic headaches, inability to multi task, skin rashes, nasal congestion, sinusitis, pneumothorax, pleurisy, diarrhea, hand numbness, twitching eyelids, eye pain, joint pain, and change in vision. He then outlines the history of your exposure to mold from January 2005 through February 2008. He provides a summary of your medical and family history and findings upon examination. He then outlines his medical opinion that your "allergic disease" is consistent with mold exposure in the Detroit Tower. He provides the definition of an antigen medicated disease and states that you appear to have suffered allergic disease including rash and sinusitis and cognitive decline and fatigue due to mold exposure in the workplace.

In a letter to Dr. Shoemaker from Dr. Chiodo dated 12/08/08 he reiterates this opinion and states that you may not return to your duties as an air traffic controller due to cognitive deficits found on evaluation by David R. Drasmin, PhD.

In his 14 page typed report dated 02/09/09, Dr. Guidice, MD indicates she evaluated you on 12/22/08 that included a detailed history, neurological mental status and neurological physical status requiring approximately 7 hours. She notes that history was obtained from you by phone on 12/18/08 and from your wife by phone on 12/27/08. She indicates that you were 40 minutes late for your appointment, neglected to bring your computer, neglected to bring your check book, and required a two hour lunch due to what you described as difficulty finding a branch of your bank. She notes that this is attributed to your cognitive deficits. She says that difficulty with these things is "quite consistent with the patient's cognitive impairment suggesting his inability to rapidly prepare, organize, and sequence information."

Dr. Guidice then reports your history of work exposure and indicates you began having cognitive deficits including an increased processing time with transmissions, inability to multitask on a rapid basis, inability to filter out transmissions that do not need a response, and errors. Dr. Guidice reports that you were medically disqualified by the FAA physician and she agrees with that assessment. She further reports similar complaints from other employees in the Detroit Tower. A plethora of your personal "post mold exposure" symptoms were further outlined. Your past medical, educational, military, work history, social history, and family histories were outlined. Positive findings upon neurological examination were reported as impaired concentration and memory, work finding difficulties, stuttering, impulsive and distractible, all based on "simple, formal" testing. Dr. Guidice stated that your judgment and problem solving is functionally impaired for daily life. Visual problems were also reported based on testing, as was bilateral heel-shin ataxia, right greater than left; slowed left hand fine coordination; mild decreased rapid alternating movements of the left hand; and mild decreased strength in the left finger abductors.

File Number: 092085516
reconvacate-D-

Dr. Guidice outlines your test results as follows: genetic haplotype of 7-2-53 susceptible to the development of toxigenic mold illness; increased Ca4 of 8581 (reference less than 2830); increased MMP-9 of 414 (reference 85-332) also consistent with mold exposures; positive ANA, nonspecific; VCS results unavailable; and magnetic resonance spectroscopy of 04/24/07 revealed abnormalities in the bilateral frontal and bilateral temporal lobes, particularly the hippocampal areas of the latter, consistent with capillary hypofusion in these areas and reduced glutamine to glutamate ratio indicating a reduction of oxygen delivery in the capillary beds of these areas of the brain, which is significant substantiation, in that these areas also delineated independently based on Dr. Drasmin's evaluation; and neuropsychological testing and interview consistent with those reported and noted in history and examination. In addition to those discussed above, Dr. Guidice also reports impaired new learning, "paraphasic errors," impaired math skills, difficulty calculating dimensions in construction business, mood symptoms, personality changes, irritability, and behavioral changes.

In the summary portion of her report, Dr. Guidice advises that you were, in fact, exposed to mold in the Detroit Tower and have a "distinctive grouping of symptoms and a distinctive grouping of signs on the clinical neurological examination" which were "strikingly and distinctively similar" to those of another cohort from the Detroit Tower examined by her in 2006. She also reports that you have absence of confounding diagnoses, including Lupus and/or traumatic brain injury; and positive blood tests consistent with mold exposure with improvement with treatment and removal from the toxic mold environment with then recurrence of symptoms and laboratory abnormalities being documented over a 3 day unprotected mold re-exposure. She outlines your diagnoses and treatment recommendations and states that you are neurologically disabled from working as an Air Traffic Controller.

It should be noted that this office has not received any evidence from Dr. Drasmin dated 11/3/08 or 08/19/08. We also have not received the VCS test results from Dr. Shoemaker. Results received include: spirometry results of 04/24/07; C3a and C4a results dated 04/24/07; and MRI Brain/Spectroscopy Without Contrast dated 04/24/07.

BASIS FOR DECISION:

The missing element in your claim was that you had not provided sufficient medical evidence to establish that you sustained an injury as defined by the Federal Employees' Compensation Act. Specifically, you had not provided any medical documentation from Dr. Shoemaker that discussed diagnostic/lab testing and results of findings upon examination to support the diagnosis of chronic inflammatory illness. Although Dr. Shoemaker had stated that you met the criteria for the diagnosis, he neglected to outline what that criterion is and how the testing and exam findings support it.

We have since received extensive medical records from Dr. Shoemaker, as well ad reports from Dr. Chiodo and Dr. Guidice. These records support that you have the clinical diagnosis of chronic inflammatory illness based on laboratory testing, diagnostic testing, and physical and neurological evaluation. Specifically, you have abnormal findings of the brain demonstrated on MR Spectroscopy; cognitive dysfunction; increased Ca4; increased MMP-9; and genetic susceptibility. Each doctor essentially stated that they believed your exposure to mold at the Detroit Tower was or likely was the cause of your condition and at the very least contributed to it. Dr. Shoemaker provides a wealth of information and argument to support his opinion, ranging from similar studies of similar patients, objective clinical data, the "rule out" of other illness, positive examination findings, and his medical expertise in this particular area of medicine. Thus, his opinion is both definitive and well rationalized.

Subsequently, the medical evidence of file now meets the criteria to establish that you have sustained an injury as defined under the Federal Employees' Compensation Act.

File Number: 092085516
reconvacate-D-

CONCLUSION:

In conclusion, the prior decision of 02/12/08 hereby vacated. Your claim has now been accepted for chronic inflammatory illness, ICD-9 code 989.7. Please see the enclosed acceptance letters.

MELISSA J. MYERS
SENIOR CLAIMS EXAMINER

File Number: 092085516
CA-1008-D-ACC

U.S. DEPARTMENT OF LABOR

EMPLOYMENT STANDARDS ADMINISTRATION
OFFICE OF WORKERS' COMP PROGRAMS
PO BOX 8300 DISTRICT 9 CLE
LONDON, KY 40742-8300
Phone: (216) 357-5100

March 30, 2009

Date of Injury: 06/20/2007
Employee: ROBERT D. HAEFNER

ROBERT DANIEL HAEFNER
50464 ELMWOOD CT
PLYMOUTH, MI 48170

Dear Mr. HAEFNER:

This is to notify you that your claim has been accepted for:

Diagnosed condition(s) and ICD-9 code(s): **CHRONIC MULTISYSTEM ILLNESS, ICD-9 CODE 989.7, DUE TO MOLD EXPOSURE AT THE DETROIT AIRPORT TOWER IN NOVEMBER 2005, DECEMBER 2006, JANUARY 2007, AND FEBRUARY 2008.**

Please advise all medical providers who are treating you for this injury of the accepted ICD-9 code(s). If this code needs to be revised, your doctor should explain in writing. Accurate coding facilitates timely bill processing.

It appears you have not returned to work with the FAA since August 2008. If your injury results in lost time from work, you may claim disability compensation using Form CA-7. Please refer to the attachment entitled "Now That Your Claim Has Been Accepted."

It is further noted that the medical evidence of file substantiates that you cannot currently return to work as an Air Traffic controller. However, it does not outline what you can currently do. Please arrange for an update of your work capabilities to be provided to this office within the next 30 days. Be advised that the goal of our program is return to work, so if we cannot return you to your prior place of employment, we need to assess your work restrictions/ limitations in order to place you elsewhere. Please be advised that second opinion evaluation will likely occur during this process.

TO EMPLOYER: IF A FORM CA-7 CLAIMING COMPENSATION FOR WAGE LOSS IS FILED, YOU ARE REMINDED THAT 20 C.F.R. 10.111(c) REQUIRES SUBMISSION OF FORM CA-7 WITHIN 5 WORKING DAYS. PLEASE SEND A COPY OF THE POSITION DESCRIPTION (INCLUDING PHYSICAL REQUIREMENTS) FOR THE JOB HELD BY THE EMPLOYEE ON THE DATE OF INJURY.

If you have any questions regarding your claim you may contact the Office at the above address. Automated information regarding compensation payments is available 24 hours per day by phoning 1-866-OWCP IVR (1-866-692-7487). All medical providers should call 1-866-335-8319 for any and all requests for authorization. For all inquiries regarding any and all bills, including claimant reimbursements, contact 1-866-335-8319 or online at http://owcp.dol.acs-inc.com. If you, your doctor, or other providers require direct contact with a customer service representative you may call 1-850-558-1818 (THIS IS A TOLL CALL).

File Number: 092085516
CA-1008-D-ACC

Sincerely,

Melissa Myers
Senior Claims Examiner

Enclosure: NOW THAT YOUR CLAIM HAS BEEN ACCEPTED

DEPARTMENT OF TRANSPORTATION
FEDERAL AVIATION ADMINISTRATION
WORKERS' COMP DIVISION-AHL-100
800 INDEPENDENCE AVENUE,SW,RM 521
WASHINGTON, DC 20591

File Number: 092085516
CA-1008-D-ACC

NOW THAT YOUR CLAIM HAS BEEN ACCEPTED

This fact sheet will answer some questions that are likely to arise. It provides information about the payment of your medical bills and compensation, and about your responsibilities in returning to work. This sheet supplements the information found in Pamphlet CA-14, which was sent to you when you first filed your claim. Feel free to access the Division of Federal Employees' Compensation web site at http://www.dol.gov/esa/regs/compliance/owcp/fecacont.htm.

MEDICAL PAYMENTS

Your file number must appear on all bills. Bills and travel vouchers must be received within the calendar year following the year in which medical service was rendered or the claim was accepted, whichever occurs later. Your acceptance letter describes the medical condition(s) OWCP accepts as work-related and only treatment for those conditions should be billed to the Office. The billing forms described below can be obtained on line at http://www.dol.gov/libraryforms/. You are not responsible for charges over the maximum allowed in the OWCP fee schedule. If a health benefits carrier has paid medical bills for your accepted condition, the carrier may submit complete, itemized billings to OWCP for consideration.

-Physicians and Other Medical Providers (Except for Hospitals and Pharmacies). Bills for your accepted condition must be submitted on the standard American Medical Association (AMA) billing form HCFA-1500, also known as OWCP-1500, to the address noted in the letterhead. The provider must itemize services for each date separately, use AMA (not state) CPT codes to describe the services performed, and provide their tax identification number (EIN). The provider must sign the form (a signature stamp may also be used).

-Hospitals. These bills must be submitted on Form UB-92. These bills must be fully itemized, and the admission and discharge medical summaries should also be sent.

-Pharmacies. These bills should be submitted electronically by your pharmacy. If this is not available, bills must be submitted on the Universal Claim Form or equivalent. The pharmacy should include the following items: the case file number, the nine-digit tax ID number, the NDC number, the prescription number, the quantity of medication prescribed, the name of the prescribing physician, and the date of purchase. Your physician's clinical notes or reports should show that the medicines prescribed were needed to treat your work-related injury. Pharmacies can obtain decisions on coverage of medications by calling 1-866-335-8319. The pharmacy will need to give your case file number, the NDC code of the medication, and the date the prescription was filled. If you, your doctor, or other providers require direct contact with a customer service representative you may call 1-850-558-1818 (THIS IS A TOLL CALL).

-Chiropractors. We will only pay for chiropractic treatment consisting of manual manipulation of the spine to correct an accepted work-related spinal subluxation demonstrated by x-ray, or if a medical doctor has prescribed physical therapy to be administered by a chiropractor.

-Reimbursements. If you have paid authorized medical expenses, you may request reimbursement by attaching Form CA-915, or a similar form, on the same required billing

File Number: 092085516
CA-1008-D-ACC

forms (such as HCFA-1500 or UB-82) specified above. In all cases, the medical provider's tax identification number (EIN) and proof of payment must be provided. Reimbursements are limited to the fee schedule amount.

-Reimbursement for Medical-Related Travel. Travel expenses should be claimed on form OWCP-957, Medical Travel Refund Request, available at http://www.dol.gov/esa/regs/compliance/owcp/forms.htm.

COMPENSATION PAYMENTS

-Claims for Compensation. Any claim for lost wages must be submitted through your employing agency on Form CA-7. Your employing agency will complete its portion of this form and forward it to the Office. In cases of intermittent wage loss, Form CA-7a is also needed.

-Claims for Leave Buy-Back. Reinstatement of leave is subject to the approval of your employing agency. Prior to using your personal leave to cover injury-related absences from work, you are urged to review the instructions for Form CA-7b. To claim a leave buy-back, you must file Form CA-7b through your employing agency, along with Form CA-7 and Form CA-7a.

-Schedule Award. A schedule award may be claimed using Form CA-7only after maximum medical improvement has been reached. A schedule award of compensation is based upon permanent loss of use of a scheduled member or function of the body due to the work-related injury.

-Penalty. Any person who knowingly makes any false statement, misrepresentation, concealment of fact, or any other act of fraud to obtain compensation, or who knowingly accepts compensation to which he or she is not entitled, is subject to felony criminal prosecution and may, under appropriate U.S. criminal code provisions, be punished by a fine of not more than $10,000 or imprisonment for not more than five years, or both.

RETURNING TO WORK

You are expected to return to work (including light duty or part-time work, if available) as soon as you are able. Once you return to work, or obtain new employment, notify this office immediately. Full compensation is payable only while you are unable to perform the duties of your regular job because of your accepted employment-related condition. If you receive a compensation check which includes payment for a period you have worked, return it to us immediately to prevent an overpayment of compensation.
-Nurse Intervention and Vocational Rehabilitation. OWCP may assign a registered nurse or a vocational rehabilitation counselor to contact you to facilitate your recovery and return to work. OWCP may suspend or reduce your benefits if you fail to cooperate with the nurse or the vocational rehabilitation counselor.

-Job Offers. You are legally obligated to accept work which is within your medical restrictions. OWCP may terminate your benefits if you refuse without good cause to accept such work.

File Number: 092085516
CA-1008-D-ACC

CONTACTING THE OFFICE

The 24-hour toll-free Interactive Voice Response line (866) 692-7487 answers case-specific concerns, such as compensation payments. You can obtain information regarding medical payments, including all reimbursements, at http://owcp.dol.acs-inc.com or by calling 1-866-335-8319. If you need a medical authorization, please call 1-866-335-8319. If you, your doctor, or other providers require direct contact with a customer service representative you may call 1-850-558-1818 (THIS IS A TOLL CALL). If you write us, please put your file number on each page.

Appendix 7

St. Bernard's Parish

RITCHIE C. SHOEMAKER, M.D., P.A.
CHRONIC FATIGUE CENTER
500 MARKET STREET
SUITE 102, 103
POCOMOKE, MD 21851
TELEPHONE (410) 957-1550
FAX (410) 957-3930

To: President, St. Bernard Parish 2/22/06
Fire Chief, St. Bernard Parish
Eddie Elovsson, General Manager M/S Scotia Prince
Larry Ingargiola, Homeland Security, St. Bernard Parish

St. Bernard Parish Mold Clinic
February 9,10,11,12
R. Shoemaker MD

Re: Results of health surveys and visual contrast testing

Dear Sirs:

EXECUTIVE SUMMARY:

Patients from five separate groups in St. Bernard Parish were screened for possible biotoxin associated illness 2/9/06-2/12/06 using a self-administered history and a non-invasive test of neurotoxicity (visual contrast sensitivity "VCS"). A formal medical history was not taken and no physical examination was performed. No laboratory data was obtained.

The pertinent findings from the screening project are summarized on the two attached tables named "St. Bernard Parish Roster" and "Katrina Cough." There is no question about the potential for illness caused by biotoxins in this population: the data are overwhelming. In the absence of known other sources of biotoxin illnesses such as that from dinoflagellates, spirochetes and cyanobacteria in these patients, and the known, massive exposure to water-damaged buildings with visible mold growth, we must consider as likely the hypothesis that ongoing exposure to toxigenic molds and water-damaged buildings is making many St. Bernard Parish residents ill. Prompt medical intervention is indicated. As my longstanding experience with over 5000 biotoxin illness patients clearly documents, the longer the illness is untreated, the worse the prognosis becomes.

Our data accumulated over the past eight years demonstrates that use of these screening modalities can confirm with reasonable medical certainty the presence of biotoxin illness, including such illness as acquired following exposure to biotoxins made by organisms residing in water-damaged buildings. The organisms that cause biotoxin-associated illness in water-damaged buildings include fungi, actinomycetes and endotoxin-forming bacteria. The

206 patients examined include an extraordinarily high percentage of affected patients compared to local (M/S Scotia Prince crew) and historical controls.

Intervention measures, including removal from exposure and further delineation of the physiologic parameters of the illness followed by definitive treatment, are strongly recommended to begin before the unprotected exposure exceeds six months. Trailers installed next to contaminated buildings, used by persons with unprotected indoor exposure to those contaminated structures cannot be considered to be a shelter strategy that provides protection from toxigenic elements, including fungi, resident in the contaminated structures.

Background:

Following an invitation from the management of the M/S Scotia Prince and Dr. Diaz, I joined Dr. Richard Lipsey, a noted toxicologist, on board the Scotia Prince in order to address two questions. First, "Is the ship a safe haven for those persons displaced by Hurricane Katrina and subsequent events?" Second, "Is there evidence that the Parish is safe for unrestricted return of its residents beginning the reconstruction process?"

As to the first question - I would note that the only "residential" location in St. Bernard Parish that did not have its "residents" (the Crew) identified with biotoxin-associated illness is the M/S Scotia Prince.[1]

In order to answer these questions, we employed an approach that assessed the potential for exposure to toxigenic elements, including fungi, on board the ship as well as in the surrounding community. Dr. Lipsey obtained multiple environmental samples in both areas. I have seen his preliminary report. First - there is no evidence of microbial contamination on the ship. Second - there is ample evidence of massive microbial growth in the community and initial laboratory analysis indicates that much of it is Stachybotrys. I will await his final report of the cultured samples. That report should then be referenced along with this report and vice versa.

In this report, I will use the term, "mold illness" to refer to those persons with a complex, multisystem, multi-symptom illness acquired following exposure to buildings with a history of water intrusion. While others may call the illness, "Sick Building Syndrome" or something else, I feel that mold illness will suffice to describe the typical biotoxin-associated illness seen in patients with exposure to water-damaged buildings (WDB) and not seen in controls. The use of the term mold illness does not imply that only molds cause the illness; indeed, other toxigenic microbes, including actinomycetes and bacteria are important toxigenic organisms in WDB as well.

As a treating physician, I am on record as insisting that the only reliable marker for a "Sick Building," is a sick person. I agree, however, that we must combine carefully obtained environmental samples with a complete health analysis to assess the potential for acquisition of adverse health effects following exposure to WDB. The assessment of health effects

[1] One crew member presented as a ciguatera case – most likely the result of eating contaminated fish several months previously at his home in the Caribbean.

involves use of a screening medical history and VCS, a neurotoxicological test of vision. Our data on over 5000 patients with biotoxin illnesses, including those caused by exposure to WDB when compared to over 4000 patients in various control groups, clearly shows the accuracy of these two tests in developing reasonable medical certainty separating those with a biotoxin illness from those without such illness. The accuracy exceeds 98.5%. I have included a graphic that demonstrates the statistical basis of this statement.

Logistic Regression Model - 8 Factor Score

Combining Symptoms: Predicting Membership in the Group of Cases or Controls

	PREDICTED CONTROLS	PREDICTED CASES	ROW TOTAL
Observed Controls	238	1	239
Observed Cases	5	277	282
COLUMN TOTAL	243	278	521
Percent Agreement			98.85

	-	+
-	Specificity	False Positive
+	False Negative	Sensitivity

Agreement Odds 515/521 Disparities 6
Standard Deviation 0.47%

Confidence Limits

2-tailed Z at α=0.05=1.960	% Agreement	Disparities
Lower Confidence Limit=	7.03	11
Upper Confidence Limit=	99.76	1

Significance Test of Agreement

Test	ChiSquare	Prob. of ChiSquare
Likelihood	656.722	<0.00000001
Pearson	497.264	<0.00000001

Ritchie C. Shoemaker MD

Such medical certainty provides us with the basis to make statements about the cohort of patients seen on the Scotia Prince, but it does not provide 100% absolute medical certainty. Our diagnostic laboratory findings and the process of differential diagnosis serve that function. Given the ongoing political and economic battles regarding the disaster relief programs, government should support an effort to expand our diagnostic accuracy using the tests that are readily available from nationally certified, high complexity laboratories such as LabCorp and Quest.

Following collection of a database of laboratory studies that will provide accurate delineation of the illness, an objective group of findings will be present that either supports the diagnosis of mold illness or not. If we find additional objective confirmation of the results of screening then we should begin to treat affected patients using protocols I have published in peer-reviewed journals and have employed for over eight years. Our protocols are both safe and effective, with a target of greater than 75% reduction of symptoms achieved in over 92% of 3000 patients with mold illness.

Case definition:

The case definition for human illness caused by exposure to toxigenic organisms, including fungi, mandates two tiers. First, there must be (1) the potential for exposure, (2) presence of multiple symptoms from multiple systems and (3) absence of confounding exposures. The second tier includes the requirement that there also must be three of six of the following elements: (1) presence of deficit in visual contrast sensitivity (VCS); (2) presence of a susceptible HLA DR genotype, as analyzed by PCR; (3) elevated levels of matrix metalloproteinase-9 (MMP9); (4) dysregulation of simultaneously measured ACTH/cortisol; (5) dysregulation of simultaneously measured ADH/osmolality; and (6) reduction of levels of MSH. This case definition was derived by identifying biomarkers present in all those patients with illness and none of control patients without illness. It was first presented in a peer-reviewed paper on September 12, 2003. To date, the case definition continues to maintain an unheard of accuracy of 100%.

For patients age eighteen and younger, the case definition necessarily excludes pituitary hormone abnormalities. We include two separate assays of autoimmunity, anticardiolipins and antigliadins as additional parameters of diagnostic significance because we have documented the unusual increase in autoimmune factors in children with mold illness compared to children without exposure and without illness. Our case definition for children must also account for the delay in maturation of the neurons involved with development of the neurologic function of vision involved in contrast detection. We include in our case definition in children five elements on the second tier, two of which must be present, including HLA DR, antigliadins, anticardiolipins, MMP9 elevation and MSH deficiency. This case definition was presented 12/14/05 at the ASTMH meetings in Washington, DC.

Results:

1. Please review the data summarized by the St. Bernard Parish Roster table. Based on population studies across various ethnic groups, we would expect that 24% of all persons with exposure to WDB would become ill once their HLA-DR based susceptibility was expressed. This "unveiling" of susceptibility follows a significant cytokine illness or massive exposure to toxigenic organisms. If the exposure to the toxigenic organisms is massive, we have seen illness prevalence that exceeds 24% in affected populations. In smaller cohorts, the expected prevalence of illness can be elevated beyond the expected 24% by reasons of chance. Larger samples (populations) won't show the effects of small sample size and then 24% becomes a criterion for consideration of potential chronic illness.

2. The finding of illness findings consistent with our case definition and previously reported logistical regression analysis of symptoms, in all subsets of patients supports our hypothesis that there are many persons with exposure to toxigenic organisms, including molds, in St. Bernard Parish. These people should undergo complete medical evaluation and treatment. Our control population, taken from ship's crew members who did not have exposure to indoor environments on land, show that the ship is a safe haven and that there is no evidence of cross-contamination of ship's

crew by individuals coming on board the ship with clothing and possessions that might be contaminated with spores, fungal fragments or other toxigenic materials. The ship employs an air filtration system that may be a factor in prevention of acquisition of illness on board.

3. We found that age, gender and race did not show any predisposition to acquisition of illness. There was only one exposure to a known biotoxin other than WDB, that being a ship's crew member who became ill following consumption of fish while in tropical reef areas suggestive of ciguatera, a dinoflagellate illness.

4. Symptoms in affected patients were no different from known cases; symptoms in ship's crew were no different from known controls. Symptoms in non-cases in the St. Bernard Parish cohort were no different from known controls.

5. Distribution of symptoms (frequency) was no different from putative cases and non-cases in this cohort compared to known cases and known controls.

6. Visual contrast (VCS) scores were no different in putative cases from St. Bernard Parish from known cases of biotoxin illnesses. VCS scores were no different in ship's crew from other known controls and were no different in putative non-cases from known controls.

7. There has been discussion in the media of the "Katrina cough," but there has been no systematic study of the origin of that persistent respiratory abnormality. Without a case definition of Katrina cough, we simply analyzed those persons who acknowledged that cough was a symptom they had on a daily basis. We have already presented materials on mechanisms involved in acquisition of cough following exposure to WDB. We know that low MSH, high MMP9, elevated C4a, low levels of VIP and low VEGF all contribute to cough in patients with normal IgE. As an aside, low levels of IgE, a reliable screening marker for allergy when elevated, are dominant in biotoxin-associated illnesses. We found that cough in the Katrina cohort was slightly higher in prevalence from cough in other biotoxin-associated illness cohorts, but that the number of associated illness symptoms had not quite reached the levels seen in patients with illness that typically is of much longer duration. Having said that, anyone with a cough and 12 other symptoms, as the Katrina cough persons report, cannot be considered to only have a respiratory illness.

Discussion:

These data support the need for more screening and case-finding in this population of mold-exposed people. Based on discussions with health care providers in the parish, there is little awareness of the diagnostic process and treatment protocols used for patients with illness from WDB. These providers expressed an interest in learning more about what to do for their patients as they are overloaded with sickened residents of St. Bernard parish at this time. The greatest risk for these patients is additional delay in proper diagnosis and treatment. The longer the inflammatory basis of this illness persists, the greater the number of additional biological cascades of "downstream" events that will occur.

I have added additional information regarding our protocols and lab testing in the enclosed Appendix "A". In an Excel workbook I have attached Appendices "B" through "L". These are various important data sets that resulted from the clinic onboard the Scotia Prince. Of particular note are the data in the first table set out below. We know from our work that there are two routes to biotoxin associated illness – (i) genetic susceptibility found in 24% of the population; and (ii) massive exposure. The St. Bernard sample size is small at 189 persons examined but the overall positive "Case" rating of 54.5%, being more than double the level due to genetic susceptibility, argues persuasively for massive exposure.

The above observations are based on a reasonable degree of medical certainty and are submitted without bias. I am willing to assist you in any way.

Sincerely,

"Ritchie C. Shoemaker"

Ritchie C. Shoemaker, MD

Location	N =	# Positive	% Positive
Firemen	31	14	45.2%
Parish Employees	16	11	68.8%
Health Workers	6	2	33.3%
Homeless Adults	126	71	56.3%
Homeless Children	10	5	50.0%
Total SB Parish	189	103	54.5%
Controls (Ship)	23	1	4.3%
Historical Controls	239	0	0

	N=	Total Symptoms	Av. Sx per person	Total Cough	% with cough	Total Sx/Cough	Sx/person with cough
Historic Mold Cases	594	10,276	17.3	345	58	6,580	19.1
Historic Normal Controls	239	645	2.7	41	17	126	3.1
Total Parish Residents							
Case	101	1,443	14.3	80	79%	1,211	15.2
Non-case	82	142	1.7	21	**25%**	54	2.6
Total	183	1585	8.7	101	55%	1265	12.5
Ship's Crew	23	33	1.43	6	26	15	2.5
Fireman							
Case	14	183	13.1	11	78	153	14
Non-case	17	43	2.52	5	29	7	1.4
Homeless							
Adult case	71	1,086	15.3	57	80	932	16.4
Adult non- case	55	83	1.5	13	23	35	2.8
Child case	5	52	10.4	4	80	44	11.0
Child non-case	5	15	3.0	3	60	12	4.0
Parish Employee							
Case	11	121	11.0	8	73	83	10.3
Non-case	5	2	0.4	0	0	0	0

Appendix A to RCS Letter February 22, 2006

Vision Tests & Analyses

All subjects who normally wore corrective lenses for near-point viewing were asked to wear them during vision testing. The visual acuity and VCS tests were administered monocularly to each eye; an eye occluder was held over one eye while the other eye was tested. All vision tests were administered under illumination from a "daylight" illuminator (fluorescent source with a correlated color temperature of approximately = 6500E K; color rendering index > 90; intensity = 1150 lux; luminance approximately 70 foot-lamberts) in a clinical unit with normal background lighting. A light meter was used to insure that luminance remained constant throughout the test sessions. A test card holder, consisting of a face rest placed just under the cheek bones or chin as comfort provided, and connected by a calibrated rod to a card holder on the distal end, was used to position the acuity and VCS test cards at a constant distance, previously standardized, from the eyes (acuity - 36 cm (14 inches); contrast sensitivity - 46 cm (18 inches)).

Near Visual Acuity

The acuity test card (MIS Pocket Vision Guide, © 1997 MIS, Inc.) contained 10 rows of numbers in which the size of the numbers progressed from a larger size in the top row to a smaller size in the bottom row. Participants were asked to first read the numbers in a middle row. Testing proceeded to the next lower row if all numbers were correctly identified or to the next higher row if an error occurred. The Snellen visual acuity of the row (20/20 or 20/30, for example) with the smallest numbers each identified correctly was recorded as the visual acuity score. Two-tailed Student t-tests 0.05 were performed, using the mean score of each participant's two eyes, to determine if scores differed significantly between cohorts.

Contrast Sensitivity (VCS)

The contrast sensitivity test card (Functional Acuity contrast Test, (FACT), Stereo Optical Co., Chicago, IL, a Gerber-Coburn Co.) contained a matrix (5 x 9) of circles filled with sinusoidal gratings (dark and light bars). Spatial frequency (1.5, 3, 6, 12 and 18 cycles/degree of visual arc) increased from top to bottom, and contrast decreased from left to right in steps of approximately 0.15 log units. The grating bars were oriented either vertically, or tilted 15 degrees to the left or right. As the investigator called out each circle from left to right, row by row, subjects responded by saying either: vertical, left, right or blank. Participants were encouraged to name an orientation if they had any indication that the bars could be seen. Participants were given the option to point in the direction to which the top of the grating was tilted if they felt any difficulty in verbalizing the orientation; none needed this assistance. The contrast sensitivity score for each row (spatial frequency) was recorded as the contrast of the last test patch correctly identified on that row following verification by repeated testing of that patch and the subsequent patch. The procedure was repeated for each row in descending order. The a priori criterion for the inclusion of data in analyses was that the eye has a visual acuity (Snellen Distance Equivalent Score) of 20:50 or better, in order to avoid confounding of the VCS results by excessive optical-refraction error. All eyes include in data analyses met the visual acuity criterion.

Data Analysis:

The units of analysis for the VCS test were the mean scores of the participant's two eyes at each spatial frequency. Standard error of the mean was calculated for each group of measurements. The VCS data were analyzed using multivariate analyses of variance (MANOVA, with the Wilks' lambda statistic) procedures suitable for repeated measures with $\pm = 0.05$. The factors in the model were group and spatial frequency. A factor for gender was not included since there aren't any gender differences in susceptibility to biotoxin-induced effects shown as yet, and no gender differences in VCS have been reported. Results that showed a significant group-by-spatial frequency interaction were further analyzed in the step-down, two-tailed Student t-tests (± = 0.05), the equivalent of a univariate ANOVA to determine which spatial frequencies accounted for the overall effect.

Laboratory:

LabCorp, Inc., Quest Diagnostics, and Specialty Laboratories, Inc., each CLIA approved, high complexity, national laboratory facilities.

MSH: alpha melanocyte stimulating hormone (MSH) is a 13 amino acid compound formed in the ventromedial nucleus (VMN) of the hypothalamus, solitary nucleus and arcuate nucleus by cleavage of proopiomelanocortin (POMC) to yield beta-endorphin and MSH. MSH exerts inductive regulatory effects on production of hypothalamic endorphins and melatonin. MSH has multiple anti-inflammatory and neurohormonal regulatory functions, exerting regulatory control on peripheral cytokine release as well as on both anterior and posterior pituitary function. Deficiency of MSH, commonly seen in biotoxin-associated illnesses, is associated with impairment of multiple regulatory functions and dysregulation of pituitary hormone release. Symptoms associated with MSH deficiency include chronic fatigue and chronic, unusual pain syndromes. Normal values of MSH in commercial labs (Esoterix and LabCorp) are 35-81 pg/ml.

Leptin: leptin is a 146 amino acid adipocytokine produced by fat cells in response to rising levels of fatty acids. Leptin has peripheral metabolic effects, promoting storage of fatty acids, as well as central effects in the hypothalamus. Following binding by leptin to a long isoform of the leptin receptor in the VMN, a primordial gp-130 cytokine receptor, a JAK signal causes transcription of the gene for POMC, which is in turned cleaved to make MSH. Peripheral cytokine responses can cause phosphorylation of a serine moiety (instead of threonine) on the leptin receptor, creating leptin resistance and relative deficiency of MSH production. Normal values in commercial labs show differences between males (5-8 ng/ml) and females (8-18 ng/ml), with levels of leptin correlated with BMI.

ADH/osmolality: abnormalities in ADH/osmolality are recorded as absolute if ADH is < 1.3 or > 8 pg/ml; or if osmolality is >295 or <275 mOsm/kg. Abnormalities are recorded as relative if simultaneous osmolality is 292-295 and ADH ≤ 2.3; or if osmo is 275-278 and ADH≥ 4.0. Symptoms associated with dysregulation of ADH include dehydration, frequent urination, with urine showing low specific gravity; excessive thirst and sensitivity to static electrical shocks; as well as edema and rapid weight gain due to fluid retention during initial correction of ADH deficits.

ACTH/cortisol: abnormalities in ACTH/cortisol are absolute if AM cortisol > 19 ug/ml or < 8 ug/ml; or if AM ACTH is >60 pg/ml or < 10 pg/ml. Abnormalities are recorded as dysregulation if simultaneous cortisol is > 15 and ACTH is > 15, or if cortisol is < 8 and ACTH <40. Early in the illness, as MSH begins to fall, high ACTH is associated with few symptoms; a marked increase in symptoms is associated with a fall in ACTH. Finding simultaneous high cortisol and high ACTH may prompt consideration of ACTH secreting tumors, but the reality is that the dysregulation usually corrects with therapy.

Androgens: total testosterone, androstenedione and DHEA-S provide measurements regarding the effectiveness of gonadotrophin secretion as influenced adversely by MSH deficiency. Normal ranges of these hormones in males are 75-205 ng/ml for androstenedione, 350-1030 ng/ml for testosterone and 70-218 ug/ml for DHEA-S. Normal values for pre-menopausal women are 60-245, 10-55 and 48-247, respectively. Post-menopausal normal ranges are 30-120, 7-40 and 48-247, respectively.

HLA DR by PCR: LabCorp offers a standard HLA DR typing assay of 10 alleles using a PCR sequence specific chain reaction technique. As opposed to serologic assays for the HLA DR genotypes, the PCR gives far greater specificity in distinguishing individual allele polymorphisms. Linkage disequilibrium is strong in these genotypes, with multiple associations made to inflammatory and autoimmune disease. These genes are part of the human major histocompatibility complex (MHC), also called the HLA complex, located on the short arm of chromosome 6. Relative risk was calculated, susceptible genotypes identified, compared within each group to location and exposure.

MMP9: matrix metalloproteinase 9 (gelatinase B) is an extracellular zinc-dependent enzyme produced by cytokine-stimulated neutrophils and macrophages. MMP9 is involved in degradation of extracellular matrix; it has been implicated in the pathogenesis COPD by destruction of lung elastin, in rheumatoid arthritis, atherosclerosis, cardiomyopathy, and abdominal aortic aneurysm. Cytokines that stimulate MMP9 production include IL-1, IL-2, TNF, IL-1B, interferons alpha and gamma. MMP9 is felt to play a role in central nervous system disease including demyelination, by generation of myelin peptides, as it can break down myelin basic protein. MMP9 "delivers" inflammatory elements out of blood into subintimal spaces, where further delivery into solid organs (brain, lung, muscle, peripheral nerve and joint) is initiated. Normal ranges of MMP9 have a mean of 150, with range of 85-322 ng/ml.

C3a and C4a: Split products of complement activation, often called anaphylatoxins. Each activates inflammatory responses, with spillover of effect from innate immune response to acquired immune responses and hematologic parameters. These short-lived products are re-manufactured rapidly, such that an initial rise of plasma levels is seen within 12 hours of exposure and sustained elevation is seen until definitive therapy is initiated. The components increase vascular permeability, release inflammatory elements from macrophages, neutrophils and monocytes, stimulate smooth muscle spasm in small blood vessels and disrupt normal apoptosis.

Anticardiolipins IgA, IgM and IgG: autoantibodies often identified in collagen vascular diseases such as lupus and scleroderma; often called anti-phospholipids. These antibodies in high titers are associated with increased intravascular coagulation requiring treatment with heparin and coumadin. Lower levels titers are associated with hypercoagulability. An increased risk of spontaneous fetal loss in the first trimester of pregnancy is not uncommonly seen in women with presence of cardiolipin antibodies. This problem does not have the same "dose-response" relationship seen with levels of autoantibodies and illness as does the anti-phospholipid syndrome. Anticardiolipins are found in over 33% of children with biotoxin associated illnesses.

Antigliadin IgA and IgG: Antibodies thought at one time to be specific for celiac disease. With the advent of testing for IgA antibodies to tissue transglutaminase (TTG-IgA), gliadin antibodies are most often seen in patients with low levels of MSH. Ingestion of gliadin, the 22-amino acid protein found in gluten (found in wheat, oats, barley and rye; often added to processed foods) will initiate a release of pro-inflammatory cytokines in the tissues lining the intestinal tract. This cytokine effect will often cause symptoms within 30 minutes of ingestion that mimic attention deficit disorder, often leading to an incorrect diagnosis. Antigliadin antibodies are found in over 58% of children with biotoxin-associated illnesses.

Vasoactive intestinal polypeptide (VIP): neuroregulatory hormone with receptors in suprachiasmatic nucleus of hypothalamus. This hormone/cytokine regulates peripheral cytokine responses, pulmonary artery pressures and inflammatory responses throughout the body. Deficiency is commonly seen in mold illness patients, particularly those with dyspnea on exertion.

Appendix B

Category Name	Symptoms (Sx) Average
Controls	2.7
JAMA Cases	18.2
Ship Crew	1.4
Fireman 0-5 Sx	2.5
Firemen 6+ Sx	12.6
Homeless Adults 0-5 Sx	2.6
Homeless Adults 6+ Sx	15.6
Parish Government 0-5 Sx	1.4
Parish Government 6+ Sx	11.8
Homeless children 0-5 Sx	3
Homeless children 6+ Sx	10.4

Appendix C

	St. Bernard Parish February 2006				
	Homeless Adults - Cases	Parish Employees- Cases	Firemen - Cases	JAMA SBS Draft Paper	Controls
Number	**71**	**11**	**15**	**288**	**239**
Fatigue	66	91	73	83	6
Weakness	48	36	33	70	5
Ache	62	64	53	68	8
Cramp	35	18	20	56	2
Unusual pain	23	9	20	51	1
Ice pick	25	18	20	41	1
Headache	66	45	66	66	9
Light	42	9	33	66	2
Red	35	18	20	48	3
Blurred	35	36	20	56	1
Tearing	41	18	20	48	3
Sinus	76	64	87	65	8
Cough	76	64	87	53	7
SOB	58	45	47	63	11
Abdominal pain	24	18	0	39	3
Diarrhea	27	9	13	39	4
Joint	58	36	53	53	11
Morning	52	18	47	44	6
Memory	51	55	87	6	2
Concentration	60	36	87	62	1
Confusion	39	9	53	57	3
Word	41	45	27	66	1
Decreased Assimilation	21	9	13	65	2
Disorientation	25	0	13	40	3
Skin sensitivity	38	36	20	ND	ND
Increased thirst	47	18	27	69	1
Increased urination	47	36	13	66	0
Static Shocks	35	9	27	41	0
Mood Swings	65	45	87	65	1
Appetite	34	18	33	58	1
Sweats (Night)	39	9	47	54	1
Reg Body Temp	24	36	13	60	2
Numbness	38	9	7	44	0
Tingling	34	36	7	51	1
Vertigo	30	36	7	48	2
Metallic Taste	21	18	7	36	0
Tremor	13	9	13	ND	ND

Appendix D

M/S Scotia Prince Crew VCS			
	Average	SD	SEM
A	71.8	24.1	14.4
B	106.3	32.8	21.3
C	117.6	36.7	23.5
D	58.2	18.3	11.6
E	44.0	11.2	8.8

Appendix E

SBS Firemen 0-5 Sx VCS Scores			
	Average	SD	SEM
A	68.0		16.2
B	97.0		23.1
C	128.0		30.5
D	60.0		14.3
E	9.0		2.1

Appendix F

SBS Firemen 6+ Sx VCS Scores			
	Average	SD	SEM
A	53.1		18.8
B	89.2		24.3
C	89.3		21.8
D	42.9		16.0
E	21.7		14.0

Appendix G

Homeless Adults 0-5 Sx VCS Scores			
	Average	SD	SEM
A	66.8	26.1	21.7
B	103.9	36.3	30.3
C	111.4	43.0	35.8
D	57.7	36.9	30.8
E	22.6	19.6	16.3

Appendix H

Homeless Adults 6+ Sx VCS Scores			
	Average	SD	SEM
A	48.6	17.6	2
B	72.7	27.8	3.2
C	62.3	33.7	3.9
D	27.8	16.8	8.0
E	10.6	8.0	0.9

Appendix I

Parish 0-5 Sx VCS Scores			
	Average	SD	SEM
A	45.8	5.6	15.0
B	84.4	17.3	28.3
C	101.4	15.2	37.5
D	61.6	19.4	17.9
E	18.6	15.2	4.2

Appendix J

Parish 6+ Sx VCS Scores			
	Average	SD	SEM
A	44.9	20.6	18.7
B	64.4	16.7	26.9
C	52.1	24.1	21.7
D	23.2	16.2	9.7
E	23.2	16.2	9.7

Appendix K

Children 0-5 Sx VCS Scores			
	Average	SD	SEM
A	52.2	19	12.9
B	101.6	23.3	45.4
C	114.0	50.8	40.0
D	56.6	33.4	12.5
E	25.2	16.3	6.3

Appendix L

Children 6+ Sx VCS Scores			
	Average	SD	SEM
A	45.2	8.4	18.8
B	69.2	36.7	28.8
C	116.6	27	48.6
D	39.4	11.1	16.4
E	14.4	7.5	6

Appendix 8

Logistic Regression Model - 8 Factor Score

Combining Symptoms: Predicting Membership in the Group of Cases or Controls

	PREDICTED CONTROLS	PREDICTED CASES	ROW TOTAL
Observed Controls	238	1	239
Observed Cases	5	277	282
COLUMN TOTAL	243	278	521
Percent Agreement			98.85

Agreement Odds 515/521 Disparities 6
Standard Deviation 0.47%

	-	+
-	Specificity	False Positive
+	False Negative	Sensitivity

Confidence Limits

2-tailed Z at α=0.05=1.960	% Agreement	Disparities
Lower Confidence Limit=	97.03	11
Upper Confidence Limit=	99.76	1

Significance Test of Agreement

Test	ChiSquare	Prob. of Chi Square
Likelihood	656.722	**<0.00000000001**
Pearson	497.264	**<0.00000000001**

Logistic Regression: Results

	Predicted Controls	Predicted Cases	Row Total
Observed Controls	37	2	39
Observed Cases	3	243	246
Column Total	40	245	285
Percent Agreement			98.2

Agreement Odds: 280/285

Standard Deviation: 0.78%

95% confidence interval:

(96.72%, 96.72%) or (1 to 9 disparities)

Test of Agreement

Test	Chi-sq	P-value
Likelihood	183.01	< .0001
Pearson	244.71	< .0001